ACOUSTICAL HOLOGRAPHY

Volume 4

ACOUSTICAL HOLOGRAPHY

Volume 1—Proceedings of the First International Symposium, December 1967, edited by A. F. Metherell, H.M.A. El-Sum, and Lewis Larmore

Volume 2—Proceedings of the Second International Symposium, March 1969, edited by A. F. Metherell and Lewis Larmore

Volume 3—Proceedings of the Third International Symposium, July 1970, edited by A. F. Metherell

Volume 4—Proceedings of the Fourth International Symposium, April 1972, edited by Glen Wade

ACOUSTICAL HOLOGRAPHY

Volume 4

Proceedings of the Fourth International Symposium on Acoustical Holography, held in Santa Barbara, California, April 10-12, 1972

Edited by

Glen Wade
Department of Electrical Engineering
University of California
Santa Barbara, California

PLENUM PRESS • NEW YORK–LONDON • 1972

Library of Congress Catalog Card Number 69-12533
ISBN-13: 978-1-4615-8215-1 e-ISBN-13: 978-1-4615-8213-7
DOI: 10.1007/978-1-4615-8213-7

Softcover reprint of the hardcover 1st edition 1972

A Division of Plenum Publishing Corporation
227 West 17th Street, New York, N.Y. 10011

United Kingdom edition published by Plenum Press, London
A Division of Plenum Publishing Corporation, Ltd.
Davis House (4th Floor), 8 Scrubs Lane, Harlesden
London, NW10 6SE, England

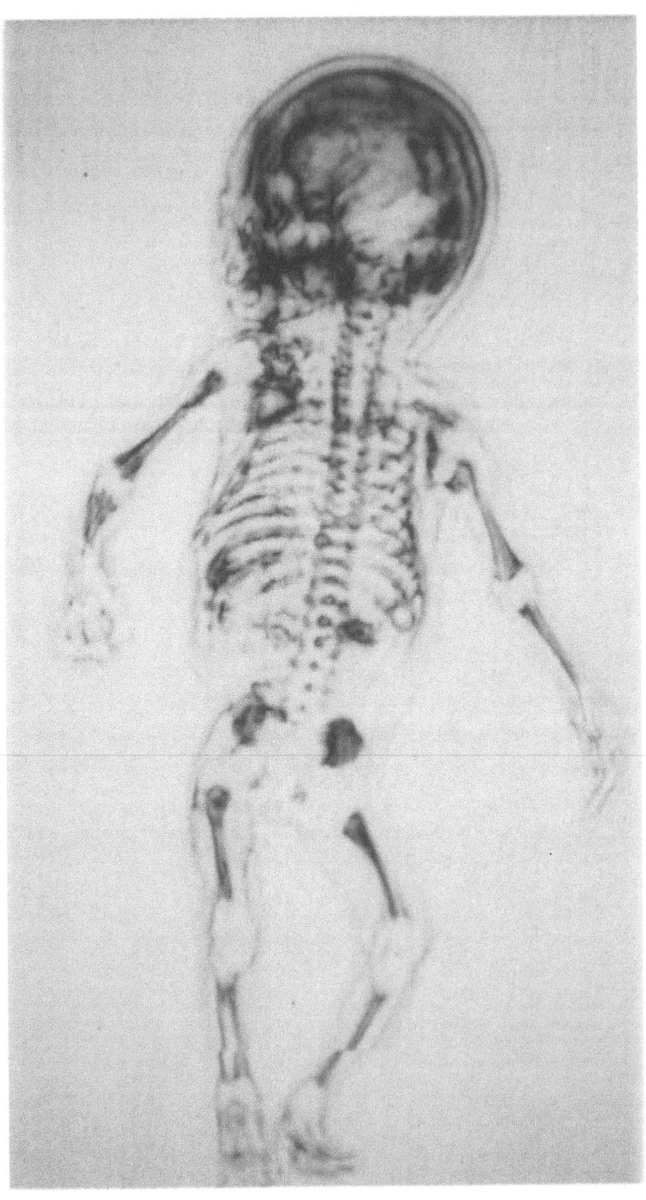

Transmission image of an aborted human fetus in approximately the 17th week. This non-holographic acoustic image was produced by mechanically scanning a 5-MHz ultrasonic transducer, focused in the fetal midplane.

PREFACE

The latest progress in acoustical holography and related research areas, generally involving imaging by means of acoustic waves, was discussed and treated in depth at the Fourth International Symposium on Acoustical Holography, held in Santa Barbara, California on April 10-12, 1972. This volume contains the proceedings of that symposium.

As the papers presented here indicate, a number of startling advances have been realized in the state-of-the-art since publication of Volume 3 of Acoustical Holography. Progress has been particularly impressive in the field of acoustical imaging. The Fourth International Symposium represents something of a landmark conference in this respect.

The scope of this volume is substantially broader than the term "acoustical holography' usually implies and encompasses the whole area of visualization, detection, and recording of sound fields whether with long wavelengths, microwaves, or with extrememly short sound wavelengths. The 37 symposium papers appear here each as a separate chapter. In general, the work reported deals mainly with experimental and theoretical developments in the above areas. This work has significant practical potential use in terms of seismic sensing, underwater imaging, non-destructive testing, real-time acoustic microscopy, and medical diagnosis.

The 37 chapters are grouped into the following 7 sections: I. Real-Time Imaging Systems, II. High-Resolution Imaging Systesm, III. Systems for Biomedical Applications, IV. Array-Imaging Techniques, V. Systems for Non-Destructive Testing, VI. Experimental Systems for Underwater and Seismic Exploration, and VII. Theory and Methods. It should be pointed out that the above sectional titles can serve only as a rough guide. Several of the chapters could logically have been placed into more than one of these sections. A reader interested in new methods of acoustical imaging, for example, will find that subject discussed in Section VII, but also in chapters from certain of the other sections as well.

The program committee for the symposium deserves much credit for the quality of the work selected for presentation at the symposium and, therefore, for inclusion in this volume. The committee members spent many hours in soliciting outstanding contributions and in making assessments concerning all of the papers which were eventually contributed. The Editor wishes to thank the following persons who served as members of the program committee: E.E. Aldridge, A.E.R.E. Harwell, England; B.A. Auld, Stanford University; H.M.A. El-Sum, El-Sum Consultants; P.S. Green, Stanford Research Institute; A. Korpel, Zenith Radio Corporation; J.L. Kreuzer, Perkin-Elmer Corporation; A. Metherell, McDonnell Douglas Corporation/Actron Industries, Inc.; R.K. Mueller, Bendix Research Laboratory; and F.L. Thurstone, Duke University. The Editor would also like to express profound gratitude to Mrs. Oneita Wilde for her substantial and effective effort in compiling this volume.

The symposium was sponsored by the Office of Naval Research*, the IEEE, and the Acoustical Society of America.

*Grant No. NONR(G)-00001-72

CONTENTS

I. REAL-TIME IMAGING SYSTEMS

II. HIGH-RESOLUTION IMAGING SYSTEMS

III. SYSTEMS FOR BIO-MEDICAL APPLICATIONS

IV. ARRAY-IMAGING TECHNIQUES

V. SYSTEMS FOR NONDESTRUCTIVE TESTING

VI. EXPERIMENTAL SYSTEMS FOR UNDERWATER AND SEISMIC EXPLORATION

VII. THEORY AND METHODS

REAL TIME ACOUSTICAL IMAGING BY MEANS OF LIQUID SURFACE HOLOGRAPHY

Byron B. Brenden

HOLOSONICS, INC.

2950 George Washington Way, Richland, WA.

INTRODUCTION

The techniques of liquid surface acoustical holography have proven to be very effective and useful in applications to industrial testing and biomedical imaging. The most useful images have been produced by focused image techniques, that is, by using acoustic lenses to focus the image into the hologram. Because liquid surface holography has usually been illustrated by use of images focused into the hologram, there has been an erroneous but growing belief that liquid surfaces are not effective in forming true holograms capable of imaging outside the hologram plane. Although published papers (1,2,3) provide evidence that this belief is incorrect, further evidence at this time may be useful and will be presented in this paper.

Focused image techniques still provide the best images. Examples of recent biomedical images obtained by focused image holography are included to illustrate the most recent results.

LENSLESS HOLOGRAPHY

Figure 1 is the schematic diagram of the liquid surface holography system as normally used with an acoustic lens to image the object into the hologram plane. When the acoustical

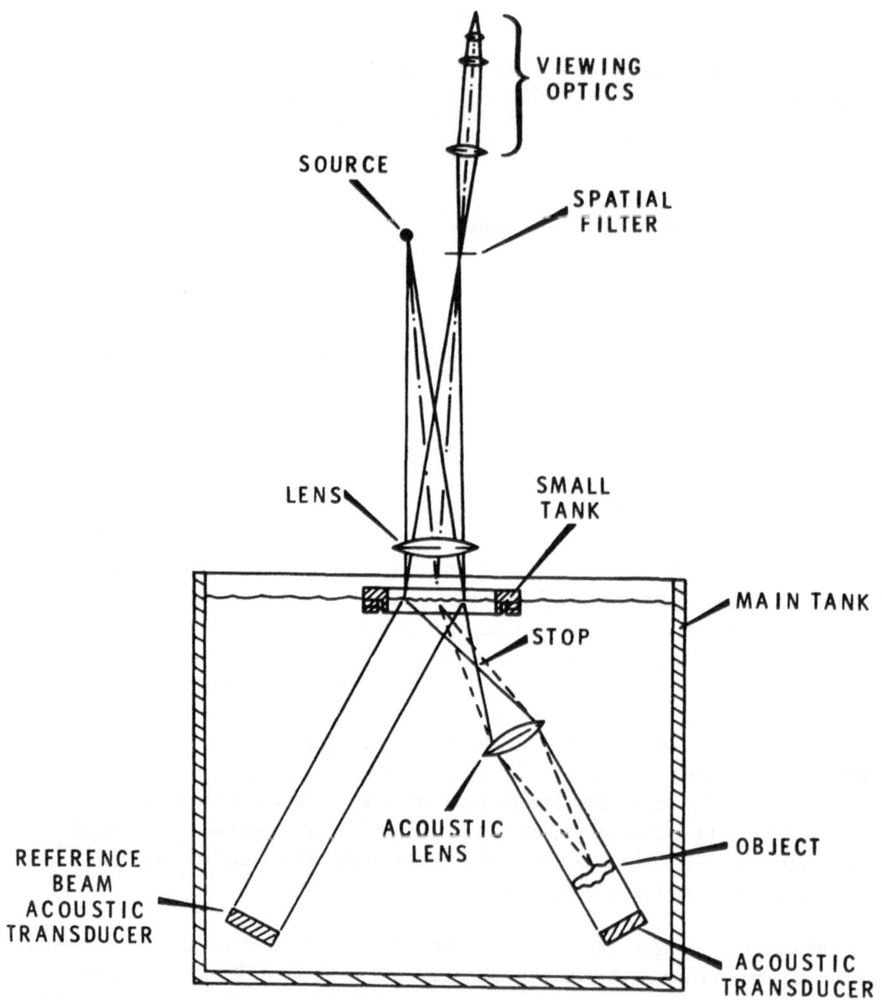

FIGURE 1: Schematic of Liquid Surface Imaging System

image exists in the hologram plane, the optical reconstruction of it also exists there. However, when the acoustic lens is removed, the optical image formed by reflecting coherent light off the liquid surface is located a great distance from the hologram. If r_1 represents the distance from the object to the liquid surface, r_2 the distance from the reference source to the liquid surface, and r_a the distance from the light source to the liquid surface, then the image will be located at a distance r_b from the surface such that (4,5):

$$\frac{1}{r_b} = \frac{1}{r_a} \pm \frac{\lambda}{\Lambda}\left(\frac{1}{r_1} - \frac{1}{r_2}\right) \tag{1}$$

where λ is the wavelength of the light and Λ the wavelength of the sound.

If the reference beam transducer generates a plane wave $r_2 = \infty$ and if the light source is in the focal plane of the optical lens $r_a = \infty$, so

$$r_b = \pm\frac{\Lambda}{\lambda}r_1 \ . \tag{2}$$

Two images are formed, one virtual, indicated by the negative sign, and one real, indicated by the positive sign. The virtual image is the true image and the real image is the conjugate image. An optical lens of focal length f will focus the undiffracted light to a point in its focal plane and will form the true and conjugate images at distances u, given by:

$$\frac{1}{u} = \frac{1}{f} - \frac{1}{r_b} \ . \tag{3}$$

Since r_b is usually very much greater than f because the ratio $\Lambda/\lambda \simeq 1000$, the true and conjugate images are located near the focal plane of the lens with the true image being somewhat further from the lens and the conjugate image somewhat nearer.

Figure 2 shows a test object and the resulting true and conjugate images both recorded in the plane of sharpest focus for the true image. The zero order, undiffracted light normally appearing between the two first order diffracted light

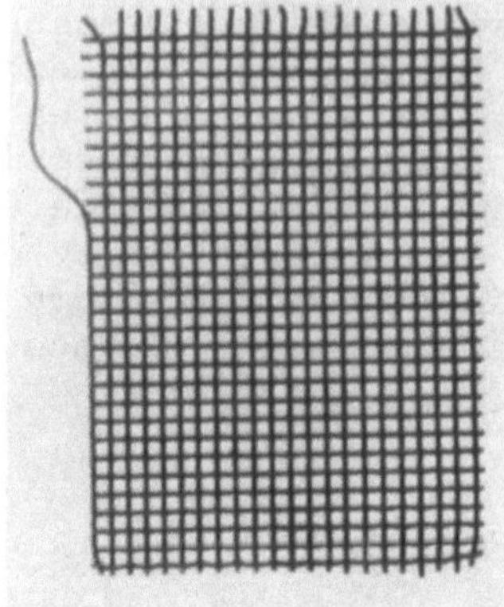

Test Object
Wire Diameter 0.050"
Wire Spacing 0.125"

Focused True Image

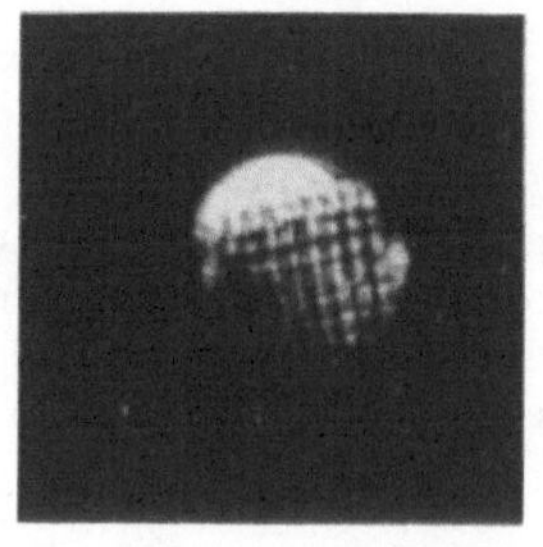

Out-of Focus Conjugate Image

FIGURE 2

beam has been blocked out. No acoustic lens was used in forming these images at 10 MHz acoustic frequency with the object to hologram distance about 13 cm.

DISCUSSION OF LIQUID SURFACE RESPONSE TIMES

Analyses (3,4,5) of liquid surface response to continuous insonification by two plane waves incident at opposite but equal angles to the liquid surface shows that the liquid surface distortion buldge exceeds the amplitude of the hologram pattern by a factor of about 100. Such distortion would indeed destroy much of the effective image forming capability of a liquid surface hologram. Such distortion is never experienced, however, because in actual practice, using acoustical wave trains of 100 μs duration or less, the distortion buldge actually has a smaller amplitude than the hologram pattern (5). This favorable result occurs because the duration of the acoustical wave train is adjusted to a quarter period of the free oscillation of the hologram pattern. The natural period of the buldge oscillation is much longer and consequently the liquid surface displacement for such short excitation pulses is very small. Analysis indicates that the ratio of effective buldge amplitude, B_e, to hologram pattern amplitude A can be given by:

$$\frac{B_e}{2A} \simeq 0.48 \ . \tag{4}$$

When the liquid surface is properly excited, the hologram pattern amplitude will be less than a quarter wavelength of light so that the phase distortions introduced by the distortion buldge B_e are not severe.

BIOMEDICAL IMAGING

For practical reasons, it is still desirable to operate in the focused image mode. Without acoustic lenses, it is difficult to place the object in a position which maintains a favorable numerical aperture. Furthermore, focused image holography permits the use of higher power levels and almost totally removes concern for maintaining a favorable $B_e/2A$ ratio. Focused image techniques are used in the examples that follow.

The real-time imaging capabilities of liquid surface systems are best illustrated by use of motion pictures, video tape or by direct viewing of the image. Swint, Yee and Godbold (6) and Clements (7) have presented examples of the application of these systems to industrial testing. Although many other examples of images related to industrial testing could be provided, I shall limit illustrations in this paper to biomedical imaging.

One of the most striking examples of the capability of liquid surface holography to delineate soft tissue structure is given in Figure 3. Figure 3 is the acoustical image of the upper arm just above the elbow. The edge contours of the humerus appear much more irregular than anticipated on the basis of x-ray images. The primary reason for the irregular contour is that ultrasound interacts with soft tissue to a much greater extent than do x-rays. Thus, rather than seeing true bone edge contours, we actually see soft tissue attachments. The major soft tissue attachments to the humerus shown in Figure 3 are those of the tendons of the muscles of the forearm, namely, the brachioradialis and the extensor carpi.

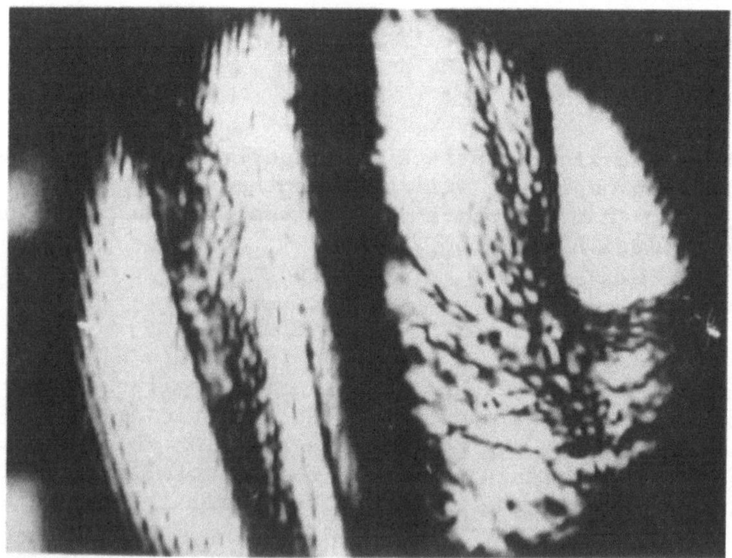

FIGURE 3: Acoustical Images of the Mucles and Muscle Attachments of the Upper Arm. Picture Courtesy of Holosonics, Inc.

In the particular view presented in Figure 3, the attachment pads are glowing with ultrasound. These large glowing areas completely obscure the edge contour of the humerus. The tendon of the biceps running perpendicular to the tendons of the brachioradialis is also quite evident.

Liquid surface holography shows considerable promise for early diagnosis of cancer of the breast. Figure 4 shows the configuration used to bring the breast into the field of view. One of the first acoustical images of a malignant tumor produced by liquid surface holography techniques is shown in Figure 5. As anticipated from earlier studies with rats (8,9), tumors absorb more ultrasound than normal tissue. Furthermore, hard tumors can readily be distinguished from cysts which transmit more energy than the surrounding tissue and therefore appear as bright areas as compared to the surrounding tissue whether that surrounding tissue be normal or diseased.

In Figure 5, the main large tumor is labeled 1 and two smaller tumors are labeled 2 and 3. Three cysts occurred in the locations 4, 5 and 7. The contour of the breast is identified by a Figure 6. The field of view was approximately 13 cm in diameter. This picture, one of the first of tumors in the breast, is due to Dr. J. Hevezi of M. D. Anderson Hospital and Tumor Institute, Houston, Texas and G. N. Langlois of Holosonics, Inc.

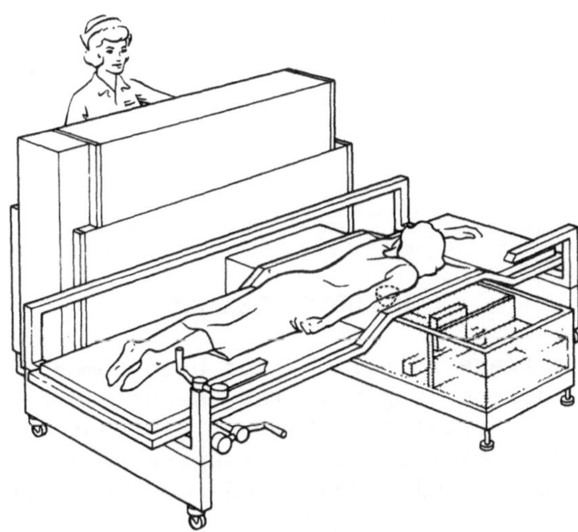

Figure 4: Machine Configuration for Mammography

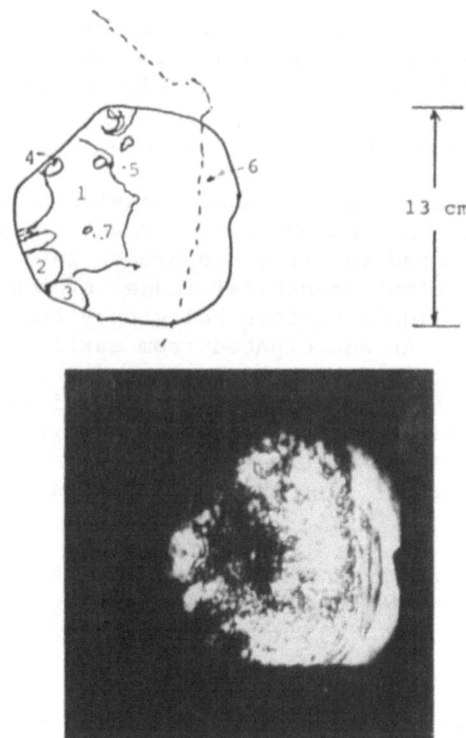

FIGURE 5: Acoustical Image of a Tumor in an Excised Breast. Picture courtesy of Holosonics, Inc., and M. D. Anderson Hospital and Tumor Institute.

CONCLUDING REMARKS

Liquid surface holography has been demonstrated to be capable of lensless imaging. The use of lenses does have great practical advantages, however, in that high quality images are produced with less stringent requirements on operating conditions. Fully developed commercial systems are now available and are being used for industrial testing and biomedical imaging.

REFERENCES

1. R. K. Mueller and N. K. Sheridan, Applied Physics Letters 9:328-329, 1966.

2. P. S. Green, Acoustical Holography, Vol. 1, A. F. Metherell, H. M. A. El Sum and Lewis Larmore, Eds., pg. 13, Plenum Press, 1969.

3. B. B. Brenden, Acoustical Holography, Vol. 1, A. F. Metherell, H. M. A. El Sum and Lewis Larmore, Eds., Chapter 4, Plenum Press, 1969.

4. B. P. Hildebrand and B. B. Brenden, An Introduction to Acoustical Holography, Chapter 2, Plenum Press, 1972.

5. B. B. Brenden, Optical and Acoustical Holography, Ezio Camatini, Ed., Plenum Press, 1972.

6. J. B. Swint, G. B. W. Yee and N. H. Godbold, "Application of Acoustical Holography to Flaw Detection", Acoustical Holography, A. F. Metherell, Ed., Vol. 3, 1971.

7. H. Clements, "Nondestructive Testing Evaluation of Graphite Epoxy Composites and Adhesive Bonded Aluminum Structures Employing Acoustical Holography", Acoustical Holography, A. F. Metherell, Ed., Vol. 3, 1971.

8. L. Weiss and E. D. Holyoke, "Detection of Tumors in Soft Tissues by Ultrasonic Holography", Surg. Gynec. and Obstetrics, 128, 953, 1969.

9. B. B. Brenden, "Acoustical Holography as a Tool for Nondestructive Testing", Materials Evaluation, 27(6) pp 140-144, 1969.

A PROGRESS REPORT ON THE LASER SCANNED ACOUSTIC CAMERA

R. L. Whitman, M. Ahmed and A. Korpel

Zenith Radio Corporation, Research Dept.

6001 W. Dickens Avenue, Chicago, Ill. 60639

In April of 1969, Korpel et al[1] described a method for rapidly forming acoustic holograms at 2.25 MHz using a laser scanning technique shown in Fig. 1. It was decided

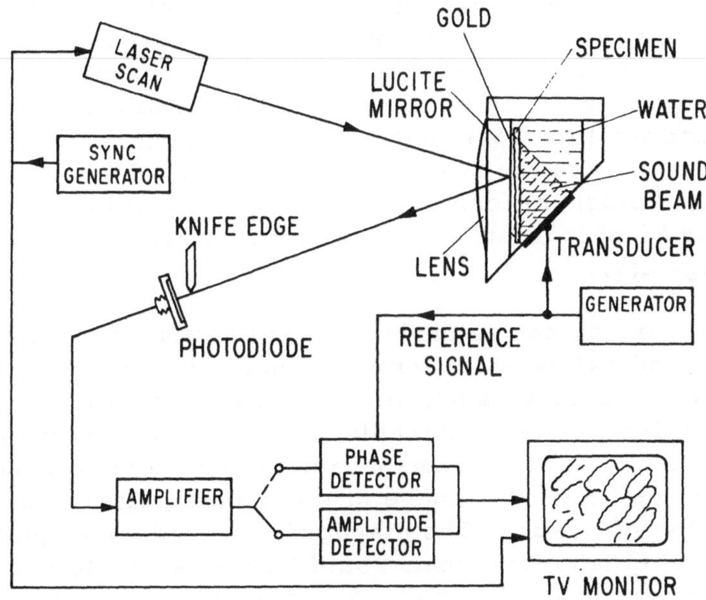

Fig. 1. Experimental setup used to form an acoustic hologram by laser scanning technique (after Korpel et al[7]).

to use a system similar to this in the development of a large aperture acoustic camera which could form acoustic images and holograms in real time. The rationale behind this, the problems encountered in the development, their eventual solution and some novel modifications will be described in this paper.

RATIONALE

The system shown in Fig. 1 requires no further development to display real time, near field acoustic holograms of an object placed in the water tank. The instantaneous acoustic displacements on the reflectively coated faceplate on one side of the water tank form a dynamic hologram of the sound field scattered by the object. The scanning laser beam is periodically deflected by an angle proportional to the slope of these displacements. This deflected beam is partially intercepted by a knife edge placed in the image plane of the laser beam scanner. Thus, the light collected by a photodiode behind the knife edge is intensity modulated and produces a diode current proportional to the laser beam deflection and hence to the slope of the acoustic surface displacements. This r.f. signal, at the acoustic frequency (Doppler shifted by the scanning motion of the laser beam), is mixed with an electronic reference signal and fed to a TV monitor which is being scanned in synchronism with the scanning laser beam. The resultant real time hologram formed [see Fig. 2(a)] on the monitor may then be photographed for subsequent reconstruction [see Fig. 2(b)].

However, since the projected application of the device was for medical examination of acoustically transparent portions of the body, such as the breast, a real time imaging capability seemed desirable. This allows the examiner to position the patient or insonifying transducer in such a way as to optimize the image of some particularly interesting section of tissue. The diagnostic procedure requires considerably less time than would be necessary to process and reconstruct a hologram. A record "in depth" may still

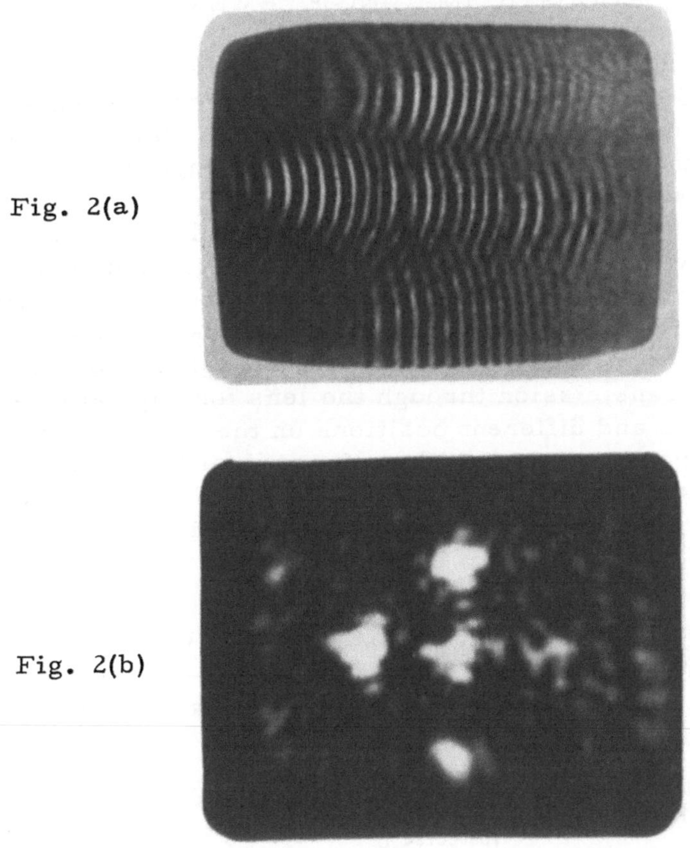

Fig. 2. (a) Hologram of five circular sound sources recorded at 9 MHz, (b) reconstruction (after Korpel and Desmares[1]).

be made by producing and photographing an image hologram on the CRT. In principle this CRT display could be produced by a computer in real time from the holographic information available in the system of Fig. 1. However, a conventional acoustic imaging system seemed to be the best approach at the present time.

ACOUSTIC IMAGING ELEMENT

The acoustic imaging element had to image a 10 cm x 7.5 cm object (the nominal size target chosen with a TV monitor aspect ratio) onto a faceplate. Since a resolution on the order of an acoustical wavelength was desired, a low F number system was necessary. One-to-one imaging was used to maintain resolution without sacrificing field of view. Plastic lenses made of plexiglass or polystyrene were considered for the prototype but were subsequently rejected because of transmission losses, and uncertainty of mode conversion effects, as well as nonuniformity of transmission through the lens for rays at different angles and different positions on the lens. A 20 cm diameter brass reflector with a 20 cm focal length was finally chosen as the imaging element, yielding the folded 1:1 imaging system configuration shown in Fig. 3. The

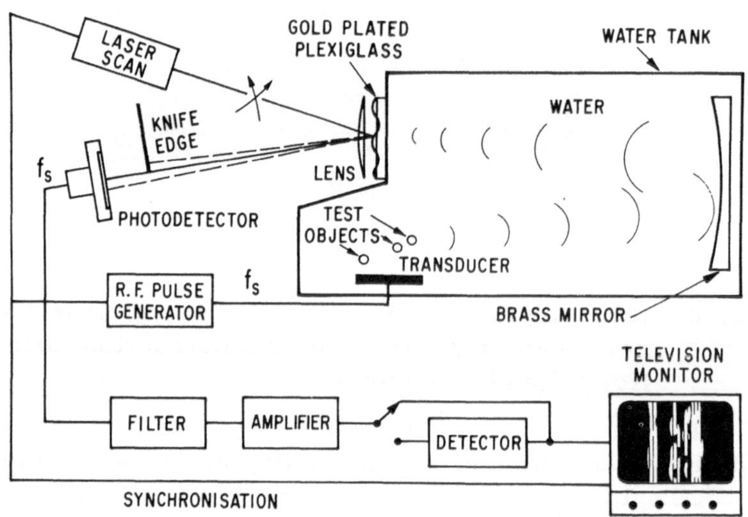

Fig. 3. Experimental setup used to form an acoustic image hologram at 2.268 MHz (after Whitman et al[6]).

off-axis nature of this system is well matched to the acoustic detection technique whose response goes to zero for acoustic plane waves normally incident on the faceplate.[2]

FACEPLATE

The reflectively coated plastic faceplate forms the liquid-solid interface which is perturbed by the acoustic field to form a dynamic hologram. The theoretical response of a Lucite (or Plexiglass) faceplate, in terms of normal surface displacement (normalized with respect to the incident acoustic plane wave displacement amplitude) as a function of the angle of incidence, has been calculated and plotted by Korpel et al,[3] yielding the results shown in Fig. 4. The response shown is that for the faceplate side

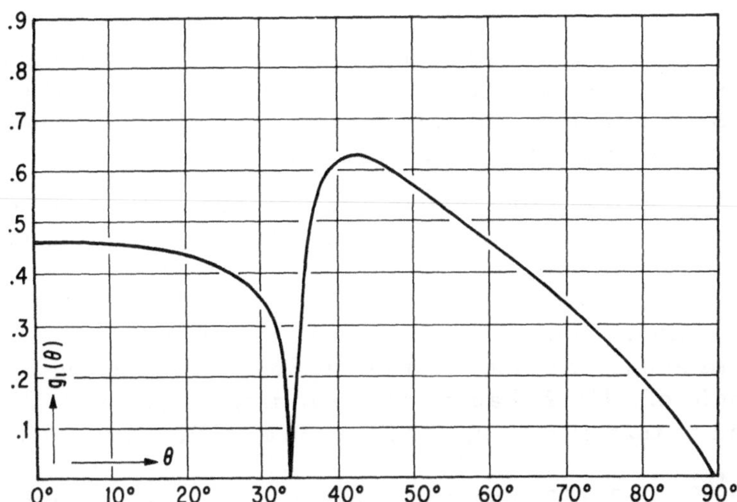

Fig. 4. Normal displacement of water-lucite interface in the absence of reflections off the solid-air boundary (after Korpel and Kessler[3]).

facing the water and assumes no interference from acoustic reflections off the solid-air interface. (This is a good assumption for a 1" thick plexiglass faceplate as is shown in the appendix which discusses resonance effects in plates.) It can be seen that the displacement goes to zero at the critical angle (34°).

The factors which determined the choice of the faceplate were:

1) the acoustic velocity in the material which specifies the critical angle;

2) the acoustic impedance Z_m of the material which gives the acoustic reflectivity $R = \frac{Z_m - Z_{water}}{Z_m + Z_{water}}$ at the interface and thereby determines the normal surface displacement amplitude A_s relative to the incident acoustic wave amplitude A_p ($A_s = A_p(1-R)$ for normal incidence).

3) the acoustic losses in the material, which indicate whether multiple reflections within the faceplate will present a problem, and also whether the faceplate may be used with the reflective coating on the air-solid interface;

4) the optical smoothness and flatness of the material surface, since it acts as a reflector for the scanning laser beam.

Plexiglass, the material used in the prototype tank; 1) has an acoustic velocity of about 2680 m/s[4] giving a critical angle of 34°; 2) has an acoustic impedance of 2.1 (normalized to that of water) resulting in a reflection coefficient $R = \frac{2.1-1}{2.1+1} = .32$; 3) has an acoustic loss of approximately 3 db/cm[4] at 2.5 MHz, which only causes a drop in displacement amplitude of the solid-air interface by a factor of .84 compared to the solid-water interface in a 1" thick plate (this small change in displacement is partly due to the fact that the incident and reflected waves add at the perfectly reflecting air-solid boundary to give a displacement twice that of a normally incident compressional wave.) The loss is large enough, however, to prevent multiple reflections within the faceplate from being significant; 4) has a surface sufficiently smooth and flat to be used as a satisfactory reflector when coated with Gold or Aluminum. It has been found that the surface at the solid-water interface soon becomes so badly distorted, due to water absorption, as to be unuseable. The acoustic camera prototype, therefore, uses the solid-air interface as the reflecting surface.

Other plastics, e.g. polycarbonate and polystyrene have lower velocities and impedances than Plexiglass, which makes them theoretically more desirable as a faceplate for reasons 1) and 2) above. The commercially available sheet stock of these materials, however, do not meet the smoothness requirement 4). It is possible that, with the correct casting techniques, the above materials could be made into superior faceplates (e.g. with a critical angle of 42° and acoustic reflectivity of .26 for polycarbonate). This has not as yet been thoroughly investigated.

Even some samples of the Plexiglass deviated from flatness enough to noticeably distort the light pattern in the image plane of the scanner's exit pupil. This distortion, coupled with imaging aberrations due to the lens in front of the faceplate, caused the reflected laser beam to move off of the knife edge in the scanner's image plane, thereby causing loss of signal on certain portions of the faceplate. This condition forced a search for an acoustic detection technique which was less sensitive to low spatial frequency deviations from flatness of the faceplate, the most common distortion encountered.

ACOUSTIC DETECTION TECHNIQUES

The spatial frequency response and pertinent characteristics of some laser probe techniques for acoustic surface perturbation detection (including the knife edge technique) have been discussed in detail previously[2]. It will be immediately obvious from Fig. 1 that the system cannot tolerate changes in faceplate flatness which would move the laser beam off the knife edge. Three techniques were developed to overcome this defect. These consisted of A) placing a diffusing material in front of the knife edge, B) substituting a wedge attenuator for the knife edge, and C) instead of the knife edge using a grating, placed in an appropriate plane.

Techniques A and B both trade off signal power for system tolerance to spot shift in the scanner image plane. The action of the diffuser, placed slightly in front of the knife edge plane, produces an enlarged diffuse spot which is periodically deflected over the same distance in the knife edge plane as is the spot in the undiffused case. The fractional periodic change in the light power passing the knife edge is inversely proportional to the spot diameter in the knife edge plane. The tolerance on spot position, however, is directly proportional to spot diameter; hence the trade-off between this characteristic and photodetector signal current. It can be simply shown that a wedge optical attenuator (i.e. a neutral density filter with a linearly varying attenuation along one dimension) achieves this same trade-off.

Figure 5(a) shows an acoustic hologram of a focused transducer point source taken using the set-up of Fig. 1 with a slightly distorted faceplate. Figure 5(b) shows the improvement in display uniformity achieved with a diffusing paper placed .5 mm in front of the knife edge.

When the knife edge is replaced by a grating placed in an appropriate plane, faceplates distorted in a manner not correctable by methods A and B above yield a uniform display with no loss in sensitivity compared to the knife edge technique. This grating technique has been used previously[2,5] as a narrow spatial frequency bandwidth detector of acoustic surface waves. It makes use of the fact that a surface having a moving sinusoidal displacement perturbation impresses a moving phase ripple on a beam of laser light reflected off it. This phase corrugation is

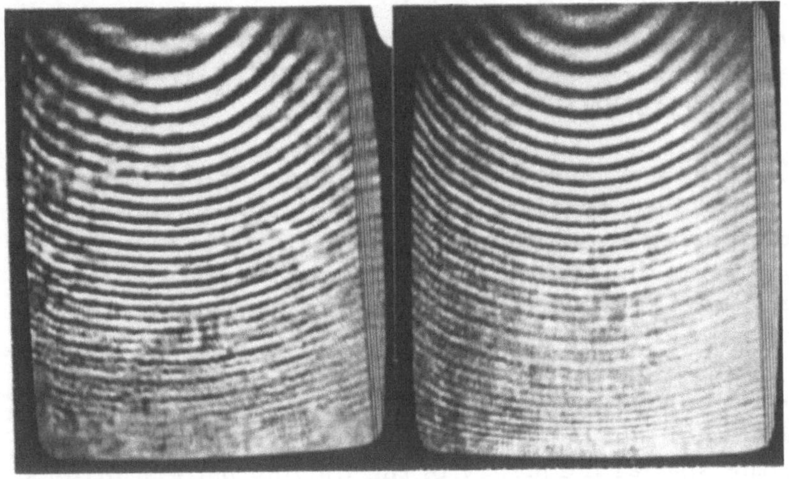

Fig. 5(a) Fig. 5(b)

Fig. 5. (a) Acoustic hologram of a focused transducer point source taken with a slightly distorted faceplate using the knife edge technique. (b) Improvement in display uniformity achieved with a diffusing paper placed 0.5 mm in front of the knife edge.

transformed into a moving amplitude ripple at planes a distance $Z_n = (n+\frac{1}{2})(\frac{\Lambda^2}{\lambda})$ from the surface. Here Λ is the surface ripple wavelength, λ is the wavelength of the light, and $n = 0,1,2 \ldots$. If an amplitude grating, having the same spatial wavelength Λ, is placed in one of these planes (sometimes called Fresnel image planes), the light passing through it will be amplitude-modulated at the acoustic frequency which caused the surface perturbation. When the light beam has a diameter d which is very large compared to the wavelength Λ, the amplitude modulation on the light passing through the grating drops sharply to zero for any surface ripple wavelength not equal to the grating wavelength Λ_g. Such a wide beam system has been the only grating type used previously. However, it can be seen that decreasing the beam diameter increases the uncertainty, 1/d, in the spatial frequency of the modulation impressed on the laser beam. This in turn leads to a correspondingly greater bandwidth in the spatial frequency detection capabilities of the system.

The effect of laser beam scanning can be extrapolated from the above description of the wide beam system by superimposing a moving aperture of size d in the grating plane. When the spatial frequency of the moving ripple pattern exactly matches that of the grating, the relative phase of the two is the same everywhere on the grating. Thus, moving the aperture to different parts of the pattern can cause no change in phase of the amplitude modulation on the light passing through the grating. If, however, the spatial frequency of the moving pattern differs from that of the grating, their relative phase at any instant changes continuously from one grating opening to the next. Here, then a uniformly moving "sampling" aperture will introduce a uniformly time changing phase (i.e. a frequency shift) to the amplitude modulation of the light passing the grating. This "Doppler shift" in detected frequency caused by the laser scanning action is similar to that experienced in the knife edge case, the only difference being that it is zero when the spatial wavelength of the acoustic perturbation matches the grating wavelength while the knife edge Doppler shift is zero when the acoustic perturbation wavelength is infinite.

A more rigorous analysis of this technique will be given in a later publication, which will also include a description of some alternate modes of operation not used here. It is sufficient for this discussion to summarize the characteristics of the grating technique as being comparable in detection sensitivity to that of the knife edge technique and as having a spatial frequency bandwidth (defined to the zero response points) of $\frac{2}{d}$ as compared to $\frac{1}{d}$ for the knife edge technique. Since the spot size is not the limiting element in the system resolution, the factor of 2 improvement mentioned above is not as significant a difference between the two systems as is the increased tolerance to faceplate distortion afforded ty the grating technique. A low spatial frequency distortion of the faceplate causes a shift in position of the light field on the grating, but no loss of detector signal current. Figure 6(a) shows a holographic display of the near field of a flat 5 cm x 5 cm transducer obtained by using the knife edge technique with a badly distorted faceplate. Figure 6(b) shows the improvement that results from the use of a grating.

Fig. 6(a)

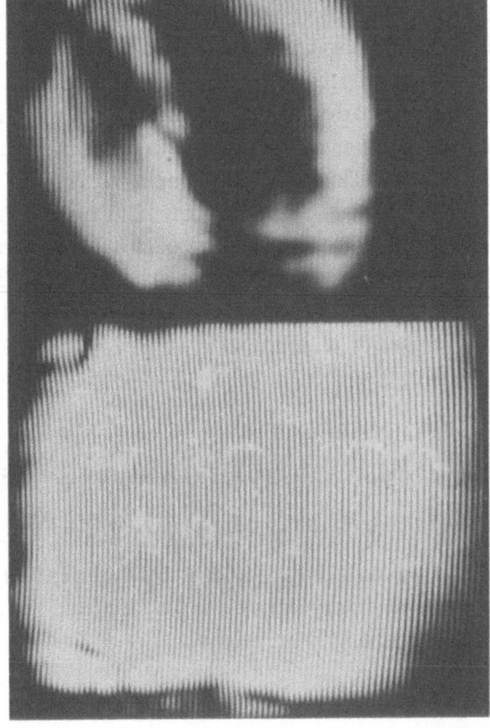

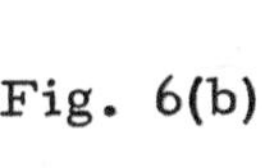
Fig. 6(b)

Fig. 6. (a) Holographic display of the near field of a flat transducer using the knife edge technique with a badly distorted faceplate. (b) Improvement resulting from the use of the grating.

SYSTEM RESOLUTION

The resolution of the total system is determined by a combination of the responses of the acoustic reflector, the faceplate and the acoustic detector effects discussed above.

The acoustic reflector of Fig. 3 subtends an angle θ of approximately 30° for any spot in the image plane. The resolution in this case would be of the order of $r = \Lambda_s/(2 \sin \theta/2) \approx 2\Lambda_s$ (where Λ_s is the acoustical wavelength) if the total angular input were usable. The corresponding spatial frequency bandwidth ΔF equals $\frac{1}{2\Lambda_s}$. It may be seen from Fig. 3, however, that while image points near the edge of the tank receive rays from the reflector making angles 0° - 30° with the faceplate normal, points near the center of the tank receive rays at ±15°. Since the response of the knife edge or grating detection technique goes to zero for acoustic waves at 0° to the faceplate normal, it can be seen that the overall system response is not spatially invariant. There are plans to correct this by the addition of two transmission type acoustic lenses, but for this system we can describe the useful spatial frequency bandwidth of the reflector as being between $\frac{1}{2\Lambda_s}$ and $\frac{1}{4\Lambda_s}$.

The faceplate response as shown in Fig. 4 goes to zero at $\theta = 34°$ where the spatial frequency $F = \frac{\sin \theta}{\Lambda_s} = \frac{.559}{\Lambda_s}$. Although the shape of this low pass response can be modified by the use of resonant faceplates, as shown in the appendix, the upper limit on faceplate bandwidth cannot exceed the figure given above.

Similar to the knife edge, the grating acoustic detection technique has a response which goes to zero at 0° incidence. When the laser beam diameter, d, is adjusted so that the upper spatial frequency limit is equal to that imposed by the faceplate, the combined faceplate-laser scan detection response is that shown in Fig. 7.

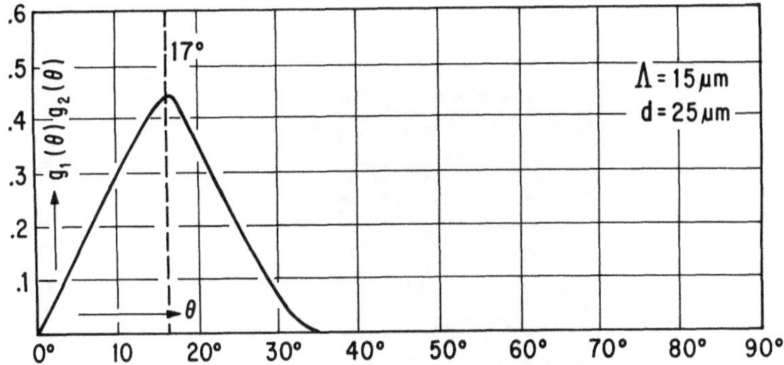

Fig. 7. Overall angular response of the combined faceplate laser scan detection system (after Korpel and Kessler[3]).

The frequency response of the electronics and CRT offer no limitation to the system and need not be discussed here.

In summary, the composite system is capable of detecting acoustic plane waves incident on the faceplate from 0°-30° near the outer edge of the tank and 0-15° near the center of the tank. With this angular aperture the system should resolve elements between $2\Lambda_s$ and $4\Lambda_s$ (i.e. between 1.3 and 2.6 mm at 2.25 MHz).

EXPERIMENTAL RESULTS

The measured system sensitivity of 2.4×10^{-5} W/cm^2 for a signal-to-noise ratio of 1, is within a factor three of that theoretically predicted for the configuration of Fig. 1[1] with a system bandwidth of 2 MHz and a laser power of 1.4 mW at the knife edge. A display of an object containing well resolved 2 mm holes on 4 mm centers indicated that the resolution predicted above is obtainable. Figure 8 shows a composite acoustic transmission picture of a hand. This was obtained from a series of photographs of the system display. A 5 cm x 5 cm quartz transducer provided the insonification at a level of 40 mW/cm^2.

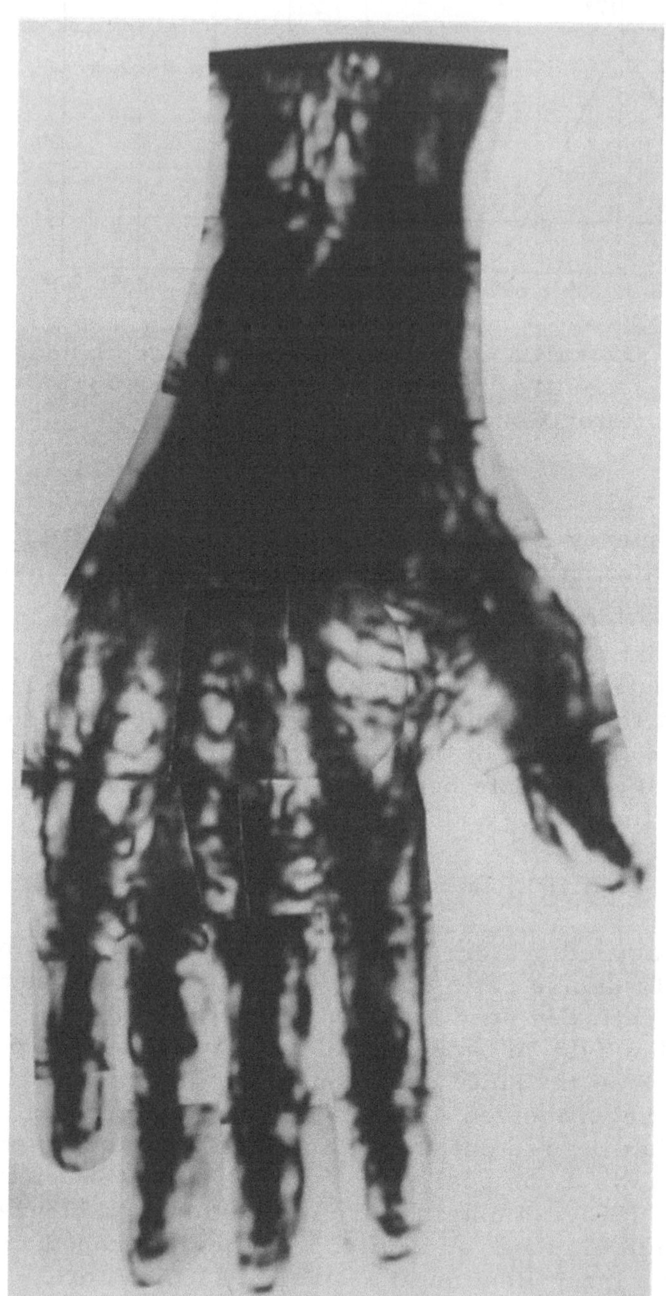

Fig. 8. A composite acoustic transmission picture of a hand, taken at 2.25 MHz.

DEPTH GATING MODE OF OPERATION

The system has been operated in a novel mode in which pulsed sidewise acoustical insonification is used. In this mode one obtains reflection images with highly improved depth resolution.[6] The pulses of sound, which occur once every horizontal laser scan time, cause an acoustic image to be formed in the faceplate plane as a moving vertical line of sound. When the detecting laser beam scans in synchronism and at the same rate as the moving sound line, it gates out sound images reflected from planes in front of or behind the desired object plane. This is because these undesired image lines arrive at the faceplate before or after the detecting laser beam has arrived. The depth resolution obtainable in this mode is approximately equal to the length of the sound pulse in water (e.g. 1.5 mm for a 1 μsec pulse). The operation is described in more detail in reference 6.

Figure 9(b) shows the reflection image of three 6.3 mm diameter brass rods sidewise insonified by .13 W/cm^2 of CW sound from a 5 cm x 5 cm transducer as shown in Fig. 9(a). The overlapping and interference of the 3 images is apparent. When the sound is emitted as 6 μsec pulses, having the same peak power as above, the three rods may be separately resolved as shown in Fig. 9(c), (d) and (e). The desired object plane is selected by adjusting the time delay between the start of the horizontal laser scan and the initiation of the acoustic pulse.

This mode of operation, which allows displays with the transverse resolution of an imaging system (about 2 Λ_s) and the depth resolution of a pulse echo system (also about 2 Λ_s), is an important new feature of the acoustic camera system. It should be noted that the gating effect in this mode does away with multiple reflection artifacts common in pulse echo systems, as well as having the property of eliminating the out-of-focus spurious images from undesired cross sections of thick objects.

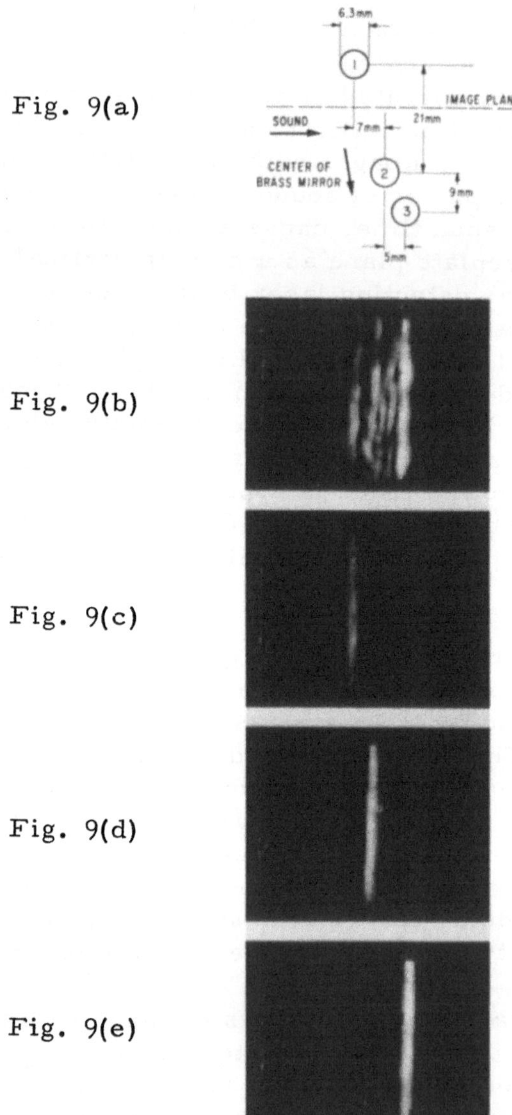

Fig. 9. Depth gating mode of operation (a) rod configuration, (b) monitor display with continuous insonification, (c), (d) and (e) display with pulsed insonification, depth gating the planes of rod (1), (2) and (3) respectively (after Whitman et al[6]).

APPENDIX

The response of the plexiglass-water interface, in terms of normal surface displacement, shown in Fig. 4, assumes that no acoustic energy is reflected back to this surface from the plexiglass-air interface of the plate. This is a good assumption when the plate is thick and has a significant attenuation constant, as is the case for example, for a one-inch thick plexiglass plate at 2 MHz. If either of these requirements are not met, the signal reflected back from the plexiglass-air interface is strong enough to interfere with the original signal, and this modifies the response characteristic of the surface. As an example, consider the theoretical response curves of the plexiglass-water surface at 2.25 MHz shown in Fig. Al,

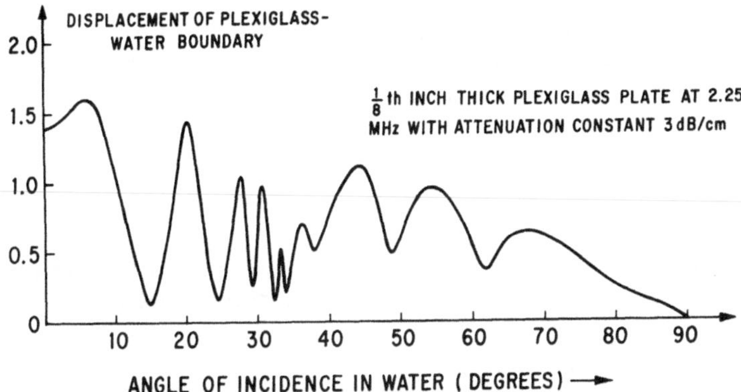

Fig. Al. Response of the plexiglass-water interface of a 1/8 inch thick plate.

for a 1/8 inch thick plate, and in Fig. A2, for a one-inch thick plate. The displacements shown are normalized to the displacement in water. An attenuation constant of 3 db/cm within the plexiglass is assumed in each case. The similarity between Figs. 4 and A2 is quite evident, especially in the region beyond the critical angle. In contrast the response of the 1/8 inch plate (Fig. Al) is

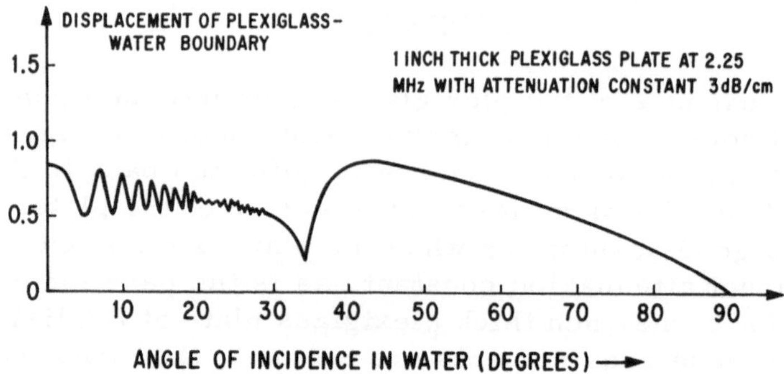

Fig. A2. Response of the plexiglass-water interface of a 1 inch thick plate.

characterized by many large resonance peaks. Figures A3 and A4 are the corresponding response curves of the plexiglass-air interface at 2.25 MHz.

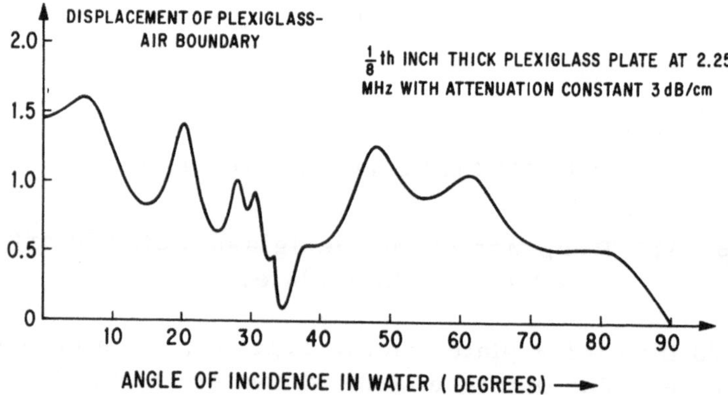

Fig. A3. Response of the plexiglass-air interface of a 1/8 inch thick plate.

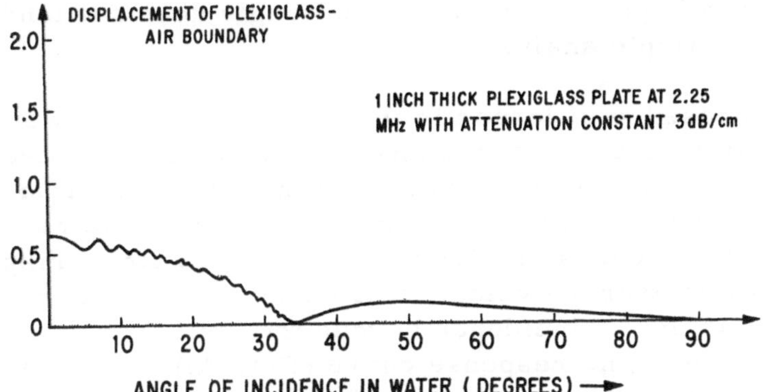

Fig. A4. Response of the plexiglass-air interface of a 1 inch thick plate.

The curves A1 → A4 are obtained from an exact analysis of the stresses and strains in the bulk and on the two surfaces of the plexiglass plate, and include the effects of internal friction and evanescent longitudinal waves within the solid. The exact analysis is too lengthy to be presented here and will be deferred to a later publication.

It is interesting to note, however, that the angles at which the minima and maxima occur correspond to those situations where the plate thickness d is given by

$$d = \frac{n}{4} \Lambda_{eff} \tag{A1}$$

where n is an integer and Λ_{eff} denotes the effective longitudinal - or shear wavelength inside the plate. This effective wavelength is defined by the intersection of the wave fronts with the normal to the surface.

In analyzing the response curves in such a manner it should be taken into account that longitudinal waves are predominant at angles of incidence below the critical angle and that only shear waves are generated beyond the critical angle. The region near the critical angle does not lend itself to a simple analysis.

The presence of losses in the plate generally lowers the quality factor, Q, of the resonant system. The peaks on the response curves therefore diminish in amplitude while the valleys increase. This explains why the response curve of a thick lossy material consists of a small ripple around some average value. Resonance due to shear waves at angles of incidence beyond the critical angle are quite evident in the response curve (Fig. A1) of a 1/8 inch thick plate with a loss constant of 3 db/cm; but the same are not visible in the response curve (Fig. A2) of the one-inch thick plate with the same loss constant.

To a certain extent these losses are desirable in a face-plate material since they cause the amplitude of the ripple on the response curve to be small. Another important feature of faceplates made from lossy material is that the displacement of the water-solid boundary is not zero at the critical angle, as can be seen from Fig. A2. The usable spatial bandwidth of the plate is thus increased which imparts to the imaging system the potential for higher resolution.

The relationship between the displacement at the critical angle and the losses within the solid can be understood by considering Fig. A5. Here the longitudinal wave travels at an angle θ_2 to the normal $\bar{n}$. The phase fronts cut the x-axis giving rise to a component of displacement u_x in the x-direction which is different (at least in phase) for different z-planes. Because of this there exists a tangential strain $\partial u_x/\partial z$ at the boundary. It is this strain, of course, that generates a shear wave within the solid at an angle γ to $\bar{n}$. At the critical angle, when θ_2 is 90°, the phase fronts of the longitudinal waves cut the boundary normally. In the absence of losses within the solid the planes of equal phase are parallel to the planes of equal amplitude and the strain $\partial u_x/\partial z$ is zero. It is clear that under these conditions no shear wave can be generated and that the displacement normal to the boundary is also

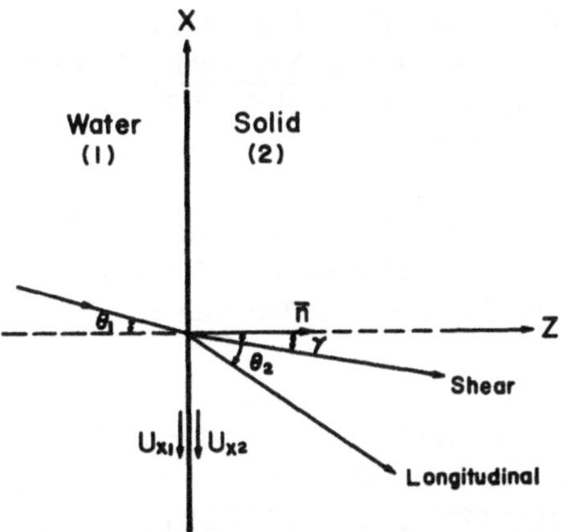

Fig. A5. Diagram used for the explanation of the critical angle response in lossy media.

zero. In lossy media, however, it can be shown, that planes of equal phase are not parallel to planes of equal amplitude. The amplitude of a plane wave in the plane of constant phase is not uniform in this case. Therefore the strain $\partial u_x/\partial z$ does not vanish at the critical angle and consequently a shear wave is generated which, in turn, causes a displacement normal to the boundary. This is the explanation of the finite critical angle response shown by all the curves of Figs. A1 → A4.

REFERENCES

1. A. Korpel and P. Desmares "Rapid Sampling of Acoustic Holograms by Laser Scanning Techniques", J. Acoust. Soc. Am., Vol. 45, No. 4, April 1969, pp. 881-884.

2. R. L. Whitman and A. Korpel "Probing of Acoustic Surface Perturbations by Coherent Light", Applied Optics, Vol. 8, August 1969, pp. 1567-1576.

3. A. Korpel and L. W. Kessler "Comparison of Methods of Acoustic Microscopy", Acoustical Holography, Vol. 3, A. F. Metherell (ed.), Plenum Press, New York, 1971, Chapter 3.

4. W. P. Mason and H. J. McSkimin "Mechanical Properties of Polymers at Ultrasonic Frequencies", Bell Syst. Tech. J., Vol. 31, No. 1, January 1952, pp. 122-171.

5. A. Korpel, L. J. Laub and H. C. Sievering "Measurement of Acoustic Surface Wave Propagation Characteristics by Reflected Light", Appl. Phys. Lett., Vol. 10, No. 10, May 15, 1967, pp. 295-297.

6. R. L. Whitman, A. Korpel and M. Ahmed "A Novel Technique for Real Time, Depth Gated Acoustic Image Holography", Appl. Phys. Lett., Vol. 20, No. 7, May 1, 1972, pp. 370-371.

7. A. Korpel, L. W. Kessler and P. R. Palermo "Acoustic Microscope Operating at 100 MHz", Nature, Vol. 232 No. 5306, July 9, 1971, pp. 110-111.

REAL-TIME RECONSTRUCTION OF IMAGES FROM HYDROACOUSTIC HOLOGRAMS

J. L. Weaver and G. C. Knollman

Lockheed Research Laboratory
Palo Alto, California 94304

ABSTRACT

Three systems capable of reconstructing images from acoustic holograms are described. In the acoustic reconstruction system the received hydroacoustic hologram is stored as a modulation pattern on a square transmitting array of piezoelectric transducers. An experimental electronic reconstruction system employs digital multiplication and analog summing to transform a 10×20 element, phase-only hologram into a 10×10-element image. In the third system a stored hologram is scanned zone by zone. Electronic analog processing of the resulting signal generates the intensity at each image point.

INTRODUCTION

Satisfactory reconstruction of images from acoustic holograms currently is achieved by optical surface-relief methods[1,2], by electrooptic devices[3], and by digital computers[4]. Three additional methods proposed in this paper are acoustic reconstruction, digital-analog processing, and zone scanning. These methods avoid the wavelength scaling problem encountered in optical reconstruction. The proposed schemes are intended for real-time construction of images containing in the order of 10^3 to 10^4 resolvable elements. In view of the slow-motion nature of underwater phenomena, "real-time" reconstruction here will include scan rates as low as 2 frames/sec.

HYDROACOUSTIC SCANNING SYSTEM

The reconstruction techniques described herein were proposed and partially developed for use with a previously described hydroacoustic scanning system[5]. This system consists of a plane, oscillatory scanning mirror and a linear array of 100 transducer elements. The acoustic mirror sweeps the field of object waves across the transducer array at a rate of 2.4 frames/sec. The array is sampled electronically at a rate of 240 lines/sec to provide holographic video output equivalent to approximately 7500 elements/frame.

Phase reference is provided electronically by modulating the received signal with a voltage having constant amplitude and a frequency identical to that of the acoustic source (2.5 MHz). When the phase of the reference voltage is constant relative to that of the source a cylindrical hologram is obtained. The cylindrical phase-reference surface results from the mirror scanning motion which sweeps the object field through an arc about the mirror axis. To reconstruct an image from a cylindrical hologram the cylindrical reference surface must be simulated and properly illuminated.

The reconstruction process can be simplified considerably if the surface of constant reference phase is a plane tangent to the cylindrical scan surface of the mirror scanning system. The required plane reference surface has been simulated electronically by phase-modulating the reference voltage in a manner which compensates for the phase difference between the cylindrical scan surface and a tangent plane arbitrarily oriented with respect to the viewing axis[6]. The simulation is valid for small angular apertures ($<20^\circ$) and for object distances much greater than the mirror scanning radius.

ACOUSTIC RECONSTRUCTION

To construct images from the simulated plane holograms obtained with the mirror scanning system an experimental acoustic reconstruction system has been designed. An acoustic image is reconstructed by transmitting hydroacoustic waves from a two-dimensional array of transducers. The transducers are driven in phase with a distribution of amplitudes which corresponds to the interference pattern detected at the hologram receiver. The required distribution of amplitudes is obtained from the multiplexed video output of the linear receiving array. Similar output from a rectangular receiving array or from a single receiving element mechanically scanned could also be used.

The square transmitting array is depicted in Fig. 1. The mozaic was cut from a 2"×2"-square of PZT4. The opposite face of the mozaic has a continuous ground electrode which is exposed to the water in the imaging tank. The 2mm × 2mm-transducers in the array are driven in phase from a 2.5-MHz source. Amplitude at each element is controlled by a field-effect transistor (FET) acting as an amplitude modulator. To distribute the holographic modulation pattern on the array of modulators, the video signal from the receiving array is demultiplexed and stored as a charge on a capacitor at the gate of each FET. The stored charge decays and is renewed with each hologram frame. A frame of only 20×20-elements is used with the present experimental array.

Each element in the array emits a hydroacoustic wave having an amplitude proportional to the amplitude at a corresponding element in the received hologram. In the far-field region the wavefront generated by a transmitting element is approximately spherical over a limited aperture and corresponds to a Huygens wavelet. Superposition of the acoustic wavefronts emitted by the array reconstructs the object wave field as received by a planar receiving array (or by the simulated plane array tangent to a cylindrical scan surface). Thus the hydroacoustic reconstruction array is analogous to an "acoustic transparency" of the hologram which is illuminated with plane coherent waves.

In the present system the wavelength for acoustic reconstruction equals that of the recording waves. Since the simulated reference and reconstructing wavefronts are plane, the acoustic image is constructed at a distance from the array equal to the object distance. This image can be displayed on an oscilloscope by scanning the image plane with a hydroacoustic image converter. Zero-order radiation at a

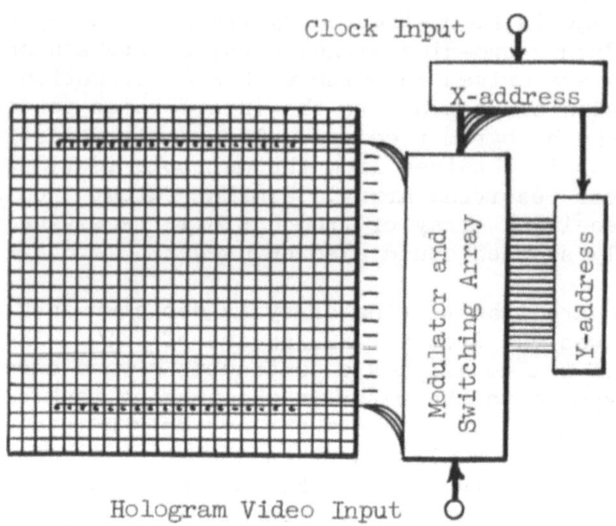

Figure 1. Transmitting array for reconstruction of an acoustic image

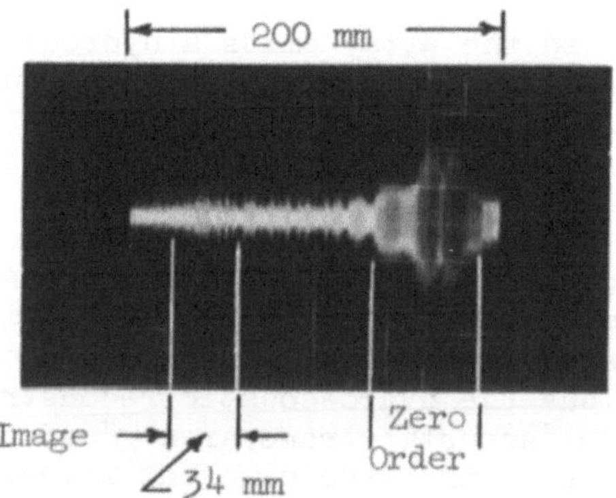

Figure 2. Line-scan through the reconstructed acoustic image of a simulated, off-axis point source

high level appears on axis but can be excluded from an off-axis image field.

Since the modulator array has not been completed only preliminary tests have been made of the acoustic reconstruction array. A spatial modulation pattern was distributed over the array by supplying excitation to the transducers through an electrode pattern which simulated a phase-only hologram of an off-axis point source. A profile through the real image of the source was scanned with a fixed linear receiving array. In the amplitude profile reproduced in Fig. 2 the reconstructed source image is visible near the left end. Resolution is degraded by the small range of spatial frequencies present in the electrode pattern. Most of the acoustic power appears in the zero-order radiation from the array. An off-axis image field which excludes the zero-order radiation can be obtained by simulating off-axis reference waves to produce the hologram.

A cylindrical reconstruction array could be used to construct images from cylindrical holograms produced by the mirror scanning system. For this purpose the transducers would be mounted on a cylindrical arc which has a radius of curvature scaled to the mirror scanning radius in the ratio of reconstruction wavelength to recording wavelength. If these wavelengths are equal, a single, linear receiving array can be used with the scanning mirror for acoustoelectric conversion of the reconstructed image as well as the hologram field. For example, by increasing the mirror scanning angle both acoustic inputs to the converter can be encompassed in a single sweep of the mirror.

A full-scale acoustic reconstruction array would contain a number of transducers and modulators equal to the number of resolvable elements in the hologram. Distortionless demagnification (to about $\frac{1}{4}$) would be obtained by scaling down both the array dimensions and the reconstructing wavelength. The chief disadvantage of acoustic reconstruction is that the image must be viewed by means of an acousto-electric converter. However, the acoustic image is constructed in real time and it is free of the type of distortion usually encountered in optical reconstruction.

DIGITAL-ANALOG PROCESSING

Our second approach to real-time image reconstruction employs digital sampling of a phase-only hologram, followed by analog summing of the weighted samples. Reconstruction of acoustic images by means of general-purpose digital computers approaches real-time display when the diffraction integral can be expressed in a form compatible with the "fast Fourier transform"[7,4]. Rapid reconstruction usually requires that the Kirchoff or Fresnel approximation be used. The analog-summing approach avoids approximations, other than quantization of phase and amplitude, and it permits rapid summation of products involved in a discrete analog of the diffraction integral. However, the method assumes a fixed image plane and therefore does not lend itself to focusing of the image or reconstruction in many image planes.

The diffraction integral can be expressed in discrete form by the double summation,

$$A_{mn} = \frac{i}{\lambda} \Delta x' \Delta y' \sum_{a=1}^{M} \sum_{b=1}^{N} H_{ab} F_{mn}^{ab} \tag{1}$$

where $k = 2\pi/\lambda$, (λ = the acoustic wavelength); A_{mn} is the amplitude at an image point located at coordinates $m\Delta x$, $n\Delta y$ in the image plane; H_{ab} is the quantized amplitude of the received signal at coordinates $a\Delta x'$, $b\Delta y'$ in a hologram containing $M \times 2N$ elements; and

$$F_{mn}^{ab} = \frac{e^{-ik\rho}}{\rho} = \frac{\cos k\rho - i \sin k\rho}{\rho} . \tag{2}$$

$\rho(a,b;m,n)$ is the path length,

$$|\bar{r} - \bar{r}'| = [(m\Delta x - a\Delta x')^2 + (n\Delta y - b\Delta y')^2 + z^2]^{1/2} \tag{3}$$

from a point $r'(a\Delta x', b\Delta y', 0)$ in the hologram plane to a point $r(m\Delta x, n\Delta y, z)$ in the image plane located at a fixed distance z from the hologram plane. Equation (1) is an approximation to the Fresnel-Kirchoff diffraction integral:

$$A(\bar{r}) = \frac{i}{\lambda} \iint_H H(x',y') \frac{e^{-ik\rho}}{\rho} dx'dy' \tag{4}$$

The real and imaginary parts of A_{mn} are

$$\text{Re } A_{mn} = \frac{k\Delta x'\Delta y'}{\lambda} \sum_a \sum_b H_{ab} \frac{\sin k\rho(a,b;m,n)}{k\rho(a,b;m,n)} \tag{5}$$

$$\text{Im } A_{mn} = \frac{k\Delta x'\Delta y'}{\lambda} \sum_a \sum_b H_{ab} \frac{\cos k\rho(a,b;m,n)}{k\rho(a,b;m,n)} \tag{6}$$

These double summations are performed in an analog summing network in the following manner. The amplitude H of a phase-only hologram is amplified to digital logic levels 0,1. H_{ab} is then fed serially into N parallel-output shift registers, each having a capacity of 2M bits. The outputs are permanently connected to two summing networks which attenuate H_{ab} in proportion to the phase factors in Eqs. (5) and (6). A matrix of required attenuation factors for $(\sin k\rho)/k\rho$ and for $(\cos k\rho)/k\rho$ is computed for the set of path lengths $\rho(a,b;\ m,n)$ between hologram coordinates $a\Delta x'b\Delta y'$ and fixed image coordinates $m\Delta x$, $n\Delta y$.

Between each M-bit line of the hologram an M-bit line of zeroes is entered into the registers until the complete frame of H_{ab} occupies the (first-M) $\times$ N addresses in the registers. The parallel outputs then are weighted by the two phase-factor matrices and summed as currents in two diode summing networks. The intensity I_{mn} at the first image element, $m = n = 1$, is obtained by squaring the analog sums in analog multipliers:

$$I_{mn} = (\text{Re } A_{mn})^2 + (\text{Im } A_{mn})^2 \tag{7}$$

As the contents of the N registers are simultaneously shifted from $m = 1$ to $m = 2,3,4,\cdots M$, the first line of the image intensity is displayed. At the end of the line H_{ab} occupies the (last-M) $\times$ N addresses. H_{ab} is then recycled into the (first-M) $\times$ (N-1) addresses. The bottom line of H_{ab} is simultaneously discarded and a new line of H_{ab} from the receiver is entered into the first M positions on line 1. Subsequent shifting of the N registers generates the intensity I_{m2} along the second line of the image display. In a similar manner the entire image is displayed at $\frac{1}{2}$ the line frequency of the hologram receiver. To compensate for the alternate blank lines during reconstruction the receiver must scan consecutive lines at $\frac{1}{2}$ the resolvable line width.

An experimental model of this system has been constructed to determine cumulative effects of errors in the summing network. The model accepts only a 10 × 20 hologram, it uses very coarse phase increments ($\Delta k\rho = \pi/2$), and it assumes a constant value for ρ in the denominator of Eq. 2. The model is shown schematically in Fig. 3. For each discrete ± value of sin $k\rho$ the corresponding attenuated outputs of the shift registers are fed through diodes to the ± inputs, respectively, of a differential amplifier, and similarly for cos $k\rho$. Images have not as yet been constructed with the experimental model, but the parameters of the model have been used to calculate the amplitude profile for an image of an off-axis point source. Although the profile has the predicted width, the amplitude is poorly defined as a result of the small number of elements in the hologram.

A full-scale reconstruction system capable of accomodating a hologram containing 100 × 100-elements would require 200 shift registers, each having 200-bit capacity. Nevertheless, images could be constructed at high frame rates, since clock rates above 10 MHz are feasible. This scheme appears to be competitive with fast-Fourier-transform methods as applied to general-purpose computers.

ZONE SCANNING

A third reconstruction scheme which we have considered is that of scanning a hologram zone by zone. A hologram may be viewed as a group of overlapping sinusoidal zone plates which correspond directly to the group of point sources comprising the object. Each zone plate diffracts and focuses the reconstructing waves upon a particular image point. The amplitude at the image point is determined by the transmittance of the associated zone plate. Other zone plates in the hologram also diffract waves to the same point but these wavefronts arrive with random phases and, when averaged over the entire hologram, do not contribute significantly to the amplitude. Thus, an image could be

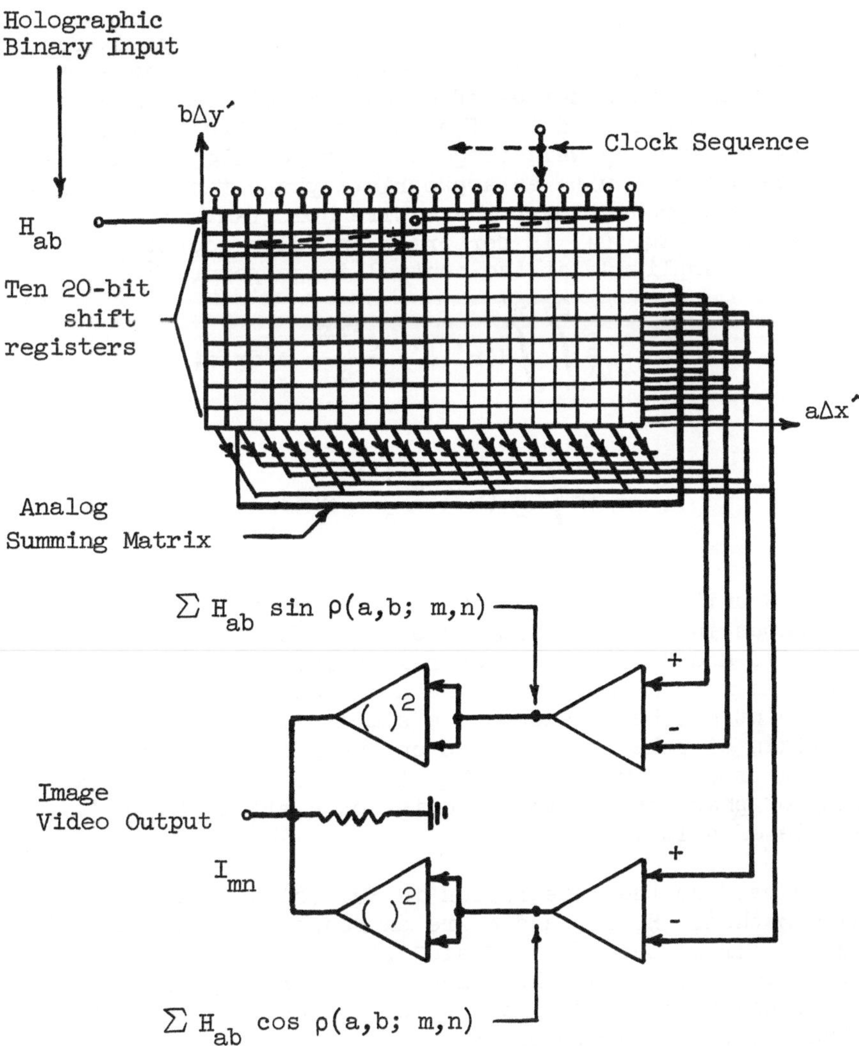

Figure 3. Schematic representation of an experimental digital-analog system for image reconstruction

constructed point by point if the hologram were scanned with a matching zone plate, i.e., one having a focal length matched to that of the hologram.

The electronic equivalent of a matched zone plate for a particular image point can be realized by scanning a stored hologram in a series of concentric circles on each of which $k\rho$ is a constant. The amplitude at the image point is obtained by weighting the average hologram amplitude on each circle with an appropriate phase factor and integrating the product over the entire hologram. When the integration in Eq. (4) is performed in polar coordinates, r', ψ', the amplitude at the central image point, $\bar{k}z$, is

$$\begin{aligned} A(\bar{k}z) &= \frac{i}{\lambda}\int_0^{2\pi}\int_0^{r'} H(r',\psi')\,\frac{e^{-ik\rho}}{\rho}\,r'dr'd\psi' \\ &= ik\int_0^{r'} \underline{H}\,(r')\,\frac{e^{-ik\rho}}{\rho}\,r'dr' \end{aligned} \tag{8}$$

where $\bar{k}$ = unit vector along the Z-axis, r', ψ' are polar coordinates in the hologram plane, ρ is the path length $|\bar{k}z - \bar{r}'|$, and $\underline{H}(r')$ is the average value of $H(r', \psi')$ on the circle r' = a constant. The amplitude $A(\bar{r})$ at any other image point can be obtained in a similar manner by shifting the origin of polar coordinates on the hologram. The zone-scanning process embodied in Eq. (8) is a two-dimensional analog of phase-sensitive detection of a signal in the presence of noise.

To perform the integration in Eq. (8) by electronic analog methods, $\underline{H}$, r', and ρ are expressed as functions of time, t. If we let $\omega t = \theta \equiv \sin^{-1}r'/\rho$, where θ is the angle between the Z-axis and the scanning vector, $\bar{\rho}$, and ω is an angular scanning speed, ρ and r' can be written in terms of ωt and the fixed parameter z:

$$r' = z \tan \omega t \tag{9}$$

$$\rho = z \sec \omega t \tag{10}$$

Since these functions are not continuous nor easily synthesized, approximate sinusoidal functions or polynomials are preferable even though their limited range of validity restricts the usable aperture.

To simplify the integration in Eq. (8) it is convenient to choose approximations which satisfy $r'dr' \sim \rho d\rho$ (or $zd\rho$), for example,

$$r' = z\omega t \tag{11}$$

$$\rho = [z^2 + (r')^2]^{\frac{1}{2}} = z(1+\omega^2 t^2)^{\frac{1}{2}} \approx z(1+\omega^2 t^2/2) \tag{12}$$

Here r' may be a triangular function of time in a range $0 \leq r' \leq R'$, where R' is the length of the diagonal of the hologram.

If the hologram amplitude $\underline{H}(r')$ on a circle of constant r' is obtained as a voltage $\underline{V}(\overline{t})$ proportional to $\underline{H}[r'(t)]$, Eqs. (11) and (12) can be used to derive an approximation to Eq. (8) suitable for analog computation:

$$A(\overline{k}z) = e^{-ikz} \int_0^{\tau} \underline{V}(t) \exp\left[\frac{-ikz\omega^2 t^2}{2}\right] (ikz\omega^2 t \, dt) \tag{13}$$

where τ is the radial scanning period, $R'/z\omega$. Equation (13) may be integrated by parts to obtain

$$A(\overline{k}z) = -C[\underline{V}(\tau) \exp\left(\frac{-ikz\omega^2\tau^2}{2}\right) - \underline{V}(0)] + C \int_0^{\tau} \exp\left(\frac{-ikz\omega^2 t^2}{2}\right) \underline{V}'(t)dt \tag{14}$$

in which C is a complex constant and $\underline{V}'(t) = d\underline{V}/dt$. The first term depends only on the initial and final values of $\underline{V}(t)$ and can be made to vanish by setting $\underline{V}(\tau) = 0 = \underline{V}(0)$

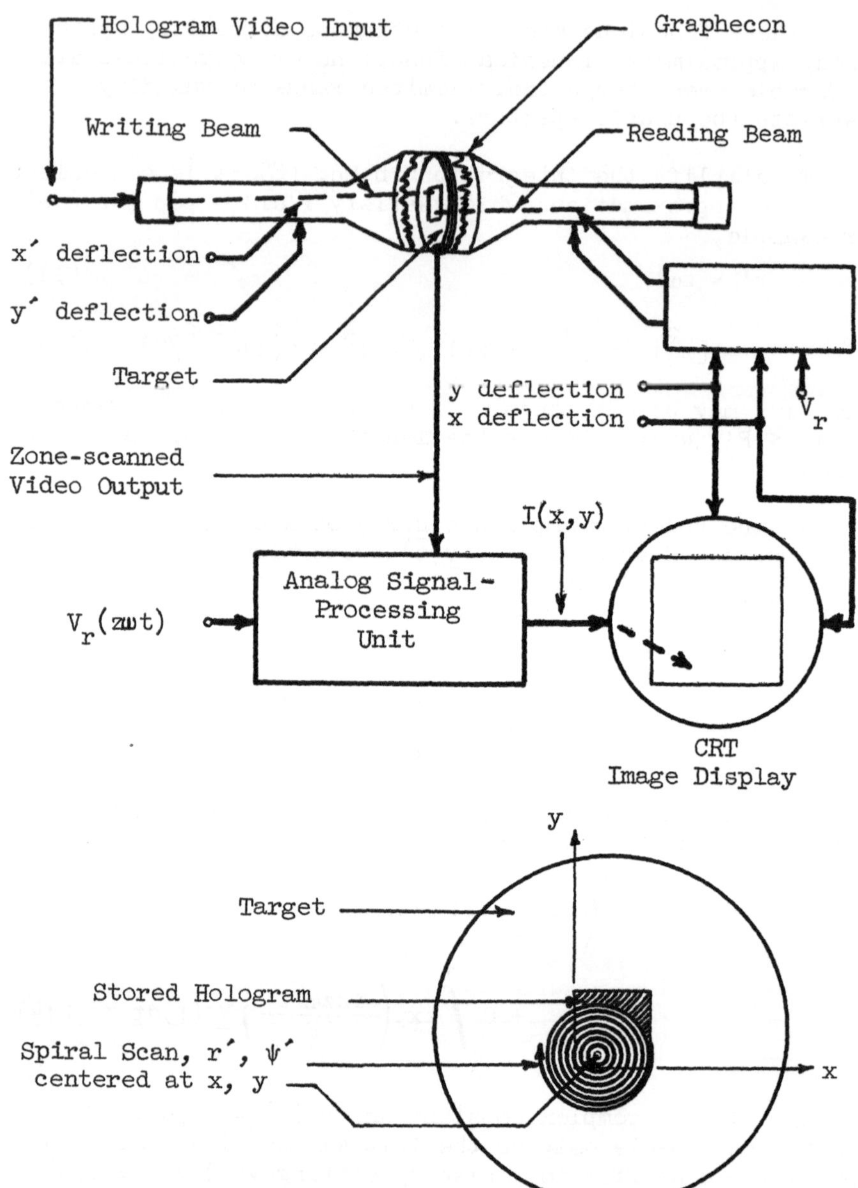

Figure 4. Zone-scanning system for image reconstruction

when the hologram is scanned. The second term may be expanded for purposes of analog computation:

$$A(\overline{k}z)/C = \int_0^T \underline{V}'(t)(\cos\,\omega_c t)\cos\left(\omega_c t + \frac{kz\omega^2 t^2}{2}\right)dt \tag{15}$$

$$+ \int_0^T \underline{V}'(t)(\sin\,\omega_c t)\sin\left(\omega_c t + \frac{kz\omega^2 t^2}{2}\right)dt$$

$$+ i\int_0^T \underline{V}'(t)(\sin\,\omega_c t)\cos\left(\omega_c t + \frac{kz\omega^2 t^2}{2}\right)dt$$

$$- i\int_0^T \underline{V}'(t)(\cos\,\omega_c t)\sin\left(\omega_c t + \frac{kz\omega^2 t^2}{2}\right)dt$$

Equations (13) and (5) are equivalent, but a carrier signal of unit amplitude and angular frequency $\omega_c >> D\omega$ has been introduced in (15) in order to employ phase modulation.

A proposed zone-scanning system is shown schematically in Fig. 4. The received hologram signal is stored and sampled in a scan-conversion tube such as the Graphecon (RCA Types 7539, 4598, or similar). An acoustic hologram is stored as a charge pattern on the target by the writing beam. To determine the intensity at an image point the entire pattern is scanned by the reading beam, which traverses the spiral path indicated in the figure. The spiral is centered at temporarily fixed hologram coordinates which are identical (except for a scaling factor) with the coordinates x, y of a particular image point.

Coordinates in the close-spaced spiral are $r' = z\omega t$ and $\psi' = D\omega t$, where z is the image distance scaled in proportion to the demagnification of the stored hologram relative to the hologram as received; $z\omega$ is the radial scanning speed, $D\omega$ is the angular speed of the spiral scan, and D is approximately equal to the number of resolvable elements on the diagonal of the hologram. Although not written as such,

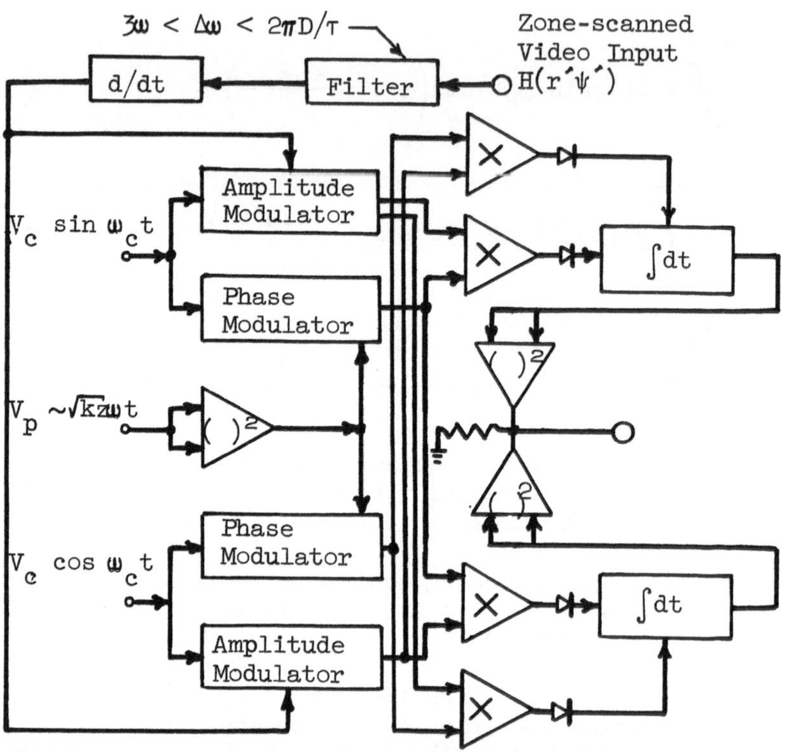

Figure 5. Block diagram of an analog signal-processing unit for zone-scanned holographic video.

the radius r' [as well as the radial sweep voltage $V_r(z\omega t)$] is a triangular function of time having a period $\tau = R'/z\omega$. The range of r' is from 0 to R', the length of the diagonal of the stored hologram.

The video signal produced by the scan-conversion tube is converted to a voltage proportional to the image intensity by means of the analog processing unit shown in Fig. 5. This unit performs the modulation and integration implied in Eq. (15). The required analog multipliers and integrators are indicated in the block diagram. The bandpass filter rejects the high-frequency spectrum which results from the high, spiral, scanning velocity. This filter also rejects low frequencies generated by overscan of the stored hologram into the surrounding "dark" area.

Integration of all radial modulation products over the hologram generates a voltage proportional to the intensity, $I(x,y)$, at the image point x,y. Amplitudes at all other coordinates in the image plane are obtained in the same manner as the center of the spiral scan sweeps through the hologram raster. Dwell time for each spiral scan is provided by a staircase waveform for the X-deflection of the reading beam.

The zone-scanning approach to image reconstruction uses analog processing and thus avoids the coarse quantization of the hologram phase and amplitude employed in the digital-analog system. If one uses the Graphecon for scan conversion, about 7 gray shades can be distinguished in the stored hologram. A 500 x 500-element hologram can be written on the target and stored for about 1/2 sec. However, the entire hologram must be sampled sequentially to obtain the intensity at one point in the image. Sequential sampling is inherently much slower than the parallel sampling realized in the digital-analog method. For example, N^2 complete spiral scans are required per frame to construct an N x N-element image from a hologram containing N^2 elements. The required spiral frequency is at least FN^3, where F is the frame rate. Thus, at a spiral deflection frequency, $D\omega/2\pi = 64$ KHz, a 30 x 30-element image could be constructed at a rate of about 2 frames/sec.

CONCLUSION

Three concepts have been presented relevant to the reconstruction of images from hydroacoustic holograms. The acoustic reconstruction and digital-analog techniques have real-time imaging capability because all hologram elements are sampled in parallel. The zone-scanning technique, being a sequential process, approaches real-time display only for about 10^3 resolvable elements. Although models have been built to explore feasibility of two of these concepts, further analysis is required in order to compare their ultimate performance with that of existing reconstruction techniques.

ACKNOWLEDGEMENTS

This work was supported by the Lockheed Independent Research Program. The authors wish to thank Mr. W. W. Walker for the concept of digital-analog processing.

REFERENCES

1. B. B. Brendon and D. R. Hoegger, "Acoustical Holography with Real-time Color Translation," Acoustical Holography, Vol. 2, Chapter 21, Plenum Press, New York (1970)

2. P. S. Green, "A New Liquid-surface-relief Method of Acoustic Image Conversion," Acoustical Holography, Vol. 3, Chapter 10, Plenum Press, New York (1971)

3. H. R. Farrah, E. Marom, and R. K. Mueller, "An Underwater Viewing System Using Sound Holography," Reference 1, Chapter 12

4. A. L. Boyer et al., "Reconstruction of Ultrasonic Images by Backward Propagation," Reference 2, Chapter 18

5. G. C. Knollman et al., "Experimental Hydroacoustic Imaging System," J. Appl. Phys., 42, 2168-2180 (1971)

6. G. C. Knollman and J. L. Weaver, "Acoustical Holography with a Scanned Linear Array," J. Appl. Phys. (to be published)

7. J. W. Goodman, "Digital Image Formation from Detected Holographic Data," Acoustical Holography, Vol. 1, Chapter 12, Plenum Press, New York (1969)

PRACTICAL HIGH RESOLUTION ACOUSTIC MICROSCOPY

L. W. Kessler, P.R. Palermo and A. Korpel

Zenith Radio Corp., Research Department

6001 W. Dickens Avenue, Chicago, Ill. 60639

INTRODUCTION

This paper describes recent progress towards the development of a practical 100 MHz acoustic microscope whose principle of operation is based upon optically measuring the localized dynamic displacements of a boundary caused by an incident angular spectrum of sound waves.[1] The applicability of this technique was first suggested in Vol. 3 of this series[2] where, in addition, several other possible methods of acoustic microscopy were compared. The term "acoustic microscopy", while not new, has traditionally referred to the visualization of detail in the millimeter range of acoustic wavelengths. The present use of this term denotes imaging detail in the micron range.

Compared with the more traditional aspects of acoustic imaging the field of acoustic microscopy is so new that it is still at the point where a practical instrument must first be fabricated before the usefulness of the technique can be assessed. Applications of an acoustic microscope are foreseen in the areas of biology and medicine as well as nondestructive testing. However, in many of these applications it will be very important to compare the acoustic patterns with the well-known optical images. The technique described here is capable of simultaneous optical and acoustic transmission imaging. This feature should be of great value in the assessment of acoustic microscopy as such.

The present 100 MHz device works in real time and demonstrates resolution of 1.3 wavelengths of sound; viz., 20 μm resolution using a sound wavelength of 15 μm in water. A specimen may be displayed either as an acoustic hologram or an acoustic image. The measured sensitivity is of the order of 10^{-3} W/cm^2 and may be further improved by increasing the frame time or the laser power as described earlier.[2]

The experiments at 100 MHz serve a twofold purpose: First, preliminary applications investigations can be undertaken and secondly, the subsequent greater understanding of the characteristics of the technique provides guidelines for the development of a much higher resolution device, e.g., 2 μ resolution is expected at 1 GHz. Here, the resolution will start to approach that of an ordinary optical microscope.

PRINCIPLE OF OPERATION

A fundamental parameter associated with acoustic wave propagation is the particle displacement within the medium. A sound wave crossing or being reflected from a non-rigid (finite acoustic impedance) boundary will distort the interface as shown schematically in Fig. 1 for an incident plane wave. The boundary displacement is dynamic and the fluctuation occurs at the acoustic frequency. The phase front of the "displacement wave" propagates to the right as shown. In practice the boundary separates water on the bottom and a plastic faceplate on top. Plastic is chosen since, compared to other solids, its acoustic impedance is reasonably close to that of the water. It is quite lossy acoustically, so that the transmitted sound effectively propagates into semi-infinite space; thus, resonance effects are avoided.

In order to detect the magnitude of the surface distortion the interface is made optically semi-reflective by coating

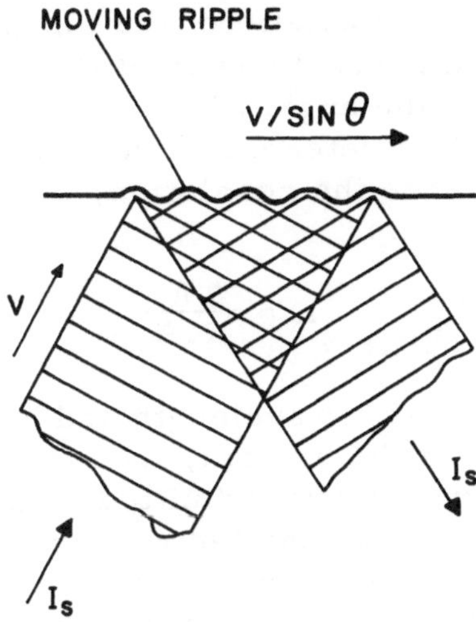

Fig. 1. Dynamic mechanical displacement of an interface resulting from an incident acoustic plane wave at angle θ.

the faceplate, and is illuminated with a focused laser beam. The reflected light beam becomes angularly modulated by the changing slope of the surface. The reflected light is imaged onto a knife edge[3] whereupon the angular modulation is converted into an intensity modulation. A photodiode collects the light passing by the knife edge and produces an electrical signal which is coherent with the local sound pressure. An object to be visualized acoustically is placed just under the boundary so that it will cast a sharply defined (shadow) ripple pattern on the interface. If the specimen or plane of interest is not close to the boundary the image will become blurred. In this situation it is necessary to record the phase of the signal by means of an electronic reference signal. This will result in the display of an acoustic hologram which may be reconstructed optically in a subsequent stage.

An important characteristic of the imaging system is the overall response characteristic of the mirror and knife edge as a function of the angle of sound incidence. It is this characteristic that determines the numerical aperture of the system and hence the resolution, Δ, as governed by the familiar relationship

$$\Delta = \frac{\frac{1}{2}\Lambda}{N.A.} \quad (1)$$

In eq. (1) Λ is the acoustic wavelength and N.A. denotes the numerical aperture of the system, i.e. the sine of half the apex angle of the cone of rays admitted by the imaging system. Figure 2 shows a calculated response curve for a particular set of light beam spot size, d, and wavelength, Λ, for a polymethylmethacrylate-water interface. The disparity in the compressional wave velocities of these materials gives rise to a critical angle at about 34°. At this angle the acoustic impedance of the boundary appears

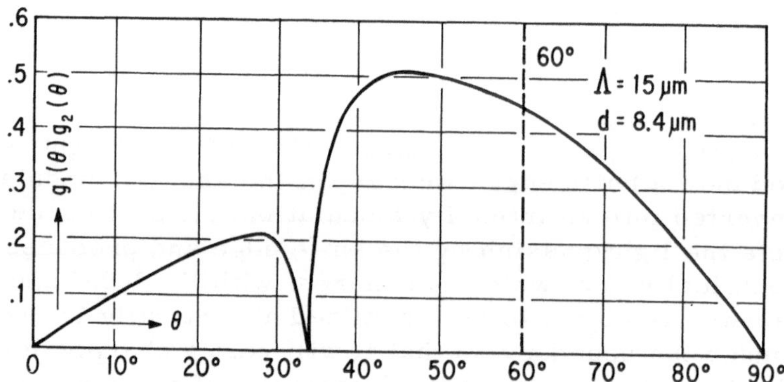

Fig. 2. Calculated response $g_1 \cdot g_2$ of the faceplate and knife edge plotted as a function of angle of incidence of sound. In this figure $g_1(\theta)$ is the normal displacement amplitude of the interface and $g_2(\theta)$ includes the effect of the light beam diameter on the knife edge response

to be infinite, hence no displacement of the interface occurs. Thus, the angular aperture is restricted to the regions between 0° and 34° or between 34° and 90° for this interface.

Depending upon the material chosen for the faceplate, various overall transfer curves are possible. The ideal response, that is one with no zero between 0° and 90°, would be obtained with a faceplate whose characteristic velocity of sound for compressional waves is less than or equal to that of water, viz., 1500 m/sec. The typical plastics with which we have been quite successful so far, are polymethylmethacrylate and polycarbonate. These have sound velocities of 2670 m/s and 2200 m/s respectively, thus giving rise to single critical angles at 34° and 43°. For these materials the shear wave velocity is less than the compressional wave velocity in water and, therefore, no second critical angle occurs. However, if a much harder substance such as glass is employed as a faceplate a second critical angle, caused by the shear wave, will restrict the angular aperture even more.

The mirror material, therefore, imposes conditions upon the arrangement for imaging. For example, the nominal angle of acoustic illumination can be placed midway between 0° and the critical angle or midway (Fig. 2) between the critical angle and 90°. Furthermore, the illumination can be directed off center instead, thereby, in effect, employing a partial single sideband arrangement to increase the effective numerical aperture. An interesting method for producing dark field acoustic imaging system is to place the nominal angle of incidence at the critical angle, where the response is zero, thus using to advantage this intrinsic "zero order stop".

The response characteristic shown in Fig. 2 can be modified optically[2], or by electronic filtering. This latter method arises because the laser probe scans over the moving ripple pattern thereby Doppler shifting the signal output from the sound frequency by an amount f_d where

$$f_d = \frac{V_{scan}}{\Lambda} \sin\theta \tag{2}$$

In eq.(2) V_{scan} is the linear velocity of the scanning light beam on the surface and $\frac{\Lambda}{\sin\theta}$ is the wavelength of the ripple pattern on the interface. Hence, spatial frequencies are translated into temporal ones which may be processed electronically. Electronic filtering is especially important in order to restrict the angular response to one side of normal incidence. This is necessary since the response characteristic is an odd function about 0°, causing a 180° phase shift of sidebands for which $\theta < 0$. Such phase reversals may severely distort the resultant image and produce false image detail.

The sensitivity of this technique is governed by several parameters, such as laser power, observation time, detector efficiency, choice of faceplate material, etc.[2] The minimum detectable signal, i.e. the level at which the signal-to-noise ratio is unity, has been derived previously.[2] Our observed sensitivity of 10^{-3} watts/cm^2 for a 1/30 sec frame time at 100 MHz agrees with the predicted value to within one-half an order of magnitude.

EXPERIMENTAL SETUP

The experimental arrangement is shown in Fig. 3. An X-Y deflection system capable of 200 x 200 resolvable spots is employed to scan the laser beam over the field to be acoustically visualized. Lenses L_1 and L_2 form a 1:1 telescope that images the exit pupil of the deflector onto the knife edge. In this manner the light beam position remains essentially independent of the instantaneous scan angle and is dependent instead only upon the surface distortion. The knife edge is placed so as to block one-half of the light beam; then as the mirror surface tilts back and forth, the reflected laser beam changes position slightly upon the knife edge thus producing an intensity fluctuation on the other side. The now intensity-modulated light is collected by a photodiode whereupon the electrical signal produced,

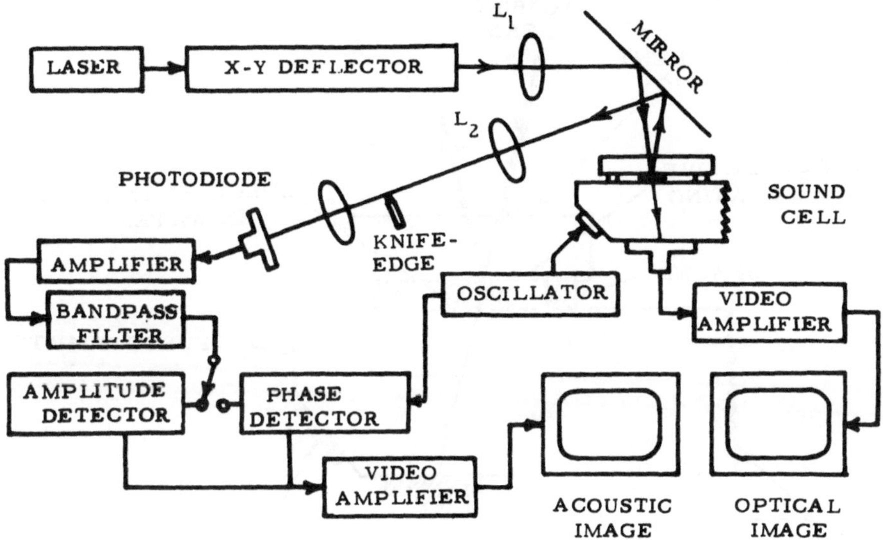

Fig. 3. Experimental setup

which is coherent with the local sound pressure, is amplified, filtered, detected and fed into a TV monitor that is synchronized with the X-Y deflection.

The function of the sound cell shown in more detail in Fig. 4 is to couple acoustic energy to the specimen at the desired angle of illumination and to mechanically support the specimen parallel to and in close proximity to the faceplate. The simplest arrangement[2] which is not shown here involves a water path coupling between a transducer and faceplate. At 100 MHz the necessary acoustic path lengths of a few millimeters pose no serious problem as regards acoustic absorption. However, at frequencies up in the GHz region the high acoustic losses in the water severely restrict the permissible geometry. Some of the experiments reported in this paper employed this simple type of sound cell in order to explore the effects of various angles of acoustic illumination but a more satisfactory arrangement

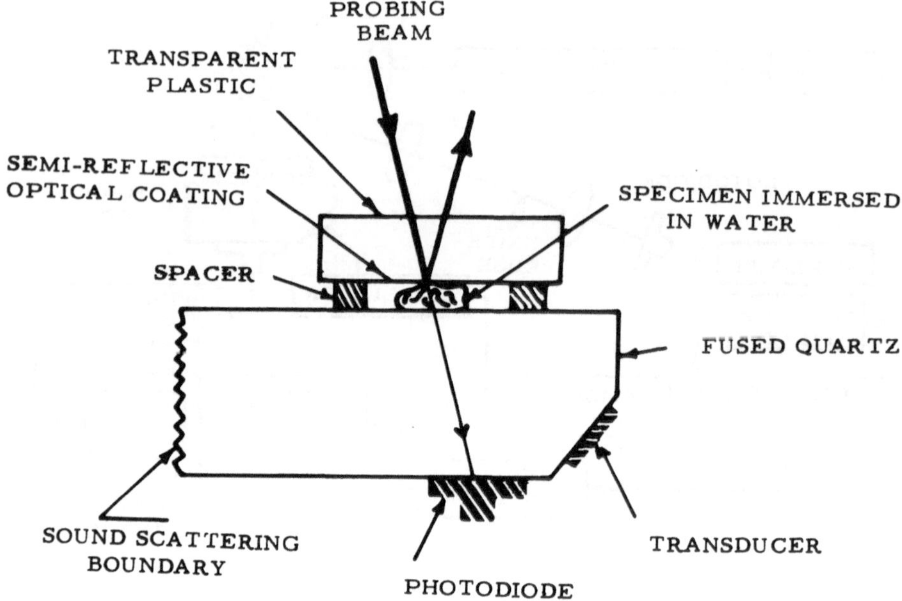

Fig. 4. Sound cell (10° acoustic illumination)

also suitable for higher frequencies is shown in Fig. 4. In order to circumvent lossy transmission paths a fused quartz block is employed as an intermediate substrate. In this arrangement the sound beam is refracted from the solid into the water at 10° due to the large ratio of acoustic velocities. The transmitted sound intensity is about 10 db below that of the incident beam and the remaining sound is scattered and absorbed within the block. A plastic mirror, with a semi-transparent coating is placed directly above the specimen sandwiching it between the mirror and quartz substrate. A fraction of the probing light beam is transmitted through the sandwich to a photodiode which receives the optical image information of the specimen. The electrical output of this photodiode, as shown in Fig. 3,is amplified and fed to a second TV monitor in order to enable side-by-side comparisons of acoustic and optical transmission properties of the specimen.

Fig. 5. Photograph of setup

Figure 5 is a photograph of the experimental apparatus displaying an optical image on the top TV monitor and an acoustic image on the bottom. The fused quartz cell is located in the lower left corner of the picture and the components on the optical bench are associated with the acousto-optic scanning system.[4] The metal box located in the upper left contains the photodiode and receiver front-end. The knife edge, a black card, may be seen in front of the photodiode obstructing one-half of the light beam.

EXPERIMENTAL RESULTS

As a first example of the capability of the 100 MHz microscope, Fig. 6 shows an acoustic image (on the left) and an optical image (on the right) of a commercially available "Honeycomb Finder Grid",[5] using the sound cell shown in Fig. 4. The grid is formed out of nickel and is used by electron microscopists for support and reference location. The grey scale is essentially binary and, in addition, the grid passes and blocks light and sound in a similar fashion.

The relative dimensions will orient the reader familiar with more traditional low frequency acoustic imaging to the

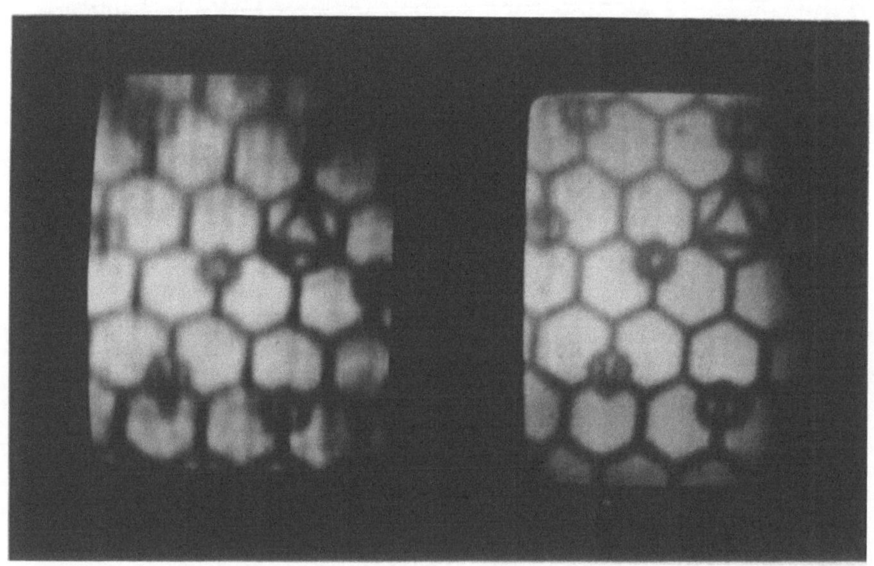

Fig. 6. Honeycomb finder grid. The acoustic image is on the left and the optical image is on the right. Acoustic illumination is at 10°

"microscopic domain". The total field of view on each TV monitor is about 1 mm x 1.3 mm. The parallel faces of the hexagons are 250 μm apart. The small circles which enclose recognizable letters of the alphabet are about 100 μm in diameter. Depending upon the particular letter being examined, it is possible to deduce that the minimum resolvable detail is about 20-25 μm. In addition, we have, in a separate experiment, been able to image a 25 μm period wire grid which confirmed resolution of ≤ 25 μm.[6] That these figures corroborate the theoretical prediction is indicated as follows: The optical resolution is determined principally by the light spot diameter employed, i.e. 20 μm. The acoustic resolution is, in addition, governed by the extent of the angular spectrum of sound diffracted by the specimen and detected by the faceplate. Assuming this "detectable" part of the spectrum to extend from the carrier at 10° to somewhat before the critical angle of 43°, say 30°, we find for the acoustic resolution, Δ,

$$\Delta = \frac{\Lambda}{2(\sin 30° - \sin 10°)} = 23\ \mu m \qquad (3)$$

An acoustic image of a periodic wire mesh insonified at 45° with an earlier version water cell is shown in Fig. 7 using a polymethylmethacrylate faceplate. This mesh is composed of 25 μm diameter wires spaced on 100 μm centers. To obtain this image the electronic filtering bracketed incident angles of sound between 34° and 90°. Note the faint hologram lines in the image which are due to an unintentional, spurious reference signal that leaked into the receiver. The other lines and blotches on the picture correspond to scratches and voids on the reflective coating of the plastic mirror. This particular image is shown to point out one of the difficulties of the technique observed thus far, namely that of achieving a microscopically good optical reflective surface on plastic. As evidenced by improvements seen in other figures presented here, however, the problem has so far been relatively tractable.

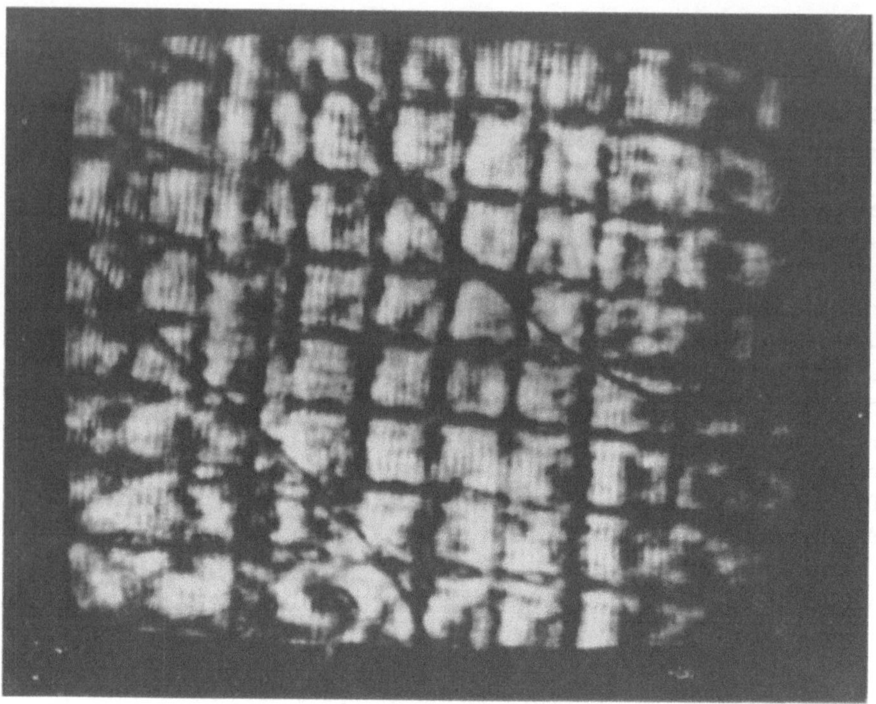

Fig. 7. Acoustic image of a 100 μm spacing, 25 μm wire diameter mesh insonified at 45°

In Fig. 8 is shown a Schlieren image of the same mesh as in the previous figure, made on a separate occasion at an angle of illumination corresponding to the critical angle. The angular aperture was unrestricted on both sides of the critical angle resulting in a numerical aperture approaching 1/2, hence the improved resolution over that in Fig. 7. Here, of course, since the zero order is intrinsically suppressed the sound scattering structures appear bright. This imaging method is of special importance for phase objects where the acoustic absorption or reflection is insufficient to cause observable contrast.

Fig. 8. Same mesh as Fig. 7 but insonified at the critical angle

Before presenting the experimental results of biological imaging it is interesting to examine acoustic absorption in tissue and water. In Fig. 9 are tabulated absorption values at 100 MHz and 1 GHz. The values given for tissue are extrapolations of data over 1-10 MHz,[7] this being the best estimate possible without making further measurements. For simultaneous optical and acoustic viewing a maximum tissue thickness of 25 μm has been chosen since thicker specimens may become increasingly optically opaque. Note that the resulting acoustic absorption through this thickness at 1 GHz is 2.5 db; at 100 MHz, however, it is essentially insignificant.

Fig. 9. Estimated acoustic absorption in tissue and water

SIGNAL LOSS DUE TO ABSORPTION	ESTIMATE FOR TYPICAL TISSUE	WATER
at 100 MHz	10 db/mm	2 db/mm
at 1 GHz	100 db/mm	200 db/mm
at 100 MHz through 25 μm thickness	0.25 db	0.05 db
at 1 GHz through 25 μm thickness	2.5 db	5 db

On the basis of these numbers, ignoring their uncertainty for the moment, we may suspect that acoustic absorption imaging at 100 MHz in thin specimens will probably not reveal any as yet unknown structural detail over what has already been found optically and histologically. Rather, acoustic microscopy at low frequencies will display interfaces which may (or may not) correlate with an optical index of refraction change. However, as will be shown later, the acoustic contrast compared to the optical may be quite significant, thus making it possible, even at low frequencies, to selectively examine certain types of structures better with acoustic energy. On the other hand, in the GHz frequency range where acoustic absorption becomes significant it should be possible to render visible new structural information.

At this juncture it is worth bringing out the notion of "mechanical staining" which may be considered roughly analogous to the optical histological techniques. By pretreatment of the specimen, even at low frequencies, it may be possible to enhance the acoustical differentiation of structures. The simplest example is the mere fixing of tissue in formaldehyde. This, as most readers are probably aware, not only preserves the sample but also "rubberizes" it, thus altering the mechanical characteristics. Thus it is reasonable to expect the development of a series of procedures for emphasizing certain structures acoustically.

The first biological specimen ever visualized with an acoustic microscope was a piece of onion skin in November 1970 and this was published in reference (8). A recent specimen (July 1971) is shown in Fig. 10. The field of view

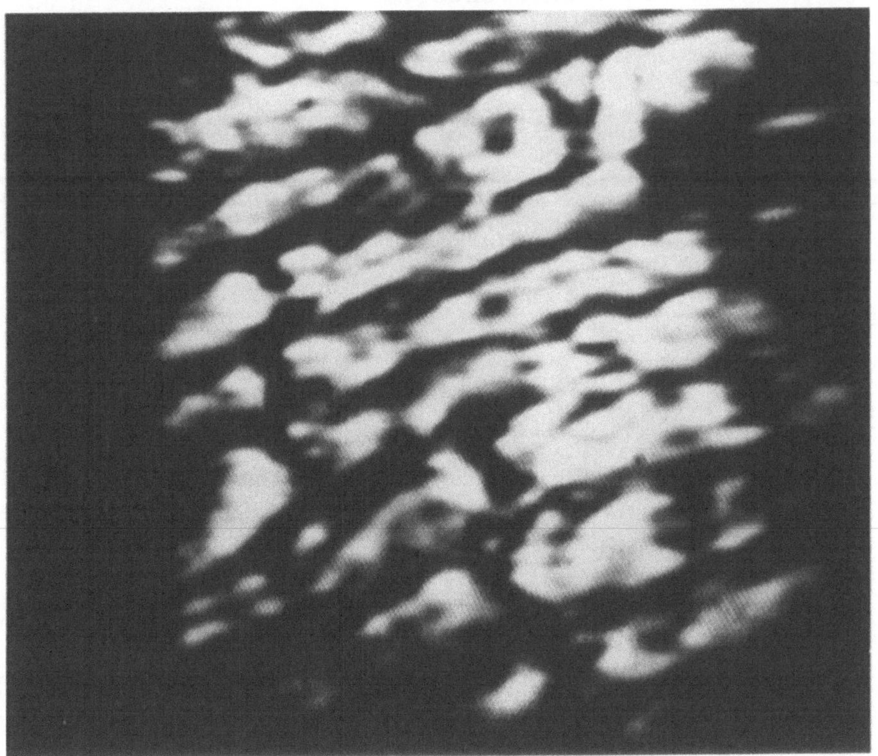

Fig. 10. Acoustic image of onion skin at 45°

is approximately 1.5 mm horizontally and the acoustic illumination is at 45°. The cell walls are easily seen since they are structured of cellulose and, therefore, acoustically different from the surrounding, aqueous-like, intracellular fluid. Although our findings are not yet consistent we feel that the dark spots within the cells may be either nuclei or vacuoles. In addition there appears to be an internal cellular architecture of as yet undetermined origin.

Figure 11 shows the simultaneous acoustic and optical imaging of the fresh onion skin. The field of view here is about 1.3 mm x 1 mm and the optical image is located on the right. This was viewed under similar conditions as Fig. 6. Ideally, it should now be possible to compare and correlate structural features of any specimen and to begin interpretation of the differences.

Fig. 11. Onion skin. The acoustic image is on the left and the optical image is on the right. Acoustic illumination is at 10°.

Figure 12 shows the acoustic and optical images of the live larva of the fruit fly Drosophila melanogaster, the optical image on top. This particular specimen is about 1 mm diameter and 4 mm long. Each TV monitor displays an area of 1.7 x 1.2 mm and, since the animal was considerably longer than the uniform region of the sound field, it was necessary to image only one section at a time. About 40% overlap has been provided, however, for better continuity. The optical image is dark field, produced by

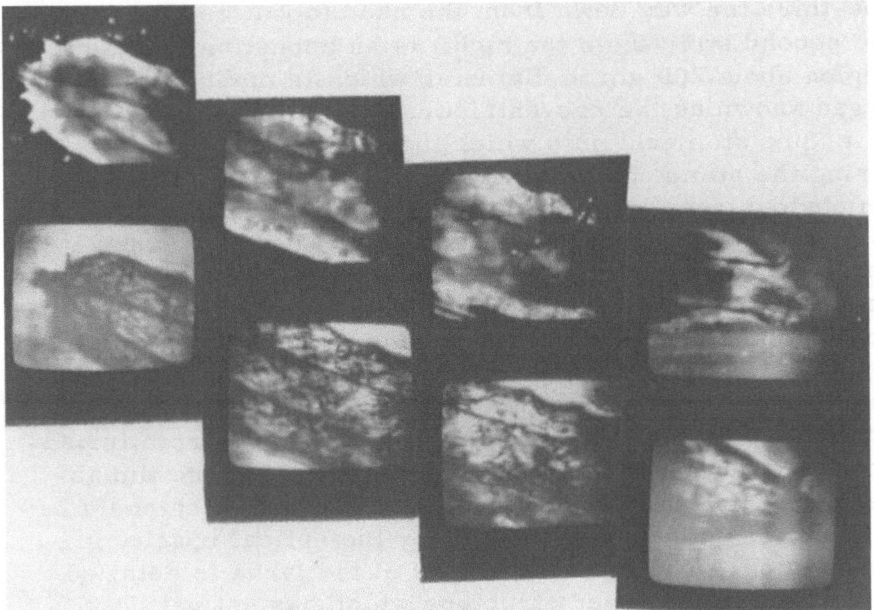

Fig. 12. Drosophila melanogaster larva. The optical image is on the top and the acoustic image (10° insonification) is on the bottom.

slightly offsetting the appropriate photodiode so as to not receive the unscattered light from the specimen.

There are a number of interesting features displayed in this figure. Most of the fine line detail in the acoustic image represents the tracheal network branching out from 2 main trachea (visible in both images) running the length of the animal. This is the system of tubes and tubules through which oxygen is delivered to the individual cells of the larva. In retrospect it could have been expected to easily visualize these air-interfaces since the acoustic impedance mismatch is great. The head of the larva is located on the right and the foremost dark structure (acoustically and optically) is the mouth armature. About one-third the way down from the head (about centered on the second strip from the right) is an acoustically opaque region about 200 μm in diameter which is most likely the organ known as the proventriculus. This organ functions as a "pre stomach" into which the food passes before entering the stomach, or more properly the "midintestine". An obvious question that arises is: If the proventriculus is so clearly visualized with sound why don't the other parts of the alimentary tract such as the intestine, and esophagus also show up? It has been found[9] that the proventriculus, unlike the other organs, consists of a chitinous layer, similar to the material which forms the shells of mollusks (such as crabs). Obviously, the acoustic impedance of <u>this</u> substance would differ greatly from impedance of tissue in the normal sense. Further note the absence of this structure in the optical image which could be a result of the general masking by the optical opacity in this region. The hind two-thirds of the larva is detailed with trachea and other structure which has not yet been interpreted.

Figure 13 is an acoustic hologram of another larva which was made by the addition of an electronic reference signal as shown in Fig. 3. The composite image was then pieced together. The lines represent equal phase contours

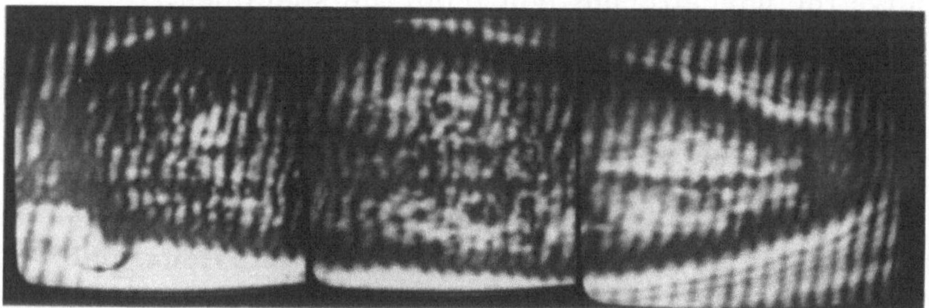

Fig. 13. Acoustic hologram of D. melanogaster at 10 ° insonification

and, if the out-of-focus regions of this animal were to be visualized it would be necessary to reconstruct this hologram with coherent light.

Acoustic microscopic visualization of non-biological specimens is of considerable interest as well; however, the dual optical and acoustic scheme, described here, may not be as useful for less optically transparent objects.

CONCLUDING REMARKS

Although at 100 MHz there appear to be significant differences between acoustic and optical microscopy, these will be magnified even further in the GHz region owing to acoustic absorption and increased resolution. High frequency acoustic examination of biological tissue can provide unique information about the physical state of the constituent structures that make up tissue. Elasticity and density are parameters accessible through acoustic impedance, whereas structural and thermal relaxations as well as the conformal states of the macromolecular components of tissue cause changes in the ultrasonic absorption.[10]

A practical and experimentally employable acoustic microscope operating in the GHz frequency range and having resolution comparable to that of an ordinary optical microscope is not yet available. The 100 MHz experiments presented here offer encouraging evidence that such a device is not only possible but will also provide relevant and useful information not otherwise obtainable.

ACKNOWLEDGMENT

The authors gratefully acknowledge the technical assistance rendered by R. Stetz, the photographic work by J. Van Roon and the typing provided by E. Litka. Additional gratitude is extended to S. G. Kessler, University of Wisconsin, for suggesting and supplying specimens of Drosophila melanogaster used in this study.

REFERENCES

1. A. Korpel and P. Desmares "Rapid Sampling of Acoustic Holograms by Laser Scanning Techniques", J. Acoust. Soc. Amer. 45, 881 (1969).

2. A. Korpel and L. W. Kessler "Comparison of Methods of Acoustic Microscopy", Acoustical Holography, Vol. 3, Ch. 3, A. F. Metherell (ed.), Plenum Press, New York, 1971.

3. R. L. Whitman and A. Korpel "Probing of Acoustic Surface Perturbations by Coherent Light", Applied Optics, 8, 1567 (1969).

4. A. Korpel, R. Adler, P. Desmares and W. Watson "A Television Display Using Acoustic Deflection and Modulation of Coherent Light", Applied Optics 5, 1667 (1966).

5. Ernest Fullam Inc., Schenectady, New York 12301.

6. L. W. Kessler, A. Korpel and P. R. Palermo "Characteristics of a Scanning Laser Acoustic Microscope", (Abstract) J. Opt. Soc. Amer. 61, 1573 (1971).

7. D. E. Goldman and T. V. Hueter "Tabular Data of the Velocity and Absorption of High Frequency Sound in Mamalian Tissues", J. Acoust. Soc. Amer. 28, 35 1956.

8. A. Korpel, L. W. Kessler and P. R. Palermo "An Acoustic Microscope Operating at 100 MHz", Nature 232, 110 (1971).

9. M. Demerec ed. "Biology of Drosophila", Hafner Publishing Company, New York (1965), Chapter 4, p. 275 "The Postembryonic Development of Drosophila by D. Bodenstein.

10. L. W. Kessler and F. Dunn "Ultrasonic Investigation of the Conformal Changes of Bovine Serum Albumin in Aqueous Solution", J. Phys. Chem. 73, 4256 (1969).

A 1.1 GHz SCANNED ACOUSTIC MICROSCOPE*

B.A.Auld, R.J.Gilbert, K.Hyllested,
C.G.Roberts, D.C.Webb

Stanford University

Stanford, California 94305

ABSTRACT

A 1.1 GHz acoustic microscope is being developed with an anticipated resolution capability of 5 μm. The microscope uses a unique scanning method which employs the photoconductive effect to locally switch a CdS piezoelectric transducer. A focused laser beam scanned across the CdS transducer sequentially deactivates resolution elements and generates the video signal. To form the image, the video signal intensity modulates a CRT display which is scanned synchronously with the laser beam.

The present system has resolved a metal grid of 50 μm wires on 275 μm centers at a 20 second frame time. Both phase and amplitude information are used to generate the video signal in the present system. However, a phase discrimination system is being developed to separate phase and amplitude information.

Signal and noise considerations are presented and compared to other imaging techniques. Problems with spurious signals, optical scattering, acoustic interference, and signal uniformity have been encountered and are discussed with proposed solutions.

* This work has been supported by The John A. Hartford Foundation, Inc.

INTRODUCTION

The interest in developing an acoustic microscope arises from a desire to form acoustic images of small structures, principally biological samples. Acoustic images present information about the elastic properties of specimens, in contrast to optical images which present index of refraction or optical absorption information. Because acoustic waves, unlike optical waves, cannot propagate in a vacuum, the object to be imaged must be placed in some material medium. The logical medium is water because it is convenient and is compatible with most biological samples. The need to image small objects forces one to operate at high frequencies. Since details on the order of microns are of interest, the acoustic frequency should be near 1.0 GHz where the acoustic wavelength is 1.5 μm in water.

Several imaging schemes have been used to form visible images of acoustic patterns: Bragg scattering systems,[1,2] laser scanned systems,[3] interferometric systems with reconstruction,[4] and transducer and/or receiver scanned systems.[5,6] The system described in this paper is a scanned system. A large area acoustic beam illuminates the object, and the video signal is generated by scanning the receiver. A light spot scans and locally deactivates the receiving transducer to generate the "negative" scan. The image information appears as an amplitude and phase modulation of the 1.0 GHz carrier.

SYSTEM DESCRIPTION

A block diagram of the system is shown in Fig. 1. A focused argon laser beam scanned by a galvanometer deflection system generates the scan raster on the receiving transducer. The 1.1 GHz signal excites a transmitting transducer to generate the acoustic beam. After passing through the water cell containing the object, the acoustic beam excites the receiving transducer which couples the energy to the signal processing electronics. The video signal from the electronics intensity modulates a synchronously scanned CRT display. In this way a visual image proportional to the acoustic image intensity is presented on the display.

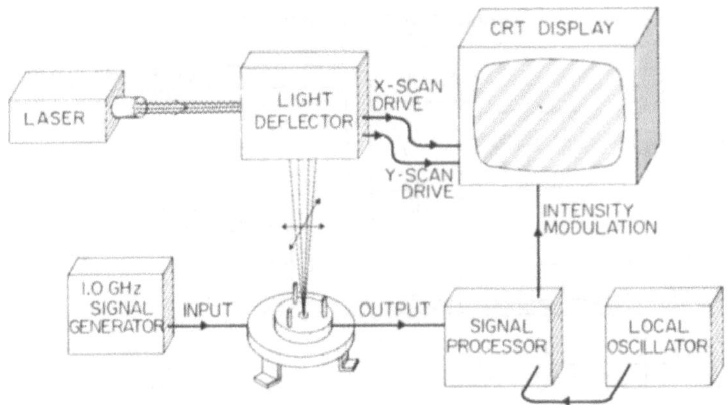

FIG. 1--Schematic of acoustic microscope system.

Figure 2 is a schematic of the acoustic microscope head. An input coupling loop excites a TM_{010} mode in the dielectric microwave cavity,[7] and the TiO_2 (rutile) dielectric rod concentrates the cavity electrical energy in the region of the transmitting transducer. A zinc-oxide transducer is rf sputtered onto acoustic impedance matching transformers on the Al_2O_3 (sapphire) rod, and the gold transformer layer is connected electrically to the cavity wall forming the counter electrode for the dielectric. The electric fields impressed across the ZnO transducer generate the acoustic signal that propagates through the Al_2O_3 rod, which is acoustically matched to the water cell by transformers. In this way a 0.75 cm diameter acoustic beam provides plane wave illumination of the object in the water cell. Scattering from and absorption by the object modify the beam as it traverses to the receiving CdS transducer, and this acoustic pattern is transmitted to the CdS receiving transducer by another set of impedance transformers. In the receiver a TiO_2 dielectric rod couples the electromagnetic energy from the transducer to the microwave output cavity.

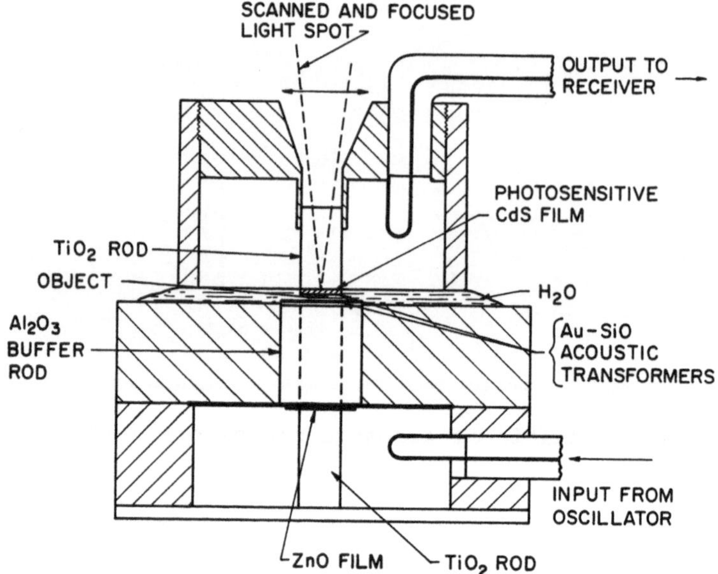

FIG. 2--Sketch of acoustic microscope head showing input and output cavities, transducers, water cell, and object plane.

Because CdS is photoconductive as well as piezoelectric, local regions of the transducer can be deactivated by illumination with a light spot.[8] The receiver receives a signal from all elements except the illuminated element, but the acoustic field distribution can be read out point-by-point when a laser spot is focused and scanned over the transducer surface. If the laser spot deactivates a resolution element that is not receiving acoustic power, the output signal is unchanged. However if the spot deactivates an element that is receiving acoustic power, the output signal decreases proportionally with the power incident on that element. This type system is referred to as a "negative receiver scan" system.

The important interaction for the microscope is the photoconductive switching of the piezoelectric transducer. This interaction can be understood by considering the equivalent circuit model in Fig. 3, where the cavity conductance (G_c) and inductance (L) are in parallel with

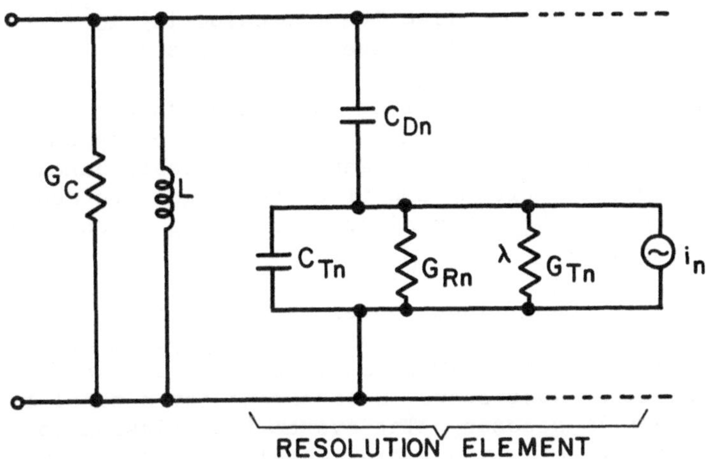

FIG. 3--Equivalent circuit for the output cavity, dielectric resonator and photoconductive piezoelectric transducer.

a parallel combination of resolution elements. Each resolution element is modeled by the capacitance of the dielectric rod (C_{Dn}) in series with the photoconductive transducer resolution element. The latter includes the geometric capacitance of the film (C_{Tn}), a radiation conductance (G_{Rn}) which describes the conversion of acoustic energy to electrical energy, the photosensitive conductance (G_{Tn}), and a current source (i_n) representing the acoustic input to the nth resolution element.[7] With no illumination, G_{Tn} is much smaller than G_{Rn} so that all the incoming signal is coupled to the cavity.* When the nth element is illuminated G_{Tn} increases and effectively shunts G_{Rn} so that the signal from this element is not converted to electromagnetic energy in the cavity. This effect provides a means for sequentially sampling the local regions of the transducer and therefore the local acoustic field. The

*The signal is, of course, subject to the conversion loss of the transducer.

conductivity range for effective switching is 10^{-3} mho/cm dark to 10^{-1} mho/cm illuminated at 1.1 GHz.[7]

A superheterodyne receiver using a balanced mixer detects the signal, which in turn is amplified by an IF amplifier to form the video signal. A variable bandwidth, single ended video preamp amplifies the signal to modulate the CRT display. No other signal processing is done in the present system.

Typically the rf input power is 60 mW which gives 20 mW/cm^2 acoustic intensity in the object plane after insertion losses. A 40 mW argon laser beam focused to a ~ 40 μm diameter spot scans 30 lines/sec for a total frame time of about 10 seconds. The system transmission loss is typically 55 dB, of which approximately one-half occurs in the receiving transducer.

Three micrometers support the receiver head on the transmitting body of the microscope. An autocollimation technique is used to align the receiving transducer parallel to the transmitting transformers on the Al_2O_3 rod. Spacing between transducers and transformer is 25 μm if consistent with object thicknesses. This spacing gives a 5 dB propagation loss through the water cell at 1.1 GHz.

SENSITIVITY CONSIDERATIONS

The signal transfer characteristics, photoconductive switching, nonuniformities in the signal generation process, system noise, and acoustic resonances are factors which affect system sensitivity. Ultimate system sensitivity is determined by thermal noise of the receiver, although one may encounter carrier modulation noise. However other effects are the limiting factors at present.

To develop the signal transfer characteristics, consider an electrically-matched transducer with a conversion efficiency η_1 and with area A , as shown in Fig. 4. Let the acoustic beam impinge on a perfect absorber with aperture of area a . The acoustic power in the transmitted beam P_B is given by

$$P_B = \eta_1 \frac{a}{A} P_{in} \quad , \qquad (1)$$

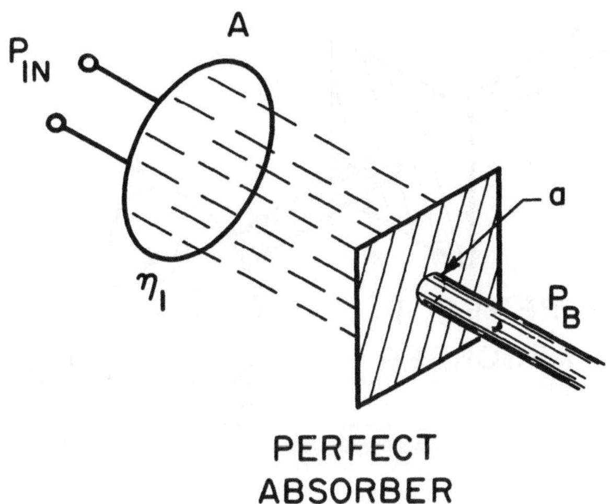

$$P_B = \eta_1 \frac{a}{A} P_{IN} = \eta_{eff} P_{IN}$$

FIG. 4--Effective conversion efficiency, η_{eff}, of a transducer of area A generating an acoustic beam which falls on a perfect absorber with transmitting area, a . The transducer has conversion efficiency, η_1, when perfectly matched electrically.

where P_{in} is the rf power exciting the transducer. By defining $\eta_{eff} = \eta_1(a/A)$, Eq. (1) can be written

$$P_B = \eta_{eff} P_{in} \quad . \tag{2}$$

Since piezoelectric transducers are reciprocal devices, reciprocity can be invoked to determine the output from a transducer of area A receiving the beam of area a , as shown in Fig. 5. If the transducer is identical to that

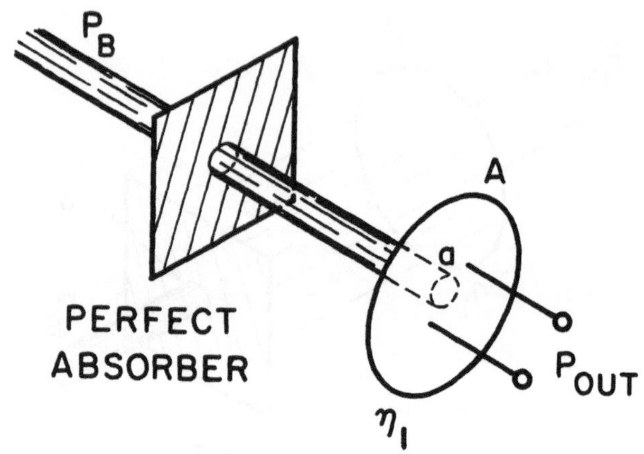

$$P_{OUT} = \eta_{eff} P_B = \eta_1 \frac{a}{A} P_B$$

FIG. 5--Effective conversion efficiency, η_{eff} , of a transducer of area A receiving an acoustic beam of area, a . The transducer has a perfectly matched efficiency, η_1 .

used in the transmission argument above; i.e., it has the same electrically-matched conversion efficiency, the electrical output power is given by

$$P_{out} = \eta_{eff} P_B \tag{3}$$

where η_{eff} is defined above. Rewriting (3) gives

$$P_{out} = \eta_1 \frac{a}{A} P_B \quad . \tag{4}$$

Combining (1) and (4) finally yields

$$P_{out} = \eta_1^2 \left(\frac{a}{A}\right)^2 P_{in} \tag{5}$$

as the relationship between input and output power when a restricting aperture defines a beam smaller than the transducer area.

To relate this to the signal output of a transducer receiving acoustic energy from an object with N resolution elements of area a and transmission τ_n for the nth element, consider Fig. 6. Let η_1 and η_2 be the electrically-matched conversion efficiencies of the transmitting and receiving transducers respectively. Then the electrical output power for a signal from the nth element is

$$P_n = \left(\eta_2 \frac{a}{A}\right) \left(\eta_1 \frac{a}{A} \tau_n P_{in}\right) \quad . \qquad (6)$$

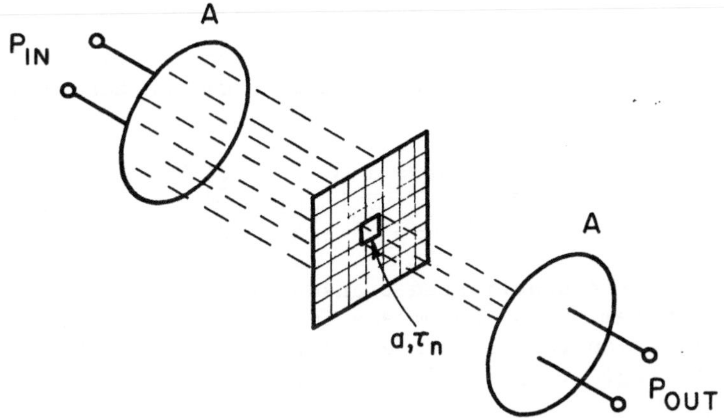

FIG. 6--Compound acoustical system showing transmitting transducer, object with resolution elements of area a and transmittance τ_n , and receiving transducer.

If this is delivered to a characteristic impedance R_o, the voltage developed by the nth element can be written

$$|V_n| = (2 R_o P_n)^{1/2} . \tag{7}$$

Combining (6) and (7) yields

$$|V_n| = \frac{a}{A}\sqrt{\tau_n}\, K \tag{8}$$

where $K = \sqrt{2 \eta_1 \eta_2 R_o P_{in}}$. In general the acoustic transmission through the n-th element experiences a phase shift as well as an attenuation, and the output voltage due to that element can be written as

$$\underline{V}_n = |V_n| \, e^{i\phi_n} \tag{9}$$

where ϕ_n is the phase angle relative to some reference and the sub-bar denotes a complex number. The total voltage output of the transducer, $\underline{V}_B$, is the sum of (9) over the N elements,

$$\underline{V}_B = \sum_{n=1}^{N} \underline{V}_n , \tag{10}$$

while the signal output voltage is just the difference between V_B and the voltage received from the deactivated element. That is

$$V_{out} = |V_B| \left(1 - \frac{\underline{V}_j}{\underline{V}_B}\right) \tag{11}$$

where j is the deactivated transducer element. For a scanning system V_{out} becomes a function of time because the transmission varies from element to element. The time varying voltage $V_{out}(t)$ is expressed

$$|V_{out}(t)| = |V_B| \left(1 - \frac{a\,K}{A\underline{V}_B} \sqrt{\tau(t)}\, e^{i\phi(t)}\right) \cos \omega t \tag{12}$$

where ω is the rf carrier frequency. Since the system is a linear detection system, the time varying modulation in (12) can be expressed as a fourier series

$$\frac{a}{A}\frac{K}{\underline{V}_B}\sqrt{\tau(t)}\; e^{i\phi(t)} = \sum_{m=0}^{\infty} \underline{M}_m \frac{aK}{A\underline{V}_B} \cos 2\pi f_m t \qquad (13)$$

where $\underline{M}_m$ are fourier expansion coefficients and f_m frequency components related to the spatial frequency of the object. Now consider the transfer relationship for one fourier component. Equation (12) becomes

$$|V_{out}(t)| = |V_B| \left(1 - \frac{a}{A}\frac{K}{\underline{V}_B} \underline{M}_m \cos 2\pi f_m t\right) \cos \omega t \;.(14)$$

As shown by Goldman,[9] the signal-to-noise power ratio for a signal of the above form is

$$(S/N)_{POWER} = \left(\frac{a}{A}\right)^2 \frac{|\underline{M}_m|^2 \eta_1\eta_2 P_{in}}{F\,kT\,B} \qquad (15)$$

where F is the noise figure of the receiver, k is Boltzmann's constant, T the temperature of the receiver, and B the electronic bandwidth of the receiver.

Equation (15) shows that the $(S/N)_{POWER}$ is inversely proportional to the square of the number of resolution elements, where $N = A/a$, and varies linearly with input power. A more useful expression for describing the acoustic microscope system is obtained when $\eta_1\eta_2$ are combined with the propagation loss into a transmission efficiency η_T for the system. The expression becomes

$$(S/N)_{POWER} = \frac{|\underline{M}_m|^2}{N^2} \frac{\eta_T P_{in}}{F\,kT\,B} \quad . \qquad (16)$$

Using Eq. (16) with $P_{in} = 30$ dBm , $\eta_T = -50$ dB , and $F = 7$ dB , one calculates the signal-to-noise ratios listed in Table I for several bandwidths and numbers of resolution elements. These calculations assume 100% switching and 100% contrast in the object. Notice that quite

System Bandwidth B \ Number of Elements N	50 × 50	100 × 100	200 × 200
1 KHz	49 dB	37 dB	25 dB
30 KHz	34 dB	22 dB	10 dB
1 MHz	19 dB	7 dB	-5 dB

TABLE I. Table of calculated signal-to-noise ratios for N resolution elements and electronic bandwidth B .

satisfactory signal-to-noise ratios are predicted for a small number of resolution elements and a narrow bandwidth. At 2500 elements-1 MHz, 10^4 elements-30 kHz, and 4×10^4 elements-1 kHz contrasts as low as 10^{-2} should be detectable, but with limited grey-shade capability. At higher numbers of elements and large bandwidths only high contrast objects would be detectable. Once a bandwidth and number of elements is defined, the shortest frame time has been determined. These ratios can be improved if the input power is increased. Because of the insertion loss at the input transducer, the acoustic intensity in the object plane is only 200 mW/cm^2 for a 1.0 watt input and should be within safe limits for most biological tissue. Any improvement in transmission loss, or receiver noise figure would also improve the expected signal-to-noise ratios.

The predicted values of S/N have been compared with experiment by using a time-modulated stationary light spot to generate a signal. Figure 7 is a plot of S/N as a function of input power for two different light spot sizes.

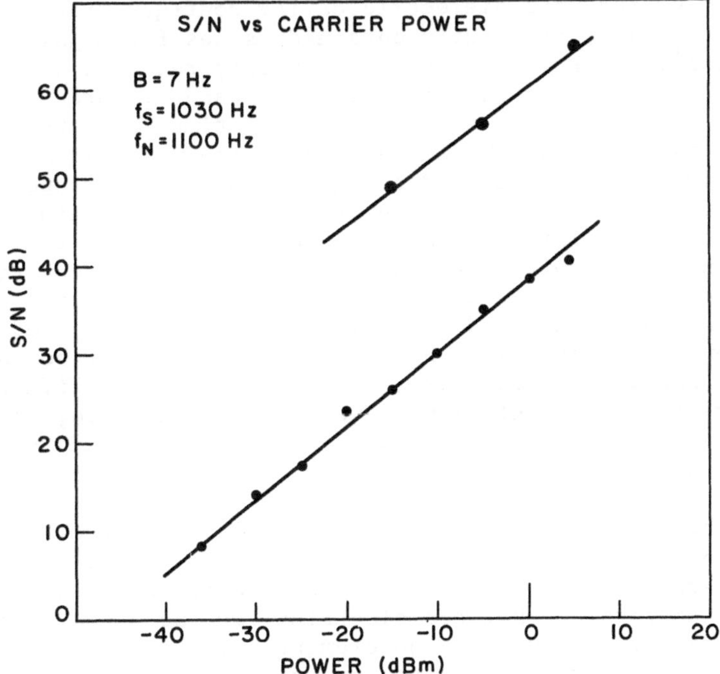

FIG. 7--Measured signal-to-noise ratio as a function of power into microscope head. Signal was generated by modulating the focused light spot. The two curves are for different spot sizes.

The S/N is obviously linear with input power as predicted, and the measured value of S/N agrees very well with calculated values. Although not as conclusive, preliminary measurements also strongly support the proportionality of S/N to $(a/A)^2 = 1/N^2$ predicted by Eq. (16).

The signal-to-noise limitation imposed by the photoconductive switching can be understood in terms of the equivalent circuit shown in Fig. 3. If the dark conductance of the film G_{Tn} is high compared to $G_{Rn} + j\omega C$, the transducer will have a poor conversion efficiency and signal will be lost in the conversion process. This requirement imposes a maximum conductivity of 10^{-3} mho/cm for the unilluminated CdS material. For effective element switching the conductivity must be increased to approximately

10^{-1} mho/cm for this dielectric cavity configuration. Normally , these are easily obtained values for highly sensitive CdS. However, high sensitivity requires a long photoconductive response time (~10 msec) and, in order to scan 10^4 elements in 1 sec, a response time less than 16 μsec is required. To achieve this, one must use a rather insensitive film and a high intensity laser spot. With sufficient laser power, photoconductive switching should not be a limiting factor. Nonuniformity in the photoconductive switching characteristics of the receiving transducer could prove to be a limitation if one scans with low light intensity. CdS material is a trap controlled photoconductor and small changes in local impurity concentration can drastically alter the photoconductive lifetime.[10] However, if sufficient light intensity is available, it is possible to saturate the photoconduction process and thereby obtain uniform switching.

Other nonuniformities in the system can severely limit detection sensitivity. Variations in transducer thickness or conversion efficiency create nonuniformities in the spot-to-spot conversions between electrical and acoustic energy. Since the ultimate resolution element size is projected to be 5 μm, the uniformity criterion must be applied to regions of this size. Thin film transducers do not characteristically have the uniformity associated with high quality bulk material and are likely to impose a significant limitation on the system sensitivity.

Variations in the acoustic impedance matching transformers also are a potential source of signal conversion nonuniformity. This source is less likely to limit performance because it is not a critical function of crystalline structure, material purity, and stoichiometry as are the transducers. With reasonable care during material evaporation, this factor should therefore not limit sensitivity.

Finally the cavity electric field distribution impresses a low spatial frequency variation radially across the transducers. This gradation is inherent to the dielectric cavity excitation. Low frequency filtering and limiting the scanned area to the central region of the rod minimize this effect.

For the signal due to nonuniformities to be less than the thermal noise, the signal generation process must be uniform to greater than one part in $(\eta_T P_{in}/N^2 FkTB)^{1/2}$ in Eq. (16). For the conditions of Table I with $N = 10^4$ and $B = 30$ kHz , this imposes one part in 12 (or 8%) uniformity.

Because the receiving transducer is relatively inefficient, much of the acoustic energy propagates through the transducer into the TiO_2 resonator. After being reflected off the resonator surface, it again excites the transducer. The round-trip propagation loss is approximately 4 dB, so that a significant amount of the energy is still available in the reflected wave. Since the microscope operates CW, the reflected pattern interferes with the incoming pattern to generate structure which is not directly related to the object transmission. This effect is another severe limitation of the present system.

In imaging operation the present system does not achieve the predicted sensitivity. Because of transducer nonuniformities and interference between incoming and reflected waves, the observed sensitivity is at least 10 dB poorer than the theoretical value.

RESOLUTION CONSIDERATIONS

The wavelength of the acoustic wave in the water cell determines an ultimate resolution limit of 1.5 μm. However other factors such as diffraction of the acoustic wave, diffusion of the photogenerated carriers, photoconductive lifetime, and optical spot size impose a more stringent limitation of 6 μm.

Because acoustic waves traveling at an angle to the transducer axis are detected less efficiently than normal incidence waves, the acceptance angle of the receiving transducer is limited to an angle which is less than that determined by the physical dimensions of the transducer. Two off-axis effects are important. First, the effective piezoelectric coupling constant decreases for waves propagating at an angle to the principal axis. Second, the phase difference across the transducer or transformer surface increases until, at some large angle, complete cancellation occurs in the transducer. This second limitation is directly related to the acoustic thickness of the transducer or transformer.[7] In the present configuration the SiO transformer layer limits the numerical aperture. This establishes the minimum spacing for resolvable objects at 6 μm. The numerical aperture of the transducer itself gives a resolution limit of 4 μm.

For an acoustic shadow to form on the transducer, the object must be less than a distance z from the transducer,[7] where

$$z \lesssim \frac{\delta^2}{4\lambda_a} \quad , \qquad (17)$$

δ is the object dimension to be imaged, and λ_a is the wavelength in water. For $\delta = 6$ μm as given above, $z = 6$ μm . This places a severe restriction on positioning the receiver relative to the object.

For diffraction limited optical systems the scanning light spot has a minimum radius

$$r = 0.61 \lambda f^{\#} \qquad (18)$$

where λ is the wavelength of light and $f^{\#}$ is the lens f-number. For f/2.5 lens $r = 0.75$ μm , which should be readily attained. However a surface free from minute scratches and good optical quality TiO_2 is necessary to prevent severe optical defocusing and scattering when the spot is focused through the TiO_2 resonator. This is a problem which has been encountered with small intense spots.

Diffusion of photogenerated carriers causes the deactivated spot on the transducer to be larger than that of the optical spot. Taking a carrier mobility of 10 cm^2/V-sec and a carrier lifetime of 10^{-6} sec, one finds that the diffusion length,

$$L_D = \sqrt{\frac{\mu kT}{q}\tau} \quad , \qquad (19)$$

is 5 μm. More typical values of mobility for thin films are 1-3 cm^2/V-sec, so that $L_D = 2$ μm . This sets the switched area as ~ 4 μm diameter.

Long photoconductive response times limit the scan rate. If the carriers do not decay within a dwell time on a resolution element, the switching process will decrease the point-to-point modulation. To overcome this factor, short response times or long frame times are required. Obviously, short response times are the desirable alternative.

The resolution limiting factors of the system are currently long photoresponse times and optical spot size. However, greater resolution is not useable at this stage of development because the effects mentioned in the sensitivity discussion already limit the signal-to-noise ratio to minimal values for currently achieved resolutions. Any decrease in spot size would further decrease the signal-to-noise ratio.

SYSTEM PERFORMANCE

Several objects have been imaged using the acoustic microscope. All have been metallic mesh and grid patterns which are high contrast objects. Most are periodic and therefore lend themselves to narrow band filtering in the electronic circuit.

Figure 8 is an acoustic image of gold bars which are 500 μm wide on 1000 μm centers. The distortion at the left of the image was caused by a scan nonlinearity. The gold foil was 25 μm thick.

A stainless steel mesh image is shown in Fig. 9. The wires are 180 μm wide and are on 430 μm centers. The openings are therefore 250 μm square. The uniform shading across the image indicates a uniform conversion for the transducers used to generate this picture. Filtering eliminated high frequency noise and caused loss of edge sharpness.

Figure 10 is the image of a stainless steel grid of 40 μm wires on 275 μm centers. This image was generated with a 15-15 kHz bandwidth. The CdS film, which was deposited by D. B. Fraser of Bell Telephone Laboratory, was highly uniform and had a short photoresponse time. Both features are apparent in the image which has good edge definition and uniform shading. A low signal-to-noise

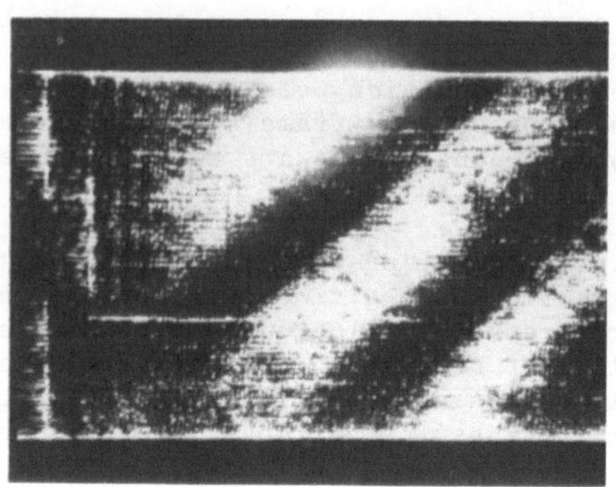

FIG. 8--Acoustic image of 500 μm gold bars with 500 μm spacings.

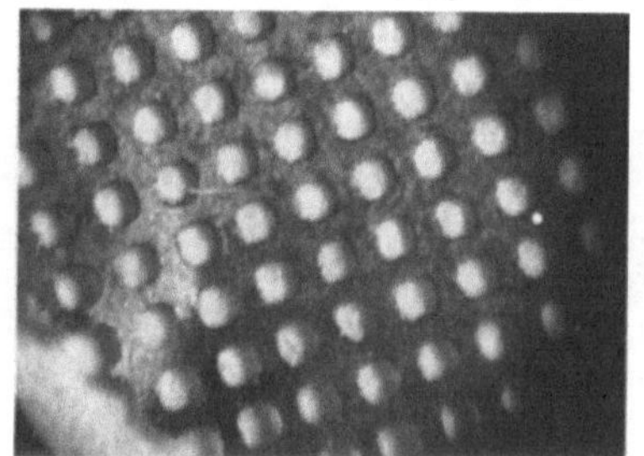

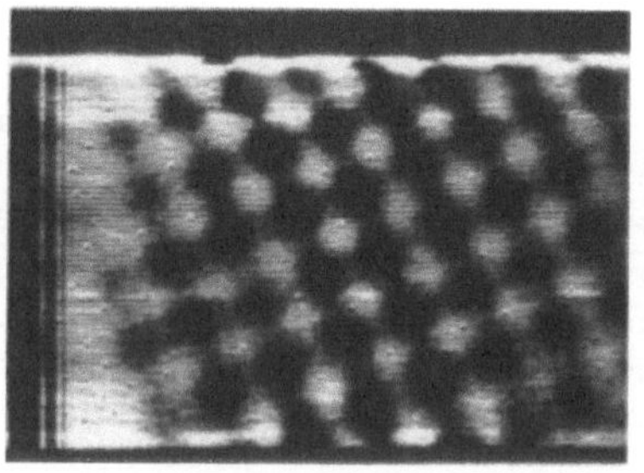

OPTICAL IMAGE ACOUSTIC IMAGE

FIG. 9--Optical image and acoustic image of stainless steel mesh of 250 μm wires on 430 μm centers.

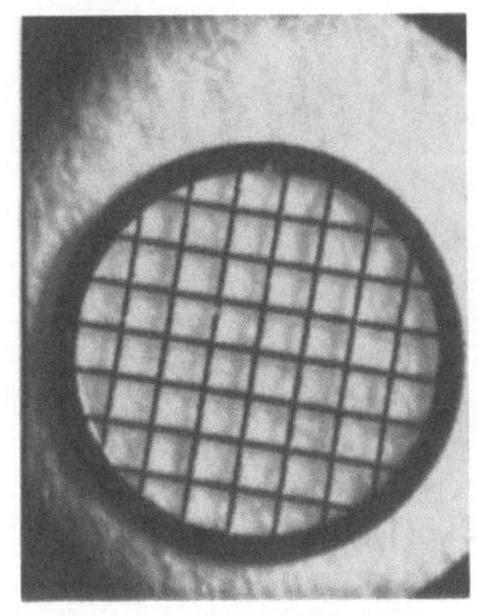

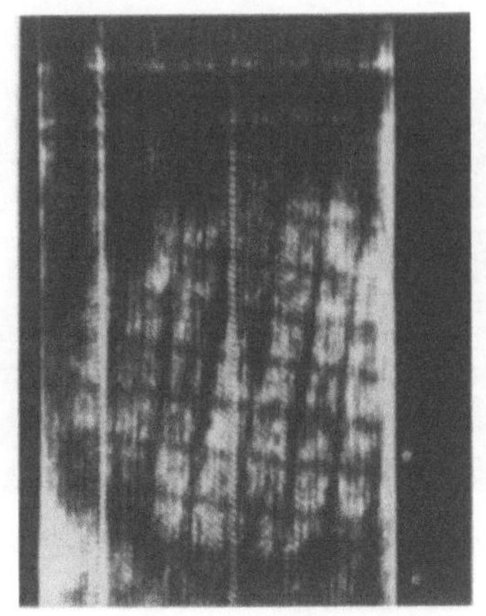

OPTICAL IMAGE ACOUSTIC IMAGE

FIG. 10--Optical image and acoustic image of stainless steel mesh of 50 μm wires on 275 μm centers.

ratio is indicated by the noise level apparent in the image.

The image of a nickel resolution chart shown in Fig. 11 demonstrates some of the effects encountered when imaging nonperiodic structures. The three-dimensional effect apparent in the image is a result of narrow-band filtering. The artifacts in the image (dots, lines, and structure not in the object) are spurious signals caused by light scattering off fine scratches on the surface and crystal imperfections in the TiO_2 resonator. The resulting modulation of switching generates a signal just as does the differences in acoustic energy on the transducer.

The image shown in Fig. 12 has a definite periodicity which obviously is not that of the object. The object is a rectangular grid of 25 μm wires on 65 μm and 285 μm centers. The acoustic image is believed to result from a superposition of the reflected wave on the incoming wave to create a moiré effect. This dramatically demonstrates the problems associated with the reflected acoustic energy in the TiO_2 output resonator.

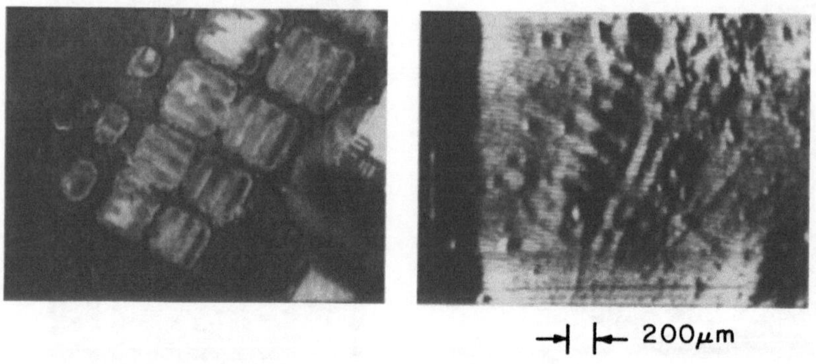

FIG. 11--Optical image and acoustic image of nickel resolution chart.

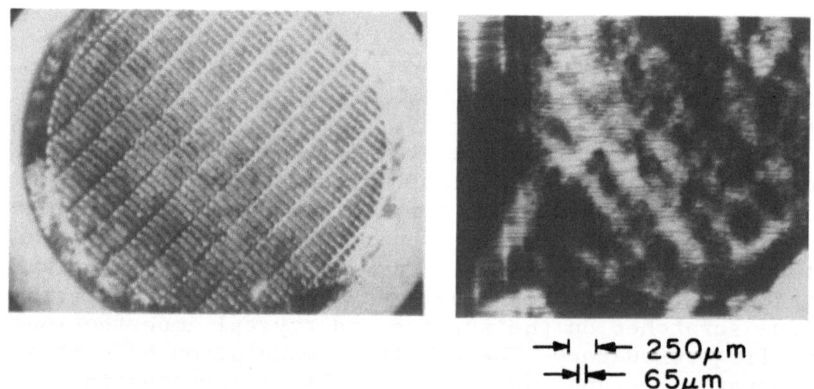

FIG. 12--Optical image and acoustic image of rectangular mesh with 65 μm and 285 μm spacings. The discrepancy in periodicity between the acoustic image and that of the mesh is believed to result from interference between the incoming wave and the reflected wave (off resonator end) at the receiving transducer.

Figure 13 shows that the system resolution capability approaches 25 μm. The grid is 37 μm wires on 115 μm centers. In this image picture quality is limited by long range nonuniformities or interference as evidenced by the shading gradation. However, the system bandwidth for the photograph was 30-30 kHz which indicates that adequate signal-to-noise ratios can be realized for useful bandwidths and numbers of resolution elements.

The above results indicate that the acoustic microscope can indeed resolve small high contrast structures. The apparent performance limitations are nonuniformities and interference from a reflected wave in the TiO_2 output resonator. Several alternative schemes have been considered for eliminating this reflected wave. A slant cut across the output resonator, which deflects the acoustic energy into the side of the rod, effectively eliminates the reflection but places additional restrictions on resonator optical surface preparation in order to achieve a

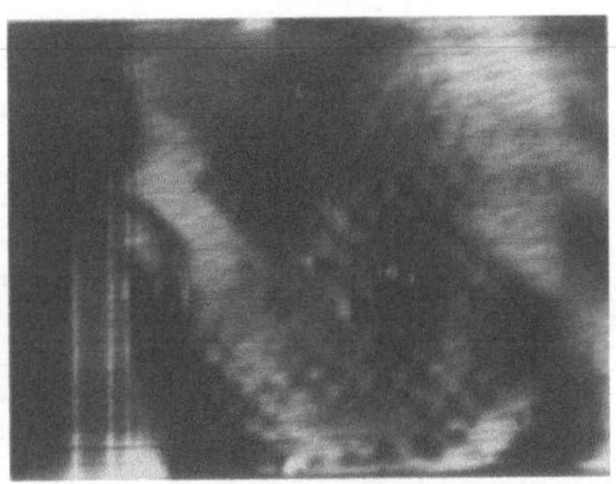

OPTICAL IMAGE ACOUSTIC IMAGE

FIG. 13--Optical image and acoustic image of copper mesh with 37 μm wires on 115 μm centers. Transducer nonuniformities and acoustic interferences cause nonuniform shading of image.

small spot in the CdS transducer plane. Because of the high index of refraction of TiO_2, these restrictions are severe enough to make this solution somewhat impractical. Use of an acoustically lossy dielectric to replace TiO_2 would be a satisfactory solution. However a material with optical properties and dielectric constant compatible with microscope design has not yet been identified. Pulsed operation of the system offers several advantages and is being considered. Another alternative is to process the electronic signal so that the phase and amplitude of the signal on each resolution element is sensed. This can be accomplished by the system shown schematically in Fig. 14. A reference signal is delayed and inserted into a differential amplifier for comparison to the signal from the microscope. The output from the differential amplifier is just the amplitude signal. A normalization circuit — the sample and hold and phase shifter-attenuator — resets the reference signal after each scan line. The phase information is available at the phase detector. With both amplitude and phase information available more elaborate signal processing can be performed.

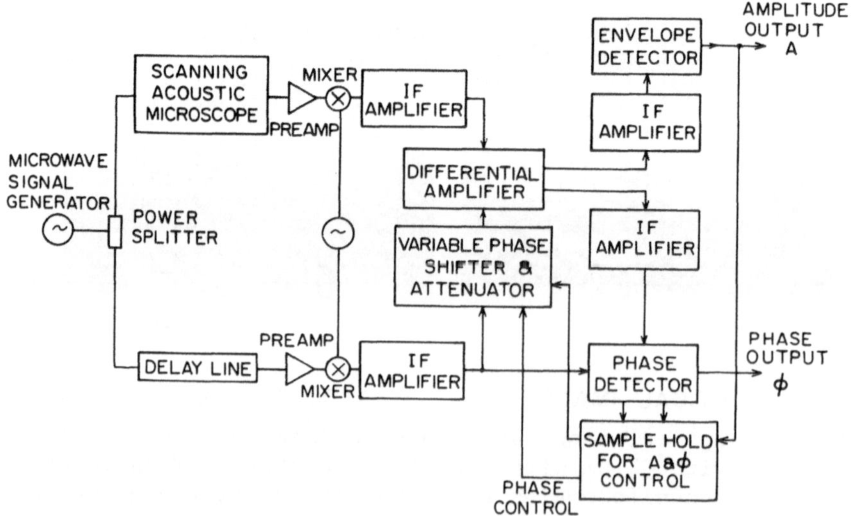

FIG. 14--Block diagram of electronics designed to separate amplitude information from phase information of microscope signal.

Finally, the results obtained with this system are encouraging. The severe loss in signal-to-noise ratio for a negative scan system ($S/N \propto 1/N^2$) can be decreased dramatically by using a positive scan system where the S/N is inversely proportional to the first power of N.[11] With the improvement predicted for a positive scan system, acoustic microscopes would offer greater potential as analytic instruments for biological samples. Positive scan receiver and positive scan transmitter schemes are presently being considered.

The authors wish to acknowledge the loan of an argon ion laser by Spectra-Physics.

REFERENCES

1. A. Korpel, Appl. Phys. Letters 9, 425 (1966).

2. J. Havlice, C.F. Quate, and B. Richardson, IEEE Trans. SU-15, 68 (1968).

3. A. Korpel and P. Desmares, J. Acoust. Soc. Amer. 45, 4, 881-884 (1969).

4. J. A. Cunningham and C.F. Quate, Acoustical Holography, vol. 4, p. 667 (Plenum Press, 1972).

5. K. Preston and J.L. Kreuzer, Appl. Phys. Letters 10, 150-152 (1967).

6. G.A. Massey, Proc. IEEE 56, 2157 (1968).

7. D.C. Webb, Ph.D. Thesis, Stanford University, Stanford, California (1972).

8. B.A. Auld, R.C. Addison, and D.C. Webb, Acoustical Holography, vol. 2 (Plenum Press, 1970).

9. S. Goldman, Frequency Analysis, Modulation and Noise (McGraw Hill, 1948).

10. R.H. Bube, II-VI Compounds, eds. M. Aven and J.S. Prener (Wiley and Sons, 1967).

11. K. Wang and G. Wade, Acoustical Holography, vol. 4, p. 431 (Plenum Press, 1972).

CONSIDERATIONS FOR DIAGNOSTIC ULTRASONIC IMAGING

Philip S. Green, Louis F. Schaefer, and
Albert Macovski

Stanford Research Institute
Menlo Park, California 94025

ABSTRACT

It is becoming increasingly apparent that ultrasonic imaging (focused and holographic) will develop into a useful tool for diagnostic medicine. In this paper we examine the benefits to be derived with this application, particularly as they compare to the results obtainable with currently-used techniques of diagnostic ultrasonics.

A list of performance requirements for a practical, general purpose diagnostic imager is set forth. Among the factors considered are frame rate, sensitivity, resolution, mode of operation (transmission, backscatter), and physical configuration.

In general, ultrasonic images formed of planes within internal organs will suffer degradation due to the inhomogeneous tissue surrounding the plane of interest. The results of an experimental study of this problem are presented. The effects examined are loss of contrast, nonuniform insonification, distortion, and the creation of false detail. Comparisons are made of images produced by transmission and by backscatter of ultrasonic waves. Images of excised organs are presented to illustrate these phenomena.

INTRODUCTION

Ultrasound, because of its nontoxic and noninvasive nature, together with its strong interaction with soft tissue, is finding increasing use in diagnostic medicine. In particular, ultrasonic imaging appears to have a promising future, as a complement to and in some cases as a replacement for conventional radiography. Although the hardware for focused imaging (and holography) has not yet been developed to the point where effective clinical evaluation can begin, another visualization method--the B-scan--has been in experimental clinical use for some time. The B-scan presents cross-sectional images of internal organs by depicting the reflectivity of tissue interfaces. As focused or holographic systems find their way into the clinic, they will certainly be compared with the B-scan as well as with radiography.

Although the B-scan has established the potential effectiveness of ultrasonic visualization in diagnosis, it has not found wide acceptance in the clinic. Most of the equipment now available provides inadequate lateral resolution and no grey scale, and it requires a lengthy period to build up a single image. Research is under way to overcome these deficiencies; however, it seems that the B-scan's most serious drawback is the image format itself. Physicians, radiologists in particular, have become used to the lateral and A-P (anterior-posterior) radiographic views; the cross-sectional B-scan images are unfamiliar to them and difficult to relate to the anatomy. This format problem is particularly unfortunate, since the physician is also faced with learning that ultrasonic waves do not interact with tissue in quite the same way as do X-rays. A focused imaging system that produces perspective images normal to the direction of propagation could greatly help to bridge this interpretation gap.

In the last few years we have seen a proliferation of the technology for focused and holographic ultrasonic imaging. Many of the imaging methods now in the conceptual or experimental stages are being considered for eventual clinical use. In the following paragraphs we set forth some of the characteristics that, in our opinion, a general-purpose diagnostic imager should have.

THE GENERAL PURPOSE DIAGNOSTIC IMAGER

One example of a focused ultrasonic imaging system is a piezoelectric transducer and lens combination that is mechanically scanned in a rectilinear fashion, in synchronization with a recording display (a "C-scanner" of this type is described in a later section of this paper). Although quite useful in the laboratory, this system lacks one characteristic that is most important for diagnostic use: the ability to produce images in "real time." A real-time focused image system--that is, one with a frame rate rapid enough to resolve motion--has a number of distinct advantages. Perhaps the most significant is the ability to resolve ambiguities as to the three-dimensional nature of the anatomy. A small motion or rotation of a real-time focused-image "camera" with respect to the object shows immediately how one position of the anatomy moves with respect to another. It thus becomes evident which structures are in front of or behind others. In our experiments with a real-time liquid-surface-relief system[1] we have found this to be extremely helpful.

In addition to resolving ambiguities about the anatomy, a real-time capability helps the observer disregard many system artifacts. Any structural noise, additive or multiplicative, of the camera itself will move with the camera past the anatomical image and will not be misinterpreted. Another undesirable structure is the quantization that results from scanning lines or matrices of receiving elements. Quantization is not only distracting, but causes an effective loss of resolution (the Kell factor). This problem is significantly reduced by a small, relative motion between the camera and the object under study (this phenomenon has long been recognized by investigators in the field of fiber optics, where a relatively coarse fiber bundle can be used to read printed matter as long as the fiber bundle is in motion).

An important advantage of real-time visualization is the ability to resolve organs in motion. The principal anatomical motions considered are those in the cardiovascular system. If a focused sonic imaging system proves successful in obtaining useful heart images, relatively high frame rates will be required, especially if valve motion is to be resolved. Observation of arteries outside the chest cavity, such as the carotid and femoral arteries, may also benefit from the use of high frame rates. In addition to the direct visualization of normal anatomical motion, such as the motions of the respiratory and cardiovascular systems, a real-time system gives the physician the opportunity to make observations in the presence of induced motions. Certain pathological conditions may be more readily detected if organs can be palpated or stressed while under observation. This is currently done in X-ray flouroscopy; however, the radiologist performing a flouroscopic examination must content himself with short observational intervals to minimize the accumulated radiation. With a real-time ultrasonic camera, this viewing procedure can be uninterrupted.

Pulse-echo methods of imaging, such as the B-scan, detect acoustic impedance interfaces but are insensitive to local variations in absorptivity. The latter, which are made visible with transmission imaging, will very likely be significant for diagnostic purposes. It would be most desirable to have within one instrument the capability of imaging in both the transmission mode and the backscatter mode.

Sensitivity, resolution, and operating frequency, three of the most important parameters of an acoustic imaging system, are tightly interrelated. The required sensitivity is determined by the maximum allowable incident ultrasonic intensity and by the absorption and scatter in the tissue. It would seem to be advisable to limit the incident intensity to 10^{-2} w/cm^2, which is consistent with current standards for diagnostic ultrasound. The absorption coefficients for most soft tissues span the range between $0.5 \times f$ and $2.5 \times f$ dB/cm, where f is the frequency in MHz.[2] From these numbers we can easily deduce that many of the imaging systems now under consideration will be of quite limited diagnostic use. Very high sensitivity would be required for examining abdominal and pelvic organs or for viewing the heart.

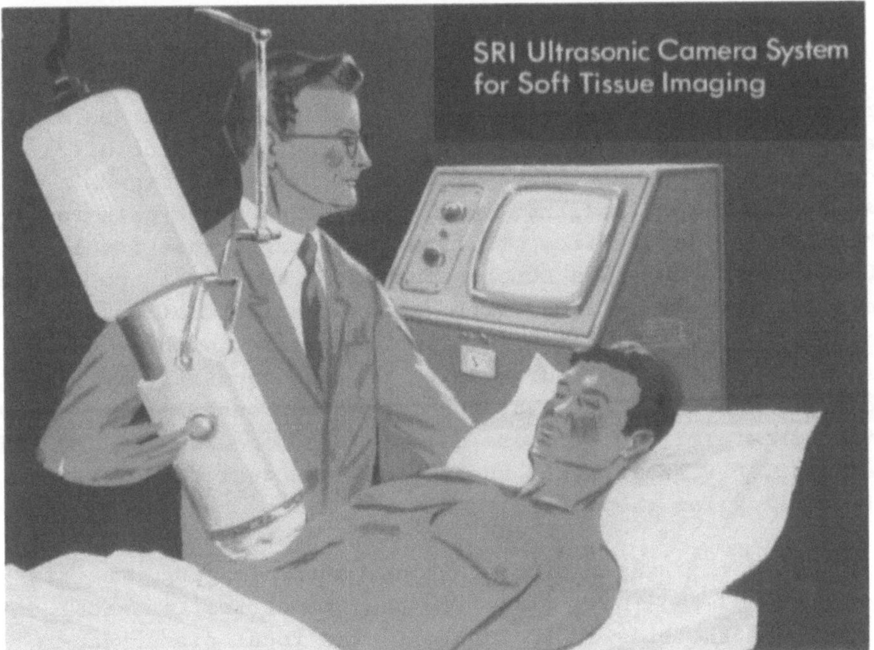

Figure 1 Artist's conception of the Ultrasonic Camera System now under development at Stanford Research Institute.

In the second part of this paper we present the results of imaging experiments with excised organs. From these experiments we conclude that high resolution and very wide dynamic range would also be essential for most diagnostic purposes.

Now under development at Stanford Research Institute is an Ultrasonic Camera System* designed to meet our criteria for a general purpose diagnostic imager. As is evident from the artist's conception of Figure 1, this system will not require immersion of the patient.

---

*The development of the Ultrasonic Camera System is being conducted under PHS Grant 1 R01 GM18780-01.

EXPERIMENTS WITH EXCISED TISSUE

Clinical evaluation of ultrasonic imaging is no longer in the distant future, and the time has come to address ourselves to a significant question: How accurately do ultrasonic images represent the structure of internal organs? Except for B-scan imaging, much of the experience gathered thus far in the practice of ultrasonic imaging has involved objects suspended in a homogeneous water bath. In the clinic, however, we will be faced with a less ideal situation, in that images of internal organs must be formed with waves that propagate through the surrounding inhomogeneous tissue. In general, we can expect that the tissue surrounding the plane of interest will be, to some extent, inhomogeneous in density, propagation velocity, and absorption, and that these properties will alter the image.

For both transmission and backscatter imaging, the tissue "behind" the focal plane (on the insonification side), as well as the tissue "in front of" the focal plane, can affect image quality. It may be useful to catagorize that inhomogeneous tissue produces (1) shading, (2) distortion, and (3) false detail. In transmission imaging, absorption and scattering in the tissue on both sides of the focal plane will result in imperfect contrast and brightness; distortion of the image can be caused by differing propagation velocities within the tissue in front of the focal plane.

The third form of degradation--false detail--might be thought of as the detail that would remain in the image if all the tissue within the depth of focus were removed and the space were filled with water. In the backscatter mode this undesired image structure can occur because of scatter from regions outside the focal depth of the system. Although in a pulse system most of this signal can be eliminated by range-gating the image converter, multiple scattering within the tissue might cause some unwanted signal to fall within the range gate. On the other hand, in transmission imaging range-gating _can_ delete multiple reflections but _cannot_ eliminate out-of-focus signals. We are accustomed to thinking of an out-of-focus image as containing only low

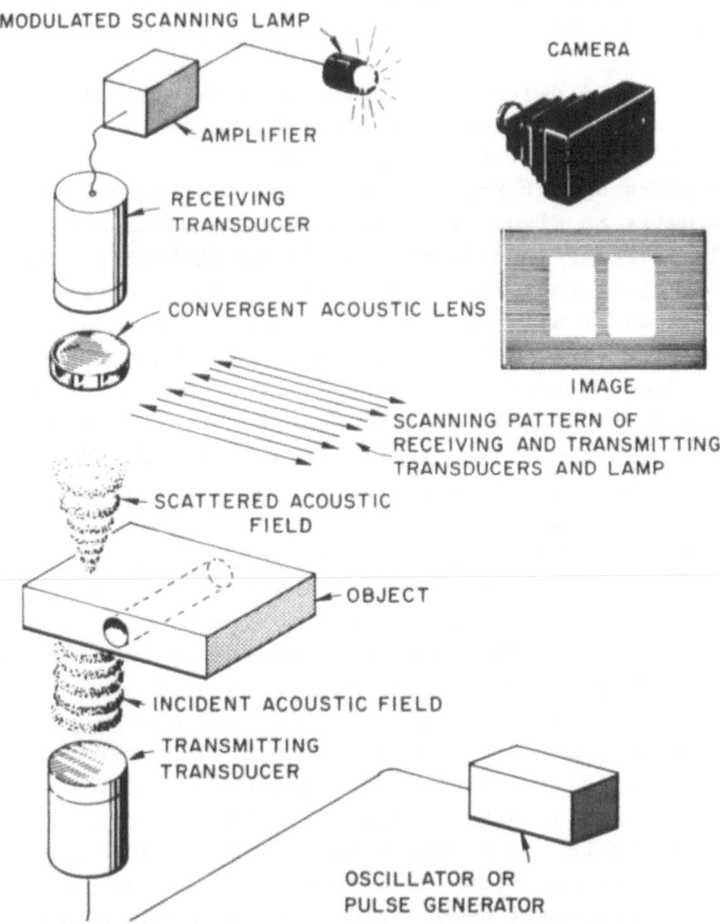

Figure 2 Arrangement for producing focused ultrasonic images of excised organs with a mechanically scanned transducer.

spatial frequencies. However, if, as in tissue, the regions in front of the focal plane contain not just point scatterers but relatively slow variations in propagation velocity as well, then we might expect to find in a transmission image extraneous structure of high spatial frequency. This phenomenon should be most pronounced in systems using coherent ultrasound, for example, in holographic systems.

Rather than defer consideration of this problem until clinical instrumentation is available, we are beginning to study it by using a laboratory C-scan imaging system developed several years ago at Stanford Research Institute.[3] Depicted schematically in Figure 2, this system consists of a single, focused receiving transducer that is mechanically scanned past the object in a raster pattern. For transmission imaging, a collimated transmitting transducer located on the opposite side of the object is scanned in synchronization with the focused receiver. A small light source, modulated by the received signal, is also scanned with the transducers and paints out the image on film in a time-exposed camera. Although producing a single image with this equipment may require a half hour or more, this method has several characteristics that make it ideal for a study of this type, in which we wish to examine acoustic image formation in soft tissue as a separate phenomenon, independent of the limitations and artifacts of a specific imaging system. For example, by using a scanning transducer focused at the desired plane in the tissue, we avoid the off-axis aberrations associated with lens/converter and holographic systems. The moving insonification transducer ensures that for transmission imaging any shading observed in the image is the result of attenuation in the tissue rather than nonuniformity of the incident field. Scan-line spacing is 1/8 mm, well below the diffraction limit associated with the 5-MHz waves and f/2.25 lens used in these experiments.

Several experiments were performed with this system, to shed some light on the shading and distortion problems. In these experiments a wire screen of about 6-mm spacing was placed in the focal plane of the lens. A layer of tissue--in this case, a fresh beef kidney of 3-cm average thickness--

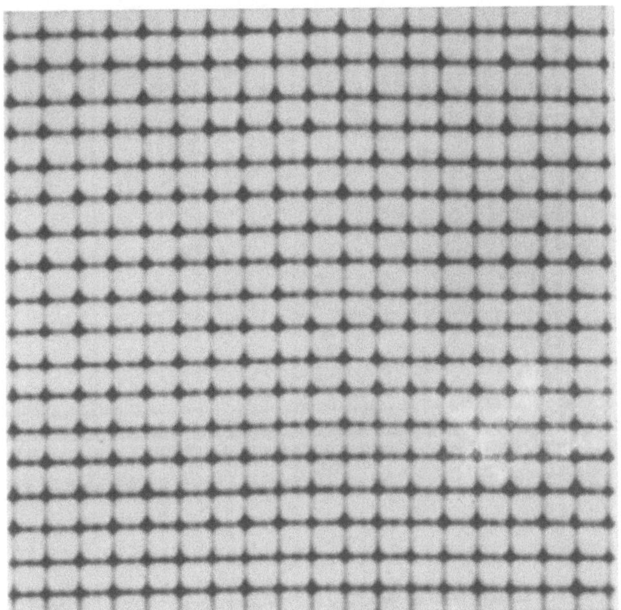

Figure 3 5-MHz transmission image of a 13-cm square wire screen of 6-mm spacing.

was placed on the screen, on the side facing the focused receiving transducer. The combination was immersed in water for imaging.

Figure 3 is a transmission image of the screen alone, showing the welds at each crossover (in this image and those following, range-gating was employed to admit only the desired pulse). Figure 4 shows a transmission image of the screen with the beef kidney in place. Three distinct regions of the image are evident: the area beyond the borders of the kidney, the lobular structure, and the hilus in the center of the kidney, where the ureter and the renal artery and vein enter. The receiver amplifier, operating in the linear mode, was adjusted to provide best contrast in the lobular region, where the tissue attenuation was about 23 dB. As a result, the detector was saturated in the region beyond the kidney and below threshold in the hilus. At best, the wires of the screen appear fragmented and nearly unrecognizable.

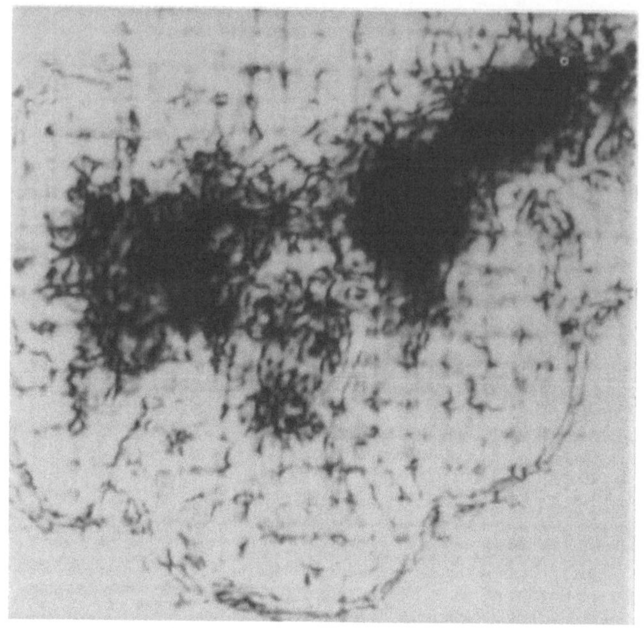

Figure 4 Transmission image of the wire screen of Figure 3, with a 3-cm thick beef kidney placed against the screen on the receiver side.

The image of Figure 5 resulted when the previous experiment was repeated with but one change: Logarithmic rather than linear amplification was used. Now, although the overall contrast is reduced, the screen is clearly visible beyond the kidney's borders, and, to some extent, within the hilus as well. The wires no longer appear broken. Examination of this image reveals that distortion of the grid is minimal, occurring principally at the borders of the kidney, where the velocity gradient is largest. Parts of the kidney appear to be in focus also; the depth of the field was about 1 cm. In the final experiment with this configuration, a backscatter image (Figure 6) was formed, the same focused transducer being used for transmitting and receiving. In this case, the attenuation differential over the image field was double that of the transmission case, and penetration of the hilus region was insufficient.

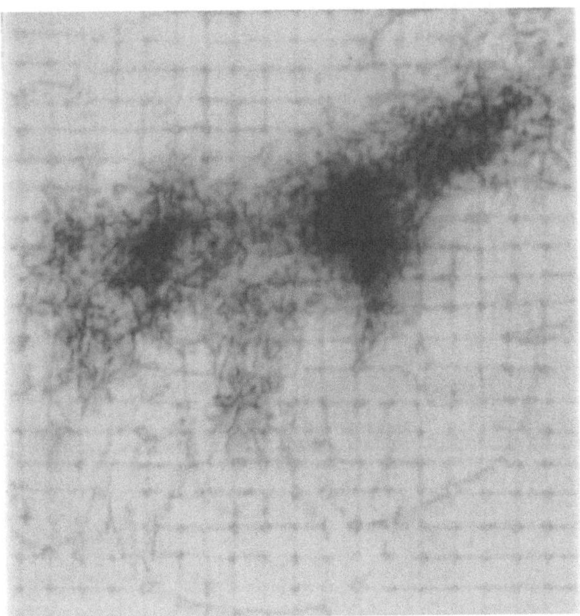

Figure 5 Same as Figure 4, except for the use of logarithmic rather than linear amplification of the received signal.

From these experiments we conclude that differential attenuation over a small region of tissue can be quite large, and that to produce images of good quality it may often be necessary to use an image converter with very wide dynamic range, followed by amplitude compression and wide dynamic-range detection and display. Distortion of the image by intervening tissue does not appear to be a serious problem, which might have been predicted, since the propagation velocities in most soft tissues are very nearly the velocity of water.[4]

A number of different organs have been imaged with this apparatus. Shown in Figure 7(a) is the transmission image of lamb kidney, in Figure 7(b) the sagittal view of a cat brain. The sagittal and coronal views of a monkey brain, shown in Figure 8, contain fine structure that may or may not correlate directly to the anatomy of the plane in which

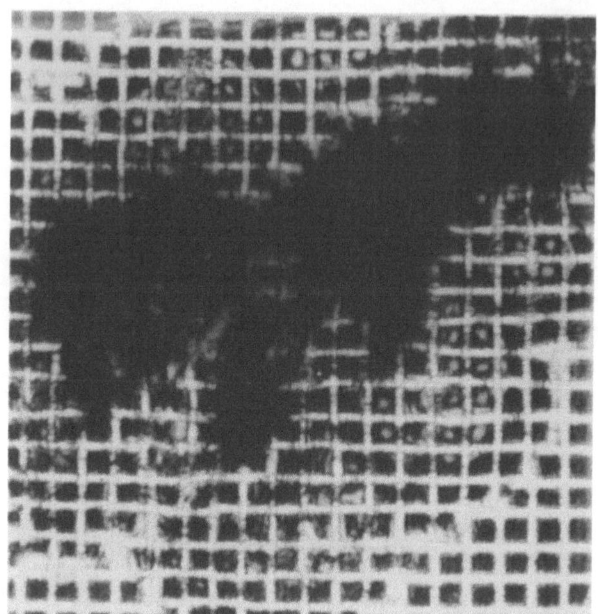

Figure 6 Backscatter image of the screen with the kidney in place.

the receiver was focused. At this time we have not yet begun to make detailed correlations. Finally, in Figure 9 we show the transmission image of a human fetus, in approximately

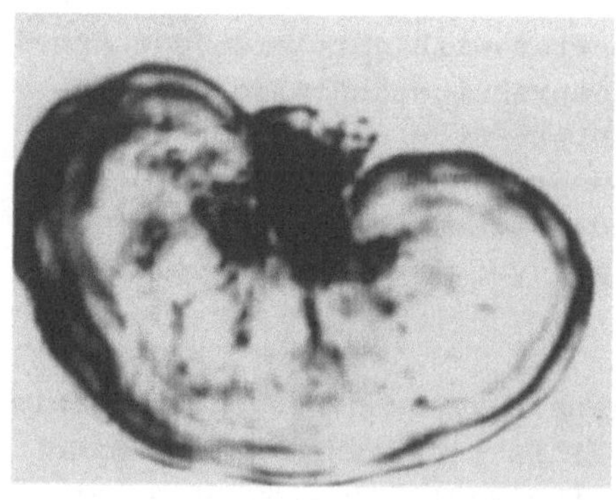

(a)

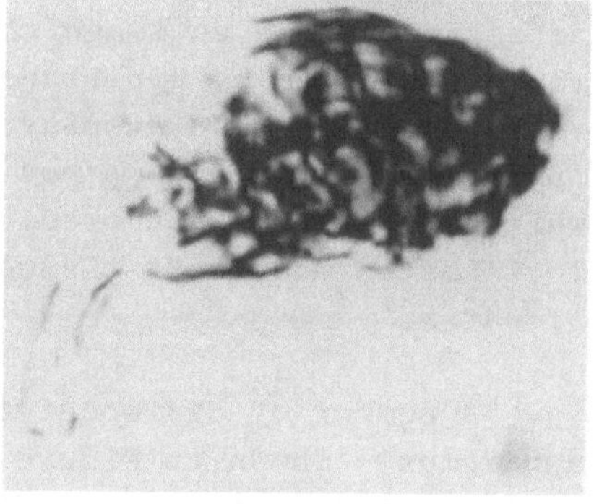

(b)

Figure 7 Transmission image of (a) a lamb kidney; (b) a cat brain.

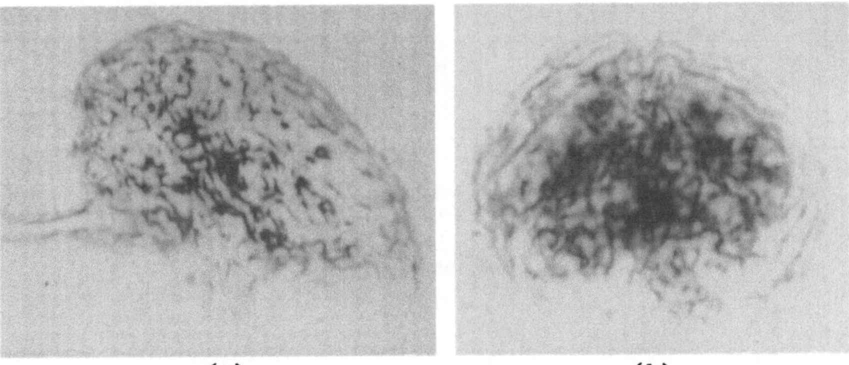

(a) (b)

Figure 8 Transmission images of a monkey brain. (a) sagittal; (b) coronal.

the 17th week.* Ossification centers of the bones are clearly revealed. By increasing the contrast internals organs can also be perceived. Hopefully, ultrasonic imaging can allow early detection of fetal abnormalities.

Several parameters that were not exercised in these experiments will be examined in future ones. One of these is the numerical aperture of the system. A greater depth of focus might sometimes be advantageous, even at the expense of resolution. The effects on image quality of changes in both center frequency and bandwidth deserve further consideration. Experiments such as these, in which excised organs are imaged under controlled conditions, should be of considerable value as a basis for interpreting clinical images produced with the real-time instruments now under development.

---

*Provided by Dr. David Holbrooke of Children's Hospital, San Francisco, California.

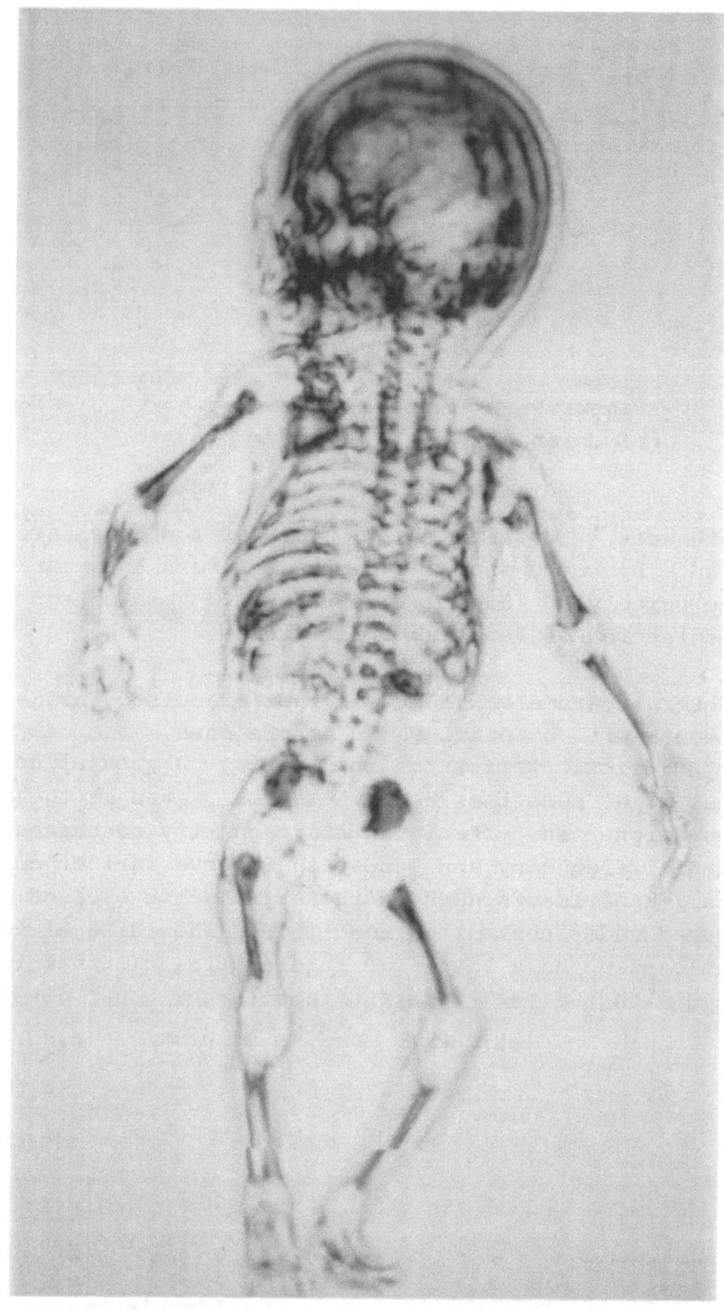

Figure 9 Transmission image of a human fetus in approximately the 17th week.

REFERENCES

1. P. S. Green, "A New Liquid-Surface-Relief Method of Acoustic Image Conversion," in Acoustical Holography, Vol. 3, A. F. Metherell, ed., Plenum Press, New York, pp. 173-187 (1971).

2. F. Dunn, "Ultrasonic Absorption by Biological Materials," in Ultrasonic Energy, E. Kelly, ed., University of Illinois Press, Urbana, Illinois, pp. 51-65 (1965).

3. P. S. Green, "Methods of Acoustic Visualization," Internat. J. Nondestr. Test., (1):1-27 (1969).

4. G.E.P.M. Van Venrooij, "Measurement of Ultrasound Velocity in Tissue," Ultrasonics, 9:240-242 (1971).

MULTI-INFORMATION RECORDING AND REPRODUCTION IN THE ULTRASONO-CARDIO-TOMOGRAPHY

Yoshimitsu KIKUCHI, Daitaro OKUYAMA and Chihiro KASAI

Research Institute of Electrical Communication
Tohoku University

Sendai, Japan

Toshiaki EBINA, Motonao TANAKA, Yoshio TERASAWA and Rokuro UCHIDA

Research Institute for Tuberculosis and Leprosy
Tohoku University

Sendai, Japan

INTRODUCTION

Medical diagnostic methods in which ultrasonic waves are utilized have developed a new field in medical science, and their applications to clinical work now involve nearly all the organs of the human body. In particular, the ultrasonic methods display full effect on the subjects hitherto not easily examined by other methods including roentgenography. The ultrasonic diagnosis in wide use today consists of the so-called pulse-echo method in Mega-Hertz frequency range. The method is classified into A-mode and B-mode according to the features of visual presentation. The sectional patterns obtained in the B-mode presentation are called ultrasono-tomograms.

It had been considered very difficult to obtain ultrasonic tomograms of the heart until the present author et al. initiated around 1963 the two methods named "Synchronized Ultrasono-Cardio-Tomography"[1,2,3] and "Ultrasono-Tomo-Kymography".[4] The reason was that the heart, unlike other organs, is in continuous pulsation and the position of every echo source repeatedly shifts along with time.

The synchronized ultrasono-cardio-tomography consists of the instantaneous control of the tomographic apparatus by electrocardiographic current (ECG) of a patient, either the synchronized control of the ultrasonic scanner at any required phase of the heart pulsation, or the synchronized control of the unblanking circuit of a cathode-ray tube in a pattern display unit.[5,6] At present, the clinical application mainly employs the latter method of control. An example of the tomograms of the human heart is shown in Fig.1, in which a certain section of the pulsating heart at a certain phase is observable. Further application of this method has made it possible to observe the movements of the heart intuitively through animated display[7,8] of a series of the stationary tomograms.

Attempts at describing the ultrasono-cardio-tomograms from different viewpoints have resulted in the ultrasono-tomo-kymography. This is a method by which both factors as the position of every echo source and its phase in relation to lapse of time are superposingly presented in one tomogram In Fig.2 is shown an example of the ultrasono-tomo-kymogram. This is considered also useful in clinical work as a tomogram which backs the synchronized ones in its clinical utility.

The time required for the clinical examination of a patient, however, is long because several leaves of synchronized tomogram for various phases of the heart must be taken of one patient. Such a long examination may impose a certain physiological burden on the patient. This situation is much more dominant in the case of obtaining "Sensitivity Graded Tomogram Pairs".[9] (A set of tomograms of a particular organ described at different sensitivity levels of an ultrasonic diagnosis apparatus: The method aims at a diagnosis of comprehensive judgement through pattern comparison.) The present authors have then developed a method[10] in which every echo from the inside of

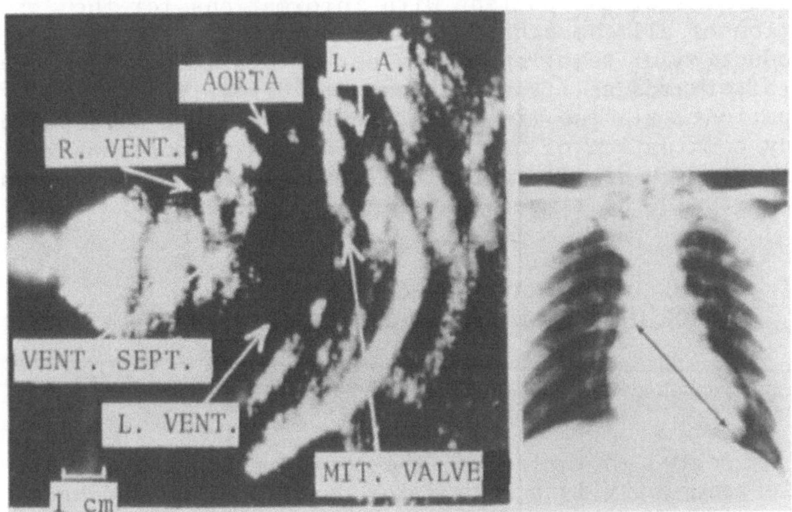

Fig.1 An Ultrasono-cardio-tomogram and the Roentgenogram in which the plane of the ultrasonic scanning is shown by the arrow.

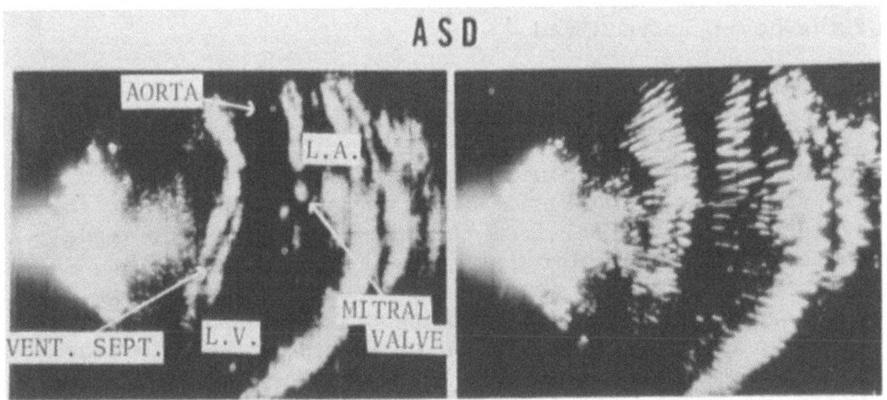

Fig.2 An Ultrasono-cardio-tomogram (left) and the Ultrasono-tomo-kymogram (right) for atrial septal defect case.

a human body is recorded on a VTR (Video Tape Recorder) at the bed side of the patient with informations for the position of all the echoes as well as ECG current. The reproduction of tomograms for a detailed examination can be made afterwards as if physicians were directly examining the patient, all the time when the reproduction is repeated at any combination of various cardiac phases and sensitivity settings. The development of this apparatus has considerably shortened the time for examining a patient.

MULTI-INFORMATION RECORDING SYSTEM FOR ULTRASONO-CARDIO-TOMOGRAPHY

1. Cardiac Synchronization

As stated in the above, the synchronized ultrasono-cardio-tomography is a method for describing the tomogram in synchronization to any desired phase of the heart pulsation by operating the apparatus for a short time in which the heart is deemed stationary. As is well known, the repetition rate of the ultrasonic pulse is limited in accordance with the time required for propagation of the sound. The repetition rate limits the scanning speed of the ultrasonic beam in order to obtain dense scanning lines. Therefore, a method similar to a panorama photographic method had to be combined in general cases when a wider sectional pattern of the heart is required.

For the synchronizing signal, the R-wave of ECG, in Fig.3 (1), is first detected, and its waveform is shaped into a pulse which gives the time origin, (2). Then, in accordance with the required phase for the stationary tomogram, a delayed signal, (3), is generated so that it may have a delay-time (T_d) from the time origin. This signal is used as the synchronizing signal for operating the apparatus; either one quick shifting of the ultrasonic beam, or the repetition of short unblanking of the cathode-ray tube in a display unit with an asynchronous and slow movement of the ultrasonic beam. In the latter method, at any single unblanking, only a partial tomogram of the heart is displayed. So the panoramic repetition of 20 to 30 times can cover the whole heart. The pattern thus obtained is a positionally patched pattern of a number of the partial tomograms for a given cardiac phase (see Fig.1). When one

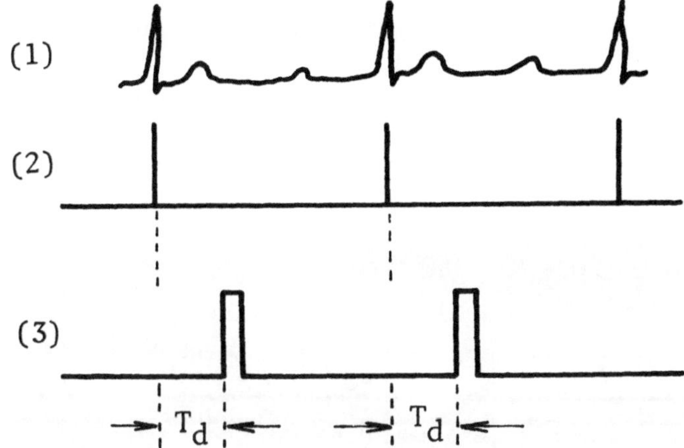

Fig.3 Time relation between ECG and the control signals for "Synchronized cardio-tomograph".

cardiac cycle is divided into nine equal intervals by changing the delay-time (T_d), a set of the stationary tomograms for one cardiac cycle is obtained as shown in Fig.4.

2. Recording system

The cause for the long clinical examination is based on the intermittent use of echo information. As a means of solving the problem of the information loss, the present authors have introduced a system of multi-information recording to the ultrasono-cardio-tomography aforementioned. In this method, entire ultrasonic echo information is recorded on a VTR together with ECG at the clinic. The time required at the clinic would be only 2 to 3 minutes no matter how many the number of phases for a cross-section may be required afterwards in the VTR reproduction. In Fig.5 is shown the composition of the apparatus.

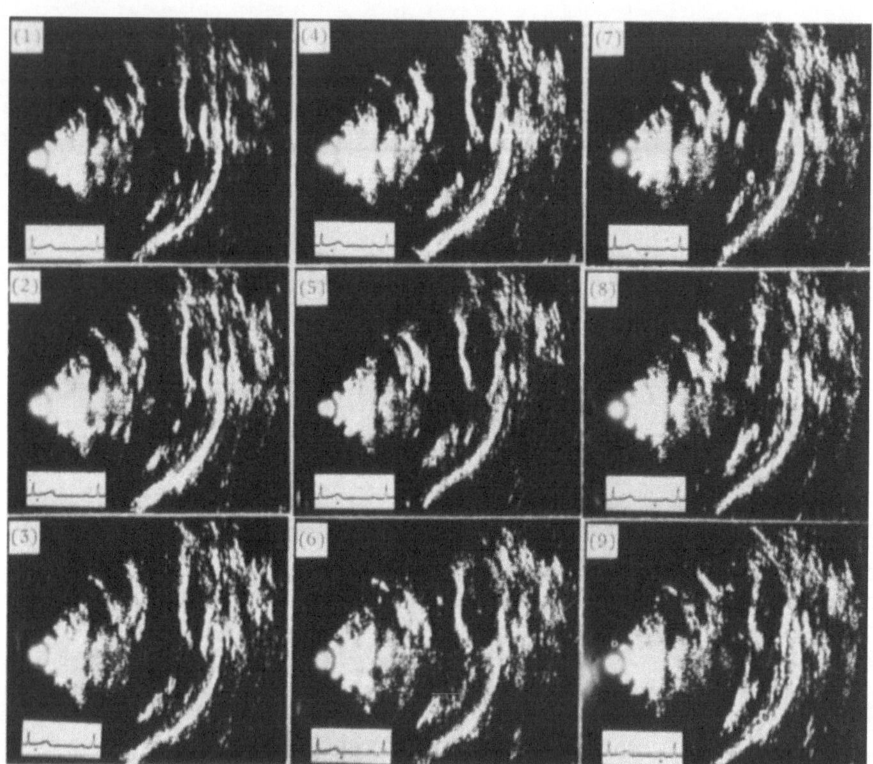

Fig.4 Ultrasono-cardio-tomograms obtained at nine different cardiac phases of one cardiac cycle.

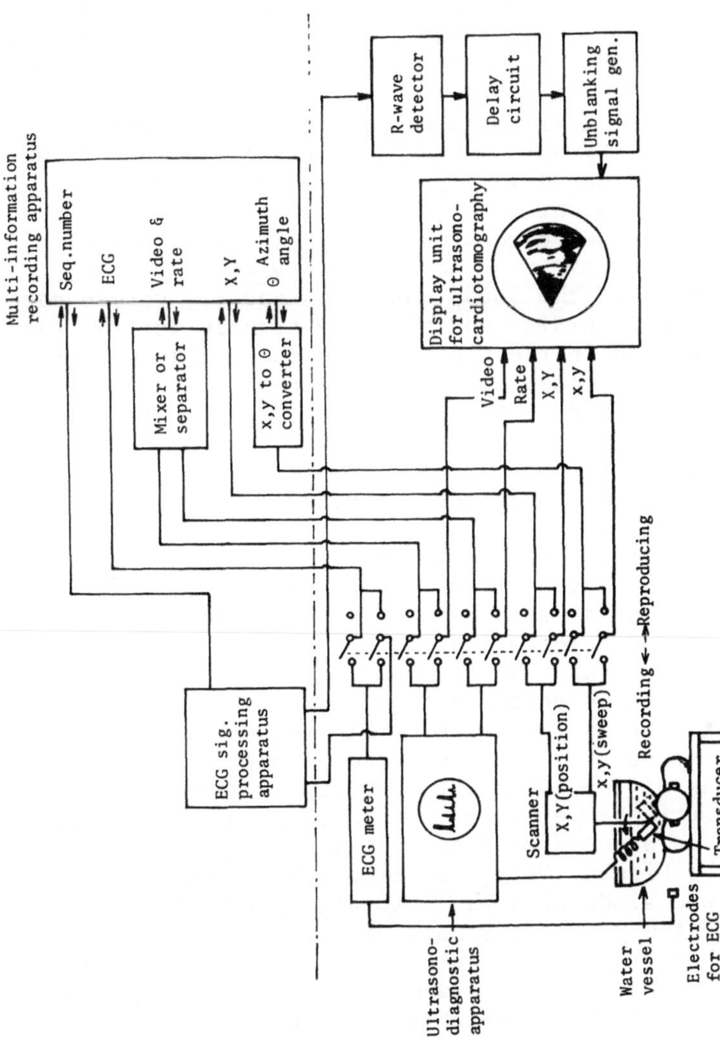

Fig.5 Schematic diagram of the equipment for the ultrasono-cardio-tomography when "Multi-information Recording System" is employed.

In the figure, the schematic diagram below the broken line is the usual apparatus for synchronized ultrasono-cardio-tomography, while the units above the broken line consist of the Multi-Information Recording Apparatus and the ECG Signal Processing Apparatus, which is shown in detail in Fig.6. Into this part, comes echo information of ultrasonic waves which are already amplified and converted into video signal. The signal is mixed with repetition rate signal of the ultrasonic beam and then recorded on the VTR video track. Of audio channels* in the VTR, one is used for recording the angular signal (Θ) of the azimuth information of the ultrasonic beam, the (Θ) being the converted result of the scanner direction (sweep signal: x and y). The other channels are used for recording the ECG, the position signals of the sound source (X and Y), voiced data related to main recording and some key signals for the tape control to be used in the playback sequence. The position signals (X, Y) are those provided for compound scanning mode, and so they are fixed in the case of simple sector scanning.

3. Reproduction System

In the playback of the VTR, the recorded signals are separated through wave-filters and demodulated to restore the original signals as shown to the right in Fig.6 . And thus the recorded information can be used as being identical with the original information, and the pattern of any tomogram is obtained on the cathode-ray tube screen of the Display Unit. It is further possible to semi-automatically repeat the reproduction of a particular part of the video tape by means of the aforementioned key signals recorded on the tape. Furthermore, if a certain program on the desired cardiac phases and the unblanking times are pre-set in the system of the Display Unit, these switchings can be automatically made along with the automatic playback and rewinding of the particular part of the video tape (an automatic photography for the cathode-ray pattern being inclusive). Through such playback operations, any number of various tomograms can be obtained after the bed-side

---

* In Fig.6, the diagram shows the case in which a frequency modulated (FM) carrier stands for one channel.

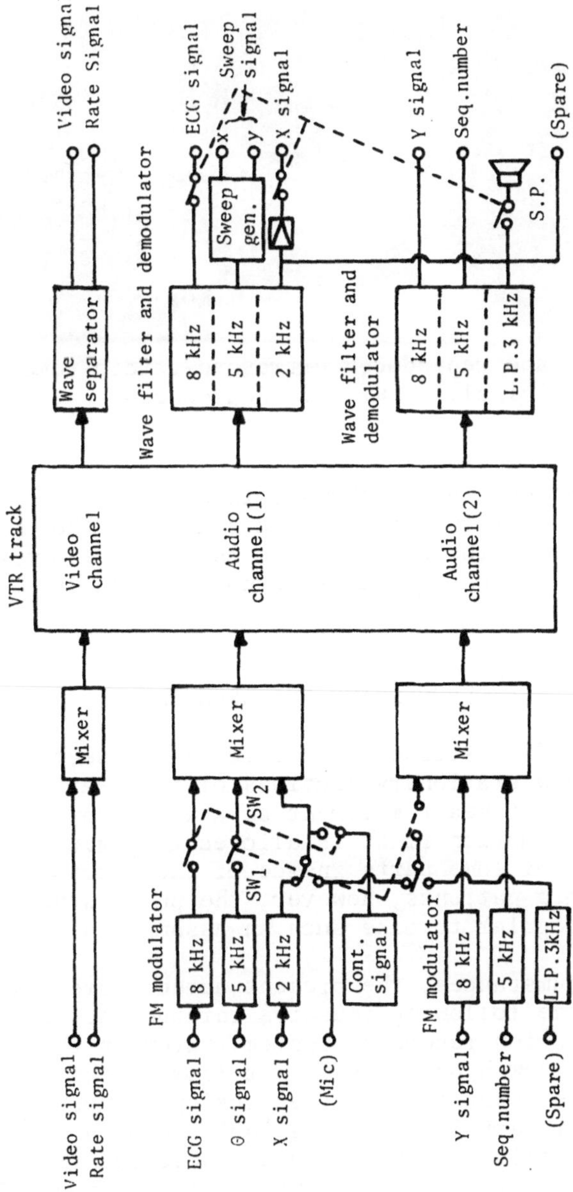

Fig.6 Detailed schematic diagram of the recording and reproduction apparatus in the "Multi-information Recording System" shown in Fig.5.

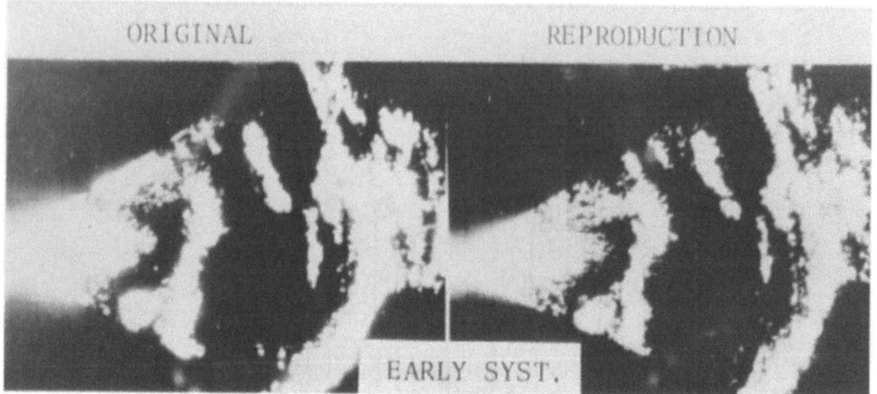

Fig.7 Original and reproduced ultrasono-cardio-tomograms. Left: Original. Right: Reproduced.

clinical examination. Fig.7 shows both examples of the original tomogram obtained at the bed side with the usual apparatus and the reproduced tomogram through the present system. It may be considered as being practically sufficient.

ECG SIGNAL PROCESSING FOR ARRHYTHMIA

In the methods described in the foregoing sections, the tomograms show stationary sections of the heart upon an assumption that each heart movement as a whole approximately repeats identical motion if it is allowed to neglect the detailed motions of some parts inside of the heart. In the case of arrhythmia patients, however, the pulse synchronization is too irregular to make such an assumption.

In the circumstances, the present authors are now trying to take the following measures through the use of the multi-information recording system: Even in the case of arrhythmia, a new hypothesis is set forward that the cardiac motion is approximately identical provided that its R-wave spacing (interval) in ECG is equal to each other. Under this hypothesis, the VTR data are selected from the recording which was taken for somewhat longer period of

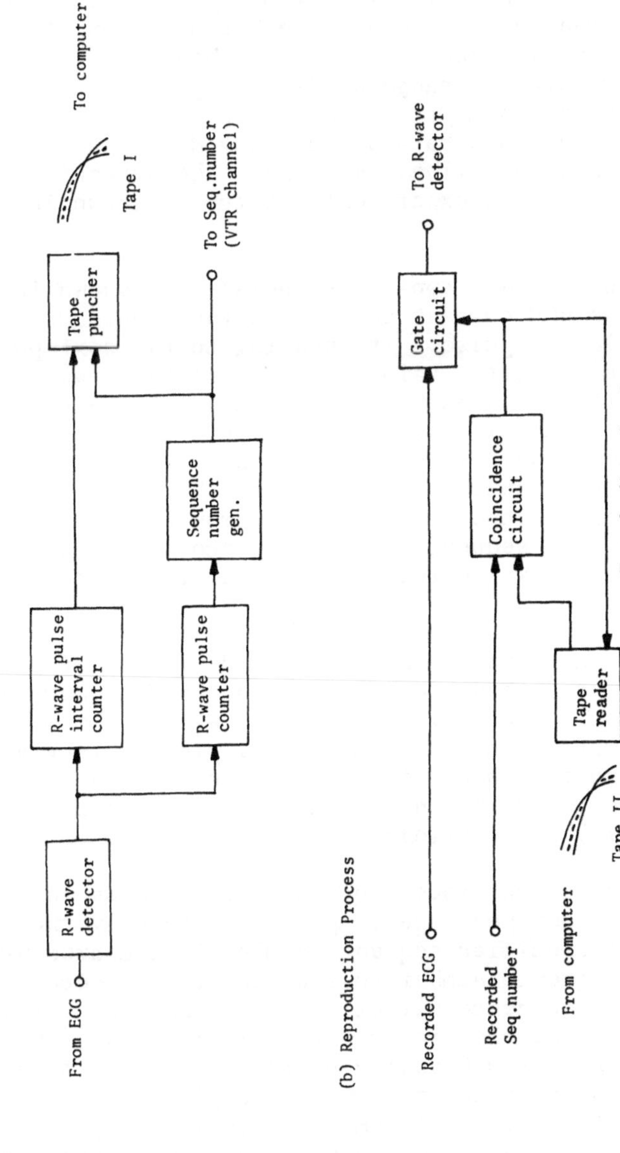

Fig.8 Detailed schematic diagram of the "ECG Signal Processing Apparatus" shown in Fig.5, provided for arrhythmia. (a) For recording process. (b) For reproduction process.

time than in the case of the normal, and only the selected data are used for the tomographic reproduction. To do this, all the R-wave spacings are first read from the recorded ECG, and next the mean value and the standard deviation are calculated by an electronic computer. Then, the medically acceptable irregularity range of R-wave spacing is pre-set in the ECG Processing Apparatus shown in Fig.5 so that only the corresponding video data are displayed on the cathode-ray tube screen to describe tomograms. The schematic diagram of a tentative experimental setup is shown in Fig.8 and its function is as follows.

In recording, this apparatus operates as shown in Fig.8 (a). The ECG current comes into the R-wave Detector to be shaped into a pulse. Its spacing to the next pulse is measured in a digital value by means of the R-wave Pulse Interval Counter. The number of the pulses in relation to time is also counted in the R-wave Pulse Counter and the pulse sequence number is generated in a certain digital serial code. Both outputs are put into the Tape Puncher to produce a tape (Tape I) for an electronic computer. At the same time, the sequence number is recorded on an audio-channel of the VTR which records the echo information.

From the punched tape (Tape I), the electronic computer calculates the mean value and standard deviation of the pulse spacing, and then specifies the sequence number of the R-wave spacings of which values are within the medically acceptable limit. A certain additional program is given to the computer so that a digit 1 is reduced from the specified sequence number, and this new sequence number is punched on a tape (Tape II) successively.

The reproduction process of the VTR is shown in Fig.8 (b). The punched tape (Tape II) is read by the Tape Reader and the specified sequence number is compared with the recorded sequence number to find the coincidence by means of the Coincidence Circuit. Every time the coincidence is found, the electronic Gate Circuit passes the recorded ECG and the unblanking of the display is made to compose a tomogram in the same way as employed for the normal. The Gate Circuit becomes closed just after the passage of the R-wave. Therefore any number of stationary tomograms for an arrhythmia patient are obtained* with various cardiac phases by repeating the playback of the VTR.

CONCLUSION

Through the development of synchronized ultrasono-cardio-tomography, it has become possible to describe the tomograms of the living human heart non-operatively. The accompanying problem of comparatively prolonged clinical examination also has been overcome by the present proposal of the multi-information recording system. The problems in arrhythmia case are now under experiment with introduction of the idea of selecting cardiac pulsations which are considered as having medically acceptable irregularity. A computer-aided processing is employed in this study.

In conclusion, the present authors would like to acknowledge close co-operation of Japan Radiation and Medical Electronics Inc., Tokyo.

REFERENCES

(1) Y. Kikuchi and M. Tanaka, "Some improvements in ultrasono-tomograph for the heart and great vessels", 1966 IEEE Symposium on Ultrasonics, J3, Cleveland, Ohio, U.S.A., 1966.

(2) M. Tanaka, S. Oka, T. Ebina, S. Kosaka, K. Unno, Y. Terasawa, Y. Kikuchi, C. Kasai, R. Uchida and Y. Hagiwara, "Ultrasono-tomogram of the heart and great vessels in living human subject", Medical Ultrasonics, 4, No.1-2, p.47, 1966.

---

* Actual tomograms for arrhythmia have not been obtained before the submission of this manuscript because of a certain trouble in the electronic circuits.

(3) Y. Kikuchi, D. Okuyama, M. Tanaka, T. Ebina, and S. Oka, "Ultrasono-Cardio-Tomography and its application to morphological measurement of the heart", Ultrasono Graphia Medica, Vol.III, (Proceedings of the 1st Congress on Ultrasonic Diagnostics in Medicine and SIDUO III, Vienna, Austria, 1969), Verlag der Wiener Medizinischen Akademie, Wien (1971), pp.423-436.

(4) Y. Kikuchi, D. Okuyama, M. Tanaka, T. Ebina, and S. Oka, "Ultrasono-Tomo-Kymography of the heart", Ibid., pp.481-487.

(5) M. Tanaka, S. Oka, T. Ebina, S. Kosaka, K. Unno, Y. Terasawa, Y. Kikuchi, C. Kasai, R. Uchida, and Y. Hagiwara, "Ultrasono-tomography by the intraesophageal method", Medical Ultrasonics, 4, No.1-2, p.48, 1966.

(6) M. Tanaka, S. Oka, T. Ebina, S. Kosaka, Y. Kikuchi, R. Uchida, and Y. Hagiwara, "Ultrasono-tomography for the heart, great vessels and other mediastinal organs", Digest of the 6th I.C.M.E. & B.E., p.294, Tokyo, 1965.

(7) Y. Kikuchi, D. Okuyama, T. Ebina, S. Oka, and M. Tanaka, "Cardiac Kineto-Ultrasono-tomography", Reports of the 6th International Congress on Acoustics, Tokyo, M-1-8, 1968.

(8) Y. Kikuchi, D. Okuyama, M. Tanaka, T. Ebina, and S. Oka, "Kineto-Ultrasono-Tomography of the heart", Ultrasono Graphia Medica, Vol.III, Verlag der Wiener Medizinischen Akademie, Wien (1971), pp.475-480.

(9) Y. Kikuchi, "Way to quantitative examination in ultrasonic diagnosis", Medical Ultrasonics, 6, No.1, 1968.

(10) Y. Kikuchi, D. Okuyama, C. Kasai, T. Ebina, S. Oka, M. Tanaka, Y. Terasawa, and R. Uchida, "Ultrasono-cardiotomography for the heart and great vessels by means of Multi-Information Recording System", (in Japanese), Reports of the 17th Meeting of the Japan Society of Ultrasonics in Medicine, p.23, 1970.

BRAGG-DIFFRACTION IMAGING: A POTENTIAL TECHNIQUE FOR MEDICAL DIAGNOSIS AND MATERIAL INSPECTION

John Landry, Hormozdyar Keyani, and Glen Wade

University of California, Santa Barbara
Department of Electrical Engineering
Santa Barbara, California 93106

ABSTRACT

Results of a recently-constructed Bragg-diffraction imaging system are shown. This new system employs a 10 × 15 cm transducer mosaic and operates at an acoustic frequency of 3.58 MHz, thus enabling it to be used with biological materials of considerable thickness. The system has a resolution capability of about 8.5 acoustic wavelengths and will produce images of bone structure in a human hand with an acoustic power-density of 100 mW/cm^2 or less.

INTRODUCTION

Experimental results during the past five years have demonstrated that ultrasonic imaging by Bragg diffraction is a useful and relatively uncomplicated method for employing acoustic energy as a means of visualizing the internal structure of optically opaque objects. Most of our early work was done at frequencies of 15 MHz or more where difficulites due to scattered light and anamorphic image distortions are minimal, as a result of the reasonably large Bragg angle.[1,2]

For many nondestructive testing purposes such frequencies are satisfactory, but for imaging biological objects 1 cm or more in thickness, the corresponding acoustic absorption would necessitate the use of unreasonably high power densities. Consequently a practical biological acoustic imaging system must operate at frequencies below about 5 MHz. Our own early attempts at using Bragg-diffraction imaging at low frequencies were unsuccessful owing primarily to a large amount of laser light being scattered into the image by minute particles in the acousto-optic interaction medium.

We describe here the design and construction of an improved Bragg-diffraction imaging system (Fig. 1) which operates at 3.6 MHz employing a transmitting mosaic consisting of six square quartz transducers, 5 cm on a side (Fig. 2). Light scattering and fluid streaming effects are

Figure 1 - 3.6 MHz Bragg-Diffraction Imaging System.

Figure 2 - Acoustic Transducer Mosaic.

reduced significantly by the use of a mylar isolation membrane between the specimen tank and the acousto-optic interaction region. Improved impedance matching techniques allow the use of incident acoustic power densities of 1.5 watt/cm^2 or more.

The new system is capable of resolving bones in the human hand at an incident acoustic power density of 100 mw/cm^2. Because of the relatively small numerical aperture of the interacting optical beam the imaging resolution is presently limited to details of about 2.5 mm spacing (8.5 acoustic wavelengths). This is a consequence of the unavailability of large-aperture cylindrical lenses suitably corrected for spherical aberration. However, from previous results at 15 MHz, we would expect a resolution commensurate with the Rayleigh criterion and can thus hope to improve the resolution by employing a cylindrical converging lens of three to four times the aperture of the one presently used.[3]

Concurrent with the development of the 3.6 MHz system, we have pursued an interest in the use of optical heterodyne detection as a means of reducing image noise due to scattered light and also as a possible technique for retaining acoustic phase in the system output. A properly designed Bragg system employing optical heterodyne detection offers the possibility of phase-contrast acoustic imaging.[4]

ACOUSTIC TRANSDUCER MOSAIC

In the design of an acoustic imaging system of relatively large active area it would, for simplicity, be desirable to use a single large transducer to produce the ultrasonic beam but considerations of cost, and the likelihood of breakage, forced us to compromise by forming the large beam with a mosaic of smaller transducers. The two main difficulties which arise from employing a mosaic are electrical impedance matching and the elimination of acoustic beam irregularities which result from destructive interference due to slight misalignment of adjacent transducers.

Impedance Matching

At resonance the quartz piezoelectric transducers which we employ do not generally have an input resistance close to

50 ohms. In fact, the resistance is typically nearly an order of magnitude higher or lower and, for this reason, impedance transformation is necessary in order to allow the use of standard electronic R.F. amplifiers and coaxial feed-line in driving the transducers.

In the case of a mosaic it would thus be desirable to connect the transducers in series when the individual impedances are low with respect to 50 ohms, and in parallel when they are high. Series connection turns out to be impractical, however, because the conductive surface on one side of each transducer is in direct contact with the water in the specimen tank and electrical shorting results from even very slight impurities. Consequently, only parallel connection is used, which in the low impedance case necessitates an impedance transformer that carries rather high current in the inductive portion of the circuit. This must be borne in mind when choosing the matching network components to insure adequate dissipation.

To arrive at a simple matching network for the x-cut quartz transducers of the system described in this chapter, a new equivalent circuit for piezoelectric transducers was used.[5]

First the equivalent circuit is simplified, for the general case of any piezoelectric material operating in the thickness-expander mode, under the assumptions that the transducer is operated at resonance and that it is interfaced with water on one side and air on the other. Then the reduced circuit is specialized to the case of an x-cut quartz transducer, and finally the matching network is calculated.

The equivalent circuit for a thickness-expander plate is shown in Fig. 3. The element values are given by

$$C_o = \frac{\ell w}{\beta^S_{33} t} = \frac{\varepsilon^S_{33} \ell w}{t} = \frac{\varepsilon_o \varepsilon'_x \ell w}{t}$$

$$Z_o = \rho \ell w v^D_t$$

$$v = v^D_t = \left(\frac{c^D_{33}}{\rho}\right)^{\frac{1}{2}}$$

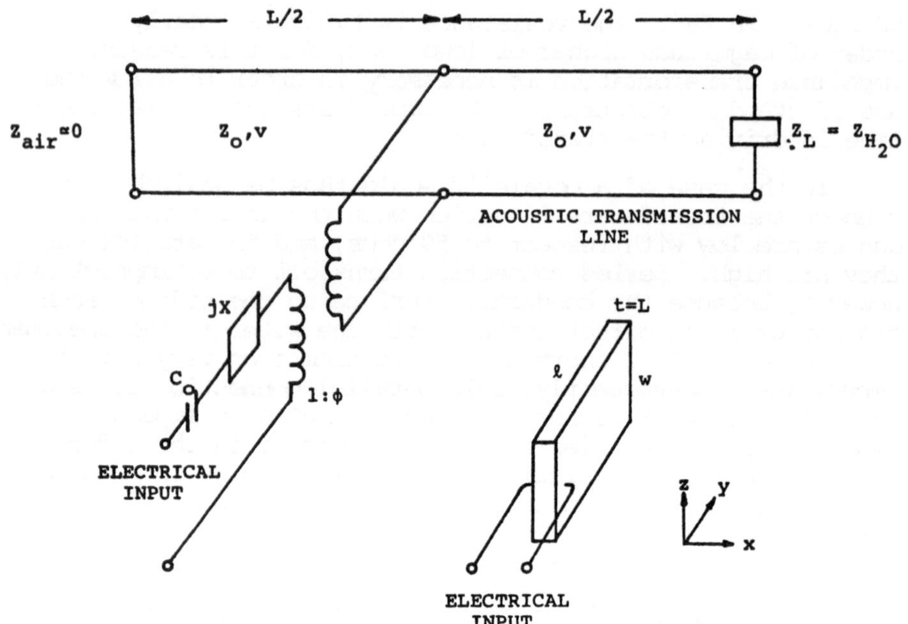

Figure 3 - The Equivalent Circuit for a Thickness-Expander Plate.

$$\phi = (\frac{1}{2} M) \operatorname{cosec}\left(\frac{t\omega}{2v_t^D}\right)$$

$$X = Z_o M^2 \sin\left(\frac{t\omega}{v_t^D}\right)$$

where $M = \dfrac{h_{33}}{\omega Z_o}$

and ℓ, w, and t are transducer dimensions as defined in Fig. 3. In these equations,

ω = angular frequency

ρ = density

β_{33}^S = dielectric impermeability

C_{33}^D = elastic stiffness

ε_{33}^S = permittivity

v_t^D = acoustic wave velocity

h_{33} = piezoelectric constant

ε_x' = clamped dielectric constant

$\varepsilon_o = 8.854 \times 10^{-12}$ farads/meter.

(For exact definition of parameters and subscripts, see Reference 6.) In the equivalent circuit, Z_o and v are the characteristic impedance and velocity, respectively, associated with the acoustic transmission line. The line length L is equal to the transducer dimension in the direction of acoustic wave propagation.

Since the transducer is operated at resonance, the length of the acoustic transmission line is $L = t = \lambda/2$, where λ is the wavelength of the acoustic wave in the transducer. Therefore, transforming

$$Z_{air} \text{ and } Z_{H_2O}$$

to the center of the transmission line reduces the equivalent circuit to that of Fig. 4a.

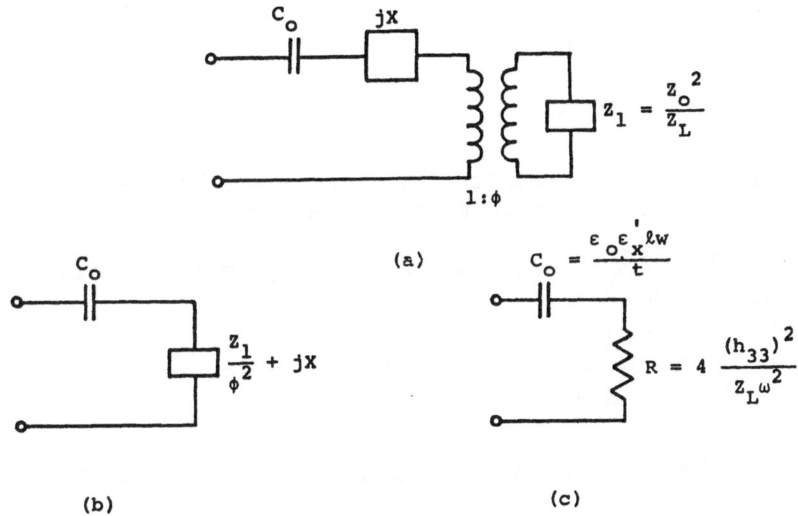

Figure 4 - Simplified Equivalent Circuits

Now, if Z_1 is reflected to the primary of the transformer the circuit is as shown in Fig. 4b.

Since

$$t = \lambda/2 \text{ and } \lambda = \frac{v_t^D}{f},$$

where f is the resonant frequency of the transducer, the expression for

$$\frac{Z_1}{\phi^2} + jX$$

becomes

$$\frac{Z_1}{\phi^2} + jX = \frac{Z_1}{\left(\frac{1}{2M}\,\text{cosec}\,\left(\frac{t\omega}{2v_t^D}\right)\right)^2} + jZ_oM^2\sin\left(\frac{t\omega}{v_t^D}\right)$$

$$= 4Z_1M^2\sin^2\left(\frac{\pi}{2}\right) + jZ_oM^2\sin(\pi)$$

$$= 4Z_1M^2$$

$$= 4\left(\frac{Z_o^2}{Z_L}\right)\left(\frac{h_{33}}{\omega Z_o}\right)^2$$

$$= \frac{4(h_{33})^2}{Z_L\omega^2} = R$$

Thus, the equivalent circuit of a thickness-expander plate at resonance is reduced to that of Fig. 4c.

Defining $Z_L = KZ_o$ and substituting into the relation for R, we obtain

$$R = \frac{4(h_{33})^2}{KZ_o\omega^2},$$ where K is a proportionality constant.

In the case of an x-cut quartz transducer we have

$$h_{33} = 4.31 \times 10^9 \text{ N/C}$$

$$Z_o = (\rho v_t^D)(\ell w) = (15.2 \times 10^6 \frac{Kg}{m^2 sec.})(\ell w)$$

Therefore R becomes

$$R = \frac{4(4.31\times10^9)^2}{K\ 15.2\times10^6 (\ell w) 4\pi f^2} = \frac{12.3\times10^{-2}}{Kf^2 A} \text{ ohms},$$

where $A = \ell w$ = the area of the transducer in m^2 and f = the fundamental frequency or any odd harmonics of the transducer in MHz.

The x-cut quartz transducer of this Bragg-imaging system has the following specifications:

$t = \frac{\lambda}{2} = 8 \times 10^{-4}$ meter

$\ell = 6"$

$w = 4"$

f = fundamental frequency = 3.58 MHz

$\varepsilon'_x = 4.45$

Thus, we get $C_o = 760\times10^{-12}$ Farad and R = 6.23 ohms.

The matching network is shown in Fig. 5. The values of L & C are obtained by setting Z_{in} = 50 ohms, and are C = 2370×10^{-12} farad and L = 3.346×10^{-6} Henry.

It should be noted that these values provide a starting point for the choice of components. To obtain an optimum match it is generally necessary to employ variable components centered on the calculated values. An RX-meter is useful for adjusting the components to the optimum setting.

R. F. Driving Electronics

The new system requires much higher transducer driving power than those systems in use previously. This is due partly to the increased transducer area and partly to the need for higher acoustic power density, since relatively large biological objects are involved. To fulfill this

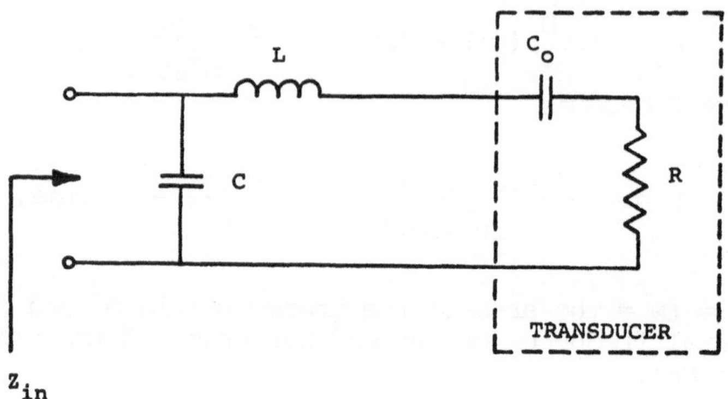

Figure 5 - Matching Network

requirement, the following electronic components were assembled.

A Marconi Instrument FM/AM signal generator is used to drive a Hallicrafters HT-40 class-C, R.F. amplifier which has a maximum output of about 50 watts into a 50Ω load. The output of the HT-40 is fed to a Swan 1200-W linear amplifier with 250 watts of maximum output. The HT-40 and the 1200-W are commercial multiband amateur-radio amplifiers and have the advantage of being rather inexpensive.

The R.F. is fed through RG-58/U coaxial cable to a Bird Thruline wattmeter and then to the impedance matching circuit and transducer mosaic in the cell. The wattmeter indicates both forward and reflected power so that the effectiveness of the impedance matching and amplifier tuning can be continuously monitored. An estimated 10% of the R.F. power is lost in the impedance transformer resulting in a usable power range of 0 to 225 watts or 0 to 1.5 watts/cm^2 in the ultrasonic beam (assuming no loss in the transducers). The maximum value causes significant heating in biological tissue within a few seconds and is well beyond the level normally considered completely safe for continuous exposure, but it can be used in a pulse mode without danger.[7]

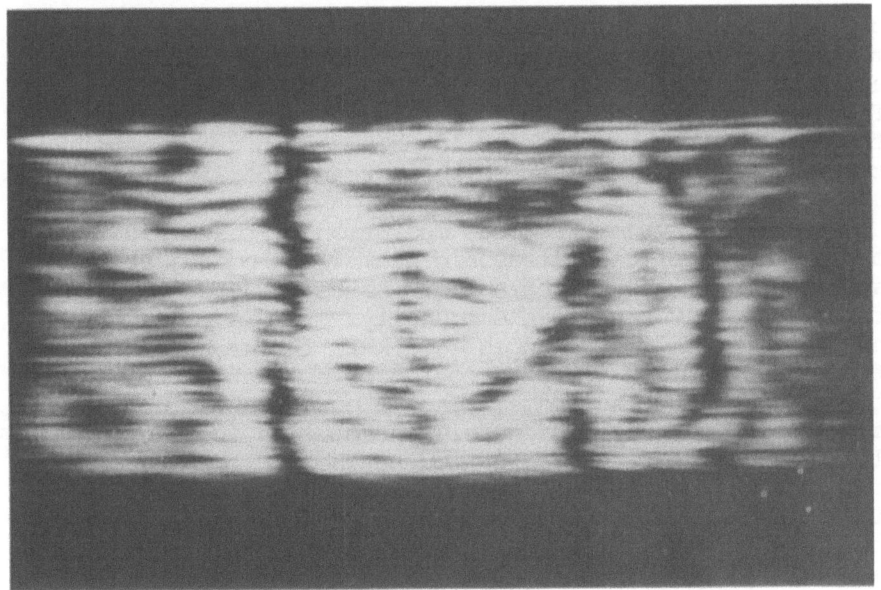

Figure 6 - Image of Ultrasonic Beam Cross-Section

Transducer Mounting Accuracy and Correction for Interference

The transducers shown in Fig. 2 were mounted in a rectangular aperture (96mm × 146mm) in a piece of 6mm thick lucite plate. A thin strip of lucite for reinforcement was placed across the middle of the mosaic in the long dimension. The edges of the transducers were joined to each other, and to the lucite, by means of RTV silicone rubber sealant. No special pains were taken to insure parallel alignment of the transducers because this would require precision machining and would have probably been ineffective as a result of transducer flexure from the hydrostatic pressure in the specimen tank.

Transducer parallelism could be insured by cementing the individual elements to a relatively thick substrate of some suitable material. This was considered, but dropped in favor of the simple arrangement because of the acoustic impedance matching problems which arise and also the difficulty of achieving a good bond between transducers and substrate. There was considerable doubt about the inhomogeneities in the ultrasonic beam which would then result from destructive interference between adjacent transducers.

Once the system was made operational, the direct optical image of the ultrasonic beam (with no object in the specimen tank) was photographed. As can be seen in Fig. 6, there is rather severe inhomogeneity. It was soon noticed, however, that slight changes (± 5 kHz) in the R.F. driving frequency resulted in rapid motion of the intensity variations throughout the image. Thus a practical solution to the problem consisted of merely introducing frequency modulation of about 10 kHz deviation at the R.F. generator. A modulation frequency of 50 kHz was used and the resulting image, considerably smoother, is shown in Fig. 7. This is a much simpler procedure than precision transducer mounting[10] and offers no disadvantages provided that the frequency deviation is small. However, at deviations of 50 kHz, or more, the image scanning due to the acoustic frequency change is sufficient to blur image detail. The fluctuation

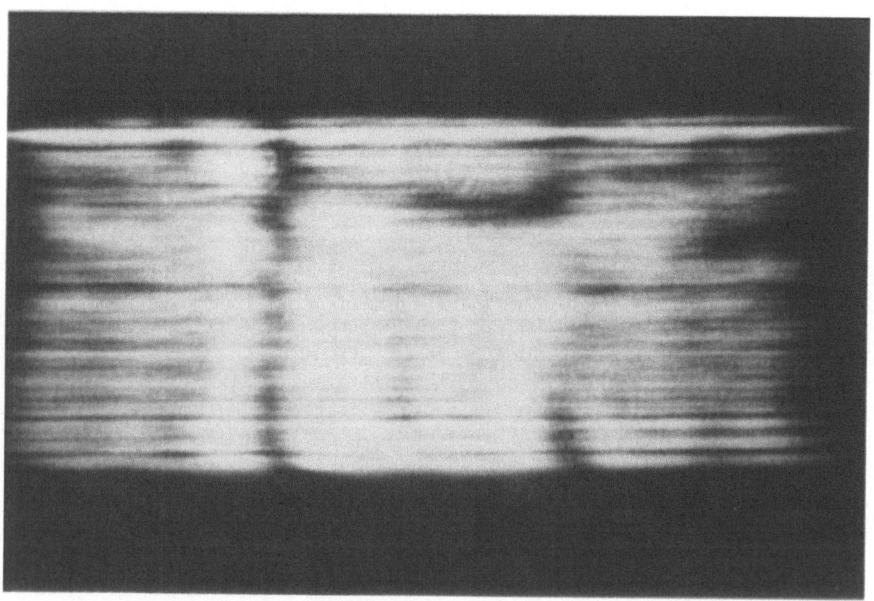

Figure 7 - Ultrasonic Beam Cross-Section with F.M.

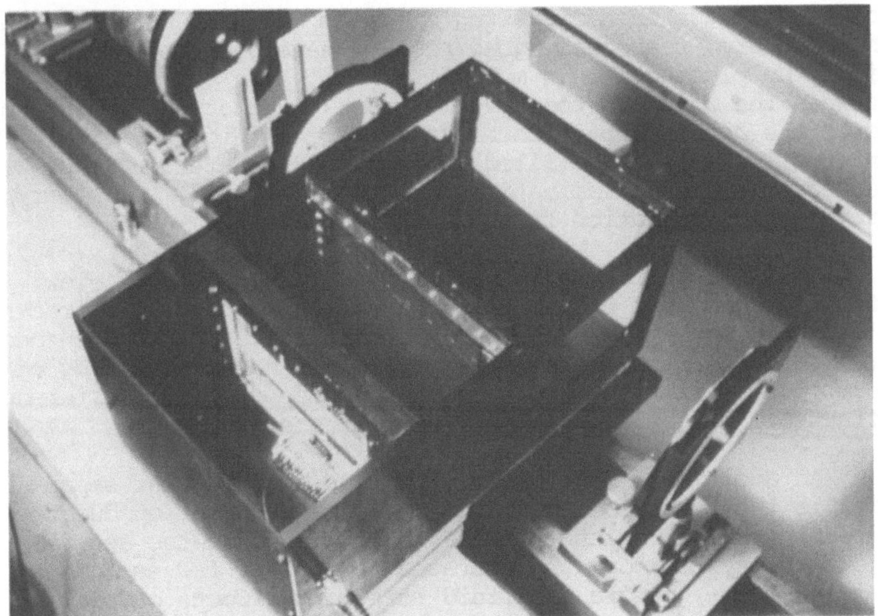

Figure 8 - Top View of Bragg-Diffraction Cell

in the optical frequency of the image, which results from this scheme, does not lead to any problems in using optical heterodyne detection for noise rejection unless acoustic phase retention is also desired.

BRAGG-DIFFRACTION CELL DESIGN

The new cell, aside from being larger than those in previous use, incorporates two additions which were made necessary by the reduced acoustic operating frequency.

Isolation of Acousto-Optic Interaction Region

Since the Bragg angle is less than 2 milliradians at 3.6 MHz, the optical image receives a large part of whatever light is forward scattered from the incident wedge of laser light. At first, it was thought that some technique of noise rejection, such as optical heterodyning, would be necessary in order to eliminate the effects due to this scattered light when operating a system at 5 MHz or less. On the contrary, reasonably good results have been obtained simply by placing a thin mylar membrane, as a separation, between the water in the specimen tank and that section which forms the acousto-optic interaction region (Fig. 8). The interaction region is filled with very pure water and then sealed to avoid contamination from the air, while the mylar prevents contamination from the specimen tank.

Mylar is particularly suitable from the point of view of ultrasonic beam transmission because its acoustic refractive index is similar to water and very little reflection occurs from the interfaces. The material in use has a thickness of about 50μm resulting in little or no tendency for interference effects to take place within the mylar sheet.

An additional benefit from the use of the mylar separation is the partial elimination of image degradation due to streaming and turbulence in the water. Without the separation, the system could be operated for only a few seconds (at 100mw/cm^2) before the fluid streaming effects completely degraded the image. With the mylar in place, streaming only becomes noticeable when the power density level exceeds about 500 mw/cm^2.

Figure 9 - Resolution Test Grid

Ultrasonic Beam Absorber

At acoustic frequencies of 15 MHz or more we have found that the attenuation in water is sufficiently high to render unnecessary the use of any absorbing material on the cell wall opposite the transducers. At 3.6 MHz, reflections from this wall are sufficiently intense, as they pass back through the coherent light wedge, so that serious image noise occurs.

Many suggestions for an appropriate absorbing material were made and several were tried in the cell. A rectangular piece of nylon plate, 8mm thick, yielded the best results and was found to be quite satisfactory for this purpose. This plate can be seen in Fig. 8 in the acousto-optic interaction section of the cell.

OPTICS

From the cost point of view the optical components of a large-aperture Bragg-diffraction imaging system cause the greatest difficulty. This applies primarily to the cylindrical elements in the system. Large-diameter spherical lenses with good corrections for aberrations are readily available, and since they need not necessarily be of large relative aperture, the cost is usually reasonable; telescope or aerial camera lenses are satisfactory.

Cylindrical lenses are another matter. The only large-scale commercial application for high-quality cylindrical elements is found in anamorphic projection lenses for the motion-picture industry and unfortunately these are free of aberration only when used in combination with spherical elements for which they are designed. The cylindrical lenses were not a difficulty with our previous system because, at high acoustic frequency, a large wedge-angle (numerical aperture) in the interacting light beam is not necessary. When used to converge a collimated, monochromatic input beam, a single-element plano-convex lens can be used down to about f/5 without serious blur in the focus which, incidentally, is due only to spherical aberration in this case. For the 15 MHz system this aperture was sufficient to privide good resolution in the image, but equivalent resolution capability in a 3.6 MHz system would require an f/1.0 converging beam. Unfortunately, no lens to provide this was available, and the best that could be obtained with sufficient size for the preliminary experiments was usable to a relative aperture of only f/9.0 (N.A. = 0.05), yielding a corresponding resolution of 8.5 wavelengths, or approximately 2.5mm.

A computer ray-trace design is currently in progress for a two-element cemented doublet which we expect to be usable at about f/3.0 (N.A. = 0.164). With the present system this lens would provide a resolution of 0.9mm. The use of a holographic correction plate in conjunction with a large-aperture single-element cylindrical lens has been suggested. This would, no doubt, yield a relatively aberration-free beam, but would be considerably less efficient than a corrected doublet.

The image-projection part of the optical system is not as subject to spherical aberration because optics of relatively high f-number can be used. On the other hand, the lens quality must be equal to, or perhaps better than, the converging optics because of the anamorphism which must be corrected in the image. The primary image is anamorphically distorted by a factor equal to the ratio of optical to acoustic wavelengths so that, to obtain an orthoscopic image, the projection optics must magnify more in one meridan than its orthogonal by a factor of Λ/λ (700 for the 3.6 MHz system). Since this can be accomplished by one or more cylindrical lenses operating at a low numerical aperture, single-element plano-convex lenses have proven satisfactory.

RESULTS

A qualitative estimation of the resolution achieved by the present system was made by using the test grid shown in Fig. 9. The grid is composed of masking tape strips, which are absorptive at 3.6 MHz, with clear gaps of 10mm × 10mm. The resulting image is shown in Fig. 10. An accurate resolution measurement would probably yield a value near the theoretically expected 2.5mm.

Figure 11 shows the image obtained when the fingers and part of one hand have been placed in the cell. There is no apparent penetration of the tissue here because the ultrasonic power density was rather low (about 20 mw/cm^2). When the power level was increased enough to insure penetration of the hand there was an excessive amount of glare in the image produced by the high intensity of light diffracted

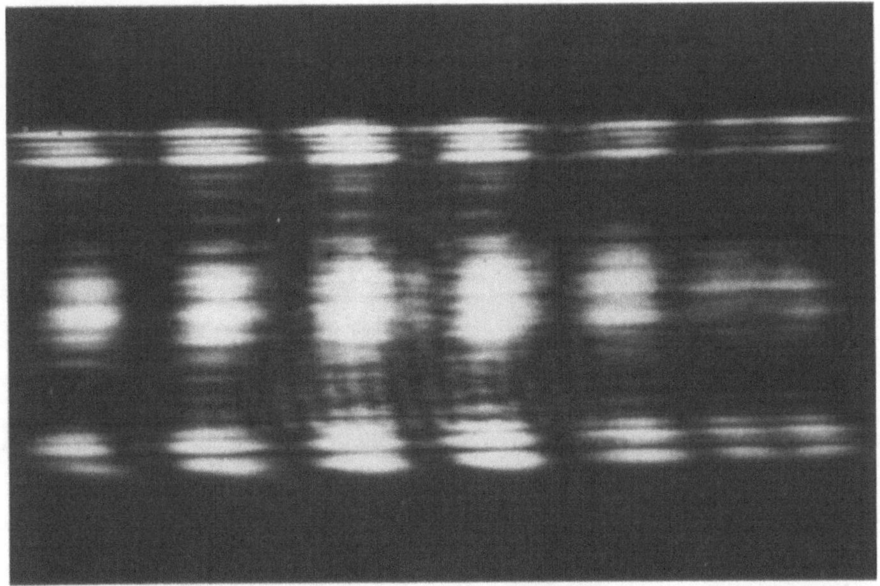

Figure 10 - Test Grid Image

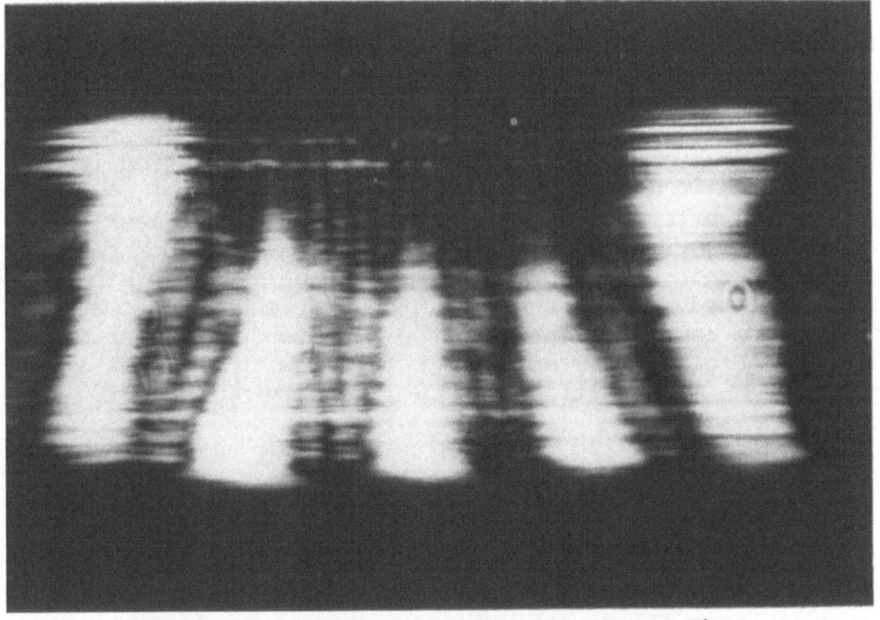

Figure 11 - Image of Hand at Low Acoustic Power

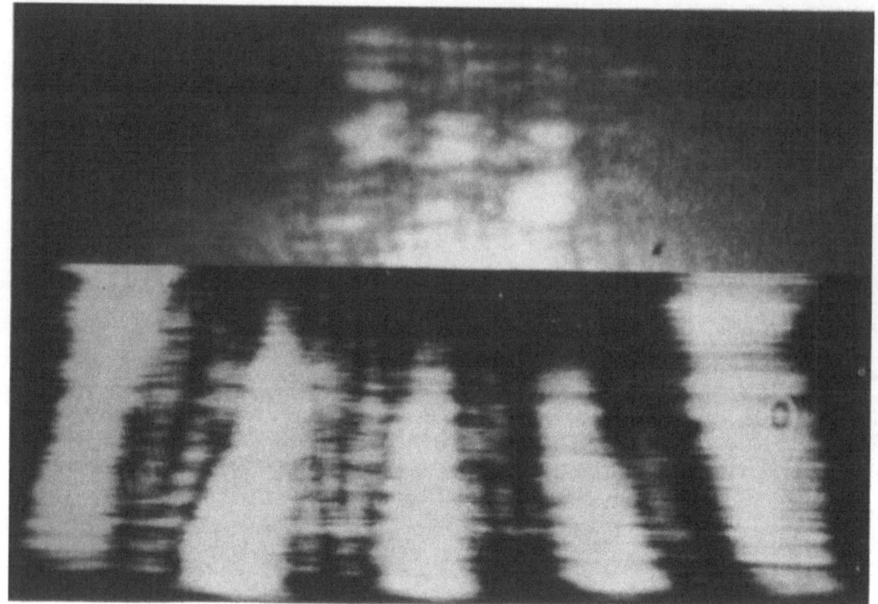

Figure 12 - Image of Bones in Palm of Hand

from the unattenuated portions of the ultrasonic beam. This has not occurred with the surface-wave deformation techniques of acoustic imaging, apparently because of their inherent non-linearity and it is, in one sense, an indication of the broad dynamic range of Bragg-diffraction imaging.

In an effort to eliminate the image glare a masking aperture of nylon plate was arranged so as to block part of the ultrasonic beam and allow only the palm of the hand to be irradiated. In this manner, the unattenuated acoustic waves were kept from entering the acousto-optic interaction region. Figure 12 is an image obtained with the mask in place at an input ultrasonic power density or about 100 mw/cm^2.

CONCLUSION

The imaging results obtained at 3.6 MHz indicate that scattered light from the undiffracted portion of the converging laser illumination is not as severe a limitation as we had previously expected it to be[8]. A mylar membrane used to isolate the acousto-optic interaction region, by preventing contamination, eliminates practically all of the previously observed scattered light. This is consistent with Smith's theoretical conclusions about the noise contributions in Bragg-diffraction imaging[9].

The predicted ultrasonic-beam irregularities due to misalignments in the transducer mosaic were rather easily overcome by frequency modulating the input R.F. at a deviation of about 0.3% of the carrier frequency[10]

The image resolution obtained with the present system is about 8.5 acoustic wavelengths. This limitation is presently set by the inability of the available cylindrical optics to exceed a relative aperture of f/9.0 without introducing severe spherical aberration. A specially-designed f/3.0 cylindrical doublet should provide resolution capability of about 1.0mm. (about 3 acoustic wavelengths).

The limitation of greatest concern at this time is the image glare which results from wide-range intensity variations across the ultrasonic wavefront as it passes through the acousto-optic interaction region. We expect to reduce or eliminate this problem by means of acoustic pulsing and range-gating of the laser beam. Since the acoustic velocity in water is typically different from that in the specimen, the unattenuated portions of the acoustic beam can effectively be separated from that part which passes through the specimen. There are several other advantages to beam pulsing which can then be taken advantage of by this scheme.

ACKNOWLEDGEMENTS

We would like to thank Sharon Scott and Mona Eyler for their effort in typing the manuscript.

This work was supported by the National Institutes of Health of the U.S. Public Health Service (Grant No. RO1 GM16474-02).

REFERENCES

1. Glen Wade, John Landry, and Alwyn de Souza, "Acoustic Transparencies for Optical Imaging and Ultrasonic Diffraction," Acoustical Holography, vol. I, Plenum Press (1969).

2. John Landry, John Powers, and Glen Wade, "Ultrasonic Imaging of Internal Structure by Bragg Diffraction," Appl. Phys. Letters, 15, 186 (1969).

3. Roy Smith, Glen Wade, John Powers, and John Landry, "Studies of Resolution in a Bragg Imaging System," Jour. Acoust. Soc. Am., 49, no. 3, 1062 (1971).

4. John Landry, Hormozdyar Keyani, and Glen Wade, "Conservation of Acoustic Phase in Ultrasonic Imaging by Bragg Diffraction," to be published in Jour. Appl. Physics.

5. Richard Krimholtz, David Leedom, and George Matthaei, "New Equivalent Circuits for Elementary Piezoelectric Transducers," Elect. Lett., vol. 6, pp. 398-9 (1970).

6. Don Berlincourt, Daniel Curran, and Hans Jaffe, "Piezoelectric and Piezomagnetic Materials and Their Function in Transducers," in Mason, W. P. (Ed.): Physical Acoustics, vol. 1, [A], Academic Press (1964).

7. Adnan Sokollu, "Irreversible Effects of High Frequency Ultrasound on Animal Tissue and Related Threshold Intensities," Acoustical Holography, vol. III, Plenum Press (1971).

8. John Landry, Roy Smith, and Glen Wade, "Optical Heterodyne Detection in Bragg Imaging," Acoustical Holography, vol. III, Plenum Press (1971).

9. Roy Smith and Glen Wade, "Noise Characteristics of Bragg Imaging," Acoustical Holography, vol. III, Plenum Press (1971).

10. It has been subsequently discovered that the inhomogeneities were largely due to multiple acoustic reflections within the cell, apparently the transducer parallelism is not critical.

BIOMEDICAL STUDIES USING ULTRASONIC HOLOGRAPHY

M. R. Sikov, F. R. Reich and J. L. Deichman

Biology and Applied Physics & Instr. Depts.
Battelle, Pacific Northwest Laboratories
Richland, Washington 99352

INTRODUCTION

The use of ultrasound to obtain information about the internal structure of the body has become an accepted clinical tool and a number of instruments are commercially available. Essentially all of these instruments and the associated techniques are based on the transmission of short pulses of ultrasound into the subject and measurement of the time lag and magnitude of the reflections or echos from interfaces within the body. There are a number of shortcomings associated with these so-called pulse-echo or echo-ranging techniques. These include high attenuation of the ultrasound, artifacts introduced by motion of the subject or inhomogeneities of the intervening tissues, and lack of gray scale in the image.

Phase information is sometimes the only way by which biological structures can be visualized in vivo. Phase-contrast optical microscopy has been found to be an important tool for the study of transparent nonabsorptive objects. By analogy, it may be anticipated that the utility of ultrasonic imaging could be enhanced by use of phase differences. Ultrasonic holography, which utilizes both phase and amplitude measurements of the radiation scattered from or transmitted through the subject, has been proposed as an alternate to the pulse-echo technique. These phase and amplitude values are measured on a plane and converted to equivalent density variations on film or to ripples on a surface. The resulting hologram is then interrogated by a coherent light beam. Interaction of the light beam with such a hologram produces a replica of the radiation field by diffraction. This replica is visible and appears to the eye or camera as an image of the subject. Although ultrasonic holography has gained acceptance as a valuable tool for the nondestructive testing of materials, only recently have biomedical applications been explored. To date, liquid-surface holography has received greatest attention because of its attractive feature of allowing real-time images to be readily obtained. On the other hand, the need for the immersion of the anatomical area of interest in the water tank presently limits the use of the technique to portions of the body, such as the limbs and the female breast, which can be readily introduced into a fluid-filled tank.

HOLOGRAPHIC IMAGING PROCEDURE

The theory and basic liquid surface technology has been described by Brenden.[1] Our laboratory unit (Fig. 1) was modified for these studies by the introduction of the milling machine controls, shown to the right, which permitted accurate and reproducible three dimensional movement of experimental subjects. A brass tube was inserted into the mill head, to which was attached a frame for supporting the specimen under study. This allowed for temporary removal of the subject for manipulative procedures and facilitated experimental manipulation while the subject was immersed in the tank. When living animals were studied a plastic cylinder, with a rubber diaphragm at the bottom, was attached to the lower end of the brass tube. In Figure 1 is shown such a cylinder attached to the positioning apparatus; the head of a living rat has been introduced into the plastic cylinder and is partially immersed in the first water tank to illustrate the technique. To allow the animal to be placed below the surface of the liquid, air was introduced into this chamber under sufficient pressure to prevent the entrance of fluid. The air flowed from the chamber through two small tubes at either side of the chamber. The air entering this chamber was first passed through an atomizer containing Penthrane (methoxyflurane, Abbott) to provide anesthesia to the experimental animal, which facilitated study for prolonged periods of time.

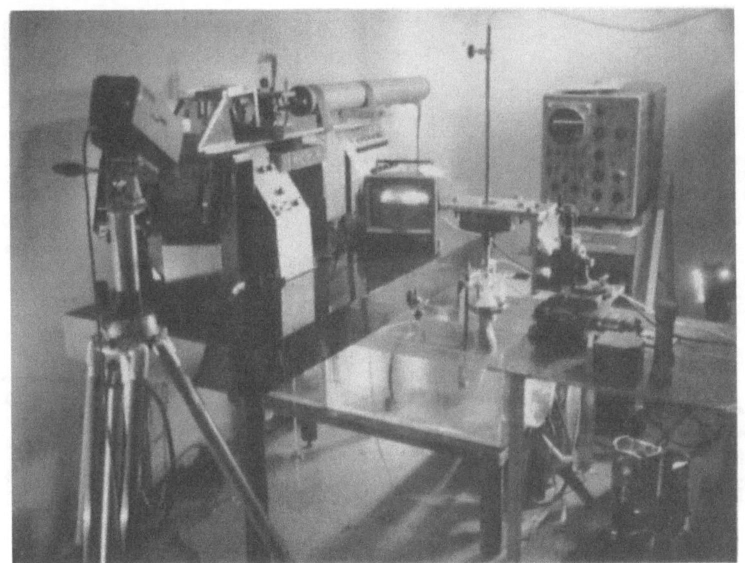

Figure 1

Ordinarily, the specimen or animal was studied using appropriate motion to bring the various planes of interest into focus and rotation to obtain several angles of projection. The images were displayed on a video monitor and video tape recordings made to allow subsequent study. Polaroid photographs of representative images were also taken to illustrate narrative material. The specimen or animal was subsequently dissected to permit comparison of anatomical relations in the subject with the images which were obtained.

EXPERIMENTAL PROCEDURES AND RESULTS

Rabbits were killed and the kidneys removed to determine the resolution obtainable under optimized conditions with this technique. Figure 2B shows one of these kidneys, which was transected after imaging. The images obtained at two focal planes through the kidney are shown in A and C. The appearance of the calyxes (a) was seen to be different at the two planes and branchings of less than 1 mm in diameter were visualized. A demarcation between the medulla (b) and the cortex (c) was apparent, although there is not so sharp an anatomic delineation between the two portions of the kidney.

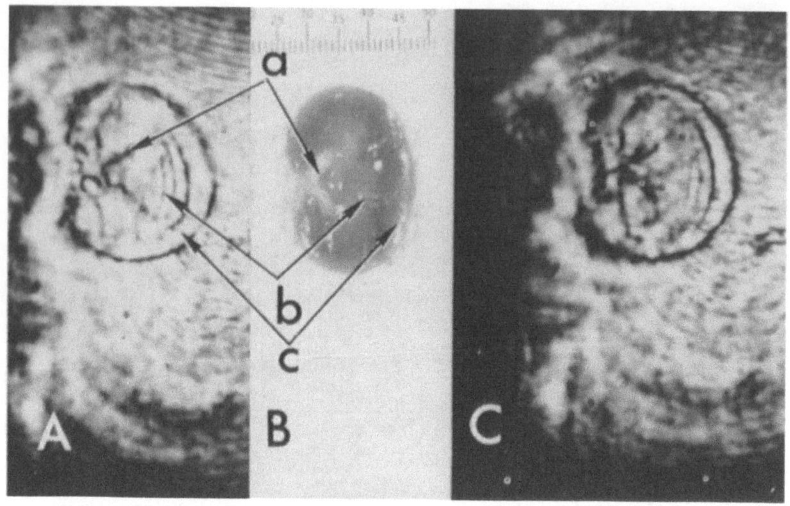

Figure 2

Mice were intraperitoneally injected with an Ehrlich tumor and imaged after the tumor had grown to substantial size. Figure 3B is a photograph of one of these mice, which was dissected following imaging. The tumor (a) is the nodular white mass lying below the liver (b), which has been lifted to show the tumor, and along the left side of the animal. The image to the left (A) was obtained through a frontal plane. The tumor was clearly imaged as lying below the liver, which appears in the image as two parallel arcs, representing the surfaces intersected by the image plane. The tumor which ran along the side of the animal also appeared in the image. The vertebral column (c) appeared as a dense structure along the long axis and a loop of intestine (d) was also clearly defined. The image to the right (C) was made after the animal had been rotated 45° about its long axis. The difference between the appearance of the liver and the underlying tumor became even more apparent in this projection.

The hind legs of several rabbits were imaged extensively, after which the animals were killed and the vessels injected with a dyed gelatin mass. Figure 4 presents the holographic image through a mid-plane at the knee (A) and the corresponding photograph (B) showing the surface vessels. Although the bones of the leg showed up quite clearly, the detail was obviously inferior to

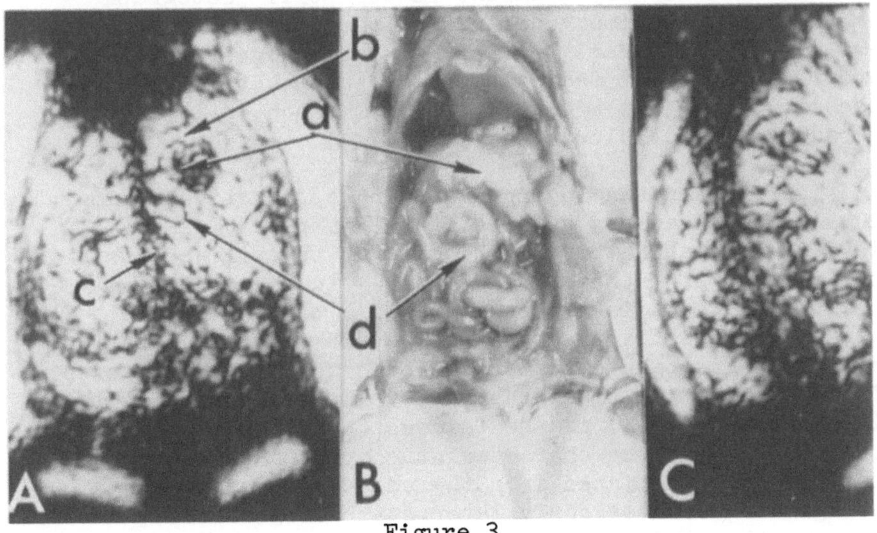

Figure 3

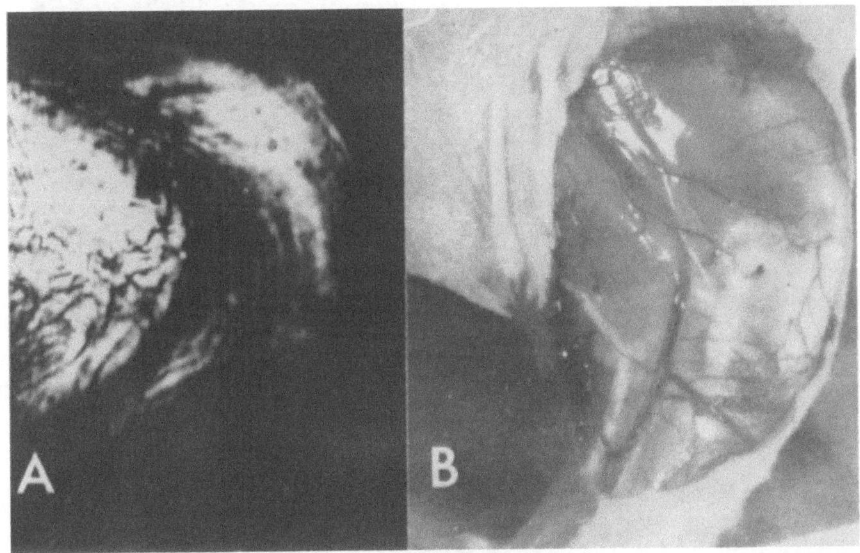

Figure 4

that obtainable with conventional X-ray radiography. Although less clearly apparent in the still photograph than in the real-time video recordings, the blood vessels were seen quite clearly. It is of particular significance for projected medical applications that it was not necessary to inject a contrast medium, such as is used in radiography of blood vessels.

There is an ongoing interest in reproductive biology and in bioengineering studies of the intrauterine device (IUD) in our laboratory.[2] This led us to examine possible applications of ultrasonic holography in these research areas. Figure 5A is a photograph of a portion of the excised uterus of a pregnant rat, at 15 days of gestation. The uterus has been cut at position 3 so that the fetus lies free with the placenta above. The corresponding ultrasonic image (B) of the fetuses (positions 1 and 2) consistently had the characteristic appearance shown and could be differentiated from the adjacent placentas. Living rats, at a slightly later state of gestation, were subsequently studied and the images (C) had a similar

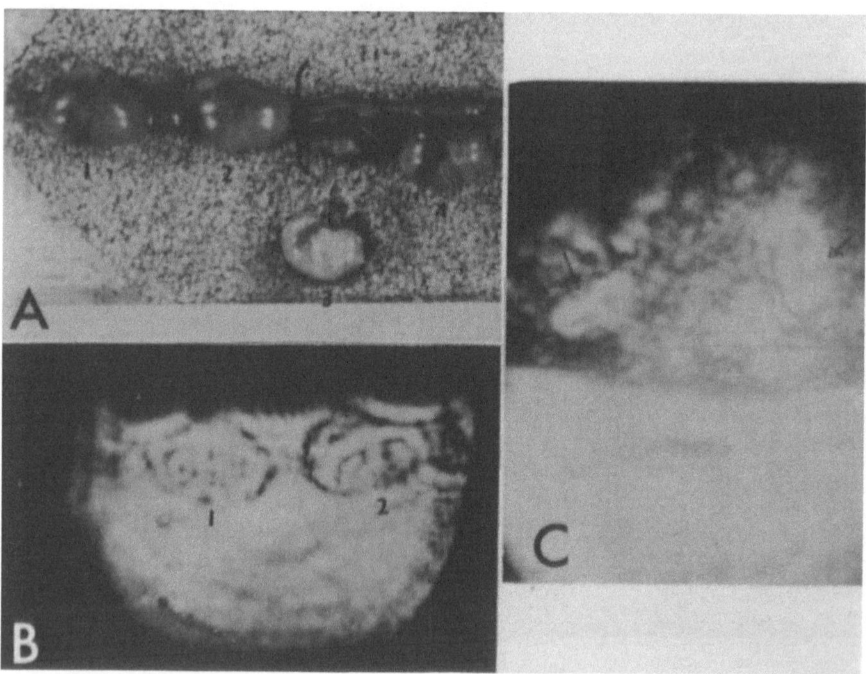

appearance. The fetuses to the left and right (arrows) are in focus and between them lies one fetus which is out of focus.

In Figure 6, a partially dissected gravid human uterus, at approximately three months of gestation, is shown to the left. This was imaged in several planes prior to dissection. Two representative (mirror) images obtained in the sagittal plane of the fetus are shown on the right. The upper image shows the abdomen, pelvis, and the upper leg and the upper portion of the lower leg. The ossified central portion of the femur and tibia are much more opaque to the ultrasound than are the ends of these bones, which are not yet calcified. The lower image was taken through the lower part of the head and shows the upper and lower jaws, the vertebral column in the neck and the ribs in the upper thorax.

The images shown in Figures 7 and 8 illustrate typical results obtained in our studies of the relationships between the human uterus and the IUD. An excised uterus, typical of those used in these studies is shown in Figure 7A. Since we wanted to introduce objects into the uterine lumen, it was positioned with the cervix upwards. The corresponding holographic image (Fig. 7B) shows the

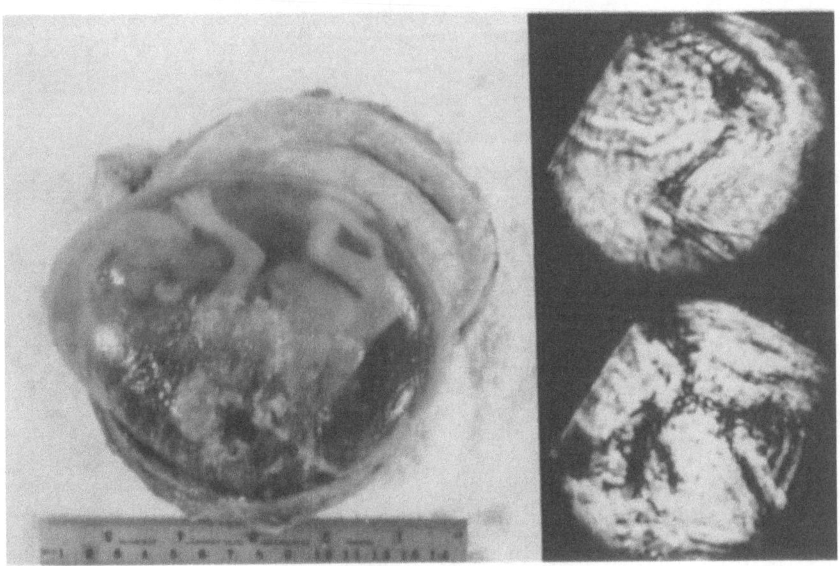

Figure 6

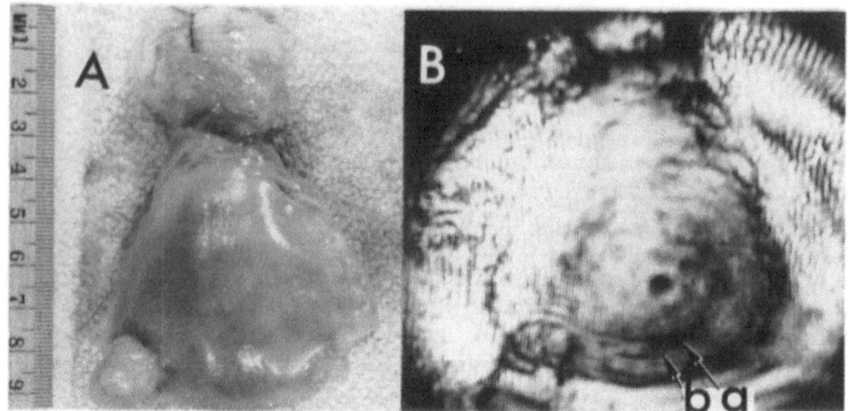

Figure 7

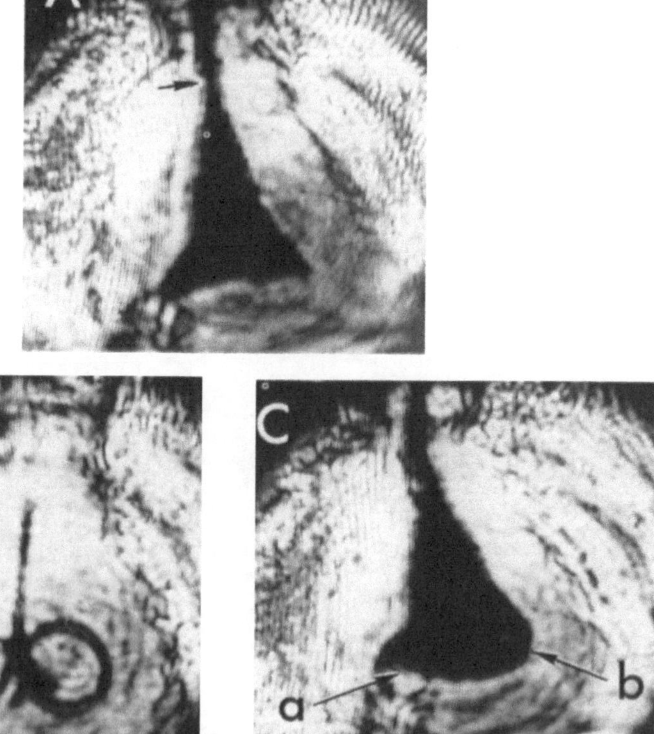

Figure 8

cervical canal and the distinction between the inner endometrium (a) and the outer myometrium (b) as well as a small fibroid tumor at the lower left of the figures.

The real-time capability of the apparatus was particularly valuable in these studies, in which contraction of the uterus was simulated. Although this aspect can not be appreciated from the still photographs, those shown in Figure 8 will convey the general findings. Air was introduced into the uterine lumen to emphasize its shape (Fig. 8A). The air was seen to pass into the utero-tubal junction below the fibroid tumor. The shape of the lumen was typical although there was a small filling-defect (arrow), which proved to be a cervical polyp on subsequent examination. A number of common IUD's were introduced to study their interactions with the uteri. The Saf-T-Coil, shown in Figure 8B, was seen to protrude into the endometrium; this was even more marked during contraction. Air was re-introduced with the devices in place (Fig. 8C). The extra-lumenal protrusion of the device was noted at (a) and the general deformation of the lumen at (b).

CONCLUSIONS

Although the photographic images are quite striking, and are useful for study, the experimental results obtained by making use of the real-time characteristics of this instrument were even more important. It is clear that this apparatus has demonstrated usefulness for biomedical research.

It also seems that this technique shows promise of becoming clinically useful, particularly in the delineation of soft tissue structures such as blood vessels. The need for immersion of the area of interest poses some limitations to the clinical usefulness of this technique. A number of minor changes of systems design might render this limitation less important.

REFERENCES

1. Brenden, B. B., A Comparison of Acoustical Holography Methods. In Acoustical Holography, Vol. I. A. F. Metherell (Ed.), Plenum Press, New York, 1969.

2. Sikov, M. R., Reich, F. R., and Deichman, J. L., Studies of Reproductive Biology Using Ultrasonic Holography, Proceedings 24th ACEMB: 262, 1971.

ULTRASONIC STEREOHOLOGRAPHY†

Bruce D. Sollish* and Isaia Glaser

Department of Electronics
The Weizmann Institute of Science
Rehovot, Israel

1. INTRODUCTION

A significant aspect of acoustical holography is the three-dimensional optical reconstruction of insonified objects. Particularly in medical diagnosis, where one-dimensional and two-dimensional imaging techniques, principally the A-scan and B-scan, are already in clinical use for the examination of internal body organs [1], a three-dimensional reconstruction would be of major importance.

However, R.W. Meier [2] has shown that in dual-wavelength holography, the resulting reconstructed image longitudinal and lateral magnification (respectively normal to and parallel to the hologram plane) are in general unequal and are related by

$$M_{long} = \frac{1}{\mu} M_{lat}^2 \tag{1}$$

where μ = reconstructing to recording wavelength ratio and $M_{lat} = m$ = hologram magnification. In addition to unequal

† This work was supported by the Israel Cancer Association.

* With the Department of Electrical Engineering and Computer Science, Columbia University, New York, N.Y., during the 1971-1972 academic year.

scaling, there are first-order aberrations, principally spherical, in the reconstructed image, unless the acoustical hologram is reduced in size by a factor $m = \mu$. Even at the relatively short acoustic wavelength of 0.1 mm typically used in diagnostic ophthalmology, a reconstruction at 0.5 μm (the green line output of an argon ion laser) gives a scale factor 1/200. An unscaled acoustical hologram of the human eye would therefore reconstruct an image, if it could be seen at all, 25 mm wide and 5 m deep! A number of solutions have been proposed to reduce scale distortion and spherical aberration in acoustical holography; see, for example, the papers by D.C. Winter and F.L. Thurstone et al. elsewhere in this volume.

Ultrasonic stereoholography is a method of three-dimensional imagery in which the problems encountered in dual-wavelength holography are entirely avoided, because ultrasonic stereoholography involves a two-step recording process. The first step is to record a sequence of two-dimensional modified B-scans, as discussed later, while the second is to synthesize a hologram from the B-scan sequence. The acoustical-to-optical conversion is completed in the first, non-holographic step. The second step is holographic but entirely optical, so that $\mu = m = 1$. Full size, undistorted images can thus be reconstructed from an ultrasonic stereohologram.

2. OPTICAL STEREOHOLOGRAPHY

The following is a brief presentation of some background material in what may be called optical stereoholography. Early this century, G. Lippmann [3] showed that a three-dimensional scene can be decomposed into an array of two-dimensional projections from which it can be reconstructed. He recorded the array of two-dimensional projections on a sheet of film placed in the back focal plane of a fly's-eye lens (an array of densely-packed spherical lenslets). The original three-dimensional scene is reconstructed by replacing the film - now developed as a positive transparency - in the back focal plane of the fly's-eye lens and illuminating the system from behind. The fly's-eye lens serves as both an encoding and decoding device that operates in white light.

In 1967 R.V. Pole [4] revived the Lippmann imaging method, but with the addition of a holographic step he was able to eliminate repeated use of a fly's-eye lens. A hologram is recorded of the fly's-eye reconstruction under coherent illumination. The three-dimensional scene is then reconstructed in the usual manner by the developed hologram or holocoder, as designated by Pole. The principal advantage of the Pole method is that the critical alignment of the fly's-eye lens and film is carried out only once.

Following the work of Pole, a number of authors including J.D. Redman [5], George, McCrickerd, and Chang [6], D.J. DeBitetto [7], and King, Noll, and Berry [8], proposed further refinements in this technique, resulting in what is now known as the holographic stereogram, or the optical stereohologram in our notation. In this method, a sequence of conventional photographs of a three-dimensional scene is recorded by moving a camera in a fixed, incremental manner between recordings. Each view, or perspective, is developed as a positive transparency and coherently projected onto a diffuse screen, as shown in Fig. 1. A recording emulsion located at a distance Z from the diffuse screen is covered by a slit mask that allows light from a transparency to reach only a certain narrow strip portion of the emulsion. The mask is shifted between exposures to uncover a fresh portion of emulsion. In this manner, each perspective is recorded as a separate subhologram. When the hologram is developed and viewed under coherent illumination with the slit mask removed, the planar images generated by the individual subholograms fuse into a single three-dimensional image. This process is examined more closely in Sec. 4.

3. PROJECTION-TYPE B-SCAN USING A CYLINDRICALLY-FOCUSED TRANSDUCER

In order to apply the method of stereoholography in ultrasonics, a suitable sequence of ultrasonically-derived transparencies must be recorded. The conventional pulse-echo or transmission B-scan recorded with a flat or spherically-focused transducer is not appropriate, because only a thin cross-section is displayed. Any particular B-scan therefore presents little information regarding the entire insonified target.

However, it is possible to record a B-scan with a cylindrically-focused transducer of a type available commercially for flaw detection. The geometry of this kind of

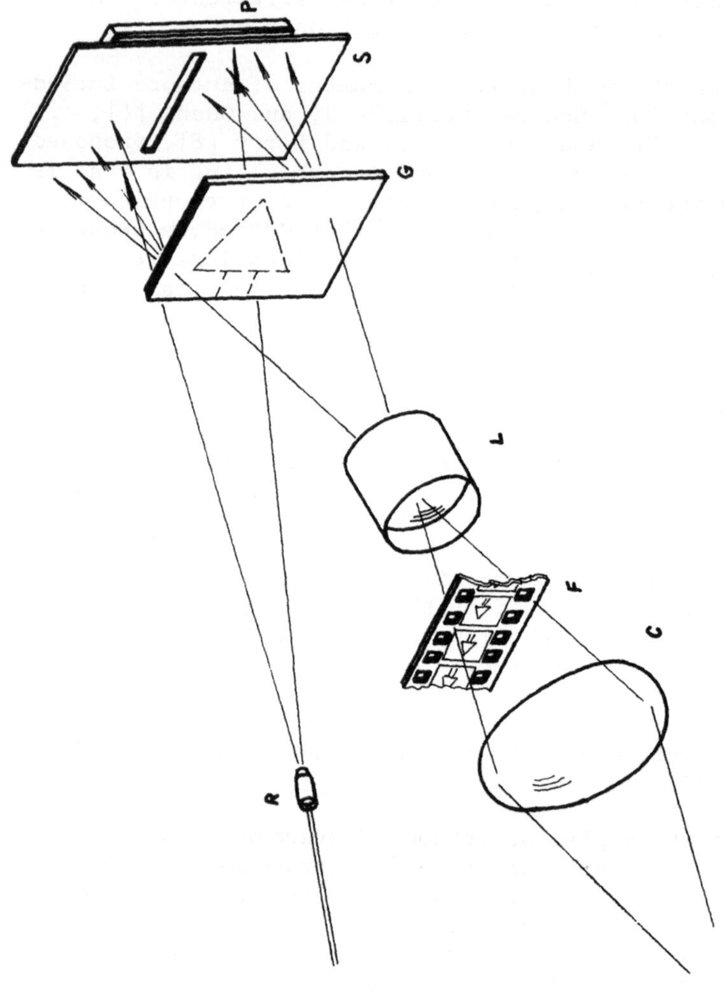

Fig.1 Recording a stereohologram (R= reference beam, C= condensing lens, F= film sequence, L = projection lens, G= groundglass screen, S= slit mask, P = high-resolution photographic plate).

linear B-scan is shown in Fig. 2. The orientation of the transducer establishes the rectangular coordinate system as indicated. The long axis of the transducer is parallel to the z-axis, and the short axis of the transducer is parallel to the y-axis. The motion of the transducer is along the y-axis. Ultrasonic pulses are propagated parallel to the x-axis. A B-scan recorded in this manner is the projection plane shown and is equivalent to the photographically-derived projection-planes used in optical stereoholography. A sequence of such projection-type B-scans, in which the target is rotated along its y-axis between scans, forms the basis of ultrasonic stereoholography.

4. LATERAL AND LONGITUDINAL MAGNIFICATION IN ULTRASONIC STEREOHOLOGRAPHY

Before proceeding to the experimental verification of ultrasonic stereoholography, we examine the lateral and longitudinal image magnification. In Fig. 2, an arbitrary point P in the arget has coordinates g_o, y_o, h_o in the body reference system. The coordinates g_o, y_o are displayed in the projection plane shown. If the target is rotated clockwise by an incremental amount $\Delta\phi$ about the y-axis, the y-coordinate in the projection plane remains the same, but the x-coordinate becomes $g_o + \Delta\phi h_o$ for $\Delta\phi << 2\pi$.

Let the two projection planes be recorded in two B-scans in which the target is rotated by $\Delta\phi$ before the second scan, and let the resultant transparencies be placed in the recording system of Fig. 1. To focus our attention on point P and its two projections consider Fig. 3. By means of an off-axis reference and slit mask (not shown), the projection of P in the first B-scan, P_1, is recorded by strip hologram H_1; the projection of P in the second B-scan, P_2, is recorded by the strip hologram H_2. The two strips are separated by ΔW. In the transparency plane, P_1 and P_2 are

$$P_1 = P_1 (m_o m_x g_o,\ m_o m_y y_o) \text{ and } P_2 = P_2 [m_o m_x (g_o + \Delta\phi h_o),\ m_o m_y y_o],$$

where m_x = ultrasonic depth magnification,

m_y = ultrasonic transverse magnification,

and m_o = optical magnification from CRT display to diffuse screen.

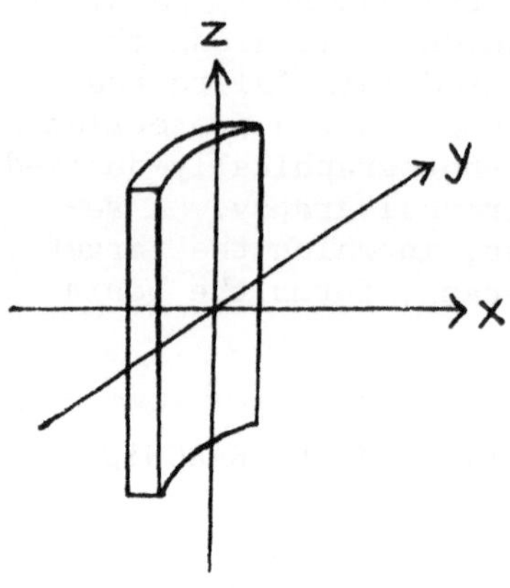

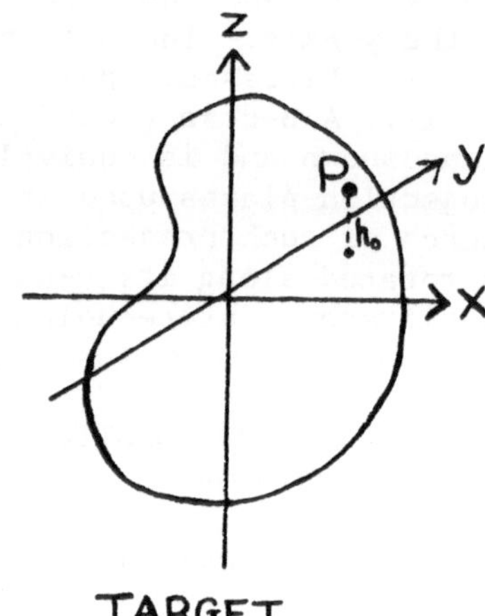

x-axis: propagation direction
y-axis: scan direction
z-axis: projection direction

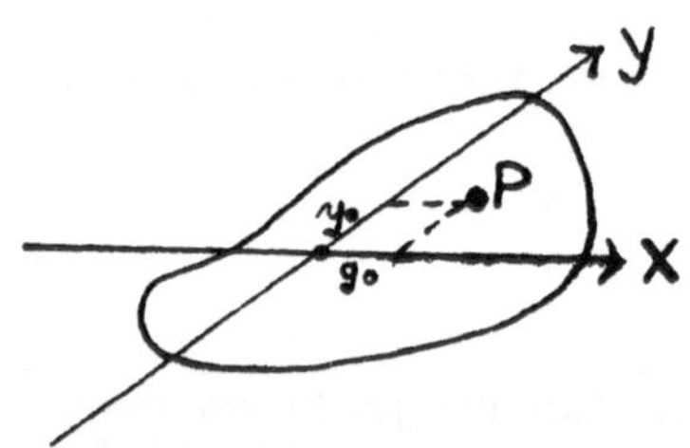

Fig. 2 Projection-type B-scan recorded with a cylindrically-focused transducer.

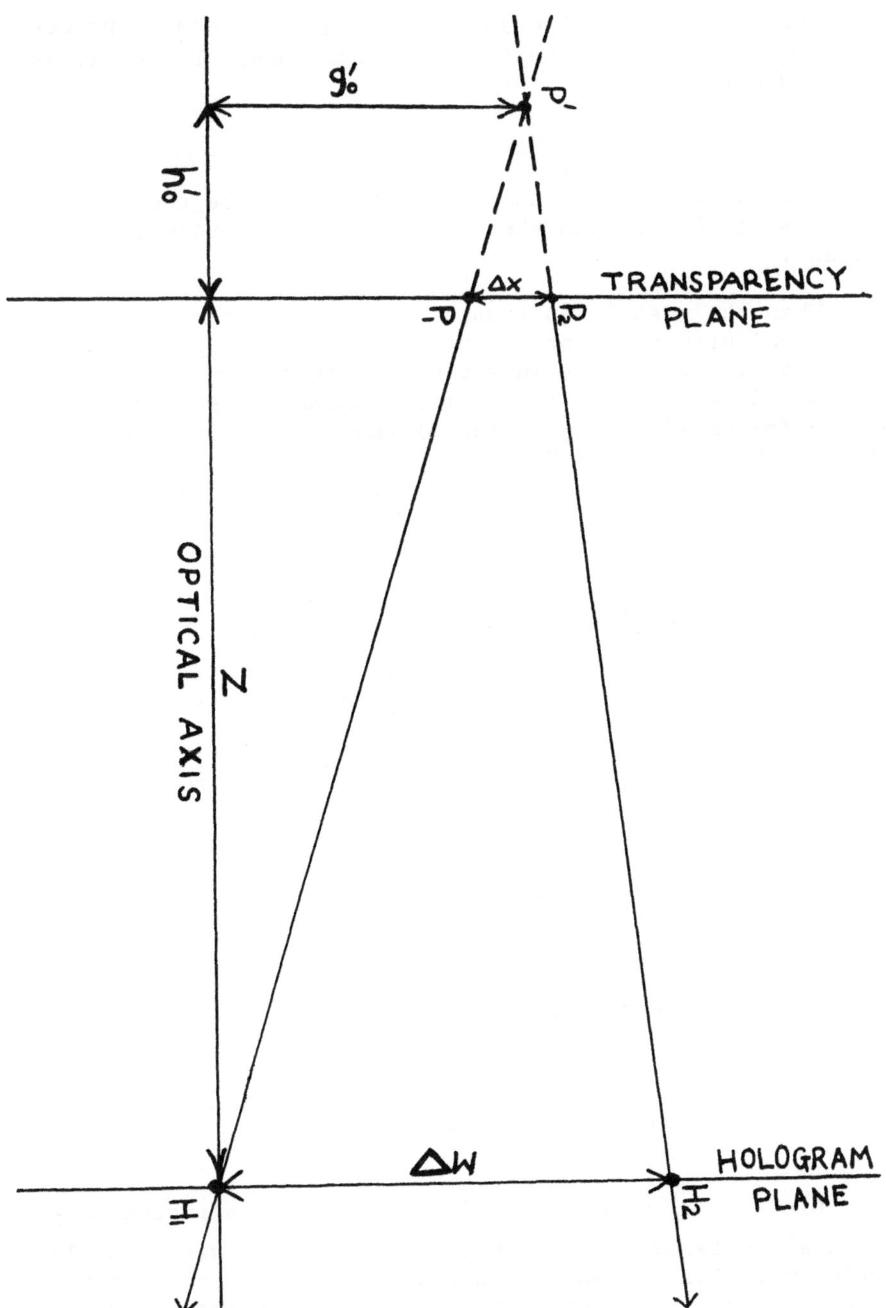

Fig.3 Stereohologram recording and reconstruction geometry.

Note that m_x and m_y are adjusted during the scanning process, while m_o occurs after the B-scan is obtained. The parallax in the x-direction is

$$\Delta x = m_o m_x \, \Delta\phi \, h_o \quad . \tag{2}$$

There is no parallax in the y-direction because there is only one axis of rotation in the recording of the projection-type B-scans.

When the unscaled ultrasonic stereohologram is developed and illuminated coherently, the points P_1 and P_2 are reconstructed simultaneously in a plane corresponding to the diffuse screen. A new image of P, denoted by P', is formed by the (extended) intersection of rays coming from H_1 and H_2 passing through P_1 and P_2. The coordinates of $P' = P'(x_o', y_o', z_o')$ are found from the geometry of Fig. 3 to be

$$\begin{aligned} x_o' &= g_o' = m_o m_x (1 + h_o'/Z) g_o \\ y_o' &= m_o m_y y_o \\ z_o' &= h_o' = m_o m_x \frac{\Delta\phi}{\Delta W} Z \, h_o \; . \end{aligned} \tag{3}$$

Under paraxial conditions, for which $Z \gg h_o'$, the image magnifications are

$$\begin{aligned} M_{lat,x} &= \frac{dx_o'}{dx_o} \cong m_o m_x \\ M_{lat,y} &= \frac{dy_o'}{dy_o} = m_o m_y \\ M_{long} &= \frac{dz_o'}{dz_o} = m_o m_x \frac{\Delta\phi}{\Delta W} Z \end{aligned} \tag{4}$$

Clearly, if $m_x = m_y$ and $\frac{\Delta\phi}{\Delta W} Z = 1$, the reconstructed image is equally-scaled in all three-dimensions and has magnification m_o. Scaling and magnification in ultrasonic stereoholography are therefore parameters that are independent of the ratio of optical to acoustical wavelength.

Fig. 4 "Phantom eye" target for ultrasonic stereoholography.

5. EXPERIMENTAL RESULTS

In order to verify the principles of ultrasonic stereo-holography, an experimental system was put together. A 10 MHz linear B-scan apparatus with binary display served as the ultrasonic imaging device. The cylindrically-focused transducer was obtained from Automation Industries, Boulder, Colorado. The piezoelectric crystal is lithium sulfate 8 mm X 38 mm operating in the fundamental thickness mode at 10 MHz. Attached to the crystal by the manufacturer is a cylindrical plastic lens with measured focal length of 31 mm in water.

The 10 MHz frequency and corresponding wavelength of 0.15 mm in water are typical values employed in diagnostic ophthalmology. A "phantom eye" is often used in ophthalmology to evaluate and calibrate ultrasonic imaging systems. The phantom eye in Fig. 4 was chosen for our experiments. It consists of eight 2 mm-diameter, 50 mm-long aluminum wires arranged in a cylindrical array on a 25 mm diameter plexiglass base.

A projection-type B-scan of this target is given in Fig. 5, which is a time-exposure Poloroid picture of the CRT

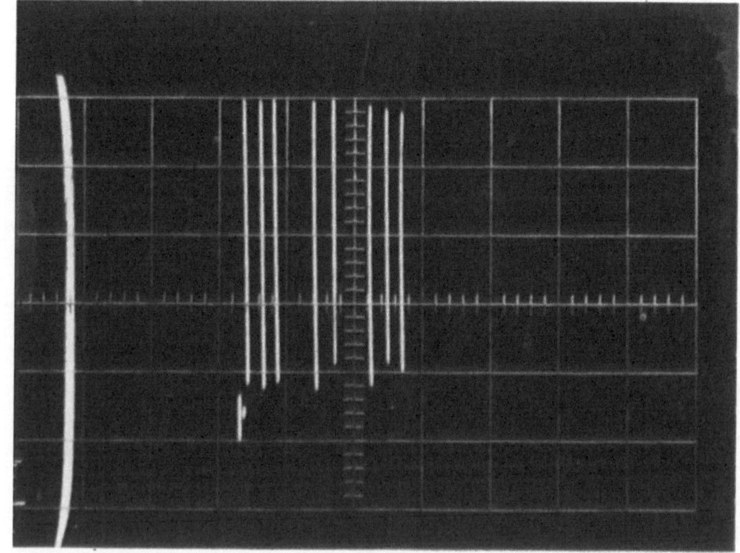

Fig. 5 Projection-type B-scan of the phantom eye

display. A single scan of this type images all eight wires simultaneously; a single B-scan of the conventional type would display only a thin vertical segment of Fig. 5. The leading edge of the plexiglass base and all eight wires are clearly shown; a conventional cross-sectional display would show only some thin vertical segment. In this particular scan, the horizontal ultrasonic magnification was $m_x \approx 1$ and the vertical magnification was $m_y \approx 1.5$ (the upper portion of the target was not submerged).

To form an ultrasonic stereohologram of the phantom eye, 33 projection-type B-scans ($m_x = m_y = 2$) were recorded. The target was rotated between scans by an amount $\Delta\phi = 1/2$ degree. The B-scan sequence was then processed in the holographic system of Fig. 1. The diameter of the slit aperture and its center-to-center displacement were each 1 mm, and the diffuse screen-hologram distance was 115 mm. The condition for equal lateral and longitudinal magnification, $\frac{\Delta\phi}{\Delta W}Z = 1$, was fulfilled. An argon ion laser provided coherent light output at 0.51 μm, a value matched to the sensitivity curve of Agfa 10E56 glass plates. An electronic shutter (Jodon ESS-100) ensured uniform exposure of the individual strip holograms.

The results obtained are shown in Figs. 6 and 7. Figure 6 is a photograph of the virtual image of the phantom eye reconstructed from the ultrasonic stereohologram. The images formed by each strip hologram have fused fairly well into eight distinct lines. The CRT grid lines appear also and can be used to determine $M_{lat,x}$ and $M_{lat,y}$.

A single plane of the reconstructed real image of the phantom eye is given in Fig. 7. This plane was chosen to coincide with two of the target wires. The CRT grid lines are somewhat out of focus because they lie in a nearby plane. The remaining target wires are decomposed into their component perspectives; these perspectives merge in other planes, in accordance with Eq. 3. The longitudinal image magnification is found by focusing first on the closest, then on the furthest, image line and noting the distance between them in image space. This was found to be $\approx$ 50 mm for the phantom eye. The measured lateral and longitudinal magnification were equal, verifying Eq. 4.

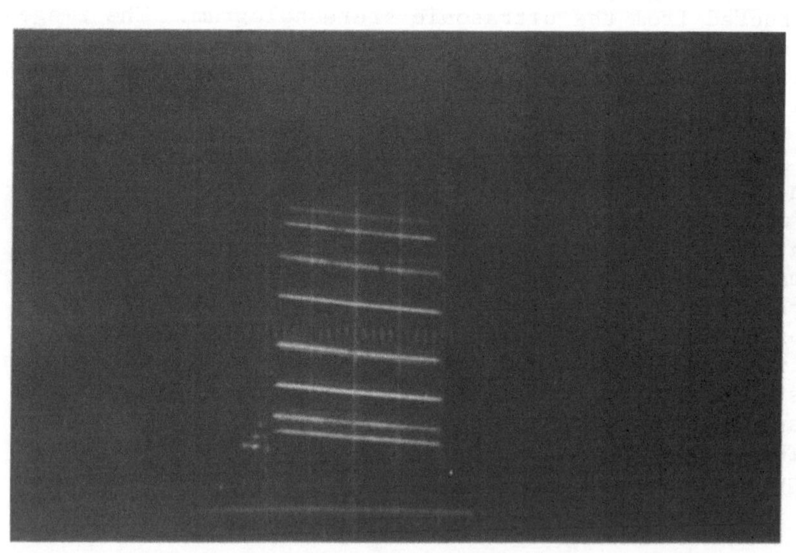

Fig. 6 Virtual image of the phantom eye reconstructed from an ultrasonic stereohologram.

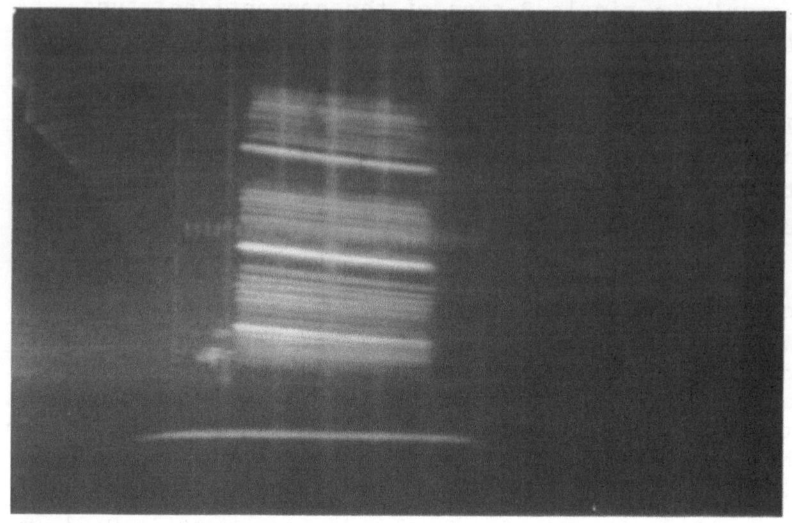

Fig. 7 Real image of the phantom eye reconstructed from an ultrasonic stereohologram.

6. WHITE LIGHT ULTRASONIC STEREOHOLOGRAPHY

Because the diffuse screen to hologram distance in ultrasonic stereoholography is the same for all transparencies, as shown in Figs. 1 and 3, all the parallax information necessary to reconstruct a three-dimensional image is contained in the single plane corresponding to the diffuse screen. As noted by King et al.[8], in such a case a second, image-plane hologram can be recorded in which the real image of the diffuse screen plane as projected by the first hologram coincides with the emulsion of the second hologram. Under white-light illumination, this image-plane hologram reconstructs a three-dimensional image.

The recording arrangement is shown in Fig. 8. A condensing lens illuminates the ultrasonic stereohologram with a conjugate reference beam, causing a real image of the diffuse screen to fall on the photographic plate. A reference beam is also brought to the plate by means of the beam-splitter and mirror. The resulting hologram is an image-plane hologram.

An image-plane ultrasonic stereohologram was recorded of the phantom eye of Fig. 4. The three-dimensional image is easily viewed in typical ambient lighting conditions when illuminated by a high-intensity desk lamp, or in darkness when illuminated by a penlight flashlight.

7. APPLICATIONS IN MEDICAL DIAGNOSIS

Although the experimental results described above were obtained for a relatively crude mechanical target, efforts are now under way at the Department of Electronics, the Weizmann Institute of Science, to image biological targets by means of ultrasonic stereoholography. We will present further results as they are achieved.

The following points can be made regarding the future application of ultrasonic stereoholography in medical diagnosis:

1. It requires only minor modifications of existing clinical B-scan equipment, specifically, substitution of a cylindrically-focused transducer and provisions for incrementing the scanning plane between scans.

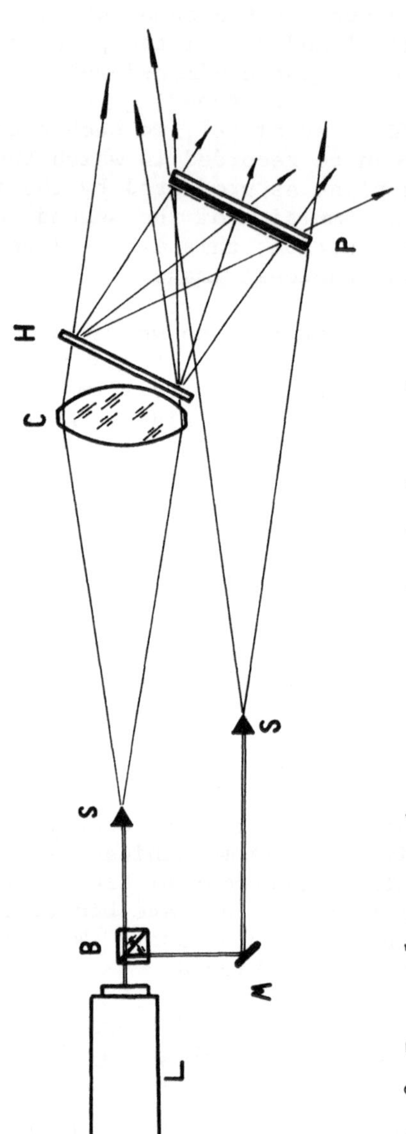

Fig. 8 Image-plane copying arrangement to obtain an ultrasonic stereohologram that reconstructs a three-dimensional image in white light (L = laser, B = beam-splitter, M = mirror, S = spatial filter, C = condensing lens, H = first hologram, P = high-resolution photographic emulsion).

2. Consequently, any advances made in conventional B-scan technology are directly applicable to ultrasonic stereoholography. These include compound scanning, time-varying receiver gain, logarithmic amplifier compression, and large dynamic-range displays.

3. The holographic step is currently carried out in the optical laboratory. We hope to design a holographic "black box" to produce a stereohologram from a transparency sequence. Then, ultrasonic stereoholograms could be prepared in a hospital's own photographic laboratory.

4. Finally, and perhaps most significantly, a physician will have available to him three-dimensional images of internal body structures, scaled to his own specifications, which he can examine in his own office under ordinary illumination.

8. ACKNOWLEDGMENTS

The authors wish to thank the following people and institutions for their assistance: G. Myers and the Biomedical Laboratory of the Riverside Research Institute, New York City; D.J. Coleman of the Eye Institute, Columbia-Presbyterian Medical Center, New York City; E.H. Frei and the Department of Electronics, the Weizmann Institute of Science, Rehovot, Israel; and the Israel Cancer Association.

9. REFERENCES

[1] See, for example, N. Lindgren, IEEE Spectrum 6, 48(1969) and P.N.T. Wells, Bio-Medical Engineering 5, 378 (1970).

[2] R.W. Meier, J. Opt. Soc. Am. 55, 987 (1965).

[3] G. Lippmann, Compt. Rend. 146, 446 (1908), and J. Phys. 1, 821 (1908).

[4] R.V. Pole, Appl. Phys. Lett. 10, 20 (1967).

[5] J.D. Redman, Holography Seminar Proceedings, Society of Photo-optical Instrumentation Engineers (SPIE) 15, 117 (1968).

[6] N. George, J.T. McCrickerd, and M.M.T. Chang, Holography Seminar Proceedings, SPIE 15, 161 (1968).

[7] D.J. DeBitetto, Appl. Opt. 8, 1740 (1969).

[8] M.C. King, A.M. Noll, and D.H. Berry, Appl. Opt. 9, 471 (1970).

FOIL-ELECTRET TRANSDUCER ARRAYS FOR REAL-TIME ACOUSTICAL HOLOGRAPHY

A. K. Nigam, K. J. Taylor* and G. M. Sessler

Bell Laboratories

Murray Hill, New Jersey 07974

ABSTRACT

The foil-electret microphone principle can be employed to construct two-dimensional transducer arrays in a relatively simple fashion. Two different designs of these arrays are described which contain N^2 microphones arranged in a N×N square matrix. The first design incorporates a backplate consisting of N×N elements and a foil electret. The array elements can be interrogated in parallel or in series. For series operation N^2 switches are required. The second design utilizes a backplate and an electret foil each with N strips of metalization arranged in an overlapping fashion. This array cannot be sampled in parallel; however, for serial sampling only 2N switches are necessary. Both designs can also be used with external biasing. Tests on 16×16 element arrays of these designs indicate usable frequency ranges of at least 70-250 kHz in air and 0.3-2 MHz in water. At 100 kHz in air, the sensitivity of elements is found to be uniform to within ±2.5 dB and, for the first design, the total interelement crosstalk is about -35 dB. Acoustic shadows of a few objects recorded at 40 frames/second by the 16×16 array in air are displayed. Larger foil-electret arrays (e.g. 200×200 element) promise to be inexpensive and rugged tools for acoustical holography.

* Presently with the Physics Department, University of Western Australia, Nedlands, W. Australia, 6009.

INTRODUCTION

A variety of methods have been used to record acoustic holograms. Examples are the mechanically scanned transducer[1], the Sokolov tube[2], Bragg diffraction of coherent light[3], deformation of liquid-gas interfaces[4] and two-dimensional or crossed linear piezoelectric transducer arrays[5]. While each of these methods has succeeded in producing acoustical holograms they are all subject to limitations such as insufficient aperture or resolution, requirement of high sound intensities, slow scanning speeds, complex data processing, or high cost of the arrays.

In the present paper two-dimensional arrays consisting of electrostatic transducers of the foil-electret type[6] are described. Such arrays can be used both in air and water over broad frequency ranges, are inexpensive, and are simple in design regardless of size and number of elements.

In the following, the basic design and performance of a foil-electret microphone is first reviewed briefly. Then the mechanical design of two arrays and a few schemes of processing the output signals are described. This is followed by a discussion of the electrical and acoustical characteristics of the arrays. Finally, examples of shadowgraphs obtained by scanned operation of a 16×16 prototype array are presented and the contemplated future development of these arrays is discussed.

FOIL-ELECTRET MICROPHONE

A cross section of an electrostatic microphone of the foil-electret type[6] is shown in Figure 1. The diaphragm of this transducer consists of a thin polymer foil, metalized on one side, and permanently charged on its polymer surface. The charged foil (foil electret) is stretched across a metallic backplate with the metal side facing out. Due to minute irregularities in the backplate surface, a shallow air gap is formed between foil and backplate. A more regular air gap may be obtained by providing ridges on the backplate surface.

A sound wave impinging on the microphone deflects the foil, generating a voltage between the metal layer of the foil and the metallic backplate. In microphones designed

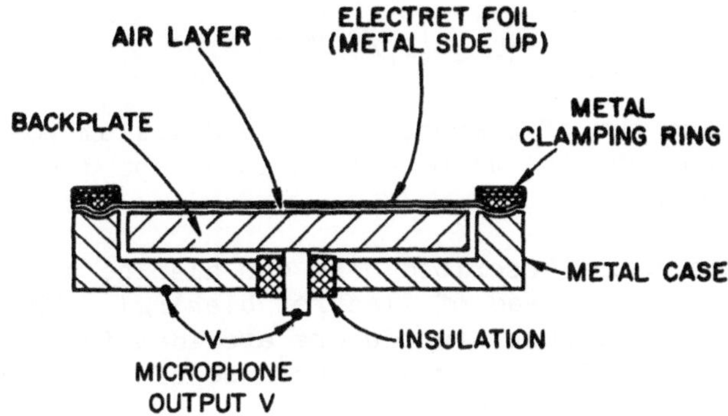

FIGURE 1. CROSS SECTIONAL VIEW OF A FOIL-ELECTRET MICROPHONE.

for high frequency operation the restoring force on the foil is primarily due to the isothermal compression in the air gap. In this case the sensitivity ρ (generated voltage/sound pressure) of the microphone well below its resonance is given by

$$\rho = [\sigma\, D\, d/\epsilon_o\, p_o(D+\epsilon d)]\, \frac{C_m}{\alpha(C_m+C_L)}\ , \qquad (1)$$

where σ is the surface charge density on the foil, D and d are the thicknesses of the foil and air gap, respectively, ϵ_o is the permittivity of free space, ϵ is the dielectric constant of the foil material, p_o is the atmospheric pressure, and C_m and C_L are the microphone and loading capacitances respectively. The factor α equals $[1+(1/\omega CR)^2]^{1/2}$ where $C=C_m+C_L$, R is the terminating resistance, and ω is the angular frequency of the acoustic signal incident on the microphone. Notice that $\alpha=1$ if $\omega \gg 1/RC$. In this case the sensitivity ρ is independent of microphone area (i.e., the capacitance of the microphone) if $C_m \gg C_L$. At the resonance frequency the sensitivity of the microphone is higher than the value obtained from Eq. (1). At higher frequencies the sensitivity decreases.

For the case $\omega \gg 1/RC$ the resonance frequency ω_r is given by

$$\omega_r = (p_o/d \cdot M)^{\frac{1}{2}}, \tag{2}$$

where M is the sum of diaphragm mass and medium loading per unit area. Notice also that ω_r is independent of microphone area.

The microphone described here can also be used with external biasing instead of electret biasing[7]. The characteristics of such transducers are identical to those with electret biasing. However, because of its simplicity, electret biasing is preferred.

The noise voltage generated in a bandwidth f_2-f_1 by an electret microphone and its preamplifier is primarily due to the resistive component R(f) of the circuit and the transistor noise. The noise voltage V_R resulting from $R(f) = R/[1+(\omega CR)^2]$, is (for $\omega CR \ll 1$ or $\gg 1$) given by

$$V_R = [4kTR(f) \cdot (f_2-f_1)]^{\frac{1}{2}}. \tag{3}$$

The transistor noise is caused by the gate leakage current. For a discussion of this noise we refer to the literature[8].

ARRAY DESIGN AND DISPLAY

Basically two array designs have been investigated. In the first design, an array of foil-electret microphones is obtained by replacing the metallic backplate of a large-area, foil-electret microphone (see Figure 1) by a subdivided backplate. Each backplate subdivision is electrically insulated from all other subdivisions and, together with the foil immediately above it, functions as an individual microphone element. Thus for a N×N element array of this design, the backplate has N^2 subdivisions. Accordingly an array of this design will be called a "sub-divided-backplate array" or alternatively a "N^2 array".

In the second design, an array of microphones is obtained by subdividing both the backplate and the metal layer of the foil along narrow strips[9]. Every foil strip partly overlaps all backplate strips with each overlap forming a separate microphone. For a N×N element array of this design, both backplate and foil have N subdivisions each.

Accordingly an array of this design is called a "subdivided-foil-subdivided-backplate array" or a "2N array".

Both designs can be used with electret or external biasing. With either biasing, the performance of the N^2 design is the same. The 2N design, however, shows higher interelement crosstalk with electret biasing (see below).

At present, 16×16 element arrays of the above mentioned designs are being studied. These investigations will determine the feasibility of larger (such as 200×200 element) arrays. In the following, the construction and signal processing for two 16×16 arrays, one of each design, are described. The display device used for recording holograms is also outlined.

Subdivided-Backplate Array (N^2 Array)

A photograph of the subdivided backplate of the 16×16 prototype array of this design is shown in Figure 2. The backplate of the array is machined from a single piece of brass by cutting a grid of narrow slots into one surface, filling the slots with epoxy and machining off the other surface of the brass so that no metal connections between the backplate sections remain. The face of the backplate is lapped to a smooth finish. Alternate methods of

FIGURE 2. SUBDIVIDED-BACKPLATE DESIGN: FRONT VIEW OF BACKPLATE OF THE 16×16 ARRAY.

backplate design utilize photoetching techniques yielding metal squares deposited on a dielectric substrate.

The overall dimensions are shown in an exploded cross sectional view of the prototype array in Figure 3. The

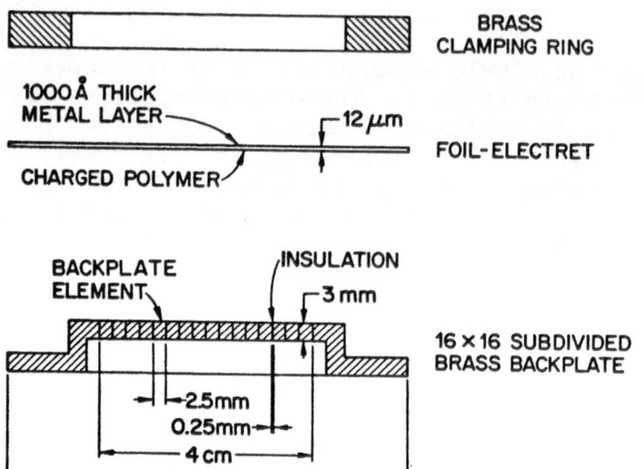

FIGURE 3. SUBDIVIDED-BACKPLATE DESIGN: EXPLODED CROSS SECTIONAL VIEW OF THE 16×16 ARRAY.

foil electret is a 12.5μm thick Teflon FEP foil charged with the electron beam method[10] to a charge level of about 2×10^{-8} C/cm^2. The metallization on the foil consists of a 1000 Å thick aluminum layer. When assembled, the average thickness of the air gap between foil and backplate is found to be approximately 10μm.

When a sound wave is incident on the array, each element of the array generates a voltage with respect to the common metal layer on the foil which is proportional to the instantaneous acoustic pressure on the element [see Eq. (1)]. Thus the array is capable of transforming sound-pressure data into electrical data in real-time. However, final reconstruction of the image conventionally requires that the electrical data first be presented in some form of a transparency which can then be illuminated with coherent light to view the image.

To obtain an optical output from the array (from which a transparency can be made), the electrical data from the elements can be read into an appropriate electro-optical device. For the case of the N^2 array, the read-in can be done either in parallel-mode, serial-mode or a combination of the two.

For an arrangement in which the read-in is in the parallel mode, we suggest a scheme in which a matrix of Light Emitting Diodes (LEDs) are directly connected to the back of the array, with one LED to each element. The electrical output (amplitude or phase) of each element would control the light output of the LED attached to the element. The attractive feature of this design appears to be its compactness --- the transducer array being a flat package with acoustic input in the front and the desired optical output (hologram or image) in the back of the array. However, it appears that each element output will have to be amplified in order to drive the LED. In the 16×16 prototype array of the present design this will require a total of 16×16 amplifiers between the elements and the LED array.

Serial data read-out, on the other hand, eliminates the need for amplifying and other data processing units for each element. For this reason, in the present design of the prototype array, serial data read-out has been employed. This is accomplished by attaching a 16×16 matrix of insulated-gate, Field Effect Transistor (FET) switches to the back of the array. By employing IC technology, the switch matrix is only 1.5cm thick and contained in the lateral size of the array (4cm×4cm) as shown in the photograph in Figure 4. A common (single) output from the switches is attained by (1) connecting the outputs of all the 16 switches in one row together and (2) connecting the 16 row outputs to a set of 16 FET switches (called row-switches) with a common output (the array output) as shown in Figure 5. The output of the array is loaded by a terminating resistance R_L=10KΩ, across which the output voltage is measured. To sample the output of a particular element (i,j) the row and column addresses, i and j respectively, are generated separately by a logic circuit and transmitted to the array. The column address j closes all the switches in column j of the array and the row

FIGURE 4. SUBDIVIDED-BACKPLATE DESIGN: REAR VIEW OF THE BACKPLATE SHOWING SWITCHES INSTALLED.

address i closes the i-th row switch. Thus element (i,j) is connected to the array output, where its output is sampled across R_L.

Thus for serial-data processing for the N^2 array a total of N×N switches are needed immediately behind the backplate elements. It should, however, be reemphasized that the N^2 array lends itself to both serial or parallel data read-out.

Subdivided-Foil-Subdivided-Backplate Array (2N Array)

A photograph of the subdivided foil of a 16×16 prototype array of this design is shown in Figure 6. The bright areas are regions where the foil has been metalized by a 1000 Å thick aluminum layer whereas the non metalized areas appear as dark lines. Also, the backplate is metalized only along parallel metal strips. In the 16×16 prototype array this was done by selective metalization of a plexiglas backplate in a manner similar to the selective metalization of the foil. The metal strips on both foil and backplate are 2.25mm wide and the gap between adjacent strips is 0.25mm. The array is assembled so that backplate and foil metal strips are mutually perpendicular and over-

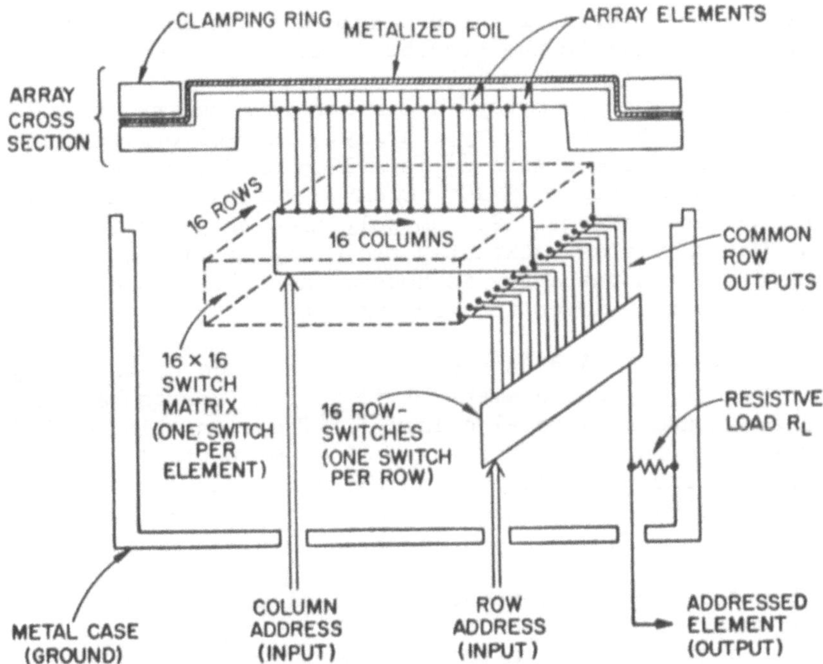

FIGURE 5. SUBDIVIDED-BACKPLATE DESIGN: LAYOUT OF ELECTRICAL CONNECTIONS FOR SEQUENTIAL SAMPLING OF THE 16×16 ARRAY. ONLY ONE ROW IS SHOWN.

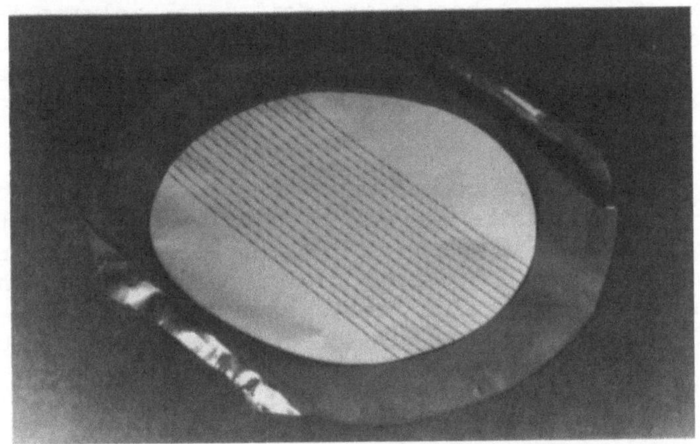

FIGURE 6. SUBDIVIDED-FOIL-SUBDIVIDED-BACKPLATE DESIGN: SELECTIVELY METALIZED FOIL OF THE 16×16 ARRAY.

lapping. Each overlap forms a single microphone element of face area equal to the overlap area.

The 2N prototype array is similar in geometry and dimensions to the N^2 prototype array. The selectively metalized foil (see Figure 6) is a 12.5 μm Teflon FEP foil charged[10] to a level of 2×10^{-8} C/cm^2. For externally biased operation, an uncharged 12.5 μm polyimide or paper foil is used. The plexiglas backplate, the clamping ring, and the casing (in which the array is housed) as well as the air gap have the same dimensions for the two prototype arrays.

When a sound wave is incident on the face of the array, the output of any one element (i,j) is read across backplate metal row i and foil metal column j. These constitute the backplate and foil, respectively, of element (i,j). In this design, data from all elements of the array can not be read simultaneously, i.e., read-out can not be in the parallel mode.

For this and other reasons outlined above, serial data read-out has been employed for the 2N prototype array. This is accomplished by attaching one switch per row and one per column (a total of 2N switches). A schematic layout of these connections is shown in Figure 7. In the externally biased version a voltage source is introduced between points A and B. Also, the foil columns not being sampled are grounded by an additional set of N switches. Connection of the switches to the foil metal columns is done by making a slot in the clamping ring and making pressure contacts between the switch inputs and the extended portion of the foil metal columns. The backplate row connections are made from the rear of the backplate.

The logic circuit constructed to operate the switches of the N^2 prototype array is employed to operate also the 2N prototype array. The row and column addresses, i and j respectively, supplied by the logic circuitry, close the corresponding row and column switches resulting in the output of element (i,j) to be connected to the array output.

Electro-Optical Display Device

For operating the arrays in the scanned mode, the row and column addressing circuitry is slaved to a system

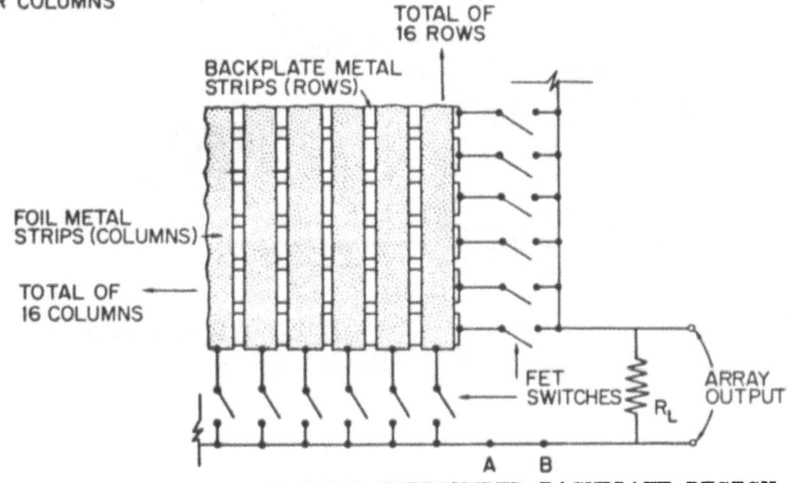

FIGURE 7. SUBDIVIDED-FOIL-SUBDIVIDED-BACKPLATE DESIGN: LAYOUT OF ELECTRICAL CONNECTIONS. EXTERNAL BIASING, IF USED, IS APPLIED BETWEEN A AND B.

clock. The system clock supplies a train of pulses at any presettable rate. With each pulse the row address is incremented by unity successively up to address 16 after which it is reset to address 1. With each row-address reset, the column address is simultaneously incremented by unity. The process is repeated until all the elements are scanned.

The time-multiplexed signal obtained in this manner at the array output is amplified and transmitted to a (remote) filtering and display device via a single channel data-link to which other data processing equipment is attached. The display device used in the present setup is a storage CRT. The array output is used to modulate the intensity of the electron beam while the position of the beam on the CRT screen is controlled by the logic circuitry. The layout of the complete instrument chain used here is shown in Figure 8.

If the face of the 16×16 array is illuminated by a plane acoustic wave and the transducer is scanned, then the

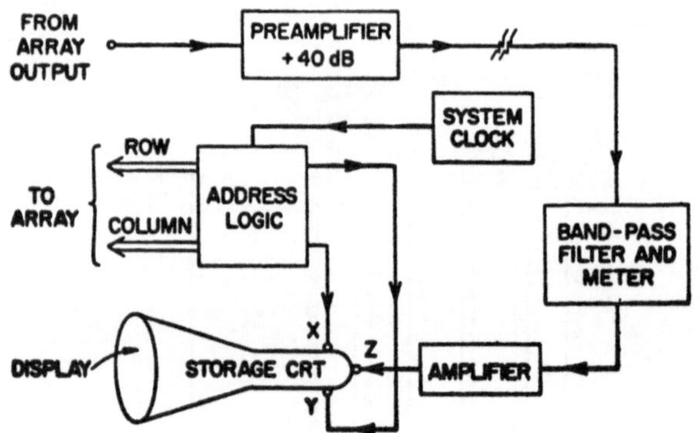

FIGURE 8. BLOCK DIAGRAM OF INSTRUMENT CHAIN USED FOR IMAGING.

display on the CRT screen is a 16×16 matrix of uniform intensity dots. If the CRT is operated in the storage mode, it is capable of displaying the accumulated intensities (as long as a threshold value is exceeded) of each of the dots. This leads to higher intensity and higher contrast displays.

CHARACTERISTICS OF THE 16×16 ARRAYS

For the 16×16 prototype arrays, the air gap is 10 μm and the mass density of the foil is 2.5×10^{-3} gm/cm^2. In air, the medium loading is negligible and from Eq. (2) the resonance frequency is approximately 100kHz. In water, due to loading effects of the medium, the resonance frequency is about 20kHz.

Below the resonance frequency, the sensitivity of elements is given by Eq. (1) and for the present arrays this amounts (for $C_m \gg C_L$) to

$$\rho = -80 \text{ dB re. } 1\text{V}/\mu\text{bar}, \quad \omega < \omega_r. \qquad (4)$$

At the resonance frequency the sensitivity is higher by about 10 dB beyond which the sensitivity decreases with increasing frequency.

The lower frequency limit of array operation is given by the frequency f_ℓ below which the array oversamples the sound field

$$f_\ell = c/2s. \tag{5}$$

Here c is the sound velocity and s the spacing of the elements. For the prototype arrays f_ℓ equals 70kHz in air and 300kHz in water.

The upper frequency limit is usually decided by requirements of field of view θ or element sensitivity of the array. The former is given by

$$\sin(\theta/2) = c/2fs, \quad (<1). \tag{6}$$

From this equation the field of view of the prototype arrays in air is 87° at 100kHz and 32° at 250kHz. In water θ equals 70° at 0.5MHz and 33° at 1MHz.

Measurements of element sensitivity at a few frequencies above resonance and extrapolation of these results indicate that the open-circuit ($C_m \gg C_L$) sensitivity of the prototype arrays at 1MHz is -120 dB re. 1V/μbar in air and -130 dB re. 1V/μbar in water (see also Reference 7). At this frequency (1MHz) the noise voltage from the arrays [see Eq. (3)] measured in a bandwidth of 100kHz is found to be -114 dB re. 1V. Thus a unity signal-to-noise ratio is attained for sound intensities I_1 of

$$\begin{aligned} I_1 &\approx 10^{-8} \text{ watts/cm}^2 \quad \text{in air} \\ &\approx 2\times10^{-11} \text{ watts/cm}^2 \quad \text{in water} \end{aligned} \quad \text{at 1MHz.} \tag{7}$$

These are the minimum detected acoustic intensities by the arrays in the respective media. The value of I_1 in water is comparable to the minimum detected intensity by piezoelectric arrays[11].

In air, attenuation in the medium prohibits[12] array operation above a few hundred kHz. In addition, the present 16×16 arrays have a relatively small value of θ above 250kHz in air. Based on these considerations, the optimum frequency range of operation of the 16×16 prototype arrays in air is 70-250kHz. From similar considerations, the optimum frequency range of operation of the 16×16 prototype arrays in water is 0.3-2MHz.

The arrays described here can be optimized for operation in any particular medium over certain frequency ranges. To achieve this, the array is designed so that its resonance frequency (where the sensitivity of elements is a maximum) lies within the frequency range of interest. This may be done by omitting the air layer between foil and backplate or replacing it by a layer of another material having the appropriate stiffness (for a particular example see Reference 13).

The values displayed in Eqs. (4) and (7) are those obtained when the array elements are not (electrically) loaded and thus represent the best presently possible values for foil-electret transducers. However, a finite loading results if C_L is comparable to C_m [see Eq. (1)]. In the prototype arrays C_m is 1.5pF. For the feasibility studies an external amplifier connected by a long cable was used which introduced a C_L of about 200pF. This resulted in a loading of approximately 45dB. The loading will be reduced to about 3pF, i.e., 10dB, by placing the amplifier directly at the output of the switches. For the 2N design, there is an additional loading caused by the capacitance of the elements not being sampled. This loading is inherent in the 2N design and for the 16×16 array (of this design) is about 20dB.

EVALUATION OF THE 16×16 ARRAYS

Of the parameters which may affect the performance of the 16×16 arrays during recording acoustic holograms, the following are of particular importance:

i. Crosstalk between elements of the array.

ii. Variation in sensitivity across the array.

In the following a brief analysis of these parameters is carried out, followed by their measured values at 100kHz in air. For brevity, analyses and results are presented only for the N^2 array. Unless stated otherwise, these results apply also for the 2N array.

Crosstalk Between Elements of the Array

In the N^2 design, the important factors that contribute to interelement crosstalk are:

a. Electrical coupling between backplate elements and among the switches (electrical crosstalk).

b. Coupling between elements due to finite mechanical transfer impedance of the shared foil (mechanical crosstalk).

The interelement capacitance is the primary cause for electrical coupling between elements of the backplate. Furthermore the capacitance between switches contributes to interswitch electrical coupling. In addition, coupling occurs also due to leakage currents across switches.

The overall electrical crosstalk between elements (i,j) and (m,n) of the array is measured in the following manner. An extremely small diameter probe (probe tip diameter 0.01cm) is inserted through the foil such that the probe only contacts backplate element (i,j). A unity voltage at frequency ω is applied to the probe and the output of element (m,n) is measured. This output represents the electrical crosstalk between elements (i,j) and (m,n). In the prototype array it was found that maximum electrical crosstalk is between adjacent elements in the same row of the array. This maximum value is -36 dB at 100kHz. In comparison, crosstalk between adjacent elements in the same column is -42 dB while for any other element pair it is less than -43 dB.

To measure the overall (electrical and mechanical) crosstalk, we employ an acoustic analog of the scheme used above to measure the electrical crosstalk. In this scheme, however, the computation for crosstalk is not as simple because it is not possible to apply an acoustic pressure on one element of the array while completely shielding all other elements. In the actual test a 5mm thick brass shield with a 2.5mm square hole is used. The shield is placed over the face of the array leaving an air gap of ≈150 μm such that the hole (in the shield) exposes only one element of the array. If this setup is now placed under a normally incident acoustic wave the exposed element is directly illuminated by the sound. However, the other elements are indirectly illuminated due to diffraction, scattering and multiple reflections between the brass shield and the foil of the array. The output V_{mn} of any element (m,n) is thus

$$V_{mn} = \rho \sum_{i=1}^{16} \sum_{j=1}^{16} p_{ij} \cdot \psi(m,n;i,j) \qquad \text{for all } m,n = 1,2,3,\ldots.16, \tag{8}$$

where p_{ij} is the acoustic pressure on element (i,j) behind the shield, ρ is the sensitivity, assumed uniform for all elements, and $\psi(m,n;i,j)$ is the overall crosstalk between elements (m,n) and (i,j) defined as

$$\psi(m,n;i,j) = (V_{mn}/V_{ij}) \text{ when only element } (i,j) \text{ is acoustically activated.} \tag{9}$$

In the present computation $\psi(m,n;i,j)$ is assumed to be a function only of the radial distance between elements (m,n) and (i,j). In this case there are only 120 unknown ψs and an equal number of unique simultaneous equations (8). These can be solved for the ψs using measured values for V and p when one element of the array is exposed (using the shield)*.

For the measurement of p_{ij} we have assumed that the motion of the foil is very small compared to the spacing between the shield and the foil. Under this assumption, p_{ij} is approximately given by the blocked-pressure distribution[14] on the surface of the array, behind the shield. The blocked-pressure distribution is measured experimentally as a function of lateral distance from the hole by a 1/8 inch B&K microphone set in a brass baffle. Some results of these measurements at 100kHz are shown in Figure 9.

To obtain some approximate results for the overall interelement crosstalk ψ an iterative procedure, which considers only the magnitudes of V and p, is employed to solve Eq. (8). The iteration is performed on a DDP-516 electronic computer. These approximate results show that, in the N^2 prototype array, the overall crosstalk is maximum

* If the assumption about radial dependence of ψ is not valid then there are, in all, $N^2(N^2-1) = 256\times255$ unknown ψs. Thus 256 separate experiments are required, each being performed by exposing a different element (using the shield) and measuring the output V_{mn} of all other elements.

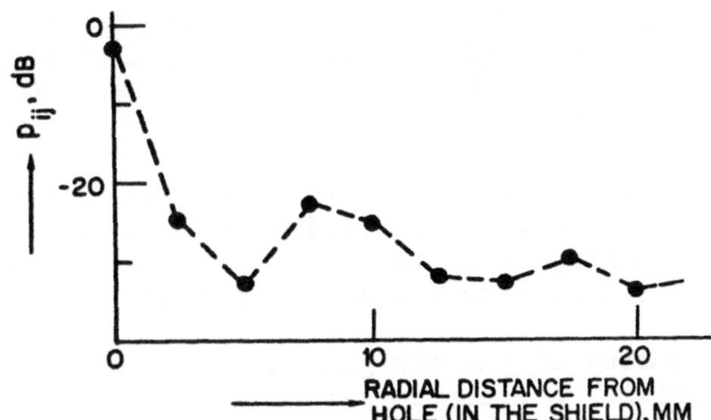

FIGURE 9. BLOCKED-PRESSURE DISTRIBUTION ON THE FACE OF THE ARRAY, BEHIND THE SHIELD, AT 100kHz IN AIR.

between adjacent elements of the array. This value is equal to -35dB. Comparison with the above figure for electrical coupling suggests that the crosstalk is primarily of electrical origin. Presently the overall interelement crosstalk is being calculated more precisely by considering both magnitude and phase of V and p in Eq. (8).

Electrical crosstalk for the externally biased 2N design is of the same order as for the N^2 design whereas the electret biased 2N design shows considerably higher crosstalk.

Variation in Sensitivity Across the Array

For successful imaging with serial data read-out, it is important that the sensitivity of all elements be essentially the same. In the present designs, variation in element sensitivities may be due to three factors. These are

a. Non-uniformity of charge density on the foil.

b. Non-uniformity in closed-switch resistances.

c. Geometric non-uniformities such as those in air gap and element sizes.

The sensitivity of an element of the array is directly proportional to the average charge density on the portion of foil belonging to that element [see Eq. (1)]. To determine the variation in average charge density across the array, charge measurements of the foil electret were made using a

2.5mm diameter probe. It was found that the variation in charge across the array (foil area 4cm×4cm) is less than ±5% (see also Reference 10). This would cause a ±0.4 dB variation in element sensitivities.

Non-uniformity in closed-switch resistances was measured at 100kHz for the 16×16 arrays (using R_L=10 kΩ). It was found that the variation in the resistance of the switches of the array causes a ±0.6 dB variation in sensitivities.

Non-uniformity of the air gap and its effect on element sensitivity can not be measured directly. However, the overall variation in sensitivity of the elements was determined experimentally by placing the arrays under an acoustic transmitter operating at 100kHz. The transmitter was found to produce an acoustic field uniform to within ±0.5 dB, as measured by a 1/8 inch B&K microphone, over the face of the array. The outputs of the elements were found to be within ±2.5 dB. Thus the overall variation in element sensitivities of the prototype arrays is at most ±3 dB. Considering the small variations in charge density and switch resistance, it can be concluded that most of this variation is due to non-uniformities of the air gap.

SCANNED OPERATION IN REAL-TIME

For imaging, the 16×16 prototype array is placed under a uniform acoustic field at 100kHz. The equipment chain shown in Figure 8 is employed. If a two dimensional object is now placed in close proximity of, but not touching, the surface of the array then the acoustic pressure pattern on the array will correspond to that of a simple acoustic shadow of the object on the array surface. This can be displayed on the CRT screen.

Experiments were done with the N^2 array using two objects (a square and the letter H). The objects were cut from 1/8 inch thick aluminum sheets and suspended at a distance of approximately 3.5mm in front of the array by means of four very fine wires (see Figure 10). The array was scanned at 40 frames/second, which corresponds to a sampling duration per element of about 10 cycles of the acoustic signal.

During recording on the CRT screen, the electron beam was defocused slightly so that the dots were each roughly 3mm in diameter. This large dot size (as opposed to a

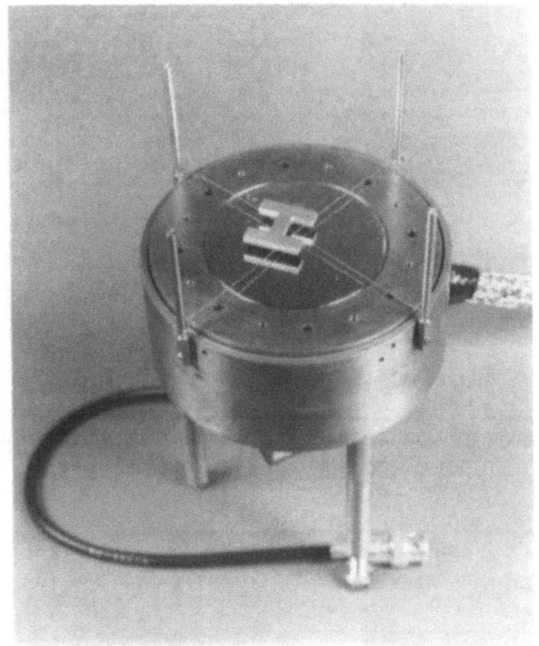

FIGURE 10. POSITIONING OF THE OBJECT IN FRONT OF THE ARRAY DURING IMAGING.

pinpoint obtained from a fully focused beam) made it easier to distinguish different intensity levels. The CRT was operated in its storage mode and a 0.5 second integration time was employed for recording the display. After recording, the electron beam was focused on the screen and the 16×16 point pattern was recorded superposed on the dot-pattern. This is done to facilitate the read-out of the row and column address of any one dot. The display is then photographed.

These photographs, along with the object for which they were made, are shown in Figures 11 and 12 (the dark areas in the top right corners of Figures 11b and 12b are due to non uniformities of the acoustic field).

SUMMARY AND FUTURE DEVELOPMENT

From the above it is seen that 16×16 transducer arrays discussed here (1) are relatively simple and inexpensive

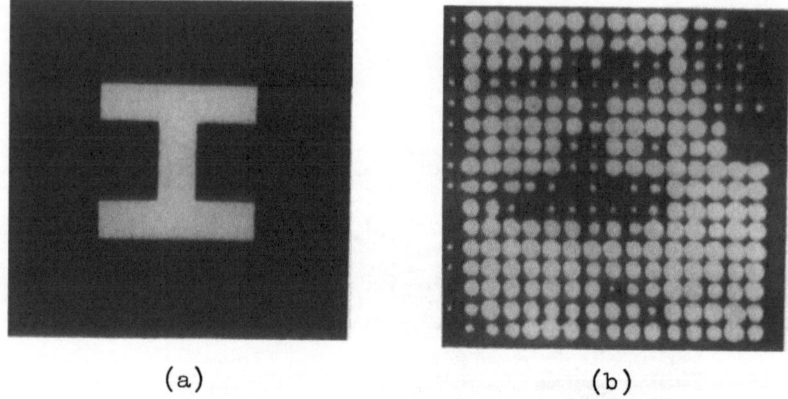

FIGURE 11. (a) OBJECT. (b) ACOUSTIC SHADOW RECORDED AT 40 FRAMES/SECOND IN AIR AT 100kHz.

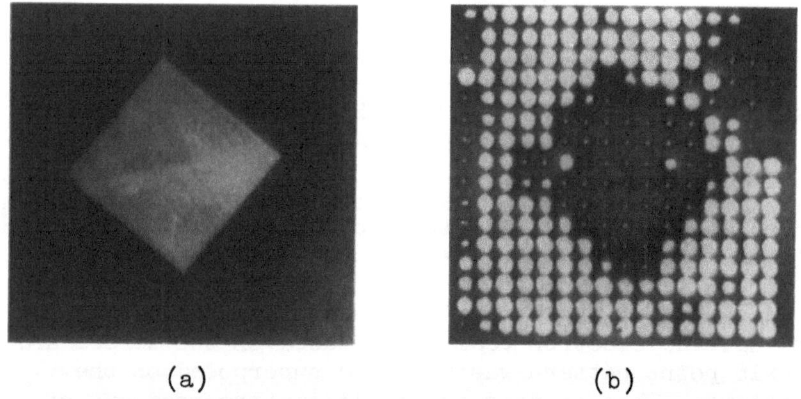

FIGURE 12. (a) OBJECT. (b) ACOUSTIC SHADOW RECORDED AT 40 FRAMES/SECOND IN AIR AT 100kHz.

to construct, (2) are capable of both air and underwater operation over broad frequency ranges in each medium, (3) have relatively large and almost uniform sensitivity across the array, (4) in conjunction with the instrument chain shown in Figure 8, are capable of scanned operation for sampling times per element of a few cycles of the acoustic signal, and (5) have overall (electrical and mechanical) crosstalk, with switches, of less than -35 dB (except for the electret biased version of the 2N design).

Larger (e.g. 200×200 element) arrays of the present designs appear to retain all of these advantages except that in the 2N design the inherent electrical loading (see above) increases with array size N. Nonetheless, for serial data read-out the 2N design with its smaller number of switches appears preferable to the N^2 design. The latter, however, is necessary if parallel data read-out is desired.

Currently the 16×16 arrays are being optimized and further underwater tests in the frequency range 0.6-2MHz are being conducted. A 200×200 element array of the 2N design is also under development. The anticipated uses of this array are in speech research for studies of the vocal tract, in underwater real-time viewing for cable laying operations, and in material testing both in air and water.

ACKNOWLEDGMENTS

The authors wish to express their thanks to J. H. Condon and J. R. Nelson for discussions on array design, to J. E. West and J. F. Puluka for charging and metalization of foils and backplates, and to J. H. Kronmeyer for assistance in array construction.

REFERENCES

1. K. Preston Jr. and J. L. Kreuzer, Appl. Phys. Letters 10, 150-152 (1967); A. F. Metherell, H. M. A. El-Sum, J. J. Dreher and L. Larmore, ibid., 10, 277-279 (1967); G. A. Massey, Proc. IEEE 55, 1115-1117 (1967).

2. S. Sokolov, U.S. Patent 2,164,125 (1930); E. Marom, H. Boutin and R. K. Mueller, J. Acoust. Soc. Amer. 42, 1169 (1967); 43 384 (1967); M. A. Plonus, Proc. IEEE Letters, 1134-1136 (1968); D. Fritzler, E. Marom and R. K. Mueller, in "Acoustical Holography, Vol. 1," Plenum Press, New York (1969) Chapter 16.

3. A. Korpel, in "Acoustical Holography, Vol. 1," Plenum Press, New York (1969) Chapter 10; H. M. A. El-Sum, in "Acoustical Holography, Vol. 2," Plenum Press, New York (1970) Chapter 2; J. Landry,

R. Smith and G. Wade, in "Acoustical Holography, Vol. 3," Plenum Press, New York (1971) Chapter 4.

4. R. K. Mueller and N. K. Sheridon, Appl. Phys. Letters 9, 328 (1966); R. K. Mueller and P. N. Keating, in "Acoustical Holography, Vol. 1," Plenum Press, New York (1969) Chapter 3; N. K. Sheridon, in "Acoustical Holography, Vol. 2," Plenum Press, New York (1970) Chapter 20.

5. W. H. Wells, in "Acoustical Holography, Vol. 2," Plenum Press, New York (1970) Chapter 8; E. Marom, R. K. Mueller, R. F. Koppelman and G. Zilinskas, in "Acoustical Holography, Vol. 3," Plenum Press, New York (1971) Chapter 11; G. L. Sackman and R. J. Larkin, ibid., Chapter 12; G. Wade, M. Wollman and K. Wang, ibid., Chapter 13.

6. G. M. Sessler, J. Acoust. Soc. Amer. 35, 1354 (1963); G. M. Sessler and J. E. West, J. Acoust. Soc. Amer. 40, 1433 (1966).

7. G. R. Schodder and F. Wiekhorst, Acustica 7, 38 (1957).

8. F. Fraim and P. Murphy, J. Audio Eng. Soc. 18, 511 (1970).

9. A. K. Nigam and G. M. Sessler, Appl. Phys. Letters (to be published).

10. G. M. Sessler and J. E. West, Appl. Phys. Letters 17, 507 (1970).

11. R. K. Mueller, Proc. IEEE 59, 1319 (1971).

12. J. deKlerk, in "Acoustical Holography, Vol. 1," Plenum Press, New York (1969) Chapter 9.

13. P. Alais and M-T. Larmande, Compt. Rend. Acad. Sc. Paris 272, 185 (1971).

14. P. W. Smith, Jr. and R. H. Lyon, "Sound and Structural Vibration," N.A.S.A. Publication CR-160, 154-156 (1965).

A SMALL SCALE MODEL FOR SEISMIC IMAGING SYSTEMS

M. G. Maginness,* G. B. Cook,† and
L. G. Higgens†
*Stanford University, Stanford, California‡
†New Zealand Post Office, Wellington, N.Z.‡

I. INTRODUCTION

This paper describes the theory and apparatus of an ultrasonic imaging system originally developed for materials inspection but now seen as a means to simulate seismic holography schemes at greatly reduced dimensions. A number of inherent features makes the arrangement particularly applicable to this purpose:

(1) The acoustic signal is entirely within solid materials and thus complete elastic wave relationships are automatically fulfilled.

(2) Transmitters and receivers are on the same surface of the material, conforming to the constraint usually imposed on the full scale system.

(3) The transmitted signals may be bursts of continuous waves, or pulses, with no alterations to the device other than the provision of appropriate electrical drive to the transmitter.

(4) One, or any number of transmitters, may be used and no accurate knowledge of their location is required.

---

‡All the authors were previously with the University of Canterbury, Christchurch, New Zealand.

(5) The receiver may consist of a two-dimensional array of detectors, or at the other extreme, a single unit moved over the receiving aperture. Given simple duplication of some portions of the recording apparatus a number of such units could be simultaneously accommodated.

(6) The transduced elastic wave signals are recorded complete, the only limits imposed being those of the recording medium capacity.

(7) Signal manipulation to reconstruct the elastic wave fields is done by digital computer program, permitting virtually any processing scheme to be followed without a commitment to specific hardware. Wide bandwidth signals may be treated in terms of their frequency components and a set of images for different frequency intervals formed. Alternatively, instant-by-instant pictures of the field structures are possible.

Although this break between signal receiving and processing was not desirable for materials testing applications it appears quite appropriate to the present use.

The description follows this break in the system. Firstly, we describe the data acquisition apparatus and the sequence of operations followed to produce, in this case on paper tape, a record of the time varying signal from each interrogated position on the receiving aperture. Other recording media (e.g., magnetic tape) or even direct input to a computer store could be substituted.

Secondly, the signal processing theory is discussed. The sequence followed in the processing algorithm is such that additional refinements may be introduced at nearly every stage.

After a brief discussion of factors affecting the relationship between the derived images and the original object characteristics, a specific example is shown.

Some details of the general system concept, of portions of the apparatus, and a brief outline of the operation may be found in References 1, 2, and 3.

II. TRANSDUCER ARRANGEMENTS

Figure 1 shows in schematic form the arrangement of transducers and the important sections of recording equipment. As drawn, a single, scanned receiving element, or an array with electrical contact made sequentially to each element is assumed.

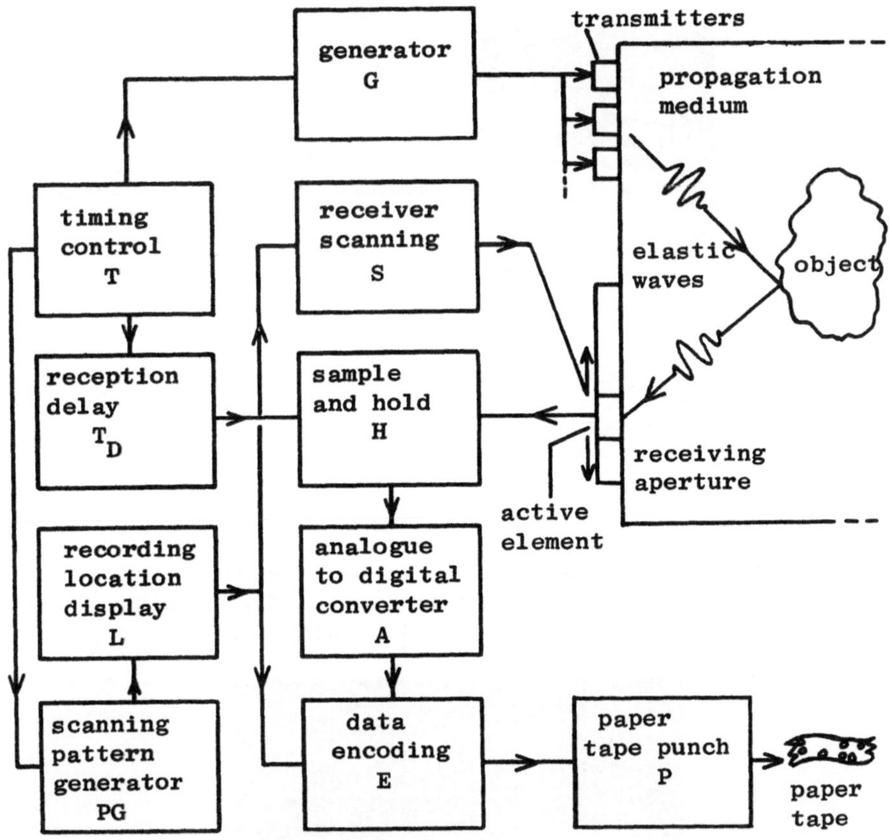

Figure 1: Block diagram of transducer arrangement and recording system.

Both receivers and transmitters were thickness mode piezoelectric transducers of PZT5A material, generally with the transmitters 5 mm diameter disks and the receiver a 15 mm disk, with nominal resonant frequencies of 5 MHz. Both transmitters and receiver were solidly bonded with very thin films of methyl-acrylate adhesive or low temperature solder to the metal samples, resulting in very heavily damped responses compared to the usual response with fluid couplants. This is illustrated by Figure 2 showing the waveforms and spectra of electrical and acoustic signals on transmission and Figure 3 showing the received signals on solidly bonded and oil coupled receivers.

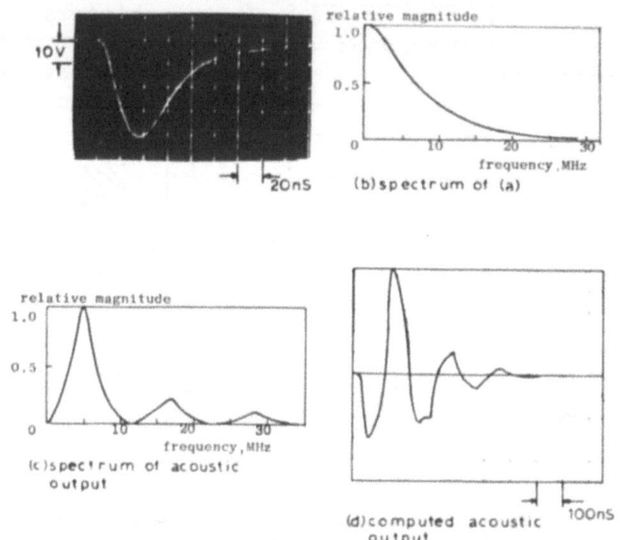

Figure 2: Waveforms and spectra of electrical and acoustic signals with a solidly bonded transmitter of PZT5A on aluminum.

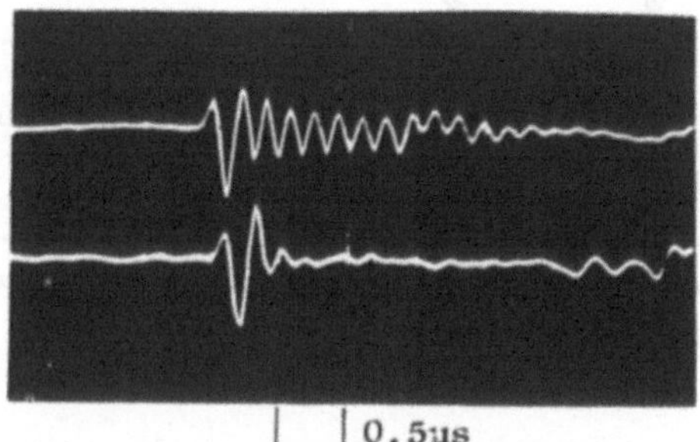

Figure 3: Effect of receiving transducer coupling. Lower; solid bonding. Upper; oil coupling. Transmitter solidly bonded.

Individual receiving elements were formed by etching the exposed receiver electrode into isolated 1 mm square plates. The high damping provided by solid bonding suppresses lateral spread of the acoustic motion.

Materials used as the propagation medium were variously aluminum or steel in the form of cylinders 15 to 20 cm in diameter and 20 to 30 cm long with the transducers on one end face. With the short pulses used having a spatial spread of only a few millimeters, no problems were encountered with standing wave patterns between the block walls and the space may be regarded as infinite.

With the type of transducer used, the system is essentially operating with longitudinal or 'P' waves. More precisely the receiver responds to particle velocity components normal to its surface and can respond to the vertical component of shear waves, but the modes can easily be separated when short transmissions are used and in the processing only 'P' waves were considered.

III. DATA ACQUISITION APPARATUS

Because of the microsecond time scale of the signals being received, the system was constructed to exploit the geometrical stability of the imaging environment over a period of time and permit the use of multiple transmissions each exactly repeating the elastic wave field. Thus, the incoming signals on any receiving element were recorded in a sequence of small sections totaling the required length of signal, one section being acquired on each transmission. This is indicated in Figure 4. The principle is similar to that used in sampling oscilloscopes.

In a similar manner the records from successive receiver sites were obtained by contacting the element and repeating the sequence of transmissions. Initially this element contact was made by manual movement of an electrical probe to the elements selected and indicated by the scanning control. Later developments used an automatically-positioned mechanical stepping mechanism.

The effective sampling rate for the system was dictated by the spectrum indicated in Figure 2(c) to a figure of 24 MHz. In processing the data it was convenient to have a

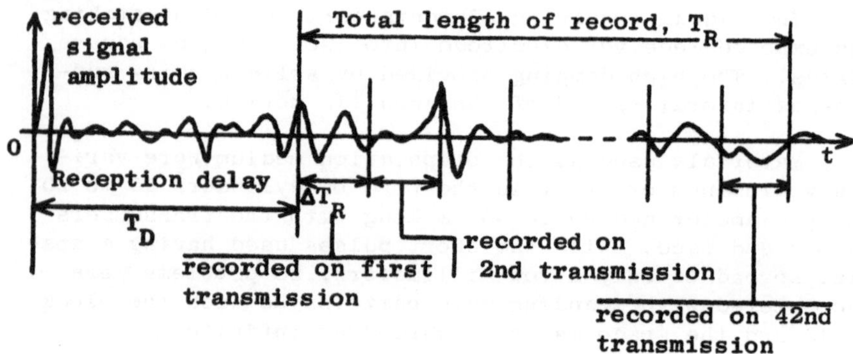

Figure 4: Recording sequence of received signals.

number of samples from each element close to a binary power and hence a total of 252 were taken, giving a figure of 10.6 μs for T_R (Figure 4). This time must accommodate the difference in arrival instants of the same event at the most widely separated elements of the array and the time differences of reflections from the most widely separated sites within the volume under examination, so that all reflections from this volume appear within T_R. With the receiver size used, the imageable volume from any record set was approximately 6 cm × 6 cm in area parallel to the transducer face and about ±1.5 cm about a preset mean depth. The precise size is a function of the transmitter location and the mean depth of the region below the transmitters. An accurate delay, T_D in Figure 4, initiated from the transmission instant was provided to permit this mean depth to be set at any value from zero to 40 cm (in aluminum).

Referring to Figure 1, operation of the apparatus follows the sequence:

(1) An array location is selected by the aperture scanning pattern generator, PG. This unit is normally set to provide a simple square raster pattern sufficient to cover all the elements used. It can accommodate arrays up to 128 × 128 elements and may be set so as to automatically skip elements not within a circular array enclosed by the scanned square. It is also possible to randomly omit elements within the scanned area, a technique [2] which, at the expense of some small image degradation, permits considerable savings in recorded material. For the 1 mm elements on the

15 mm diameter receiving transducer a pattern accessing 148 elements was generated. Random omission was not used, this being more useful for much larger arrays.

(2) Manual or automatic indication of correct element selection increments the timing control T to trigger the transmitting generator G and excite the transducers.

(3) After a delay T_D the high speed sample and hold circuit H [4] is activated to capture a portion ΔT_R of the incoming signal. This interval contains 6 of the 252 samples.

(4) An analog to digital converter circuit A converts each of the 6 held values into digital form and holds these in a temporary store. In the apparatus, quantization was done to 2 bits or 4 levels, set to indicate sample values between zero and about one third of the expected peak values, and those above this level, in both positive and negative polarities [5].

This coarse quantization was forced by the storage limitations of paper tape recording but there is some justification in that the usual target echo to reverberation level, even in a low attenuation material such as aluminum, is such that the echo information per sample does not exceed 3 to 4 bits at most. The accuracy of the sample and hold circuit is ±5% and it would thus be possible, given a recording medium of greater storage density, to obtain meaningful records of at least 4 bits with straightforward additions to the digital sections.

(5) The six samples are coded and punched over six tracks on two successive positions of the tape (Sections E and P).

(6) On completion of this punching (taking 18 ms), an extra delay equal to ΔT_R is added to T_D and the sequence from (3) onwards repeated. The punching delay provides more than adequate time for all echoes caused by the previous transmission to decay.

(7) After a total of 42 such cycles, with successive multiples of ΔT_R added each time (i.e., on the 42nd transmission a delay of 41 ΔT_R has been added to T_D) the record from this receiver element is complete. The x and y

index values for the element position are recorded in simple binary code over 7 tracks of the tape and the aperture scanning generates the next location to be interrogated. The whole cycle is then repeated, until all receivers have been contacted.

It will be seen that this is a relatively slow process. Even without allowance for element selection, interrogation of each receiver takes 0.76 seconds and the 148 element array some 115 seconds. With mechanical indexing to the element position this time increases to some 10 to 15 minutes. However, no difficulty was encountered in maintaining a stable acoustic path over these times without any special temperature stabilization. Measurements on an aluminum sample showed a change in propagation time of less than 2×10^{-4} per °C or about 5 ns in a typical 30 μs path.

Electrical timing stability is ensured by deriving all critical delays from a count of crystal oscillator cycles. With low frequency seismic signals much of this serial data acquisition can be eliminated in favor of multiple active receiver elements each connected to a separate magnetic tape recorder track. It may still be some advantage to digitize the signal and this would permit multiplexing of several records onto each track to the point where a single transmission may suffice. This is obviously of importance where the excitation cannot be exactly repeated (e.g., an explosion). The overall complexity may approach that of the present system but the required operating speeds are much less.

IV. SIGNAL PROCESSING

The end product of the recording process is a length of tape containing 252, 2 bit samples of the incoming signal from each receiver element with position information identifying each of the (in the example) 148 locations. The computer program may proceed on this information alone but the end result will need scaling to the particular dimensions and frequency range involved. Thus it is usual to provide values for the element spacing, the effective sampling frequency, and the 'P' wave propagation velocity of the material. Further, one needs to know approximately the mean depth of the examined volume so that sensible values may be assigned for the field reconstruction location.

It is not necessary that the receiver elements be on a rectangular grid. The two 7 bit position records permit unique identification of approximately 16,000 locations if regarded as a single number and one may provide a table giving the actual receiver coordinates corresponding to each number. If this is done, some extra calculation is needed as the basic algorithm requires signals from a regularly spaced grid pattern and one must first interpolate the necessary values.

It is also possible to provide a subroutine to correct for transducer response provided one initially characterizes this by some suitable test measurement. In the present program neither of these options were incorporated.

The operation of reconstructing the acoustic field (and hence perturbations caused by discontinuities) from the aperture record may be regarded in two ways corresponding to regarding the total field as composed of superpositions of plane or spherical waves, respectively.

In analytical terms these produce identical results, the distinction lies in computational efficiency, with plane wave relationships having the advantage of being representable as Fourier transforms. Taking this approach we consider the field at any location (x,y,z) and any time t as a superposition of 'P' waves having a particle displacement vector $\underset{\sim}{u}$ described by;

$$\underset{\sim}{u} = (l,m,n)u_o \exp\{-jk(lx + my + nz - ct)\} \quad (1)$$

l, m and n are direction cosines defining the propagation direction under the constraint $l^2 + m^2 + n^2 = 1$, c the P wave velocity and k the wave number $(j = \sqrt{-1})$. The complex factor u_o contains the displacement amplitude and phase at some space-time origin. Given c, waves are completely specified by quoting u_o as a function of two cosines, say l, m, and k; thus, $u_o(l,m,k)$.

The total field $\underset{\sim}{U}(x,y,z,t)$ is found by integrating over the u_o spectrum, i.e.,

$$\underset{\sim}{U}(x,y,z,t) = \iiint_{-\infty}^{\infty} \underset{\sim}{u}\, dl.dm.dk \quad (2)$$

and in particular if we define the plane containing the receiving device as $z = 0$, the displacement normal to this plane, U_n, is;

$$U_n(x,y,0,t) = \iiint_{-\infty}^{\infty} [u_o(l,m,k)(1 - l^2 - m^2)^{1/2}].$$

$$.\exp\{-jk(lx + my - ct)\}\, dl.dm.dk \qquad (3)$$

This function (or some portion of it) is the data available to the reconstruction process. The important feature is that (3) is in the form of a Fourier transform of the square bracketed term for which we write $u_o^1(l,m,k)$. Thus as the first step in reconstruction, apply an inverse transformation to U_n, giving;

$$u_o^1(l,m,k) = \frac{ck^2}{8\pi^3} \iiint_{-\infty}^{\infty} U_n(x,y,0,t)\, \exp\{jk(lx + my - ct)\}.$$

$$.dx.dy.dt \qquad (4)$$

Subject to $l^2 + m^2 \neq 1$ (this condition excludes waves near glancing incidence to the measurement plane), u_o may be extracted and the complete field distribution formed using equations (1) and (2). Failing this, one can at least get the normal or z directed component for;

$$U_n(x,y,z,t) = \iiint_{-\infty}^{\infty} u_o^1(l,m,k)\, \exp\{-jkz(1 - l^2 - m^2)^{1/2}\}.$$

$$.\exp\{-jk(lx + my - ct)\}\, dl.dm.dk \qquad (5)$$

Inspection of (5) shows that if we take reconstructions in planes of z = constant (i.e., parallel to the measurement surface $z = 0$), then the operation required is again that of Fourier transformation.

The field reconstruction procedure thus consists of:

(1) Measure the spatio-temporal field variations over some plane.

(2) Fourier transform this three dimensional function to derive a spectrum describing the field as a distribution of monochromatic plane waves present in such relative magnitudes, phases and propagation directions at each frequency as to provide the observed field.

(3) Multiply this spectrum by a 'propagation function' $(\exp\{-jkz(1 - l^2 - m^2)^{1/2}\})$ representing the effect of propagation between $z = 0$ and some other parallel plane.

(4) Fourier transform again to reconstruct the field over this plane.

A single application of step (2) can provide for any number of such plane slices by repeating steps (3) and (4) with altered z values.

Material attenuation may be accounted for by inclusion of a real exponential term in the propagation function.

While the above formulation is in terms of particle displacement, differentiation of all displacement terms with respect to time will accommodate the more usual velocity records without altering the essential form of the equations.

This procedure involving spectrum manipulation on the measured values, and final transformation, may be replaced by a direct convolution of the measurements with the spatial 'impulse' or point source response of the space between the measurement and reconstruction regions. This procedure corresponds to regarding the field as composed of a superposition of spherical waves converging to 'sinks' on the transducers and has the advantage of not requiring the restriction to parallel planes. However, if the receivers are arrayed on a regular grid over the $z = 0$ plane and reconstruction in z = constant slices suffices, then the above relationships convert directly to a discrete form immediately manipulatable by 'fast' Fourier transform (FFT) algorithms with a large saving in computing time.

It is necessary to have the basic spacing of the sampling elements sufficiently fine to accommodate arriving waves up to the maximum angle to the normal expected. With the 1 mm element spacing used, arrival angles, $\theta = \sin^{-1} (l^2 + m^2)^{1/2}$ greater than 18° for 10 MHz components cannot be unambiguously resolved. However, this increases to 37° for 5 MHz components where most of the energy is concentrated and the transmitter fields have restricted angular spreads, reducing the energy at large θ values.

That part of the infinite distribution U_n received by the system sets the resolution at any frequency in accordance with the usual formulae relating resolution to aperture size for a focussed system. With the 14 mm aperture at a wavelength λ of 1.2 mm (5 MHz in steel) the angular resolution (1.22 λ/A) is 6°.

The size of computer store available limits the size of record and thus the aperture size and frequency bandwidth that can be used. In the work here it was convenient to implement a discrete form of the equations in two parts, first computing the frequency spectrum of the 10.6 μs T_R record from each receiver element and then for each frequency component doing a two dimensional calculation for the field over some z = constant plane. The total energy within a defined bandwidth was then aggregated for the final images. Processing time on a FORTRAN programmed IBM 360/44 was 16 seconds for each frequency component.

The program as written does direct computation for receiver arrays of up to 32 by 32 elements but techniques are available for piecewise computation of any size [6].

Many aspects of the system theory have been much more fully considered in [7].

V. IMAGE CHARACTERISTICS

Apart from the basic limitation of system resolution, two further effects greatly affect the degree of correlation between calculated field perturbations and the geometry of viewed objects. Firstly, information regarding the objects is always a product of the transmitter field at the reflecting surfaces and secondly these surfaces are often shaped so that specular reflections dominate. This latter effect is accentuated when the recording arrangement suffers from a limited dynamic range or when random scattering in the bulk material gives a background return from the whole interrogated volume. Both of these features exist in the example shown later.

Mirror-like reflections give rise to images containing accurate information only about those portions of the object so orientated as to deflect energy directly into the receiving aperture. Further, the angular spectrum (u_o) available from the aperture measurements is truncated at the angle subtended by the visible portion of the objects and this results in large ripples superimposed on the reproduction.

Figure 5(b) illustrates a theoretical calculation showing these points. The geometry taken was that of a flat reflector 40 mm away from a 40 mm square aperture with the insonification consisting of a single plane wave of $k = 5\ mm^{-1}$ incident at an angle of 20° to the surface. Figure 5(a) shows this in cross section through the $y = 0$ plane. The approximately correct portion of the image is confined to an extent matching the aperture width and is offset from the z axis by 15 mm, corresponding to the 20° arrival angles. (The small ripples are the normal sidelobe response of a focussed aperture.)

Calculations for different k values show fluctuations in the large ripples such that a superposition of images from greater than an octave band would result in a greatly smoothed response. Similarly broadening the angular spectrum of the insonification gives a reproduction of increasing portions of the complete reflector. These points are stated to emphasize the importance of employing the maximum bandwidth, in the most general sense, of insonification possible, and were a main reason for working toward a system capable of recording such fields. Within the constraint of applying transducers to only one side of the volume examined, it is however difficult to approach the spatially diffuse insonification desirable for image fidelity.

Figures 6(a), (b) and (c) show the arrangement and field patterns obtained from two transmitters excited firstly with a long burst of a 5 MHz wave and secondly with a very short pulse. It will be seen that the increased bandwidth in the second case causes a filling in of the

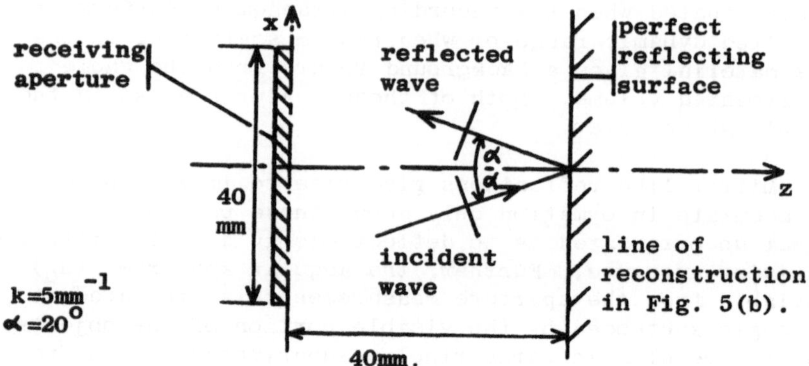

Figure 5(a): Reflector and receiver geometry for calculated specular reflection example.

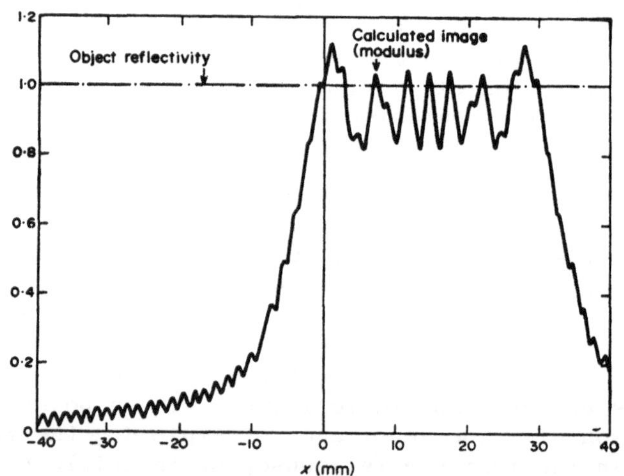

Figure 5(b): Reconstructed field section in y=0 plane along the line z=40 mm.

smaller nulls but interference in the region between the sources remains along with the usual humped distribution. Addition of a further transmitter midway between the original two fills in the central region but creates two null regions between the sources instead of one.

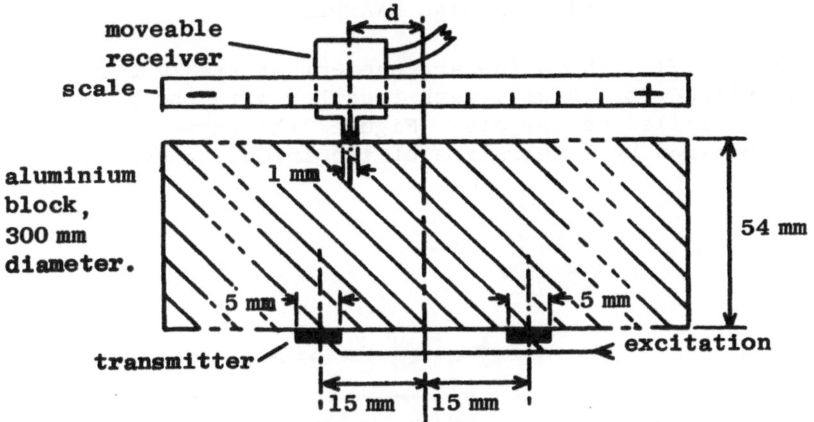

Figure 6(a): Cross-section through experimental set-up.

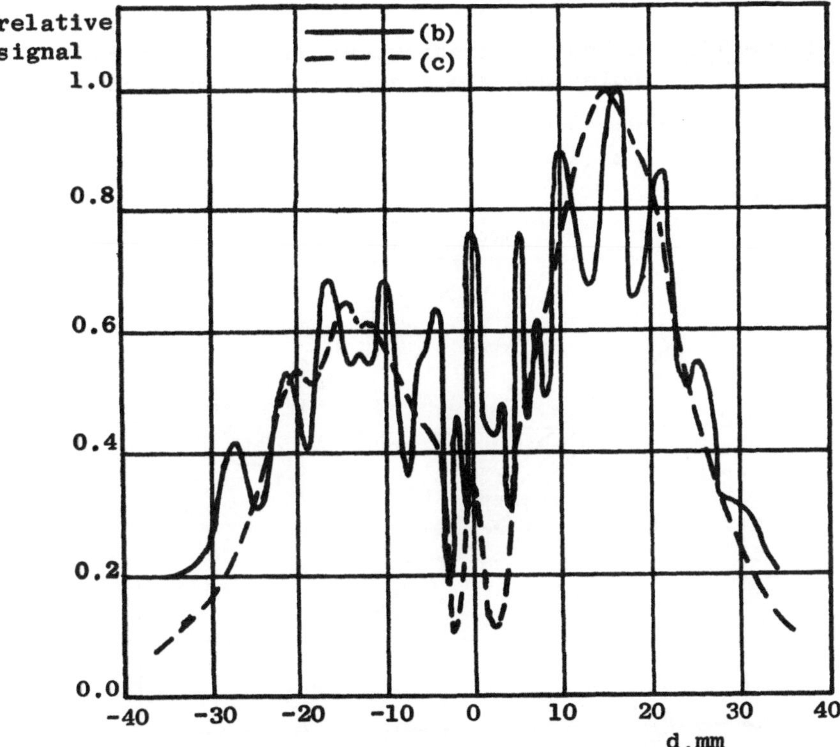

Figure 6(b),(c): Signal level in a plane 54 mm from transmitters. Excitation; (b) 4μsec burst of 5MHz; (c) 0.1μsec pulse.

VI. IMAGING EXAMPLE

Figure 7(a) shows the arrangement of transducers on a steel cylinder of 15 cm length together with the position of holes drilled as targets. Figure 7(b) shows a resultant image obtained by integration of the intensity over a 1 MHz

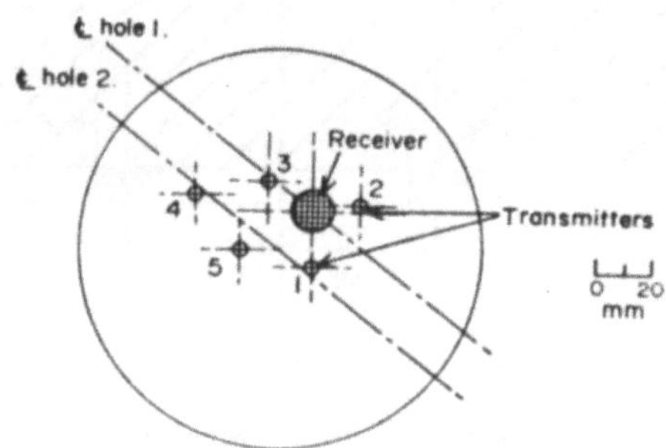

Figure 7(a): Layout of transducers on steel cylinder. Holes are 3 mm diameter and 76mm.deep.

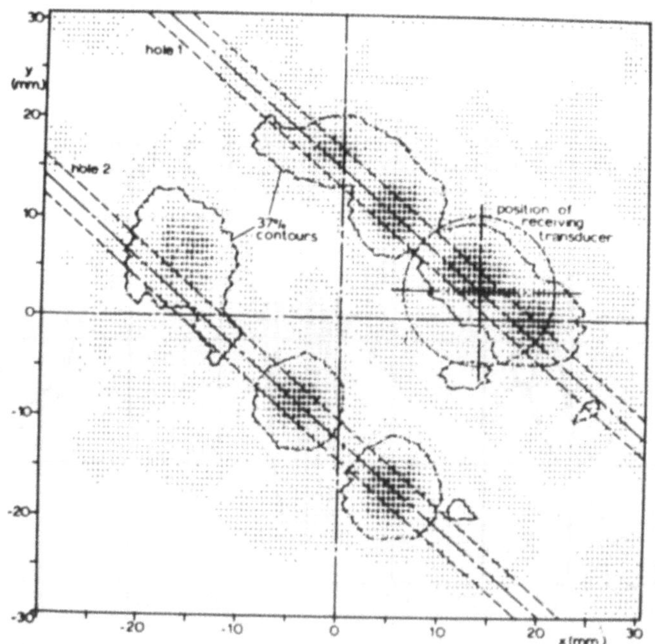

Figure 7(b): Image of two holes in steel cylinder. Darker regions indicate stronger returns.

band centered on 5 MHz. This was calculated for the known depth of the holes but with the relatively narrow bandwidth one may shift the focal plane over a range of depths without altering the essential features. All the effects discussed above are apparent in this picture. Specular scattering predominates with the image consisting of a number of 'spots' along each hole representing reflection from just those portions of each target favorably located relative to a transmitter and the receiver. Measured to the 1/e or 37% contour implied in the resolution formula used before, the diameter of the isolated spots is 8 mm, agreeing excellently with the 6° angular resolution figure found earlier for this diameter receiver.

A considerable background intensity is apparent in the picture from grain scatter in the steel and as shown by Figure 8 the echo signals on any individual receiver element are nearly lost in this 'noise'. Only the coherent form of signal recording and processing used permits extraction of the image in useable form.

A variety of other situations were tested in aluminum where the attenuation and scatter are much lower. These gave much stronger target reflections relative to material scatter but the domination of specular returns remained. Attempts to use brass and other large grained materials failed as these metals transmit virtually nothing above a few hundred kilohertz over any significant distance.

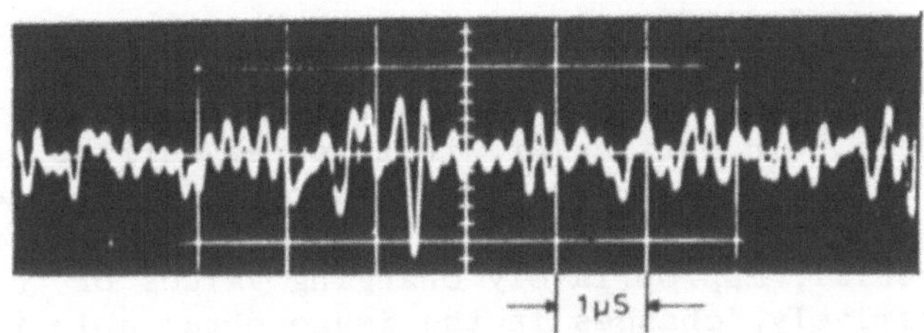

Figure 8: Typical received signal from steel block.

VII. CONCLUSIONS

A larger receiving array would improve image quality. The present 148 element array is very small relative to the signal wavelength and corresponds to an optical system with an aperture of only 7 microns. Increase of aperture size to the full 32 × 32 elements accommodated in the computer program involves no extra spatial processing since the present arrangement merely inserts zero values in the unused locations, but the time required by the time-frequency transforms is directly proportional to the number of active transducers. A full sized array renders mechanical scanning impractical from the time viewpoint and present aims are toward an electronic switching matrix in integrated circuit form, mounted directly on the piezoelectric array.

Nowhere in the formulation (Section IV) is there any reference to the original field source. Practical operation of the apparatus does require some knowledge of transmitter location so that an appropriate delay may be inserted before recording a portion of the returning echoes, but this is purely a function of the finite recording period provided. In principle, any set of signals on the receiving array that are identifiable as originating from a single source or time correlated distribution of sources, can be used provided the recording period embraces the spread of events discussed in Section III. The critical parameters are timing of relative arrival times over the receiver array and accurate location of the elements. Permissible tolerances on these are about ±1/4 of the period and wavelength, respectively, for the highest operating frequency.

Although the image shown in Figure 7 was obtained from objects in a relatively homogenous medium it will be obvious how more complex propagation paths can be simulated by building up layers of different materials. From the single set of received signals, equation (5) provides a relation for computing a corresponding set of normal components over any other surface. In particular, if this new surface is an interface between differing materials one may use the calculated field at the interface to compute the field in the second material, appropriately changing values of c and k. Alternatively, changes in the image obtainable from a given processing algorithm with discontinuities in the object-receiver path can be practically investigated.

Overall the main feature of this simulation apparatus is the relative ease with which one may vary geometrical parameters, material properties and operating frequency, or locations and shapes of objects, while still retaining the correct propagation conditions of full sized experiments. Typically, the scale factor might be 10^5, 5 MHz corresponding to 50 Hz, and 1 mm to 100 m on the ground.

REFERENCES

1. M. G. Maginness and L. Kay, "Signal Processing for Acoustic Imaging Systems," Paper K-5-3, 6th Intl. Congress on Acoustics, Tokyo, 1968.

2. G. B. Cook and D. A. H. Johnson, "Pseudo Random Selection of Elements in a Multi-Element Array," Radio and Electron. Eng., Vol. 38, pp. 82-84, Aug 1969.

3. M. G. Maginness and L. Kay, "Ultrasonic Imaging in Solids," Radio and Electron. Eng., Vol. 41, pp. 91-93, Feb 1971.

4. P. M. Cashin and D. A. H. Johnson, "A Novel Method of Analyzing Singly Occurring Pulses with Nanosecond Resolution," Radio and Electron. Eng., Vol. 38, pp. 1-4, Jul 1969.

5. H. S. Heaps and P. W. Willcock, "The Use of Quantizing Techniques in Real Time Fourier Analysis," Radio and Electron. Eng., Vol. 29, pp. 143-148, Mar 1965.

6. B. M. Gold and C. M. Rader, "Digital Processing of Signals," New York, McGraw-Hill, 1969.

7. M. G. Maginness, "The Reconstruction of Elastic Wave Fields from Measurements over a Transducer Array," J. Sound and Vib., Vol. 20, pp. 219-240, Jan 1972.

ACKNOWLEDGEMENTS

The authors wish to thank the New Zealand University Grants Committee and the National Research Development Corporation (U.K.) for their support of this work.

SOLID-STATE ACOUSTIC IMAGE SENSOR

N. Takagi, T. Kawashima, T. Ogura and T. Yamada

Toshiba R. & D. Center, Tokyo Shibaura Electric Co., Ltd.
1 Komukai Toshiba-cho, Saiwai-ku, Kawasaki
210 JAPAN

INTRODUCTION

Some ultrasonic imaging systems with lens system for underwater use have been studied[1,2,3]. An ultrasonic imaging system utilizing a piezo-electric transducer array converter has been developed[4].

The converter contains 80 x 80 piezo-electric material elements arranged at a 2 mm pitch. Each element is equipped with a diode and a register for switching. X and Y common electrodes which combine the diodes and resistors are scanned by a scanning circuit. An ultrasonic image focused on the converter is converted into electric signals and modified by a signal processing circuit made up of a filter circuit and a sampling circuit. Then the signals are displayed on a cathode-ray tube at the rate of 30 frames/second as a visible image. The frequency of the ultrasonic waves used here is 1 MHz and the sensitivity of this system is about 5×10^{-7} W/cm^2.

IMAGE SYSTEM PRINCIPLES

Piezo-Electric Element Array

Figure 1 shows a schematic diagram of the piezo-electric element array. S_{11}∿Smn denote piezo-electric elements, D_{11}∿Dmn denote diodes, R_{11}∿Rmn denote bias resistors, X_1∿Xm and Y_1∿Yn denote X and Y common electrodes, respectively. S1 and S2 are X and Y scanning circuits, which generate the scanning signals. The signal read-out method is as follows: In order to read out the element S_{11} output voltage, a positive voltage is applied to the base of transistor Q_1 from scanning circuit S1, which results in Q_1 becoming conductive. It may be thought that a conductive Q_1 state would allow the signal currents from all

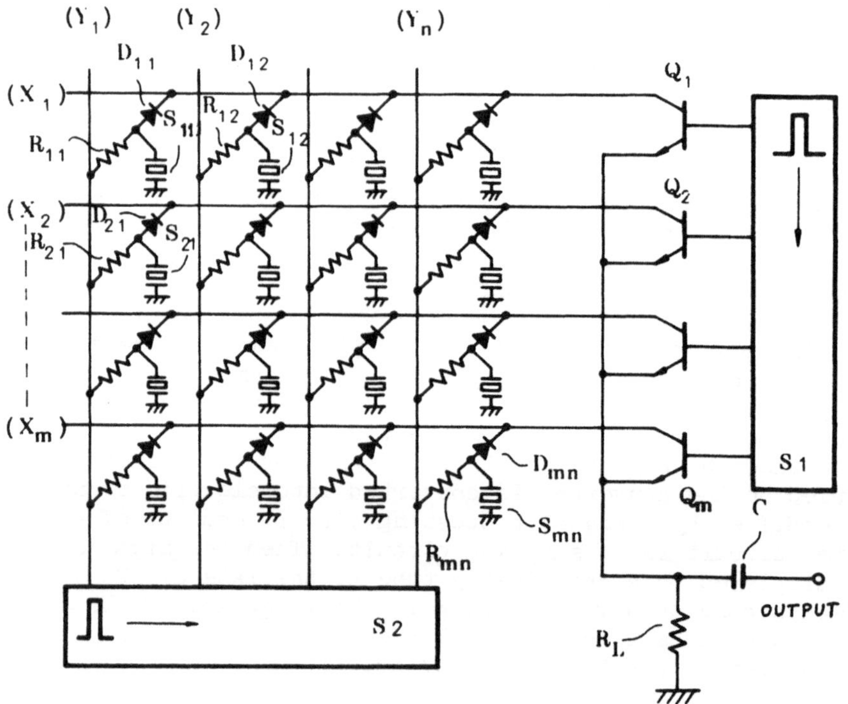

Figure 1. Schematic diagram of the piezo-electric element array

piezo-electric elements connected to electrode X_1 to appear at the output terminal through the diodes. However, the output voltage of each element irradiated with ultrasonic waves is in the order of tens of mV, which is much smaller than the threshold voltage required to induce forward current through an ordinary silicon diode. Therefore, merely the fact that, transistor Q_1 is in a conductive state is not sufficient to allow the signal from any piezo-electric elements to appear at the output terminal. A positive voltage, high enough to induce forward current through the diodes, is applied to electrode Y_1 from scanning circuit S2. This voltage induces a forward current through diode D_{11} and reduces the resistance of the diode to a very low value. Therefore, the output voltage from S_{11} appears at load resistor R_L and the signal is sent to the signal processing circuit. Positive voltage is also applied to all the diodes connected to electrode Y_1, as well as to D_{11}, but, because of the high resistance of transistors $Q_2 \sim Q_m$, no diode except for D_{11} becomes conductive. Therefor,no output voltages from any piezo-electric elements except S_{11} appear at the output terminal of load resistance R_L. In this way, by sequentially scanning X and Y electrodes with scanning circuits S1 and S2, signals from all the elements can be read out sequentially.

Electric System

Figure 2 shows a block diagram of the electric system. The output signals from the piezoelectric elements are applied to a high pass filter, where only the image signals of 1 MHz will be passed, and DC current and low frequency variations, mentioned below, will be suppressed. The low frequency variations would occur at a frequency corresponding to the read-out speed, due to the lack of uniformity of scanning pulse height, bias resistors, diode forward resistance and the resistance of transistors. This low frequency variations are not so small that they can be neglected against signals, so they have to be suppressed by a filter.

The sampling circuit is used for removing the spike noise which appears on the front and rear edges of the signal of each element and for sampling out only the desired signal components.

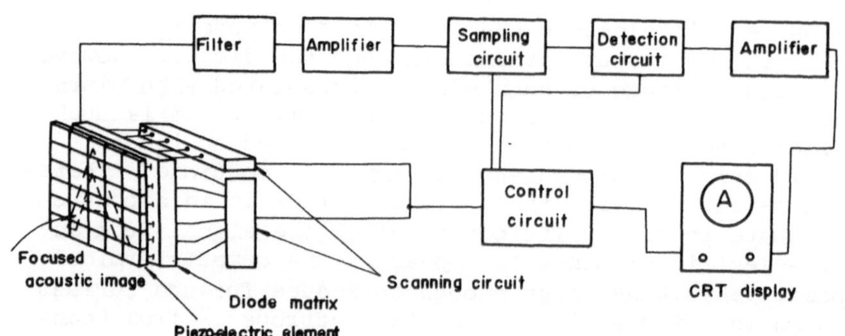

Figure 2. Block diagram of electronic arrangements

There are two causes for spike noise. One is the diode capacitance and base-emitter capacitance of a transistor, which cause some narrow spike pulses to appear at the front and rear edges of the read-out signal of an element. The other cause is the transient of the above mentioned low frequency variations. Such a transient contains higher frequency components than about 1 MHz, which cannot be suppressed by the filter circuit.

Since the image signals are in the form of 1 MHz carrier the ultrasonic wave frequency amplitude modulation, the signals are sent to a detection circuit and the envelopes of the image signals are detected to obtaine the video signals. Video signals of a few volts are then amplified to a few tens of volts, which are applied to the CRT cathode to modulate the CRT spot brightness.

The control circuit generates saw-tooth waveform voltages to be applied to the X and Y CRT deflection plates in synchronization with the horizontal and vertical scanning circuits. In this way, an ultrasonic image focused on the piezo-electric converter can be entirely reconstructed on a CRT as a visible image.

Output Voltage of Piezo-Electric Element Array

The output of the array is determined by the output

voltage from the piezo-electric element and the electric circuit constant connected to it. The fundamental piezo-electric vibrator equations are as follows:

$$\dot{F} = -A\dot{V} + (\dot{Z}o + \dot{Z}_1)\dot{v} \tag{1}$$

$$\dot{I} = (\dot{Y}o + \dot{Y}_1)\dot{V} + A\dot{v} \tag{2}$$

$\dot{F}$: External force
$\dot{V}$: Alternating voltage
$\dot{Z}_1$: Mechanical impedance of the vibrator
$\dot{v}$: Vibrational velocity
$\dot{Y}_1$: Electrical damped admittance
A : Force factor
$\dot{Z}o$: Mechanical impedance of the force
$\dot{I}$: Alternating current
$\dot{Y}o$: Electrical internal (load) admittance

To obtain the output voltage of the piezo-electric vibrator let $\dot{I} = 0$ in Eqs.(1) and (2), and $\dot{v}$ is obtained, as follows.

$$\dot{v} = -\frac{A^2/\dot{Z}_1 \cdot \dot{F}/A}{A^2/\dot{Z}_1 + (\dot{Y}_1 + \dot{Y}o)} \tag{3}$$

The damped admittance is shown by the following equation.

$$\dot{Y}m = \frac{A^2}{\dot{Z}_1} \tag{4}$$

Free admittance is shown by the following equation.

$$\dot{Y}f = \dot{Y}_1 + \dot{Y}m \tag{5}$$

From Eqs.(4) and (5), Eq.(3) is :

$$\dot{v} = -\frac{\dot{Y}m}{\dot{Y}f + \dot{Y}o} \cdot \frac{\dot{F}}{A} \tag{6}$$

where P: pressure applied on the resonator per unit area and S: resonator area

$$\dot{V} = -\frac{\dot{Y}m}{\dot{Y}f + \dot{Y}o} \cdot \frac{PS}{A} \tag{7}$$

In a thick resonance vibrator, it is expressed as :

$$A = \frac{2S}{t} Y_T^E \, \varepsilon_T^T \, g_T \tag{8}$$

wherein, t: resonator thickness
Y_T^E: Young's modulus in the direction of thickness when polarization is 0.
ε_T^T: dielectric constant in the direction of thickness when stress is 0.
g_T: electric distortion constant in the direction of thickness.
Substitute Eq.(8) into Eq.(7).

$$\dot{V} = -\frac{\dot{Y}m}{\dot{Y}f + \dot{Y}o} \; \frac{t}{2Y_T^E \, \varepsilon_T^T \, g_T} \; P \tag{9}$$

Let ρ be the density of the resonator and C be the sound velocity in the resonator,then,between the R.M.S. value of the sound pressure $P(N/m^2)$ and acoustic power density $I(W/m^2)$, there is a relation of

$$I = \frac{P^2}{\rho C} \tag{10}$$

Hence, Eq.(9) becomes

$$\dot{V} = -\frac{\dot{Y}m}{\dot{Y}f + \dot{Y}o} \; \frac{t}{2Y_T^E \, \varepsilon_T^T \, g_T} \sqrt{\rho C I} \tag{11}$$

Values calculated by putting in actual values and experimentally measured values are referred to in the "experimental results and discussion" section.

Sensitivity and S/N Ratio of Piezo-Electric Element Array

The S/N ratio of the piezo-electric element array is determined with the following equation.

$$S/N = S/(Ne + Nm + Na) \tag{12}$$

where S is the piezo-electric element read-out signal. Ne is an electrical cross-talk noise that is a leakage signal due to the electric current passing through the diode during it's off state, as will be discussed later. Nm is a mechanical cross-talk noise which is caused by the mechanical coupling of the bibrator with circumferential elements. Na is pre-amplifier noise. In an usual detection system, S/N ratio is approximately S/Na, so the larger the S, the better the S/N ratio. However, in this equipment, as the signal becomes larger, the cross-talk noises also become larger (Ne, Nm). Therefore, S/N ratio approaches a constant value.

Electrical Cross-Talk

In this equipment, since many elements are connected to a common electrode, the summation of slight leakage electric current from each element lowers the S/N ratio as a whole. The S/N ratio with leakage electric current taken into consideration is discussed below. The equivalent converter circuit of the piezo-electric element array is shown in Fig.3. In this figure, $\dot{Z}r$ denotes the internal impedance of the resonant piezo-electric element, $\dot{Z}x$ direct current bias resistor, $\dot{Z}_L$ load resistor at the output terminal, $\dot{Z}c$ diode impedance under threshold level, $\dot{Z}o$ internal impedance of scanning circuit S2 in "ON" state, $\dot{Z}o'$ internal impedance of the scanning circuit in the "OFF" state, and $\dot{V}o$ open-circuit voltage of the piezo-element. As is understood from this equivalent converter circuit, the first factor causes a signal to appear at the output terminal from the piezo-electric elements which are not scanned can be considered as the electric current passing through load resistor $\dot{Z}_L$ through $\dot{Z}c$, because $\dot{Z}c$ has a limited resistance. The second factor is that the internal impedance of the scanning circuit S2 in an "ON" state is not zero and, therefore, the current flows through $\dot{Z}o$ through $\dot{Z}x$ and $\dot{Z}r$ due to the output voltage of the piezo-

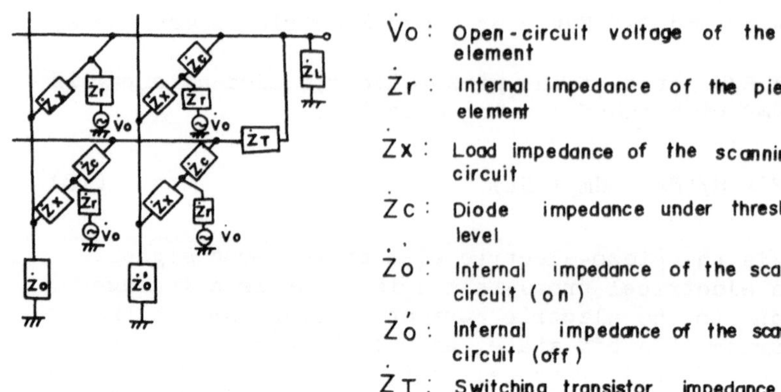

Figure 3. Equivalent converter circuit

-electric element, which produces an electric potential at the terminal of Zo. This electric potential is superimposed on the scanned piezo-electric element signal and appears at the output terminal. If the number of piezo-electric elements connected to the common electrode is small, leakage voltage due to such causes is small in comparison with the real signal. However, if the number of elements is large, this value cannot be ignored. If the S/N ratio is decreased by this leakage voltage, the contrast of the image decreases and shadows appear in the image. The limited impedance of $\dot{Z}c$ is mainly because of the diode capacitance. This is about 1 to 2 pF and corresponds to an impedance of 80 to 160 kΩ at 1 MHz. $\dot{Z}o$ is explained because of the semi-conductors used in the scanning circuit not being completely conductive. Actually, its resistance is 5 to 10Ω.

In order to design equipment having a suitable resolution and S/N ratio, the relation between the number of piezo-electric elements and S/N ratio is obtained. First, from the Fig.3, real signal voltage $\dot{V}s$ is obtained by the following equation.

$$\dot{V}s = \frac{\dot{V}o \cdot \dot{Z}x \cdot \dot{Z}_L'}{\dot{Z}r(\dot{Z}x + \dot{Z}_L) + \dot{Z}x\ \dot{Z}_L'} \qquad (13)$$

Let the leakage voltage due to the limited impedance of $\dot{Z}c$ be $\dot{V}nx$, and $\dot{V}nx$ is obtained as follows.

$$\dot{V}nx = Mx \frac{\dot{Z}x}{\dot{Z}r + \dot{Z}x} \cdot \frac{(\dot{Z}_L /\!/ \dot{Z}r /\!/ \dot{Z}x)}{\dot{Z}c + (\dot{Z}_L' /\!/ \dot{Z}r /\!/ \dot{Z}x)} \dot{V}o \qquad (14)$$

Mx stand for the number of piezo-electric elements connected to one common X electrode. In order that the above equation might be meaningful, it is assumed that all piezo-electric elements connected to a common electrode are irrdiated with ultrasonic waves at constant strength and in the same phase. Then, the S/N ratio is obtained on the assumption of $|\dot{Z}c| \gg |\dot{Z}_L'|$.

$$\frac{\dot{V}s}{\dot{V}nx} = \frac{1}{Mx} \frac{\dot{Z}c(\dot{Z}r + \dot{Z}x)}{\dot{Z}x \cdot \dot{Z}r} \qquad (15)$$

The S/N ratio with the leakage voltage due to $\dot{Z}o$ is obtained in a similar manner. Let $\dot{V}ny$ be the amount of leakage voltage due to $\dot{Z}o$, and $\dot{V}ny$ is obtained as follows.

$$\dot{V}ny = \frac{1}{My} \cdot \frac{\dot{V}o \cdot \dot{Z}o}{\dot{Z}x + \dot{Z}r + My\dot{Z}o + Zo} \cdot \frac{\dot{Z}r \cdot \dot{Z}_L'}{\dot{Z}x\dot{Z}r + \dot{Z}x\dot{Z}_L' + \dot{Z}r\dot{Z}_L'} \qquad (16)$$

My stands for the number of piezo-electric elements connected to a common Y electrode. Also, in this case, the same assumption is made as in the case of $\dot{V}nx$. $\dot{V}s/\dot{V}ny$ is obtained as follows.

$$\frac{\dot{V}s}{\dot{V}ny} = \frac{\dot{Z}x}{\dot{Z}r} \left(\frac{\dot{Z}x + \dot{Z}r}{My\ \dot{Z}o} + 1 + \frac{1}{My} \right) \qquad (17)$$

From Eqs.(15) and (17), it is understood that both leakage voltages due to the two causes increase according to the number of piezo-electric elements connected to one common electrode. Therefore, the number of piezo-electric elements connected to one common electrode is limited by the S/N ratio requirements.

PROTOTYPE EQUIPMENT PRODUCTION

Prototype Piezo-Electric Element Array Production

Since it takes a long time and is troublesome to arrange a large number of small piezo-electric plates into an array, ditches were made on one piezo-electric plate, as shown in Fig.4, so that it may both mechanically and electrically have the same effect as the arrangement of a large number of small piezo-electric elements. The piezo-electric plate used was PZT having a 1 MHz resonant frequency. Four piezo-electric plates, whose size was 80 x 80 mm, were combined and an array of 80 x 80 elements was made up.

Circuit constants of an array are as follows. 1 kΩ and 1.9 kΩ resistors are used for $\dot{Z}x$ and $\dot{Z}_L$, respectively. The values of $\dot{Z}c$, $\dot{Z}o$ and $\dot{Z}r$ are, according to the

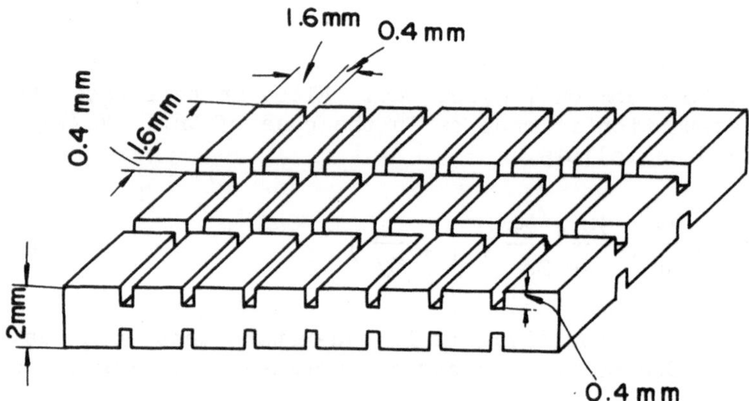

Figure 4. Appearance of the piezo-electric element

measurement, 60 kΩ for $\dot{Z}c$, 0.8Ω for $\dot{Z}o$ and 100 kΩ for $\dot{Z}r$, on the average. Substituting these values into Eqs.(15) and (17), S/N ratios with regard to electrode X and electrode Y are calculated, that is, S/Nx = 0.77 and S/Ny = 16. The S/N ratio with regard to electrode X is not satisfactory. In order to increase the value of S/Nx, either make the value of $\dot{Z}x$ smaller or make the value of Mx smaller. If the value of $\dot{Z}x$ is made smaller, the output voltage becomes smaller which causes sensitivity decreases. Therefore, it was decided to decrease the number of piezoelectric elements connected to one common electrode, that is to make the value of Mx smaller, and, in turn, to increase the number of common electrodes. If Mx = 20, S/Nx = 3 and the over-all S/N ratio becomes S/N = 2.5. Although this is not such a good value, it is considered that the actual value will be better on the average, because the worst condition is assumed where the ultrasonic wave is applied to all the element at the same strength and in the same phase. Therefore, the constant and the number of piezo-electric elements connected to the X electrode were determined based on this value for the prototype production. The whole assembly is shown in Fig.5, where diodes and resistors are connected to PZT piezo-electric elements and the lead wires of each element are wired in the form of an X and Y matrix.

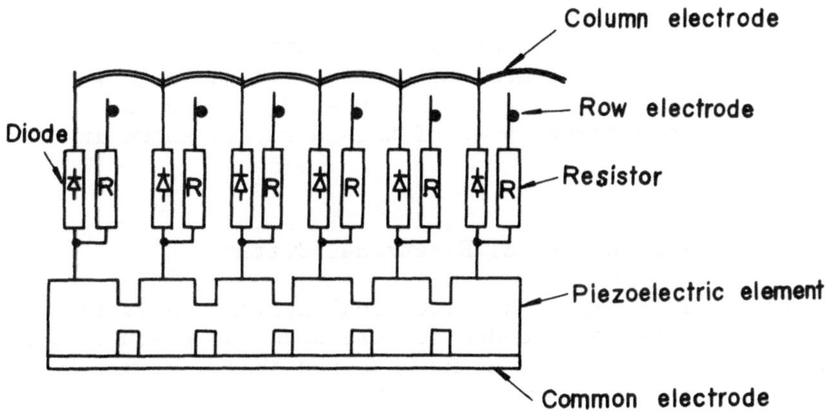

Figure 5. Appearance of assembled converter

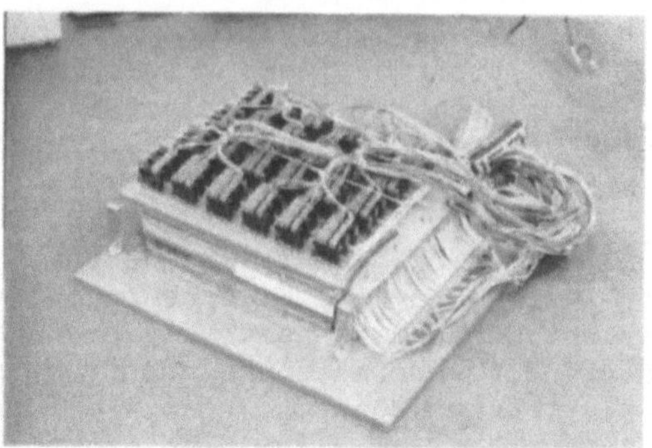

Figure 6. External view of array

Since one of the purposes for the equipment was to see the image of a moving object, the rate of frames was decided to be the same as that of a commercial television, that is, 30 frames/sec and 60 fields/sec with 2:1 interlaced scanning. In this case, let the read-out time per elements be T,

$$T = \frac{1}{f \cdot Mx \cdot My} \qquad (18)$$

where f: the rate of frames, Mx: the number of horizontal elements and My: the number of vertical elements. By substituting f = 30, Mx = 80 and My = 80, T becomes 5.2 μsec. The external view of this array is shown in Fig.6.

Manufacture of Electrical System

No problems exist the electrical system. However, since many parts were needed for the scanning circuit, ICs were used to minimize it as much as possible. Figure 7 shows actual signal waveforms. Figure 7(a) shows a waveform after passing through the filter. The spike noise are observed in front of and behind the signal. Figure 7(b) shows a sampled waveform of the signal after the spike noises are eliminated and Figure 7(c) shows the video signal waveform.

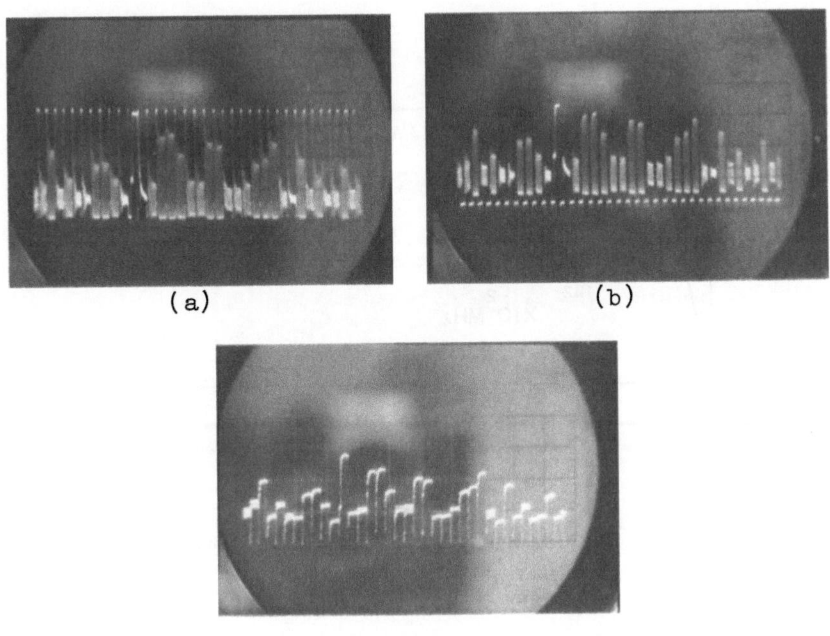

(a) (b) (c)

Figure 7. Signal waveforms: (a) waveform after passing through the filter; (b) sampled waveform of the signal after the spike noises are eliminated; (c) video signal waveform

RESULTS OF EXPERIMENTS AND INVESTIGATION

Piezo-Electric Element Characteristics

Figure 8 shows measured examples of the free admittance of the piezo-electric element alone, isolated by ditches. Regarding the resonant frequency, although the resonator has ±0.02 MHz resonant frequency centered around 1 MHz before ditched, it is noted that the resonant frequency shifts about 0.1 MHz to the higher side after ditched.

Damped capacity Cd is calculated in consideration of the sizes of the element and of the characteristic chart of the piezo-electric element used.

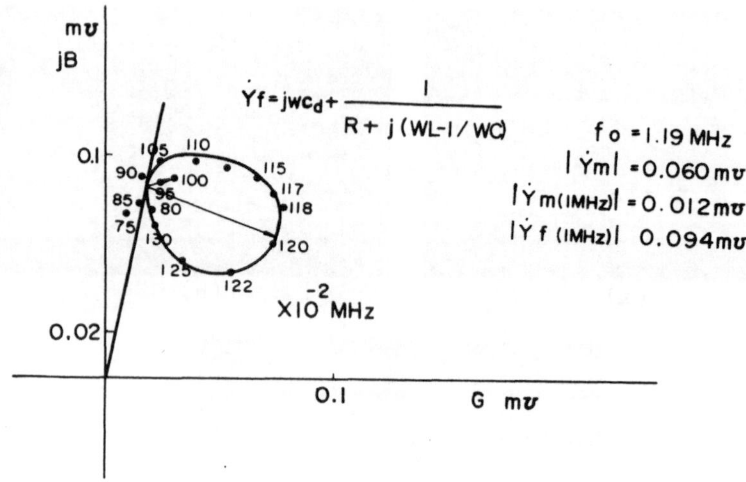

Figure 8. Free admittance characteristic of a piezo-electric element

$\varepsilon/\varepsilon o = 510$

$S = 2.53 \times 10^{-6}\ (m^2)$

$t = 2.657 \times 10^{-3}\ (m^2)$

From the above, Cd = 5.6 pF is drawn. The above internal capacitance was also measured by the following equations. From $\dot{Y}f = \dot{Y}_1 + \dot{Y}m$

$$|\dot{Y}_1| = \sqrt{|\dot{Y}f|^2 - |\dot{Y}m|^2} = \omega Cd \tag{19}$$

$$\therefore Cd = \sqrt{\frac{|Yf|^2 - |Ym|^2}{2 f}} \tag{20}$$

The average value of several measured elements was Cd = 11 pF. This deviation from the calculated value seems to be caused by the stray capacitance of measuring electrodes and by the difference between actual and nominal values of ε.

Measurement of Mechanical Cross-talk

Mechanical cross-talk is one of the important factors of determining S/N ratio of this array. Measurements were made on a cross ditched plate without switching circuits. The output voltages of an element were measured with or without masking, after all the elements were adjusted to center on 1 MHz irradiating ultrasonic waves. The mask is made of ultrasonic absorbing rubber including tungsten powder. Two sizes 4 mm square and 7 mm square, were used. The 4 mm square mask covers one element and was used to study the effects of mechanical coupling with eight elements surrounding it. Since a 7 mm square mask can cover the center element and surrounding eight elements, the effects of the second ring of elements around the central element were measured.

Using the 4 mm square mask, a S/N ratio of 5:1 was obtained. In the case of the 7 mm square mask, a S/N ratio of about 15:1 was obtained. Mechanical cross-talk is comparatively large, since a ratio of more than 30:1 was obtained in the measurement for a single 2 mm square element with 4 mm square mask.

Piezo-Electric Element Output Voltage

It is important to anticipate how much output voltage will be obtained from a piezo-electric element, in order to foresee the sensitivity and design the amplifier. The measured output voltage was compared with the calculated value of some elements of this converter. Figure 9 shows the measured value and the calculated value. The output voltage measuring methods are as follows. The 2mm square small transducer is used for sound source. Output voltage is measured by putting the transducer close to one element of an array. In order to obtain the output power of the ultrasonic wave, a microphone calibrated with a calorymeter was used. The actually measured value shown in the figure is an average of the values of fifteen points measured at random. From Eq.(11), the calculated value will be :

$$|\dot{V}| = \frac{|\dot{Y}m|}{\sqrt{|\dot{Y}f|^2 + |\dot{Y}o|^2}} \cdot \frac{t}{2Y_T^E \cdot \varepsilon_T^T \cdot g_T} \cdot \sqrt{\rho CI} \qquad (21)$$

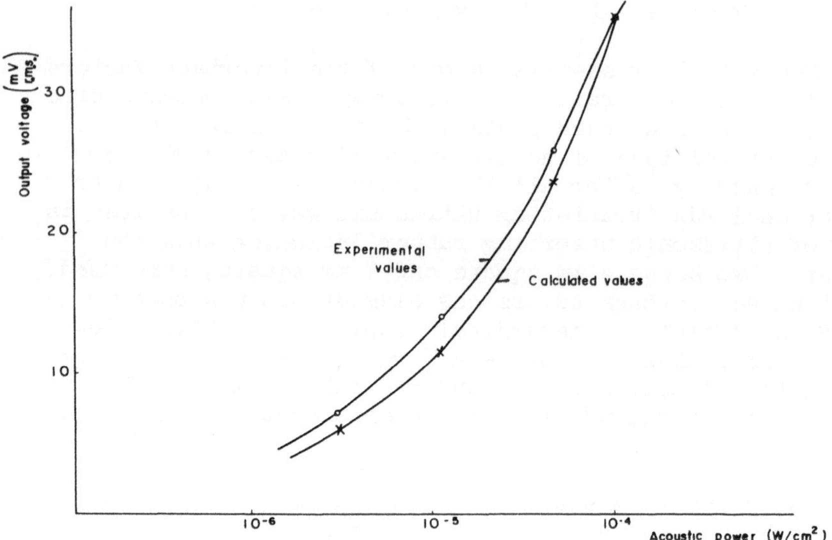

Figure 9. Converter output voltage characteristic curve vs acoustic power

The following are the average values taken from the measurement of $|\dot{Y}m|$ and $|\dot{Y}f|$.

$$|\dot{Y}m| = 10 \times 10^{-6} \mho$$

$$|\dot{Y}f| = 89 \times 10^{-6} \mho$$

Since $|\dot{Y}o|$ is a load admittance, from Rx = 1 kΩ, R_L = 1.9 kΩ

$$|\dot{Y}o| = 1.53 \times 10^{-3} \mho$$

And, the characteristic of the piezo-electric elements ;

$$\frac{t}{2Y_T^E \varepsilon_T^T g_T} = 1.33 \times 10^{-4}$$

$$\sqrt{\rho C} = 3.9 \times 10^3$$

By inserting these amounts into Eq.(21),

$$|\dot{V}| = 3.37 \times 10^{-3} \sqrt{I}$$

The output voltage was calculated, after the power density I was measured with the calibrated microphone, by substituting it into the equation.

The results are almost in agreement with those of the actually measured amount. Therefore, it was known that the output voltage could be calculated, if the characteristics of the elements are known. The reason for inconstancy of the sensitivity is variation of the output voltage. The measured output voltages at fifteen point at random, showed ±60 percent in distribution, which is considered to be caused from the inconstancy of $\dot{Y}m$, variations of Rx etc. When the ultrasonic waves are irradiated on the entire piezo-electric element, electrical and mechanical cross-talk will be superimposed onto signals, so the variations of output voltages are different in appearance.

SENSITIVITY AND S/N RATIO

In order to determine the sensitivity of this prototype equipment, the relations between the applied ultrasonic power and S/N ratio were measured. The entire piezo-electric element array was uniformly irradiated, and the ratio of output voltages for one element, with and without 4 mm square mask, was measured at various ultrasonic waves power densities.

An example of the results is shown in Fig.10. The S/N ratio increases linearly until the irradiation power reaches 10^{-7} to 5×10^{-6} W/cm^2. This is because, since the input power is small, the noise due to mechanical cross-talk or electrical cross-talk is smaller than the thermal noise of the pre-amplifier. In this range of power, the noise can be considered to be practically constant, so the S/N ratio increases as the signal increases. In the range of 5×10^{-6} to 10^{-4} W/cm^2, the S/N ratio remains almost a constant value. In this case, most of the noise is caused from mechanical cross-talk and electrical cross-talk. In more than 10^{-4} W/cm^2, the S/N ratio begins to decrease. This is because the signal

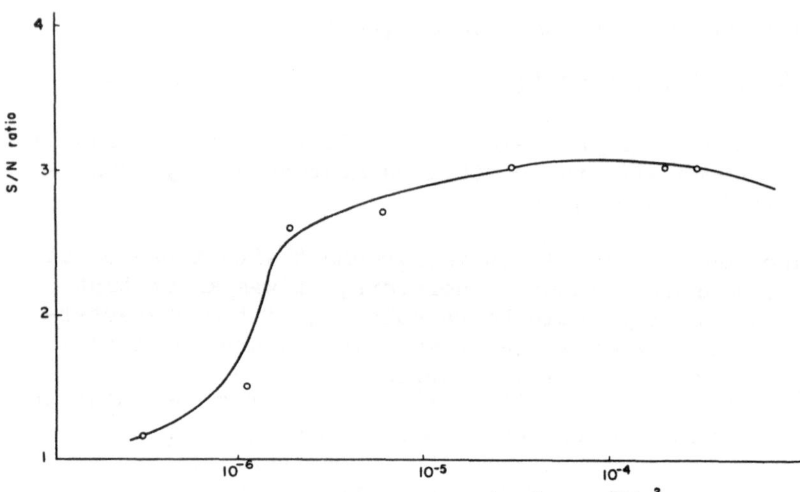

Figure 10. Characteristic curve showing S/N ratio of the converter vs acoustic power

amplified at the pre-amplifier becomes saturated as the input signal becomes larger, but the noise, being smaller than the signal, would not be saturated.

Measurements on several elements showed the variation of S/N ratio was considerably large; that is S/N = 2 to 4, at the saturation point of S/N ratio. If the mechanical cross-talk is assumed to be constant around 5, the variation of S/N ratio are considered to be attributed to electrical cross-talk. Since the entire S/N ratio is expressed by Eq.(12), the electrical cross-talk is calculated by ignoring S/Na, and substituting S/Nm = 5. In case of S/N = 4, S/Ne is 20 and in case of S/N = 2, S/Ne is 3.3. This results shows that the variation of electrical cross-talk seems considerably large.

IMAGING TEST IN WATER TANK

In order to study the imaging characteristics of the equipment, an ultrasonic image was detected using lens system shown in Fig.11. Some examples of objects and images are shown in Figs.12, 13, 14 and 15. Figure 12(a)

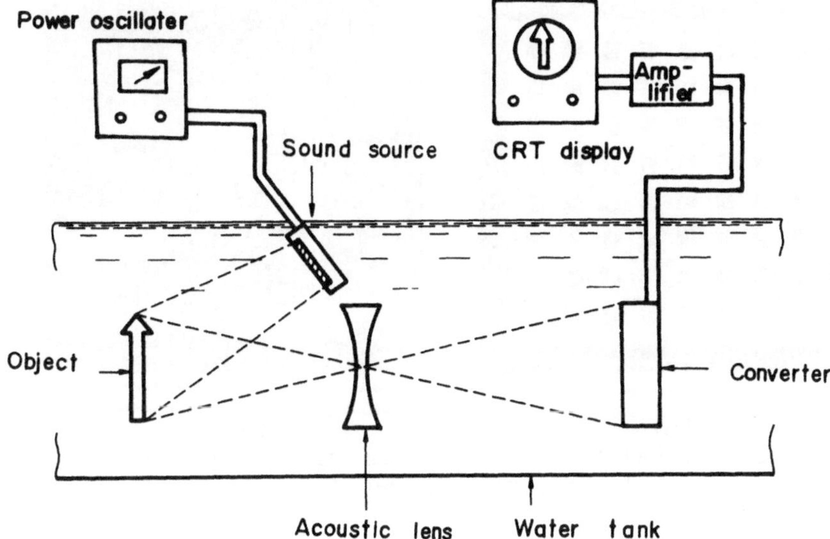

Figure 11. Reflection type acoustic imaging system

shows a 40 mm pitch aluminum lattice plate, Fig.12(b) shows a reflection image of a lattice plate. Figure 13(a) shows a 140 mm square brass resolution test plate, Fig.13(b) shows a reflection image of a resolution test plate. Figure 14(a) shows a 170 mm length Toshiba logo, Fig.14(b) shows a reflection image of the logo. Figure 15 shows a transmission image of a human hand.

The acoustic lens was made from polystyrol, with a focal length of f = 400 mm and a lens diameter of d = 280 mm. As the acoustic source, four 10 mm diameter PZT plates were used for the reflection image and one for the transmission image. The power from the acoustic source was 2 W/cm^2 on the average immediately in front of the source in water. The distance from the object to the converter was about 1.6 m when the magnification was one.

According to the theory, the resolution of the reflection image, including the lens, will be about 60 lines in the field of view. However, the experimental results showed

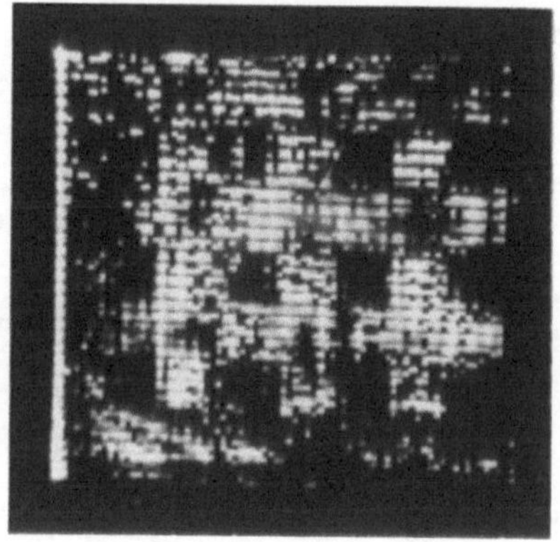

(a) (b)

Figure 12. (a) 40 mm pitch aluminum lattice plate; (b) reflection image of a lattice plate

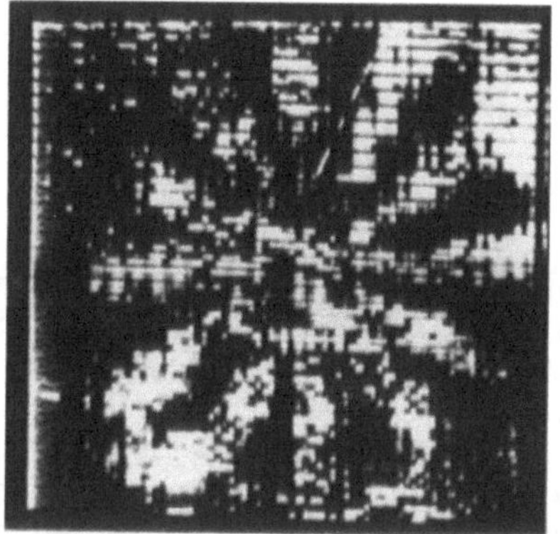

(a) (b)

Figure 13. (a) 140 mm square brass resolution test plate; (b) reflection image of a test plate

about 40 lines. The reasons for the lowering resolution are considered to be that there are many defective elements which have practically no sensitivity due to the imperfect connection of lead wires with elements owing to unskillful production techniques, that the strength of the acoustic source is not uniform and that the arrangements of ultrasonic source are not accurate and so on. Further development may solve these problems.

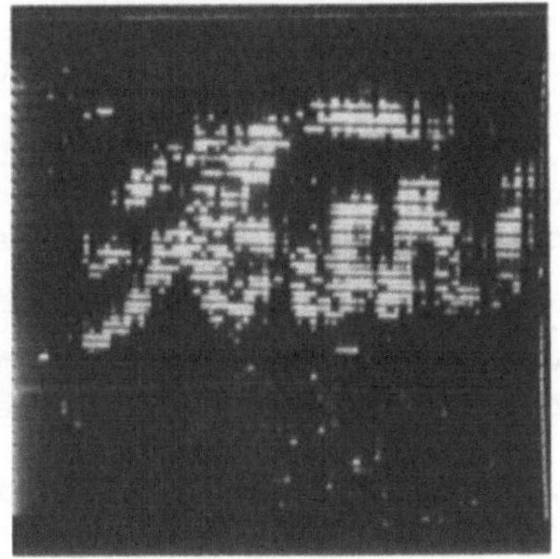

(a) (b)

Figure 14. (a) 170 mm length Toshiba logo; (b) reflection image of the logo.

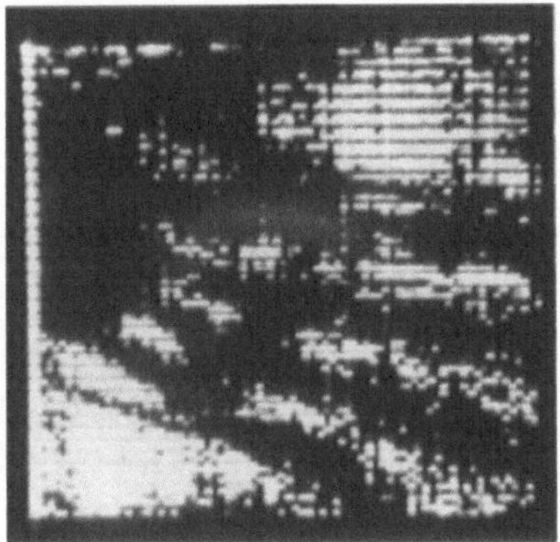

Figure 15. Transmission image of a human hand

ACKNOWLEDGEMENT.

The authors are grateful to Dr. Glen Wade who gave us the opportunity and encouragement to offer this paper at the 4th International Symposium on Acoustical Holography.

REFERENCES

1. T.Ogura, T. Kojima, and N. Uesugi, "Ultrasonic Image Converter", Japan Electronic Engineering, No.24, p.24, Dempa Publication, Tokyo, Japan, (November, 1968).

2. T. Yamada, "Development of Ultrasonic Image Converter System", Industria, Vol.1, No.2, p.95., Diamond Lead Co., Ltd., Tokyo, Japan, (December, 1971).

3. N. Uesugi, Y. Sato, M. Mishima, T. Tanaka, and T. Ogura, "Trial Construction and Experiments in the Sea of an Ultrasonic Image System". Reports of the 1972 Spring Meeting of The Acoustical Society of Japan, p.331.

4. T. Kawashima, N. Takagi, T.Ogura, and T. Yamada, "Solid State Acoustic Image Sensor", Reports of the 1971 Autumn Meeting of The Acoustical Society of Japan, p.375.

ACOUSTICAL IMAGING BY ELECTROSTATIC TRANSDUCERS

P. ALAIS

Institut de Mécanique Théorique et Appliquée

Université de PARIS VI. FRANCE

This paper describes a new way of obtaining acoustical holograms. The hologram is given by a matrix of transducers which are of the electrostatic type and may be interrogated in a way which simplifies considerably the subsequent circuitry.

Electrostatic transducers have been proved to be of interest not only in the audiorange but also in the ultrasonic range of frequencies. They may be built in a very simple way by setting a thin dielectric film between two electrodes : one is rigid and the other is thin and transparent to ultrasonic radiation.

A matrix of 256 x 256 such transducers has been realised with two printed circuits of 256 linear electrodes one horizontal and the other vertical, each transducer being located at the crossing of one line and one column. The horizontal electrodes are polarized one after the other so that the matrix is operated line by line. The vertical electrodes are connected to 256 amplifiers and synchronous detectors which deliver the holographic information corresponding to the polarized line. A multiplexing device allows this information to be obtained as a conventional video signal.

In the present state of the art, the hologram formation requires 50 ms and the first results are quite encouraging. This holographic matrix works in the range of frequencies :

0,3 to 2 Mhz and the observed sensitivity at 0,5 MHz is about 1 m W/cm^2 .

THE ELECTROSTATIC TRANSDUCER

The electrostatic transducer considered in this chapter is similar to devices which have been studied for ultrasonic transduction in air[1] and water[2,3] . A capacitor is obtained (fig.1) with a thin dielectric film (b) disposed between a rigid electrode (a) and another thin electrode (c) which may be formed by a metallization of the film itself or the superposition of another metallized plastic film. A polarizing voltage V_0 is set across the capacitor and the electrostatic pressure exerted at the interface dielectric-electrode is balanced by elastic forces developed in the dielectric material.

This equilibrium may be perturbed by a sound wave of frequency ω and the capacity C is modified according to $\Delta C = \Delta C^* e^{j\omega t}$. The device is equivalent to a generator of current $i = j\omega V_0 \Delta C^* e^{j\omega t}$ and transmits the signal

$$v = i/[Z + 1/jC\omega]$$

where Z is the impedance of the connected amplifier.

Alternately, this device may be used as an ultrasonic emitter : the electrostatic pressure may be electrically modulated by an external signal and a part of it is responsible for acoustical emission. The modulation of the electrostatic pressure may be written $p = \varepsilon E_0 E$ where ε is the dielectric constant of the dielectric material, E_0 the static field and E the perturbation field.

It is obvious that the higher the static field E_0(i.e V_0), the better the efficiency of the transducer. Some thin dielectric films are available today which have a very high dielectric rigidity and can support electric fields of a few hundred volts per micron which generate electrostatic pressures of several bars. The transducers considered here work typically with an electric field of 100 V/μ and we shall see that there is some departure from the behaviour of more classical devices like condenser microphones or electret microphones which work with typical fields of 10 V/μ . A simple model of the electrostatic transducer which takes into account the following points may be easily computed.

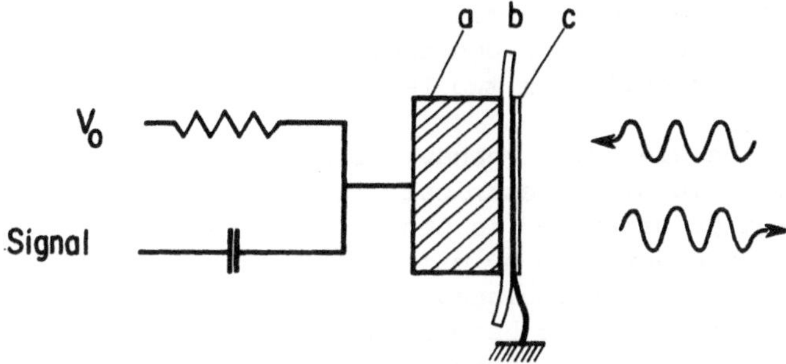

Fig.1 The electrostatic transducer.

- The dielectric film is a perfect linear dielectric material.

- The rear electrode is perfectly rigid and the forces which balance the electrostatic pressure and the sound pressure in the propagating medium are only due to the elastic compression and the inertia of the dielectric material which is submitted only to stationary longitudinal modes.

According to this model, in the emitter case, and when the inertia of the front electrode is neglected, the theoretical response versus frequency of the electrostatic transducer is,

$$p = \frac{\varepsilon E_0}{e\left(j\alpha \operatorname{cotg} \frac{\omega e}{c_1} - 1\right)} V, \quad \alpha = \frac{\rho_1 c_1}{\rho_2 c_2} \qquad (1)$$

where e is the dielectric film thickness, $\rho_1 c_1$ and $\rho_2 c_2$ are the acoustical impedances of the dielectric material and of the propagating medium.

In the receiver case, with the same assumptions and neglecting the effect of the amplifier impedance ($Z = \infty$) one has

$$V' = \frac{2j E_0}{\omega \rho_2 c_2 \left(j\alpha \operatorname{cotg} \frac{\omega e}{c_1} - 1\right)} p \qquad (2)$$

Experiments have been carried out [4,5,6] to check the validity of this model and have shown appreciable departures in the experimental results for two reasons :

a) First, the electrical behaviour of the dielectric film submitted to high electrical fields is not simple. The apparent polarization is obtained not only by molecular orientation but also by migration of charges in the dielectric volume (heterocharges) and by a transfer of charges at the interface between metallic electrode and dielectric film. This last phenomenon brings in charges of opposite sign (homocharges) and has a depolarizing effect. So, it is responsible for a reduction of the efficiency of the transduction. These processes have been recognised by different authors in studying electrets [7,8]. However, it must be emphasized that in the case of high electrical fields, the relaxation times observed are much shorter and may be evaluated in minutes or even in seconds. Besides, in our case, these electret type effects must be avoided. Different dielectric plastic films have been studied. Polyester films (Mylar) give good electrets but cannot be used for the transduction considered here because the depolarizing effect is important even at relatively low electrical fields. The transducer efficiency η no longer remains proportional to the polarizing field E_o. Figure 2 shows the departure of the efficiency η from the theoretical value η_o (identified to E_o or V_o). Negative values of η may be obtained when the polarization voltage

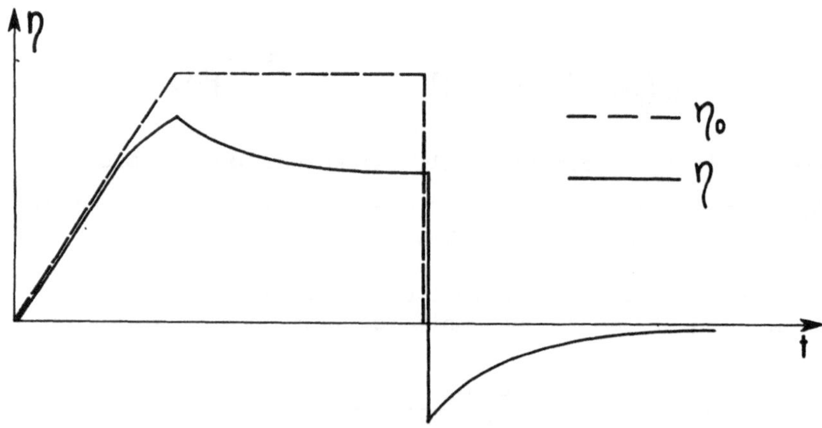

Fig.2 The efficiency of the electrostatic transduction

is removed due to homocharges which create a reversed polarization. The material which gives the best results is the polyimide film (Kapton). With this film, transducers of great stability have been realized working with high electric fields ($E_0 = 100\ V/\mu$) with a small residual transduction when the polarization voltage has been removed.

b) The mechanical behaviour of the sandwich rear electrode - dielectric film - front electrode is not simple. Imperfections in the flatness of electrodes, dust, gaseous cushions are responsible for flexural modes as well as longitudinal modes in the dielectric film and in the front electrode. Figure 3 shows the theoretical response of a delay line consisting of two electrostatic transducers which may be written from (1) and (2)

$$\left|\frac{V'}{V}\right| = \frac{2\ \varepsilon\ E_0^2}{\omega\ \rho_2 c_2 e \left[1 + \alpha^2 \cot g^2 \frac{\omega e}{c_1}\right]}\ , \qquad (3)$$

and the experimental response. The theoretical maximum is attained for a frequency which corresponds to a quarter wave-length for the dielectric thickness, i. e 25 MHz for a 12 μ Kapton film but the experimental peak corresponds to lower frequencies, near 1 MHz for our exemple. However, the theoretical maximum and the experimental peak have the same order of magnitude.

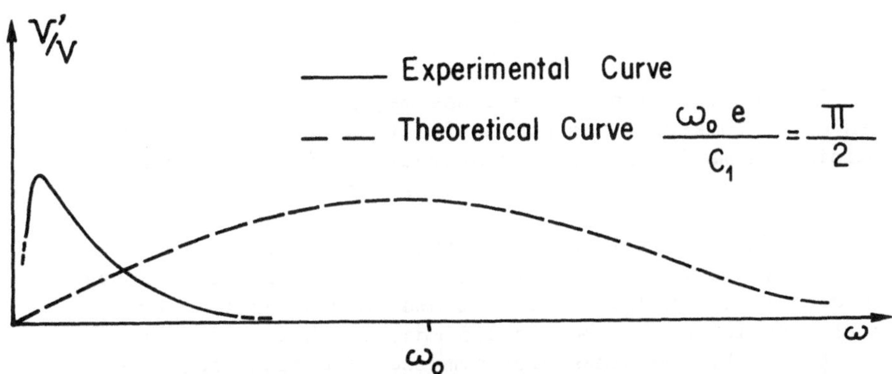

Fig.3 The efficiency versus frequency of our electrostatic delay line.

THE ELECTROSTATIC ACOUSTICAL RECORDING SURFACE.

The technology of the electrostatic transducer offers the possibility of building very easily complex matrices of transducers with printed circuits : these have linear parallel electrodes. The front circuit is a thin circuit which is as transparent as possible to the acoustical radiation and the rear circuit has a rigid support. Two such circuits, with n metallized lines on the first, and n columns on the second, and a thin sheet of dielectric material between them constitute a matrix of n^2 transducers.

But in this case the interest offered by the electrostatic conversion is not only the simplicity of technology but essentially the new possibility which it gives of exploring the information delivered locally at each crossing of one line and one column : in order to operate, the electrostatic transducer needs to be polarized, and this fact permits new kinds of commutations which could not be realized with classical piezoelectric transducers. Figure 4 shows the principle which has been adopted. Each column is connected to a receiving circuit. The lines are connected to a multiplexing circuit which polarizes them sequentially. Only the transducers corresponding to the polarized line operate and only the acoustical signal delivered at the location of this line is transmitted to the receiving circuits. Of course, an important simplification is obtained from the fact that no electronic circuit is individually associated to each transducer so that only n receiving circuits and n polarizing circuits may carry the information from n^2 transducers.

The polarization of the lines may be commuted at frequencies up to 10 kHz. It is not useful to work faster because a minimal time is required to extract the holographic signal by a synchronous detection type operation.

The first experimental realization has just been achieved in our laboratory; the matrix has 256 x 256 transducers. Its area is 256 x 256 mm^2 and it has been built to operate at frequencies of 0,5 MHz to 2,5 MHz. The Kapton film is 12,5 μ thick. The lines are commuted at a frequency of 4 kHz, so that 16 pictures/second may be delivered, and real time viewing of the hologramm is obtained.

Another method would consist in sequentially commuting each column to a unique better performing receiving circuit after each line exploration. This method is simpler, may be more sensitive, but is slow and would require in our case 16 s for a complete picture.

SYSTEM TIMING AND SIGNAL PROCESSING

During the time in which a line is polarized, the receiving circuits, for each column, amplify the acoustical signal which is mixed with the electronic reference signal used for exciting the illuminating transducer, and then integrated. They must also give the corresponding holographic information to a multiplexing circuit which performs a fast interrogation of the n circuits and furnishes the video signal. This last operation is effected after the first one has been achieved to avoid parasitic effects induced by the multiplexing circuit, so that each circuit must store the holographic information as a DC level until the first exploration has been done. Figure 5 shows the scheme of these operations :

The signal delivered by the column is amplified by an amplifier whose band width covers the relevant frequencies 0,3-3 Mhz. The mixing operation is obtained with a chopper which transmits the amplified signal only during 64 intervals of 250 ns which are initiated in phase with the reference signal. This technique permits the holographic information to be obtained with comparable levels at any frequency. 135 μs are allowed for this operation. This time corresponds to the lower frequency (500 kHz). The resulting signal is integrated and this operation furnishes the holographic signal as a DC level which is stored by means of a condenser.

The fast interrogation of the 256 charged capacitors is then operated in near 100 μs and a corresponding video signal is delivered in the same time. The polarization of the next line is then triggered but about 20 μs are used to discharge the storage condensers and to wait for parasitic signals induced by the polarization commutation to disappear.

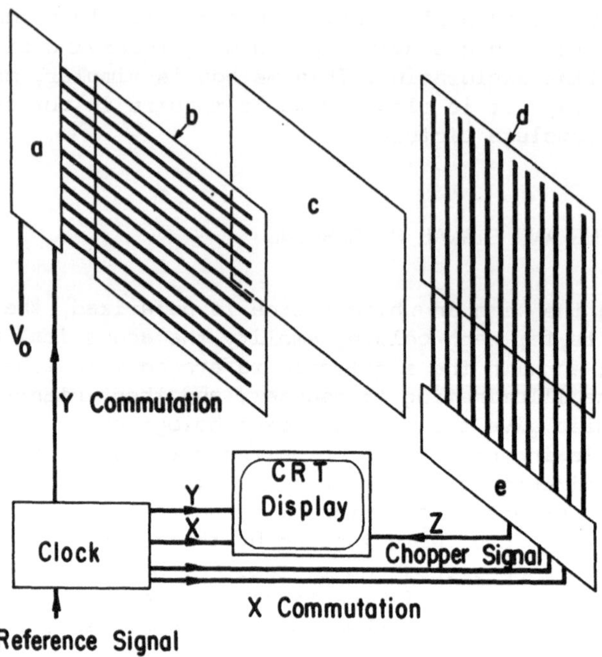

Fig.4 Simplified block diagramm of acoustic holography system.

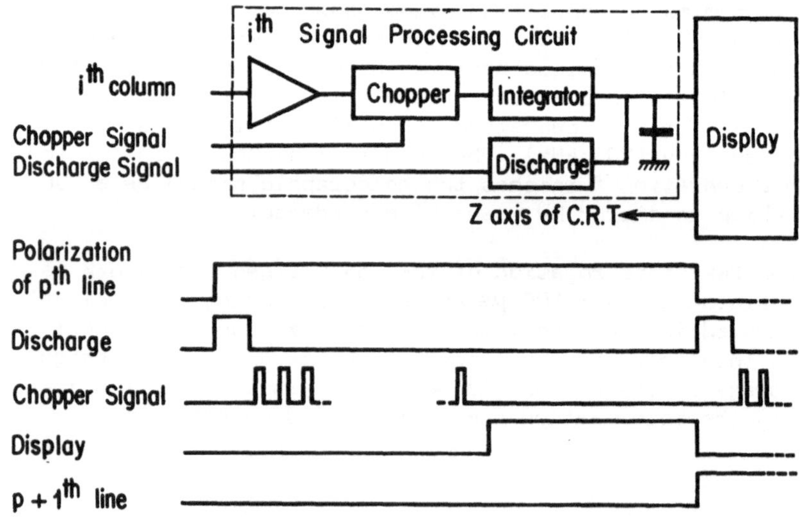

Fig.5 Signal processing circuitry and timing waveforms

Fig.6 Experimental arrangement.

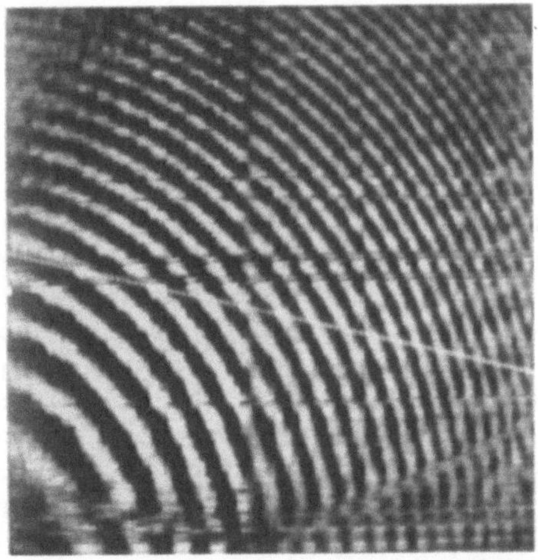

Fig.7 Acoustical hologram of a 3 mm wavelength point source located 60 cm from the recording surface.

EXPERIMENTAL RESULTS

The experimental arrangement has just been achieved (fig.6). Holograms have been obtained from an acoustical illumination at 500 kHz and at 1,5 Mhz. The observed sensitivity is much better at 500 kHz than at 1,5 MHz. Figure 7 shows fringes obtained at 500 kHz from a quasi point source located at 60 cm from the holographic recording surface. The intensity of illumination is of the order of 1 mW/cm^2. At this time, the sensitivity of the device is limited by the noise brought in by the multiplexing circuit which delivers the video signal. Figures 8 and 9 show the hologram obtained with 10 cm high plexiglass letters placed near the recording surface and the corresponding reconstruction. An image defect may appear when the holographic signal varies very slowly along one column and is located in the neighbourhood of vertical fringes (fig. 10). This phenomenon is caused by the residual transduction of unpolarized transducers which is an electret-type effect. The residual transduction gives at each column a signal which is added to the useful signal of the polarized transducer; this unwanted signal represents the integral of the holographic information along the column weighted by the factor of residual transduction. This factor decreases rapidly to a constant value. But the transducers which have just been polarized give a more important contribution which may be constructive when they all furnish an information of the same sign.

CONCLUSIONS

These first experimental results are encouraging, in spite of the fact that the actual device does not operate in ideal conditions. We hope to raise the observed amplitude sensitivity by a factor of three or four : the 12 μ Kapton film may be conveniently polarized up to 800 volts instead of the 200 volts actually imposed for simplifying the Y commutation. Besides, the front printed circuit which has been used (75 μ Mylar + 17,5 μ Cu) is too thick and too stiff to permit a good fitting, and this fact could explain the observed decrease of sensitivity with increasing frequency. In fact, this decrease is higher than for the other electrostatic transducers which we have tried.

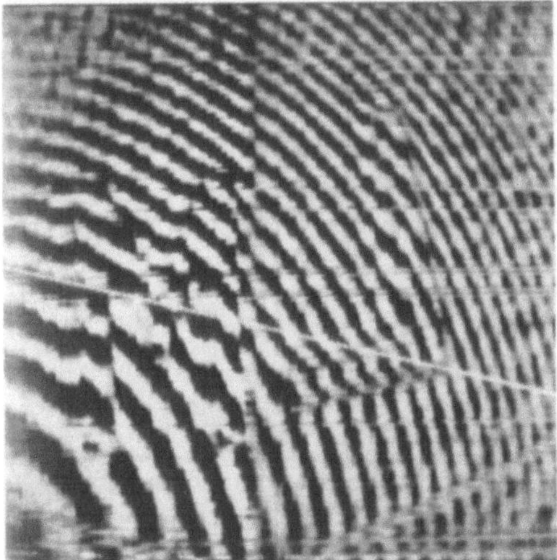

Fig. 8 Acoustical hologram of 10 cm high plexiglass letters " L " and " M " placed near the recording surface.

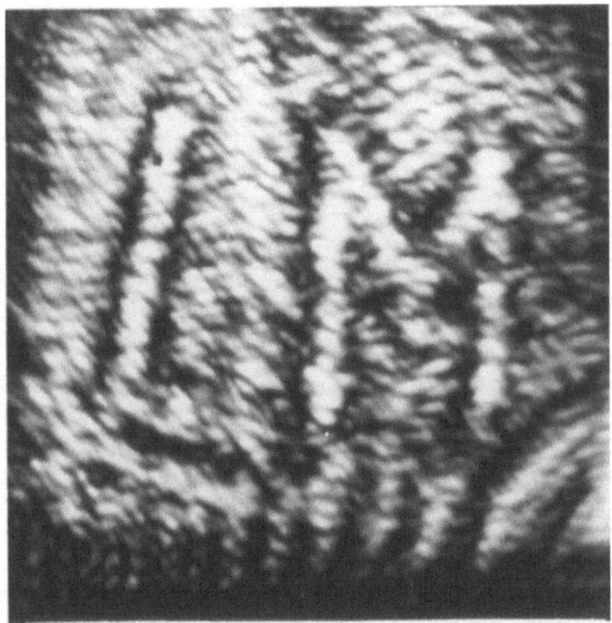

Fig. 9 Optical reconstruction with 6328 A converging laser beam of the hologram of fig. 8 after it was demagnified to 6 x 6 mm.

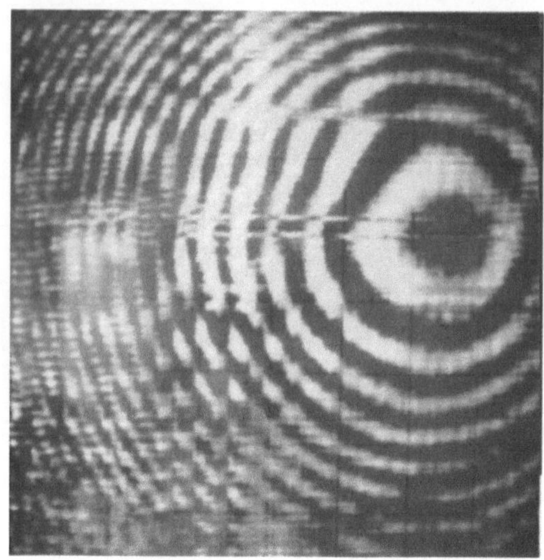

Fig. 10 Image defect due to residual electret type transduction.

The next step, of course, will be to try to improve the sensitivity of this device for frequencies up to 2 or 2,5 MHz, for medical applications. Besides, the capability of real time display is interesting only if a real time restitution may be done in conjunction.

ACNOWLEDGEMENTS

The author wishes to thank R. LALIMAN for assistance given in this work.

REFERENCES

1. W. KUHL, G.R. SCHODDER and F.K. SCHRÖDER "Condenser transmitters and Microphones with Solid Dielectric for Airborne Ultrasonics".Acustica,4,1954,p.519

2. G.R. SCHODDER and F. WIEKHORST, "Electrostatic Transducers with Solid Dielectric for Waterborne Sound", Acustica, 7, 1957, p.38.

3. J.C. MORRIS, "Broad-Band Constant Beam Width Transducers",
Journal of Sound and Vibration, 1, 1964, p. 28.

4. P. ALAIS and M.Th. LARMANDE,
"Etude d'un Transducteur Ultrasonore Electrostatique",
C.R. Ac. Sc. PARIS, t. 272, p. 185-188, Janvier 1971.

5. M.Th. LARMANDE
"Transducteurs Electrostatiques",
Thèse de 3ème Cycle, Juin 1971, Université de PARIS VI.

6. D. LEGROS.
"Etude d'un Transducteur Ultrasonore de type Electrostatique",
Thèse de 3° Cycle, Juin 1971, Université de PARIS VI.

7. G. CROSS and J. de MORAES.
"Polarization of the Electret",
J. Chem. Phys., 37, 1962, p. 710.

8. G.M. SESSLER and J.E. WEST.
"Foil Electret Microphones",
J. Acoust. Soc. Amer., 40, 1966, p. 1433.

CONVENTIONAL AND WEAK SIGNAL ENHANCEMENT HOLOGRAPHY IN THE PRESENCE OF MEASUREMENT ERRORS

P. N. Keating, R. K. Mueller, and R. R. Gupta

Bendix Research Laboratories

Southfield, Michigan 48076

The effects of phase and amplitude errors on the dynamic range in the images obtained by reconstruction by both the conventional method and by a new enhancement method are examined. It is shown that the dynamic range in enhancement imagery is little affected by phase errors. It is also shown that, if the width of the strong image is aperture-limited, the dynamic range in conventional holographic imagery increases as N in terms of intensity, where N is the number of sampling points. This increase continues until N is large enough so that the image width is no longer aperture-limited.

INTRODUCTION

Much of the interest in acoustical holography involves the use of sampled holograms. This type of hologram has a number of technical advantages, including sensitivity in detection and flexibility in processing. One example of the utility of the processing flexibility is the possibility of employing a recently-described weak-signal enhancement technique[1] which is particularly appropriate to the case of digital processing of sampled holographic data.

The weak-signal enhancement technique (WSET) is a method of processing the data by using some *a priori* knowledge about the object field to enhance a weak image in the presence of a strong virtual source. For example, a weakly-scattering target above or below a strongly reflecting interface can be enhanced relative to the interface reflection if the illumination is coherent.[1]

The present article consists of a study of the processing of sampled holograms and, in particular, a study and comparison of the effect of phase and amplitude errors on (a) conventional and (b) weak-signal enhancement holographic processing of sample holograms. It is shown that if a weak image obtained by conventional holography is drowned in phase noise from a strong image, then the WSET will give a considerable improvement in signal-to-noise. The effect of the number of sampling points on the smoothing of errors in holography is also shown.

The next three sections of this paper consist of a comparison of the effect of random phase errors on conventional and weak-signal enhancement holography. This comparison consists of (a) an analytical study of the effect of phase errors on far-field holograms and (b) a computer study of the effect on near-field holograms. The fifth section consists of a comparison of the effect of amplitude errors; the sixth consists of an examination of the problem of non-planar arrays, where non-random phase errors occur. For simplicity, the study is restricted to one dimensional holograms and images.

EFFECT OF RANDOM PHASE ERRORS

We consider far-field holograms of a function g(x) in one dimension. The field is sampled at the kth point (where u = ka) in the hologram. The resulting hologram in the absence of errors is

$$\hat{G}(u) = \sum_k G_k \left[\theta(u-(k-1/2)a) - \theta(u-(k+1/2)a) \right] \tag{1}$$

if the value at the kth point is placed in the range $k-1/2 \leq u/a \leq k+1/2$, where

$$G_k = \int g(x) e^{2\pi ikax} dx \tag{2}$$

and θ is the usual step function. Such a hologram is obtained in a current holographic underwater viewing system.[2,3]

Phase Errors in Conventional Reconstruction

Under conventional reconstruction, the image obtained from a hologram of the type described by Eq. (1) is

$$\hat{g}(x) = a \operatorname{sinc}(ax) \sum_k \tilde{G}_k e^{-2\pi ikax} \tag{3}$$

where $\operatorname{sinc}(x) = (\sin(\pi x)/\pi x)$ and where $\tilde{G}_k$ is the sampled field value complete with errors. In the case of phase errors,

$$\tilde{G}_k = G_k e^{i\Delta_k} = G_k + G_k(e^{i\Delta_k} - 1) \tag{4}$$

where Δ_k is the phase error at the kth sampling point. We shall assume that the distribution of Δ_k is symmetric, i.e., $\langle\Delta\rangle = o$. Thus

$$\hat{g}(x) = a \operatorname{sinc}(ax) \sum_k G_k(1+D_k) e^{-2\pi ikax} \tag{5}$$

where

$$D_k = e^{i\Delta_k} - 1 \tag{6}$$

If we write

$$d(x) = \sum_k D_k e^{-2\pi ikax} = \sum_k (e^{i\Delta_k} - 1) e^{-2\pi ikax} \tag{7}$$

$$\hat{g}(x) = a \operatorname{sinc}(ax) \int dx' g(x') \Big[Nf(x-x') + d(x-x') \Big] , \tag{8}$$

where

$$f(x) = \frac{1}{N} \cdot \sum_{k=-\frac{(N-1)}{2}}^{+\frac{N-1}{2}} e^{2\pi ikax} = \frac{\mathrm{Sin}(N\pi ax)}{N\ \mathrm{Sin}(\pi ax)} \tag{9}$$

Thus, the effect of phase errors is to produce an additional contribution which is the convolution of the true image with the spectrum of the distribution $\{D_k\} = \{e^{i\Delta_k}-1\}$.

In the case where $\Delta_k <<1$, $-id$ is just the spectrum of $\{\Delta_k\}$. In this case, the value of $|d|$ expected from the distribution $d(x)$ is equal to

$$<|d|> = N^{1/2}\gamma\Delta_{rms} \tag{10}$$

in terms of the variance Δ_{rms} expected in the set Δ_k; γ is a constant in order of unity which depends on the nature of the statistics.

We now consider the case of two point objects, one of which is weak (by a factor α) and one strong:

$$g(x) = \delta(x) + \alpha\delta(x-x_1). \tag{11}$$

For a "signal-to-noise" ratio of unity for the image of the weak object to be observed in the presence of the noise produced from the strong object by the phase errors, we require the average value of $|d(x)|$ to be

$$<|d|> \sim N\alpha f(o) = \alpha N \tag{12}$$

In other words, we can observe a point object down by a factor α in amplitude (α^2 in intensity) from another point object provided α is greater than α_c, where, using Eqs. (10) and (12),

$$\alpha_c = \gamma\Delta_{rms}/N^{1/2}. \tag{13}$$

The result, given by Eq. (13), shows that, _if the width of the images of the two objects is aperture-limited, amplitude differences can be handled which are appreciably larger than the variance in the phase errors would suggest_, even if N, the number of sampling points, is only 64, for example. This result is due to integration of the noise which occurs in the reconstruction process. It is very similar to the integration of shot noise, where the signal increases as N, but the shot noise increases only as $\sqrt{N}$, giving a $\sqrt{N}$ improvement in signal-to-noise in terms of amplitude. In terms of intensity, the signal-to-noise increases as N until N is large enough so that the image width is no longer aperture-limited.

RANDOM PHASE-ERROR EFFECTS IN ENHANCEMENT RECONSTRUCTION

The enhancement technique[1] is applicable to the case where a strong (virtual) source is present in addition to a weak-scattering target source which is coherent with it.

Such a situation exists, for example, with an active holographic system when a strongly-reflecting interface is present. The _a priori_ information needed by the enhancement technique is that the strong virtual source is fairly localized spatially. Under these conditions, the enhancement technique consists of (a) formation of the intensity distribution, (b) high-pass spatial-filtering of this latter, and (c) reconstruction of the filtered intensity with the original field.

Error-Free Enhancement Reconstruction

In the absence of phase errors, one obtains for the high-pass filtered intensity

$$H_k = |G_k|^2 - \frac{1}{N} \sum_{k'} |G_{k'}|^2. \tag{14}$$

This form of filtering is adequate for far-field holograms, although wider-band rejection is normally necessary in the case of near- and intermediate-fields. Thus

$$H_k = \int dx'dx''g(x')g^*(x'')\left[e^{2\pi ika(x'-x'')} - f(x'-x'')\right]. \quad (15)$$

The final image is then

$$g_w(x) = a\ \mathrm{sinc}(ax) \sum_k G_k H_k\ e^{-2\ ikax} \quad (16)$$

$$= Na\ \mathrm{sinc}(ax) \int dx'dx''dx'''\ g(x')g^*(x'')g(x''') \times h(x'''-x, x'-x'') \quad (17)$$

where

$$h(x'''-x, x'-x'') = f(x'''-x+x'-x'') - f(x'''-x)f(x'-x''). \quad (18)$$

For example, if

$$g(x) = \delta(x) + \alpha\delta(x-x_1) \quad (19)$$

we have

$$G_k = 1 + \alpha\ e^{2\pi ikax_1} \quad (20)$$

$$H_k = 2\alpha\left[\cos 2\pi kax_1 - f(x_1)\right] \quad (21)$$

$$\approx 2\alpha\ \cos 2\pi kax_1 \quad (22)$$

since $f(x_1) << 1$ in most cases of interest. Hence

$$g_w(x) \approx N\alpha a\ \mathrm{sinc}(ax)\left[(f(x-x_1)+f(x+x_1) + \alpha(f(x)+f(x-2x_1))\right] \quad (23)$$

The first term is the enhanced image at $x = x_1$ and the third is the attenuated image at $x = o$; the second and fourth terms are new terms introduced by the processing, which are separated from the main image, as shown in Fig. 1.

Phase Errors in Enhancement Reconstruction

In the presence of phase errors, we note that, although

$$\tilde{G}_k = G_k(1+D_k), \tag{24}$$

as before, we have

$$|\tilde{G}_k|^2 = |G_k|^2 \tag{25}$$

and

$$\tilde{H}_k = H_k. \tag{26}$$

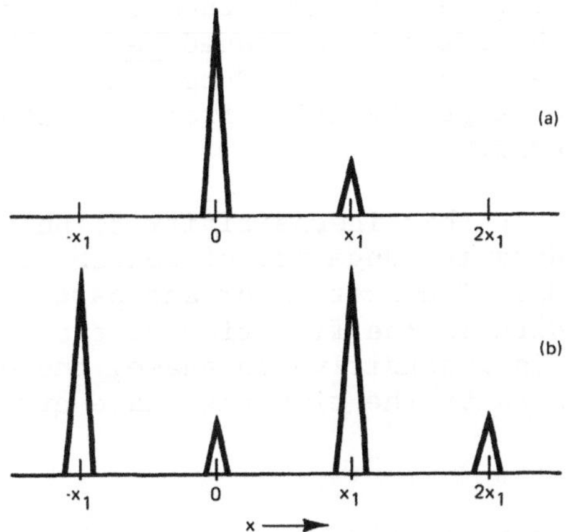

Fig. 1. (a) Object Distribution (b) Enhanced Imagery of (a).

Thus, Eq. (16) becomes

$$\tilde{g}_w(x) = a \operatorname{sinc}(ax) \sum_k G_k H_k (1+D_k) e^{-2\pi ikax} \tag{27}$$

If we again neglect $f(x_1)$ as being small, we obtain

$$\tilde{g}_w(x) = g_w(x) + \alpha a \operatorname{sinc}(ax) \Big[d(x-x_1)+d(x-x_1) + \alpha(d(x)+d(x-2x_1)) \Big] \tag{28}$$

where g_w, given by Eq. (23), is the image obtained in the absence of phase-errors.

We note that the enhanced image is degraded in the same way in the weak-signal enhancement technique as the strong image is degraded in conventional holography. However, it is the <u>enhanced</u> image which is degraded, so that the unwanted signal may be obscured by "scattering" of the enlarged, desired signal, but this is not, of course, a problem. In other words, because the enhancement involves the intensity (and thus is insensitive to phase errors), the image is enhanced <u>before</u> the phase error effects are encountered. Thus, the object of interest is not submerged by noise from the unwanted object, but only vice versa.

The reason for this insensitivity to phase error limitations on dynamic range is, of course, due to the fact that $\tilde{H}_k = H_k$. The most important part of the enhancement procedure is the filtering of the intensity and, since this is insensitive to phase, the weak signal is already enhanced by the time the phase errors have any effect.

NEAR-FIELD COMPUTER RESULTS

The above analysis was carried out for far-field holograms. It is of some interest to examine the more general case of near- and intermediate-zone holography, which is best handled by computer calculation. We now describe such calculations, which substantially agree with the far-field analytical results.

The model (Fig. 2) used for the computer calculations is very similar to that used for the weak-signal enhancement method previously.[1] It differs from it only in that the weak-scattering target was displaced upwards so that the two images are in focus in the same plane. It provides quite a stringent test for the enhancement method since the presence of both longitudinal and shear waves, together with mode conversion, ensures that the strong virtual source is a compound object. The calculations were carried out with $N = 2^7 = 128$.

The reconstruction by conventional holography of the strong and weak images ($x_1 = 25\lambda$) is shown with and without random phase errors, respectively, in Figs. 3 and 4.

The random phase errors were incorporated by means of a random number generator in such a way that (a) the maximum phase error is $\pm \pi/4$ and (b) the expected probability for a given error to arise in this range is constant. In Fig. 3, the noise from the phase errors is at least as large as the weak image and the latter is not detectable with any confidence in the presence of the former. The weak signal, however, is readily observable in the absence of noise (Fig. 4) for the same value of $\alpha \simeq 0.12$.

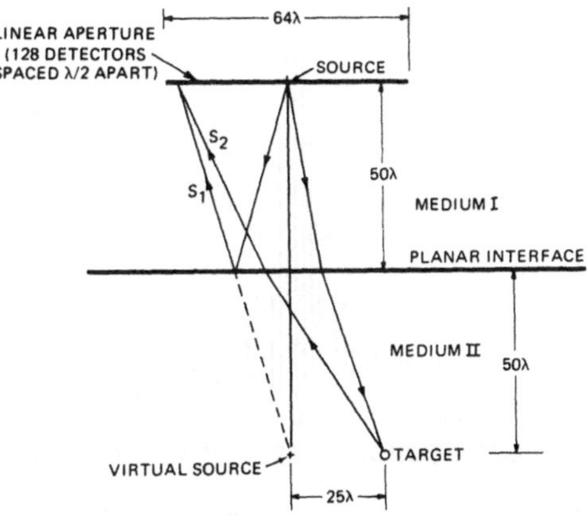

Fig. 2. Model used for near-field computer calculations.

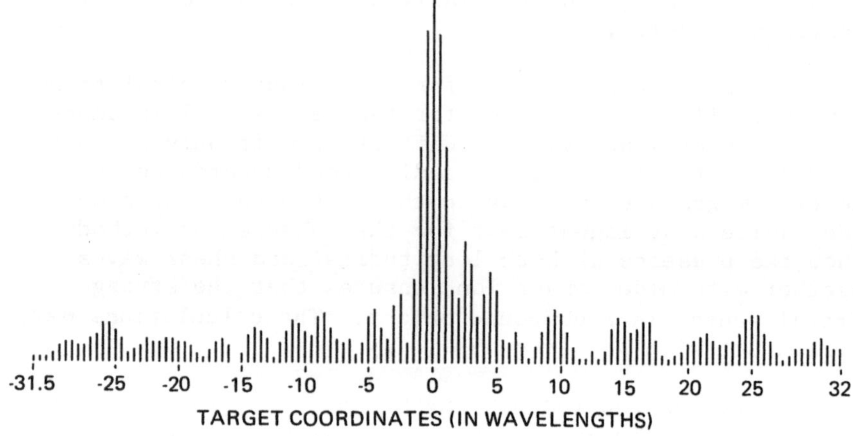

Fig. 3. Conventional reconstruction with random phase errors.

Equation (13) is now

$$\alpha_c = \bar{N}^{1/2} \langle |D| \rangle \tag{29}$$

where we have interpreted $\gamma\Delta_{rms}$ as the average value of $|e^{i\Delta}-1|$ since Δ is no longer small. Hence, $\langle |D| \rangle \simeq 0.4$,

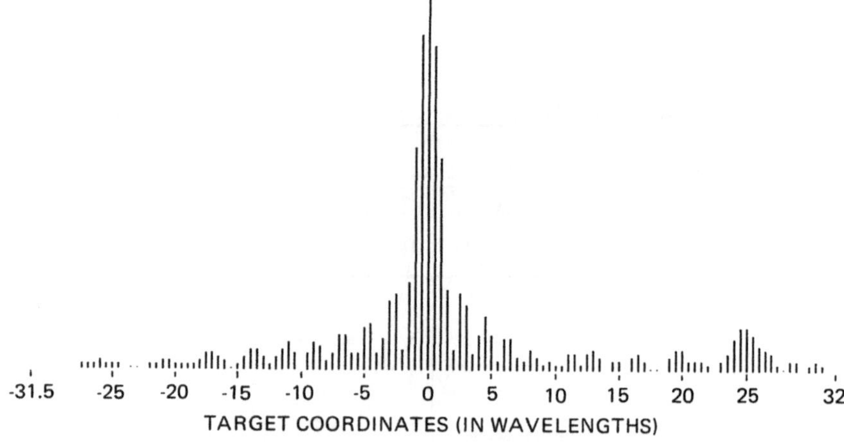

Fig. 4. Conventional reconstruction without deliberate phase errors.

$$\alpha_c \approx 0.05 \tag{30}$$

which is in reasonable agreement with the near-field results shown in Fig. 3, in view of the fact that the image widths are not aperture limited.

The weak-signal enhancement reconstruction of the same object is shown without and with random phase errors of the same magnitude, respectively, in Figs. 5 and 6. We note that, in contrast with the conventional reconstruction, the enhanced desired signal at $x_1 = 25\lambda$ is plainly visible, and there is no serious noise problem in the presence of the substantial phase errors which are also seen in Fig. 3.

EFFECT OF RANDOM GAIN ERRORS

The effect of random amplitude errors on conventional reconstruction is readily carried out by the same procedure as above. We represent the sampled holographic data with errors as

$$\tilde{G}_k = G_k(1+\lambda_k) \tag{31}$$

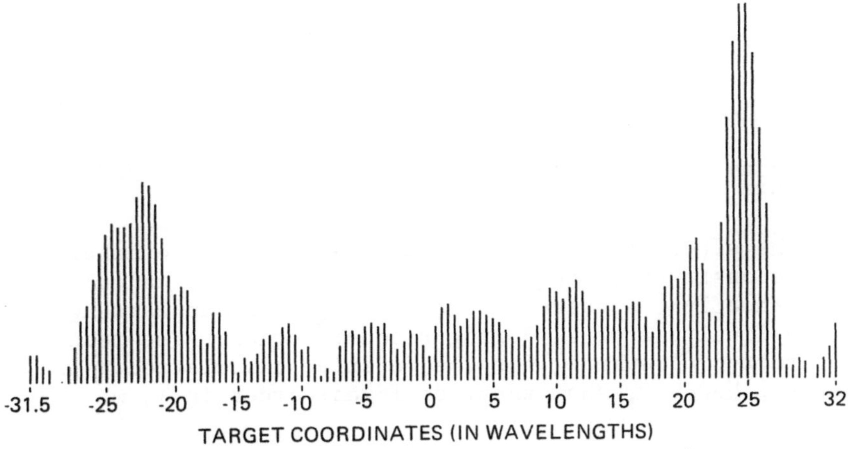

Fig. 5. Enhancement reconstruction without deliberate errors.

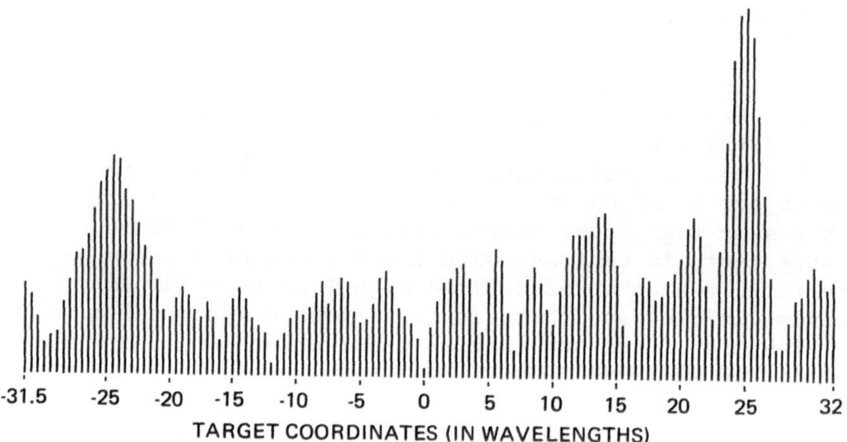

Fig. 6. Enhancement reconstruction with phase errors as in Fig. 3.

where λ_k represents the fractional error in the kth channel. We obtain a result essentially the same as Eq. (8) except that d(x-x') is replaced by

$$\Lambda(x-x') = \sum_k \lambda_k e^{-2\pi ika(x-x')} . \qquad (32)$$

Furthermore, Eq. (13) becomes

$$\alpha_c = \gamma N^{-1/2} \lambda_{rms} \qquad (33)$$

where λ_{rms} is the variance in the λ distribution (again we assume $\langle\lambda\rangle = o$).

The effect of random amplitude errors on the enhancement reconstr''ction requires a more complicated calculation since $1+\lambda_k$ appears squared in $|G_k|^2$ and appears again in the final use of G_k during reconstruction. The reader is referred elsewhere[4] for the detailed calculation. The results, however, show that, to lowest order in λ, we obtain[4]

$$\alpha_c = 2\gamma N^{-1/2} \lambda_{rms} , \qquad (34)$$

which is twice as large as in the conventional reconstruction. In other words, the effect of amplitude errors is twice as bad in enhancement reconstruction because the intensity is used. The appearance of λ_k again via G_k during the final step has a negligible effect because the weak signal has already been enhanced by the time this effect arises. This compares with the reason for insensitivity to phase errors noted above.

We have again carried out computer calculations for the near-to-intermediate zone and again find substantial agreement[4] with the far-field results.

NON-PLANAR ARRAYS

If holograms determined by field sampling over a non-planar surface are reconstructed as if obtained by planar sampling, then non-random phase errors will occur. Clearly, the previous analysis cannot be applied directly to systematic phase errors. In the present section, we shall present a few basic aspects of the reconstruction of holograms sampled over non-planar surfaces.

If a computer is used to reconstruct, it is clear that the necessary corrections may readily be included. If optical reconstruction is used, the same (demagnified) non-planar surface could be used to hold the holographic information. However, this is not an attractive possibility since the accurate production of complex surfaces is difficult. Therefore, the following points are of interest:

(a) In the Fresnel and Fraunhofer zones, the phase correction can be applied during data acquisition.

(b) The linear and quadratic parts of the sampling surface variation cause phase errors which are trivial in the sense that they either give trivial effects or can trivially be corrected.

The first point can readily be demonstrated. Using the geometry shown in Fig. 7, the path from a point $\underset{\sim}{u}$ on the object to a point $\underset{\sim}{x}$ on the sampling surface (where $\underset{\sim}{u}, \underset{\sim}{x}$ are two-dimensional vectors) is

$$\ell(\underset{\sim}{x},\underset{\sim}{u}) = [\ (R+y(\underset{\sim}{x}))^2 + (\underset{\sim}{x}-\underset{\sim}{u})^2\ ]^{1/2} \qquad (35)$$

where $y(\underset{\sim}{x})$ is the equation of the sampling surface. In the Fresnel approximation, this becomes

$$\ell(\underset{\sim}{x},\underset{\sim}{u}) \approx R+y(\underset{\sim}{x}) + \frac{(\underset{\sim}{x}-\underset{\sim}{u})^2}{2R}. \qquad (36)$$

Since there are no cross-terms between y and $\underset{\sim}{u}$ in this approximation, the phase correction corresponding to the $y\ (\underset{\sim}{x})$ term in Eq. (36) is the same for all image points and can be applied during the sampling process.

The second point (b) above can readily be ascertained by carrying out a Taylor series expansion of $y\ (\underset{\sim}{x})$:

$$y(\underset{\sim}{x}) = \underset{\sim}{x}\cdot\underset{\sim}{\Delta}y_o + 1/2\ \underset{\sim}{x}\ (\underset{\sim}{\Delta}\underset{\sim}{\Delta}y_o)\cdot\underset{\sim}{x} + \ldots \qquad (37)$$

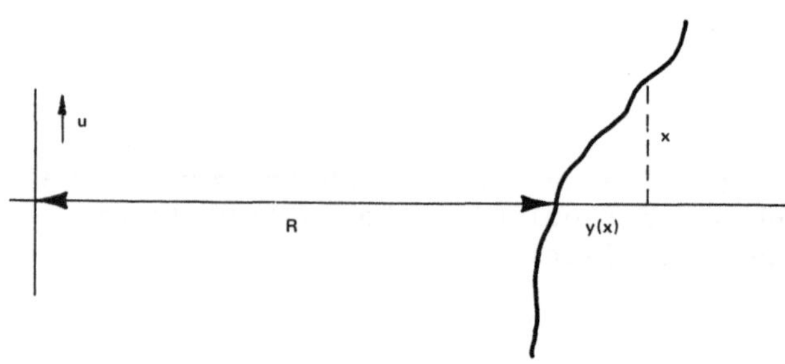

Fig. 7. Coordinates used in the discussion of non-planar arrays.

where $\underset{\sim}{\Delta}\underset{\sim}{\Delta}\, y_o$, for example, represents a second rank tensor with α, β component

$$\left[\frac{\partial^2}{\partial x_\alpha \partial x_\beta} y(\underset{\sim}{x}) \right]_{\underset{\sim}{x}=o} \qquad (38)$$

Analysis readily shows that the first term merely shifts the lateral position of the image and is therefore trivial. The second term merely defocusses the image and is also trivial since it can readily be corrected either by looking at a slightly different image plane or by the use of weak spherical or cylindrical lenses.

REFERENCES

1. R. K. Mueller, R. R. Gupta, and P. N. Keating, J. Appl. Phys., 43, 457 (1972).

2. H. R. Farrah, E. Marom, and R. K. Mueller, Acoustical Holography, Vol. 2, 173 (Plenum Press, New York, 1970).

3. E. Marom, R. K. Mueller, R. F. Koppelmann, and G. Zilinskas, Acoustical Holography, Vol. 3, 191 (Plenum Press, New York, 1971).

4. P. N. Keating, R. R. Gupta, and R. K. Mueller, J. Appl. Phys., 43, 1198 (1972).

SPIRAL SCANNING IN LONGWAVE HOLOGRAPHY

N. H. Farhat, W. R. Guard* and A. H. Farhat

The Moore School of Electrical Engineering
University of Pennsylvania
Philadelphia, Pennsylvania 19104

ABSTRACT

A general diffraction theory treatment of sampled longwave holography is presented. The results are used to demonstrate the properties of the spiral scan and other circular sampling formats in longwave (microwave or acoustical) hologram recording. Conclusions drawn from the theory are verified experimentally employing analog computer generated spiral scan formats and Fourier optics techniques. Both the uniform and the linearly "chirpped" spiral scan formats are considered. High quality millimeter microwave images of reflecting objects recorded employing the spiral scan are presented to illustrate the capabilities of this sampling format.

INTRODUCTION

Direct extension of recording techniques used in optical holography to suboptical microwave frequencies and to acoustic waves runs into difficulties stemming directly from the lack of a suitable area recording device analogous to the photographic plate for use in the recording step.

*W. R. Guard is presently with the Newark College of Engineering, Newark, New Jersey.

Several methods have been investigated to realize such a recording device. For example, the recording of microwave holograms has been demonstrated employing liquid crystals [1-3], preconditioned photographic emulsion [4-6] and more recently employing photochromics [7]. At present a major shortcoming of these methods is the high exposure levels required.

The development of more sensitive area detectors for use in acoustical holography has been more successful. A variety of methods are currently in use [8-10]. However, these methods appear to be limited when larger recording apertures are required.

The recording of longwave holograms over large apertures appears at this time to be best performed through the use of sensitive discrete sensors. Two modes of recording the object wavefront are possible: a) by discrete sampling employing arrays of stationery sensors discretely allocated in a prescribed manner over the hologram recording aperture constituting thus a "hologram recording array". The output of the individual elements can be interrogated electronically to derive the hologram data; b) by continuous scanning employing a single discrete detector scanned over the hologram recording aperture in a prescribed format to obtain in time a mapping of the hologram data over this aperture. Both modes yield a sampled version of the hologram data, the amplitude and phase distributions in the object wavefront over the hologram recording aperture.

The finite size of discrete sensors used in sampled longwave holography must be smaller or equal to the smallest fringe separation in the intensity distribution being mapped in order to avoid loss of information. This requirement is easily fulfilled at microwave and acoustical wavelengths in contrast to optical holography where the high fringe density precludes the use of practical discrete detectors.

For economical reasons, non-real-time longwave holography has been best accomplished in the laboratory when low irradiance levels are involved with the latter mode of scanned recording. The conventional raster scan is almost universally used.

The aim of this chapter is to present the results of a theoretical and experimental study that shows the advantages of the spiral scan format in longwave non-real-time holography.

A general theory of sampled holography is presented in order to determine the general effect of sampling on the retrieved image. This shows that the most desirable sampling formats are those whose Fourier transforms approximate as closely as possible a single delta function. The properties of the spiral scan are then discussed in this context. Experimental verification of these properties using analog computer generated spiral scan formats and Fourier optics techniques is given next. High quality millimeter microwave images of reflecting objects such as a wire, rectangular plate and a toy gun are then presented to demonstrate the advantages of the spiral scan.

GENERALIZED THEORY OF SAMPLED HOLOGRAPHY

To record a hologram using long wavelengths the complex field amplitude (CFA for short) of the wavefield scattered by the object is determined over the hologram recording plane with the aid of a square law detector in the presence of a coherent background reference wavefield. We will refer to the generalized recording geometry of Fig. 1 where we consider a fully or partially reflecting object illuminated with coherent electromagnetic or acoustic radiation of wavelength λ. The field scattered by the object will produce in an adjacent plane P_o referred to hereafter as the object plane situated an arbitrarily small distance ΔZ_o from the object, an object CFA $O(\bar{\rho}_o)$ where $\bar{\rho}_o = x_o\bar{1}_x + y_o\bar{1}_y$ is the position vector of a point in P_o.* Here and in the following $\bar{1}_x$, $\bar{1}_y$, and $\bar{1}_z$ designate unit vectors along the coordinates being considered. Similarly $\bar{1}_n$ is the unit normal to P_o. Because of the nature of the scattering phenomena at acoustic and microwave frequencies the object CFA will be produced by the scattering centers of the object only. A scattering center is defined here as a reflecting point or region of the object that contributes to the CFA in P_o and

*The vector notation of position in the various planes allows more concise expression of subsequent formulas that prove otherwise inconveniently lengthy.

therefore, as will be shown below, to the CFA in the hologram recording surface S_H or the recording plane P_H. We adopt the view that retrieval of the object CFA $O(\bar{\rho}_o)$ or a scaled and in practice diffraction limited, replica of $O(\bar{\rho}_o)$ in an imaging system constitutes the coherent imaging operation desired. In this fashion we avoid and need not be concerned directly with the complex problem of scattering of the illumination by the object. A retrieved image of the object will actually be an image of the scattering centers or so called "highlights" of the object that give rise to $O(\bar{\rho}_o)$. This is in direct contrast to imaging at the considerably shorter optical wavelengths where each point of an ordinary rough surface object scatters the incident light diffusely in all directions and will therefore generally contribute to the net field in P_o and the recording plane P_H which is situated at distance $Z_o >> \lambda$ from P_o and parallel to it. The position of a point in P_H is designated by the position vector $\bar{\rho}_h = x_h\bar{1}_x + y_h\bar{1}_y$.

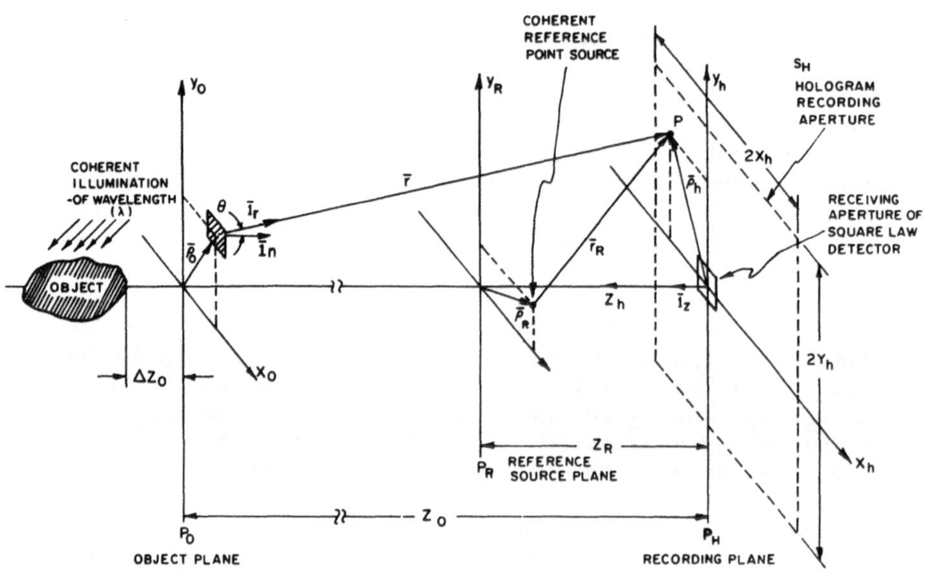

FIGURE 1 GENERALIZED RECORDING GEOMETRY OF SAMPLED HOLOGRAM

The object CFA over the hologram recording surface S_H is recorded indirectly by mapping the total intensity distribution over this surface with the aid of a scanned square law sensor in the presence of a coherent background reference wavefield of wavelength λ. This reference wavefield is produced by a point source located in the reference plane P_R at a lateral position given by the position vector ρ_R. The detector output is utilized to generate a hologram transparency record suitable for optical reconstruction using the usual CRT display methods. The resultant hologram is usually demagnified M times in size in comparison to the actual size of the recording surface S_H. Usually, in order to eliminate longitudinal image distortion, M is chosen to equal the ratio (λ/λ_L) of the recording wavelength λ to the optical reconstruction wavelength λ_L.

Optical reconstruction of the resultant optical transparency replica of the sampled (scanned) longwave hologram is performed using the generalized reconstruction or visible image retrieval geometry shown in Fig. 2.

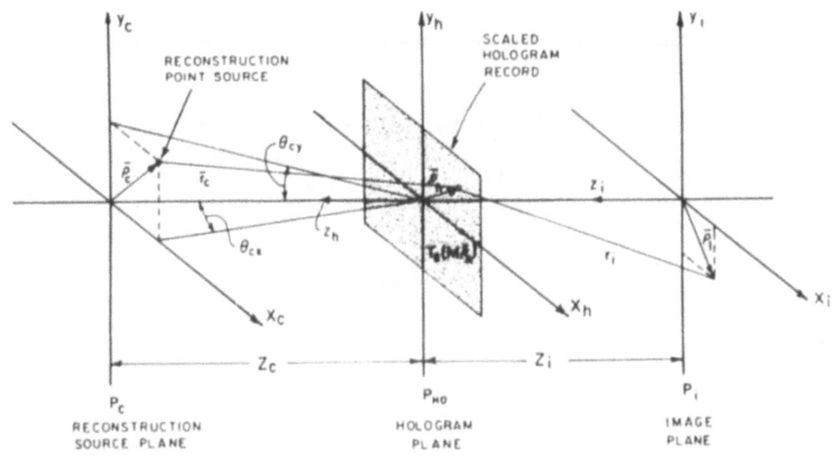

Fig. 2 GENERALIZED RECONSTRUCTION OF SAMPLED HOLOGRAM

We proceed next to determine analytically the effect of sampling on the retrieved image.

The value of the CFA $\Psi(\bar{\rho}_h)$ at the field point P on S_H produced by the propagating object field can be obtained from the object field $O(\bar{\rho}_o)$ by making use of the Fresnel-Kirchhoff diffraction integral

$$\Psi(\bar{\rho}_h) = \frac{j}{2\lambda} \int_{P_o} O(\bar{\rho}_o) \frac{e^{-jkr}}{r} (1 + \bar{1}_n \cdot \bar{1}_r) \, d\bar{\rho}_o \quad (1)$$

where $k = 2\pi/\lambda$, $d\bar{\rho}_o$ is used to designate an element of area in the object plane P_o, and $r = |\bar{r}|$ is the magnitude of the vector

$$\bar{r} = \bar{\rho}_h - \bar{\rho}_o - Z_o \bar{1}_z \quad (2)$$

directed from the element of area in P_o to the field point P on S_H as shown in Fig. 1.

In practice we are interested in the case when Z_o is much greater than $(2X_h, 2Y_h)$ and than the extent of a region of linear dimensions roughly equal to the lateral extent of the object over which $O(\bar{\rho}_o)$ is defined. Under this condition $\bar{1}_n \cdot \bar{1}_r \simeq 1$ for all points in P_o for which $O(\bar{\rho}_o)$ is not zero. Accordingly, equation (1) assumes the simpler form,

$$\Psi(\bar{\rho}_h) = \frac{j}{\lambda} \int_{P_o} O(\bar{\rho}_o) \frac{e^{-jkr}}{r} \, d\bar{\rho}_o \quad (3)$$

where

$$r = (\bar{r} \cdot \bar{r})^{1/2} \simeq Z_o + \frac{1}{2Z_o} (\rho_o^2 + \rho_h^2 - 2\bar{\rho}_o \cdot \bar{\rho}_h) \quad (4)$$

is to be used in the exponent and $r \simeq Z_o$ in the denominator of the integral. This leads to the following explicit form of (3)

$$\Psi(\bar{\rho}_h) = \frac{j}{\lambda Z_o} e^{-jkZ_o(1+\frac{\rho_h^2}{2Z_o^2})} \int_{P_o} O(\bar{\rho}_o)\, e^{-j\frac{k}{2Z_o}\rho_o^2}\, e^{j\frac{k}{Z_o}\bar{\rho}_o\cdot\bar{\rho}_h} d\bar{\rho}_o \tag{5}$$

The reference CFA over the hologram recording plane produced by the spherical wave generated by the reference point source can be expressed as

$$R(\bar{\rho}_h) = R_o' \frac{e^{-jkr_R}}{r_R}$$

$$\simeq R_o\, e^{-j\frac{k}{2Z_R}(\rho_h^2 + \rho_R^2 - 2\bar{\rho}_h\cdot\bar{\rho}_R)} \tag{6}$$

where $R_o = (R_o'/Z_R)\exp(-jkZ_R)$ is a complex quantity proportional to the reference source intensity.

The total CFA in the hologram recording aperture will, therefore, be

$$H(\bar{\rho}_h) = \Psi(\bar{\rho}_h) + R(\bar{\rho}_h) \tag{7}$$

The scanned square law sensor will provide a sampled version of the intensity distribution $I(\bar{\rho}_h) = HH^*$ over the hologram recording aperture S_H which we will designate by

$$I_s(\bar{\rho}_h) = S(\bar{\rho}_h)\, I(\bar{\rho}_h) = S(\bar{\rho}_h)[I_o(\bar{\rho}_h) + I_p(\bar{\rho}_h) + I_c(\bar{\rho}_h)] \tag{8}$$

where

$$I_o = |\Psi|^2 + |R|^2$$

$$I_p = R^*\Psi$$

$$I_c = R\Psi^*$$

give rise to the zero order light, the primary image, and the conjugate image, respectively, in the reconstruction process. $S(\bar{\rho}_h)$ is the sampling function representing the nature of the sampling format and the characteristics of the receiving aperture of the detector used.

Image retrieval from the sampled hologram data (8) involves the generation of an optical transparency record whose CFA transmittance is linearly proportional everywhere to a scaled replica of the sampled intensity distribution (8). This leads to a transparency with amplitude transmittance

$$T_s(M\bar{\rho}_h) = \beta\, I_s(M\bar{\rho}_h) \tag{9}$$

where β is a real constant dependent on the film characteristics and development. Since the intensity I_s is real, T_s is a real function.

Images are retrieved from the scaled sampled hologram (9) by placing it in the generalized image retrieval geometry shown in Fig. 2 where the transparency is assumed to be illuminated by an off-axis coherent reconstruction point source of wavelength λ_L located a distance Z_c in front of the hologram plane. The CFA produced in the hologram transparency plane by the reconstruction point source may be expressed as

$$C(\bar{\rho}_h) = C_o' \frac{e^{-jk_L r_c}}{r_c} \simeq C_o\, e^{-j\frac{k_L}{2Z_c}(\rho_h^2 + \rho_c^2 - 2\bar{\rho}_h\cdot\bar{\rho}_c)} \tag{10}$$

where

$$r_c = |\bar{\rho}_h - \bar{\rho}_c - Z_c\,\bar{1}_z| \simeq Z_c + \frac{1}{2Z_c}(\rho_h^2 + \rho_c^2 - 2\bar{\rho}_h\cdot\bar{\rho}_c) \tag{11}$$

is the distance between the reconstruction point source and a point on the hologram transparency. Here

$$C_o \simeq \frac{C_o'}{Z_c}\, e^{-jk_L Z_c} \tag{11a}$$

and C_o' is a factor dependent on the strength of the reconstruction source and $k_L = 2\pi/\lambda_L$. By allowing Z_c to become infinite we can represent a point source at infinity which produces a reconstructing plane wave obliquely incident on the hologram transparency. By changing θ_{xc} and θ_{yc} in Fig. 2 the obliquity angle of the reconstructing wavefront can be changed. The geometry of Fig. 2 is, therefore, quite general and covers a variety of cases.

Referring to the reconstruction geometry described, the CFA in the plane P_i obtained by applying the Fresnel-Kirchhoff diffraction integral will be

$$G(\bar{\rho}_i) = \frac{j}{\lambda_L Z_i} e^{-jk_L Z_i \left(1 + \frac{\rho_i^2}{2Z_i^2}\right)} \int_{-\infty}^{\infty} T_s(M\bar{\rho}_h)\, C(\bar{\rho}_h)\, e^{-j \frac{k_L}{2Z_i} \rho_h^2} \times e^{j \frac{k_L}{Z_i} \bar{\rho}_i \cdot \bar{\rho}_h}\, d\bar{\rho}_h \ . \qquad (12)$$

We will concentrate in the following on the effect of sampling on the primary image arising from the term $I_p(\bar{\rho}_h)$ in (8) only. The conclusions drawn are applicable, however, also to the conjugate image caused by $I_c(\bar{\rho}_h)$. The primary image portion of (12) can, therefore, be written as,

$$G_p(\bar{\rho}_i) = \frac{j\beta}{\lambda_L Z_i} e^{-jk_L Z_i \left(1 + \frac{\rho_i^2}{2Z_i^2}\right)} \int_{-\infty}^{\infty} S(M\bar{\rho}_h)\, R^*(M\bar{\rho}_h)\, \Psi(M\bar{\rho}_h)\, C(\bar{\rho}_h)\, e^{-j \frac{k_L}{2Z_i} \rho_h^2}\, e^{j \frac{k_L}{Z_i} \bar{\rho}_i \cdot \bar{\rho}_h}\, d\bar{\rho}_h \ . \qquad (13)$$

Substituting for $\Psi(\bar{\rho}_h)$, $R(\bar{\rho}_h)$ and $C(\bar{\rho}_h)$ from equations (5), (6), and (10) we obtain,

$$G_p(\bar{\rho}_i) = \beta C_1 C_2 R_o^* C_o \, e^{+j\frac{k}{2Z_R}\rho_R^2} \, e^{-j\frac{k_L}{2}\left(\frac{\rho_i^2}{Z_i} + \frac{\rho_c^2}{Z_c}\right)} \iint_{-\infty}^{\infty} S(M\bar{\rho}_h)\, O(\bar{\rho}_o)$$

$$e^{-j\frac{k}{2Z_o}\rho_o^2} \, e^{j\frac{kM}{Z_o}\bar{\rho}_h\cdot\bar{\rho}_o} \, e^{-j\frac{kM^2}{2Z_o}\rho_h^2} \, e^{j\frac{kM^2}{2Z_R}\rho_h^2} \, e^{-j\frac{kM}{Z_R}\bar{\rho}_h\cdot\bar{\rho}_R}$$

$$e^{-j\frac{k_L}{2Z_c}\rho_h^2} \, e^{j\frac{k_L}{Z_c}\bar{\rho}_h\cdot\bar{\rho}_c} \, e^{-j\frac{k_L}{2Z_c}\rho_h^2} \, e^{j\frac{k_L}{Z_i}\bar{\rho}_h\cdot\bar{\rho}_i} \, d\bar{\rho}_o \, d\bar{\rho}_h \qquad (14)$$

where $C_1 = je^{-jkZ_o}/\lambda Z_o$ and $C_2 = je^{-jk_L Z_i}/\lambda_L Z_i$.

When the sampling function $S(\bar{\rho}_h)$ is real and non-focusing*, i.e., it cannot be expressed in terms of quadratic exponential such as $\exp(\pm j\alpha\bar{\rho}_h)$ or $\exp(\pm j\beta\bar{\rho}_h^2)$, where α and β are real constants, we can rewrite (14) by collecting exponent terms in $\bar{\rho}_h$ and ρ_h^2 obtaining

$$G_p(\bar{\rho}_i) = F \iint_{-\infty}^{\infty} O(\bar{\rho}_o)\, e^{-j\frac{k}{2Z_o}\rho_o^2} \, S(M\bar{\rho}_h)\, e^{-j\left[\frac{kM^2}{Z_o} - \frac{kM^2}{Z_R} + \frac{k_L}{Z_c} + \frac{k_L}{Z_i}\right]\rho_h^2/}$$

$$\times \; e^{j\left[\frac{kM}{Z_o}\bar{\rho}_o - \frac{kM}{Z_R}\bar{\rho}_R + \frac{k_L}{Z_c}\bar{\rho}_c + \frac{k_L}{Z_i}\bar{\rho}_i\right]\cdot\bar{\rho}_h} \, d\bar{\rho}_o \, d\bar{\rho}_h \qquad (15)$$

where

$$F = \beta C_o R_o^* C_1 C_2 \, e^{+j\frac{k}{2Z_R}\rho_R^2} \, e^{-j\frac{k_L}{2Z_c}\rho_c^2} \, e^{-j\frac{k_L}{2Z_i}\rho_i^2}$$

is a complex factor dependent on the positions of the reference and reconstruction point sources and on the position vector $\bar{\rho}_i$ in the image plane.

---

*Examples of a real focusing sampling function are the sinusoidal Fresnel pattern $S(\bar{\rho}_h) = 1 - \cos\beta\, \rho_h^2$ and the Fresnel zone pattern.

The evaluation of this integral can be greatly simplified by making

$$\frac{kM^2}{Z_o} - \frac{kM^2}{Z_R} + \frac{k_L}{Z_c} + \frac{k_L}{Z_i} = 0 \qquad (16)$$

a relation that can be satisfied for a given Z_o by proper choice of the parameters λ, λ_L, Z_i, Z_R, and Z_c. Since (16) relates the distance Z_i of the image plane P_i from the hologram transparency plane to the other recording and reconstruction parameters it is generally called the image "focusing condition".

In view of (16) the integration with respect to $\bar{\rho}_h$ in the remaining double integral (15) can now be evaluated yielding,

$$\int_{-\infty}^{\infty} S(M\bar{\rho}_h)\, e^{-j\bar{\omega}_i \cdot M\bar{\rho}_h}\, d(M\bar{\rho}_h) = (2\pi)^2 M\, \tilde{S}(\bar{\omega}_i) \qquad (17)$$

where $\tilde{S}(\bar{\omega}_i)$ is the Fourier transform of the sampling function $S(M\bar{\rho}_h)$ and

$$\bar{\omega}_i = -\frac{k}{Z_o}\bar{\rho}_o + \frac{k}{Z_R}\bar{\rho}_R - \frac{k_L}{MZ_c}\bar{\rho}_c - \frac{k_L}{MZ_i}\bar{\rho}_i \qquad (18)$$

Substituting (17) into (15) we obtain,

$$G_p(\bar{\rho}_i) = (2\pi)^2 MF \int_{-\infty}^{\infty} O(\bar{\rho}_o)\, e^{-j\frac{k}{2Z_o}\rho_o^2}\, \tilde{S}\Big[-\frac{k}{Z_o}\bar{\rho}_o + \frac{k}{Z_R}\bar{\rho}_R - \frac{k_L}{MZ_c}\bar{\rho}_c - \frac{k_L}{MZ_i}\bar{\rho}_i\Big]\, d\bar{\rho}_o \qquad (19)$$

We consider now the case of a point object CFA, i.e., the case when $O(\bar{\rho}_o) = \delta(\bar{\rho}_o)$ where δ is the Dirac delta "function". The general result (19) yields then the impulse response or the point spread function of the linear holographic imaging process considered, namely,

$$D_p(\bar{\rho}_i) = (2\pi)^2 MF \int_{-\infty}^{\infty} \delta(\bar{\rho}_o)\, e^{-j \frac{k}{2Z_o} \rho_o^2}$$

$$\tilde{S}\left[-\frac{k}{Z_o}\bar{\rho}_o + \frac{k}{Z_R}\bar{\rho}_R - \frac{k_L}{MZ_c}\bar{\rho}_c - \frac{k_L}{MZ_i}\bar{\rho}_i\right] d\bar{\rho}_h \tag{20}$$

which yields by the sifting property of the delta "function"

$$D_p(\bar{\rho}_i)\ (2\pi)^2 MF\, \tilde{S}\left[-\frac{k_L}{MZ_i}\bar{\rho}_i - \frac{k_L}{MZ_c}\bar{\rho}_c + \frac{kM}{Z_R}\bar{\rho}_R\right] \tag{21}$$

where

$$\tilde{S}(\bar{\xi}) = \frac{1}{M} \int_{-\infty}^{\infty} S(M\bar{\rho}_h)\, e^{-j\bar{\xi}\cdot M\bar{\rho}_h}\, d(M\bar{\rho}_h) \tag{21a}$$

with

$$\bar{\xi} = -\frac{k_L}{Z_i M}\bar{\rho}_i - \frac{k_L}{Z_c M}\bar{\rho}_c + \frac{k}{Z_R}\bar{\rho}_R \tag{21b}$$

Thus, the CFA in the primary image of a point object observed in a plane P_i whose distance Z_i from the hologram transparency is given by the focusing condition (16) is within a complex constant the Fourier transform of the scaled sampling function.

Since the holographic process considered is spatially invariant and linear, the primary image for an arbitrary object function $O(\bar{\rho}_o)$ would be given, according to linear system theory, by the convolution of the primary image $O(\bar{\rho}_i)$ of this object function under ideal (non-sampled) recording conditions, with the point spread function $D_p(\bar{\rho}_i)$ of (21). This general result is seen to be independent of the hologram type (i.e., Fresnel or Fraunhofer) and the geometry used.

Since the convolution of the image CFA function obtained under ideal non-sampled recording conditions with a delta function yields back the function itself unaltered, we arrive at the important conclusion that the most desirable sampling formats are those whose Fourier transforms approach a single delta function. Evidently, the ideal image CFA is obtained when $S(\bar{\rho}_h) = 1$ for all values of $\bar{\rho}_h$ since the Fourier transform of this sampling function is a delta function.

When the transform of the sampling function possesses a single peak approaching a delta function, the image obtained using such a sampling function would be accordingly centered at

$$\bar{\rho}_i = M \frac{k}{k_L} \frac{Z_i}{Z_R} \bar{\rho}_R - \frac{Z_i}{Z_c} \bar{\rho}_c \tag{22}$$

in the image plane. Thus, for example, when normally incident reference and reconstruction plane waves are used respectively in the recording and reconstruction steps we would have $\bar{\rho}_R = \bar{\rho}_c = 0$ and $Z_R = Z_c = \infty$ representing point sources at infinity. Then the retrieved image would be centered at $\bar{\rho}_i = 0$, the origin of the image plane.

CIRCULAR SAMPLING FUNCTIONS

The results of the preceding analysis will be used in this section to account for the effect of continuous sampling when a spiral scan format is used in the recording of longwave holograms.

Consider the case of an idealized uniform spiral scan (with constant scan line separation) illustrated in Fig. 3. The pattern shown could be synthesized by moving a discrete transducer (with aperture function $g(\bar{\rho}_h)$) over the hologram recording plane utilizing constant angular velocity ω and a constant radial velocity v_r such that

$$\frac{2\pi}{\omega} v_r = \Delta$$

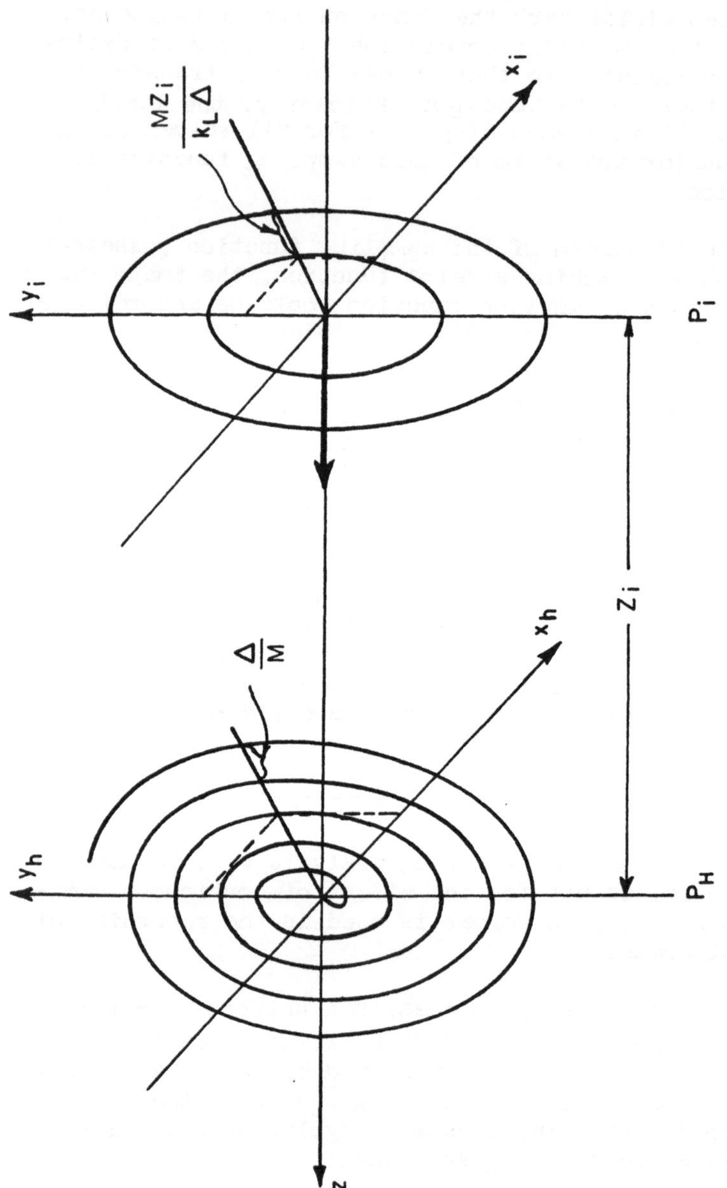

Fig. 3 THE SPIRAL SCAN PATTERN AND ITS APPROXIMATE FOURIER TRANSFORM.

where Δ is the separation between two adjacent scan lines. Because of its continuous and fly-back free nature, the spiral scan format is easy to implement in practice; however, it is quite difficult to develop analytically its Fourier transform.

To simplify the analysis, we will assume the pitch $(\frac{\Delta}{M})$ of the scan to be small and represent the spiral by a series of concentric circles so that the sampling function can be expressed as,

$$S(M\bar{\rho}_h) = g(M\bar{\rho}_h) * \sum_{n=1}^{N} \delta[(\bar{\rho}_h \cdot \bar{\rho}_h)^{1/2} - n \frac{\Delta}{M}] \tag{23}$$

where N approaches infinity for an unbounded scan pattern. The Fourier transform of (23) is

$$\mathfrak{F}\{S(M\bar{\rho}_h)\} = \mathfrak{F}\{g(M\bar{\rho}_h)\}\ \mathfrak{F}\{\sum_{n=1}^{N} \delta[(\bar{\rho}_h \cdot \bar{\rho}_h)^{1/2} - n \frac{\Delta}{M}]\} \tag{24}$$

where $\mathfrak{F}\{S(M\bar{\rho}_h)\}$ is defined by (21a). For the case of $\bar{\rho}_R = \rho_c = 0$, the Fourier transform of the concentric circle scan pattern can be shown to be [11],

$$\mathfrak{F}\{\sum_{n=1}^{N} \delta[(\bar{\rho}_h \cdot \bar{\rho}_h)^{1/2} - n \frac{\Delta}{M}]\} = \sum_{n=1}^{N} \frac{k_L n\Delta}{MZ_i} J_o(\frac{k_L n\Delta}{MZ_i} \bar{\rho}_i) \ . \tag{25}$$

The resulting summation of zero order Bessel functions in (25) will determine the properties of the Fourier transform of the spiral scan pattern. Since the index of the summation n is contained in both the argument of the Bessel function and the multiplying factor, the terms involving large n will be weighted very heavily in the result. The Poisson approximation [14] for large and positive argument can be used, therefore, to evaluate approximately the variations of the CFA along the radial direction ρ_i in the Fourier transform plane (the image plane in the holographic process), through the large argument approximation

$$J_o(\eta_i) = \left(\frac{2}{\pi\eta_i}\right)^{1/2} \left[\cos(\eta_i - \frac{\pi}{4})\cdot P(\eta_i,0) + \sin(\eta_i - \frac{\pi}{4})\cdot Q(\eta_i,0)\right] \qquad (26)$$

where $\eta_i = k_L n\Delta/MZ_i\ \rho_i$, $P(\eta_i,0)$ and $Q(\eta_i,0)$ are asymptotic Poisson series [14].

The periodic nature of the expansion in (26) shows that for an unbounded scan pattern ($N \to \infty$), the CFA will consist of an impulse function at the origin with an amplitude proportional to $N^2/2$. Surrounding the origin at multiples of the interval which corresponds to the fundamental frequency of the scan (i.e., at radial distance $\rho_i = \alpha\ MZ_i/k_L\Delta$) in the image plane are concentric rings of zero width whose amplitudes are proportional to $N^{3/2}\ \alpha^{-1/2}/2\pi$, where α is the order of the diffraction term (i.e., $\alpha = 1$ is the first ring, $\alpha = 2$ is the second, etc.). Therefore, the ratio of the amplitude of the central peak to the amplitude of the concentric rings is $\pi\,\alpha^{1/2}\ N^{1/2}$.

In view of the fact that the retrieved image is the convolution of the Fourier transform of the scan pattern with the retrieved image obtained without sampling, the above results show that the signal to noise ratio in the image due to the spiral sampling action is a function of both radial position in the image (represented by the order α) and the number of scan lines utilized (represented by N).

To evaluate (25) for finite N and thus determine the effect of concentric scan and spiral scan sampling, equation (25) was computed utilizing a Spectra 70 digital computer for the specific case of a circular aperture ($D = 1$) which was sampled by ten concentric scan lines ($N = 10$). A plot of the normalized radial CFA distribution obtained shown in Fig. 4b clearly demonstrates the effects of the spiral scan at the origin (PM_o) and around the positions (PM_1) and (PM_2), respectively, corresponding to the first and second diffraction rings ($\alpha = 1$ and $\alpha = 2$). The Fourier transform of an unsampled circular aperture of diameter ($D = 1$) is shown plotted in Fig. 4a for comparison.

In the neighborhood of $\rho_i = MZ_i/k_L\Delta$ and $\bar{\rho}_i = 2MZ_i/k_L\Delta$

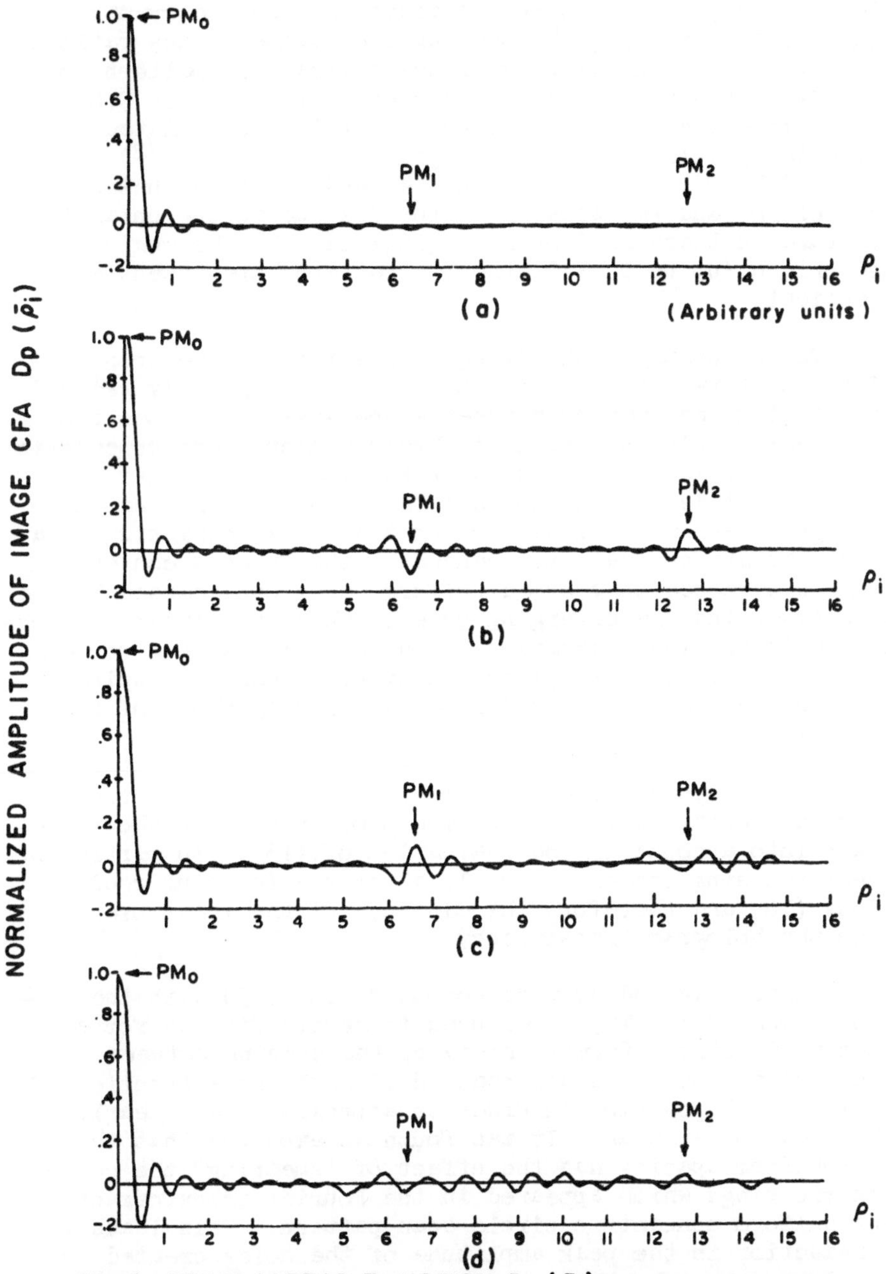

Fig. 4 POINT SPREAD FUNCTION $D_p(\bar{\rho}_i)$ OF A CONCENTRIC CIRCLE SCAN FORMAT WITH N = 10 AND D = 1.

(PM_1 and PM_2), the pattern has additional secondary maxima and minima. From Fig. 4b the measured value of the ratio of the CFA at the central peak of the diffraction pattern to the CFA at the first and second concentric rings is 9.8 and 13.7, respectively. The theoretical values calculated using the previously derived expression ($\pi \alpha^{1/2} N^{1/2}$) are 9.94 and 14.05. Comparing Figs. 4a and 4b in the neighborhood of PM_1 and PM_2 shows that the CFA due to the unsampled aperture in these regions is negligible. In the region near the origin (up to $\rho_i = 5$) the two patterns are essentially identical.

The secondary peaks (rings) appearing in the transform of the uniform circular scan pattern can be greatly reduced by changing the spacing between successive scan lines progressively during the scan to obtain a nonuniform concentric scan pattern resembling somewhat the familiar Fresnel zone pattern. It is well known that the sinusoidal Fresnel pattern possesses an optical Fourier transform consisting of a single bright central spot caused by the focused converging order superposed on a broad uniform background caused by the zero order and the diverging order. Thus, by linearly changing the scan line separation in the spiral scan one would expect to realize a sampling function, a "chirpped spiral scan," whose transform would resemble closely those of the Fresnel zone.

It is important to emphasize, however, that chirpped sampling formats possess focusing properties that should be taken into account in the evaluation of (15). In particular, these focusing properties would alter the focusing conditions (16) and therefore the distance of the image plane from the hologram transparency.

Figures 4c and 4d were computed from (25) with the same conditions of $N = 10$, $D = 1$ used to derive Fig. 4b for a concentric circle format, however, the spacing between successive scan lines was changed linearly as a function of their radial position in order to approximate a linearly chirpped spiral scan. It was found as expected that the non-uniform spacing has the effect of "smearing" the concentric rings which appeared in the Fourier transform of the uniform concentric circle scan pattern. This leads to a reduction in the peak amplitude of the noise created in the image plane by the scanning action. In Fig. 4c the rate of chirpping, the variation of the chirpped scan pattern from the constant scan, was very slight and therefore the basic ring structure is still evident. However, an improvement in peak to maximum sidelobe of about 2.5 dB has been realized. In Fig. 4d a considerably higher rate of chirpping was used so that the discrete ring structure is no longer distinguishable. The Fourier transform pattern now consists of several rings of approximately equal amplitude whose peaks are considerably less than that of the single ring for the uniform concentric scan case. In this particular case an improvement in peak to maximum sidelobe ratio of 9.0 dB was realized.

OPTICAL VERIFICATION OF THE PROPERTIES OF SPIRAL SCAN SAMPLING FUNCTIONS

Spiral scan formats were generated using an analog computer (Systron Donner Model SD 80). The simulation programs used are depicted in Figs. 5a and 5b for the uniform and the linearly chirpped spirals, respectively. The equations being simulated are also included. Graphical plots of the spirals were obtained by displaying y_1 versus y_2 on an x-y recorder or a CRT. Transparency records with the desired size were obtained from the display using ordinary photographic reduction methods.

For the two kinds of spirals the analog computer was programmed first to produce on the display a repetitive circular pattern. Then the respective x and y voltages driving the display were amplitude modulated with a ramp voltage by a process of analog multiplication. In the case of a uniform spiral the circular pattern was obtained by simulating the harmonic differential equation $\ddot{x}_1 + \omega^2 x_1 = 0$ subject to the initial condition $x_1(0) = k$. The solution, $x_1 = k \cos\omega t$, was integrated with a proper scaling factor to yield $x_2 = k \sin\omega t$. These two voltages are recognized as the parametric description of a circle. Their multiplication by a ramp voltage, $t/30$, before being displayed resulted in a plot of a uniform spiral defined by $y_1 = (Kt/30)\cos\omega t$, $y_2 = (Kt/30)\sin\omega t$. The number of scan lines per unit radial distance of the resultant pattern was set to a convenient value by choosing the angular frequency to be $\omega = 4\pi$ rad/s. The same procedure essentially was followed in the case of the linear chirpped spiral except that here the angular frequency increases linearly with time according

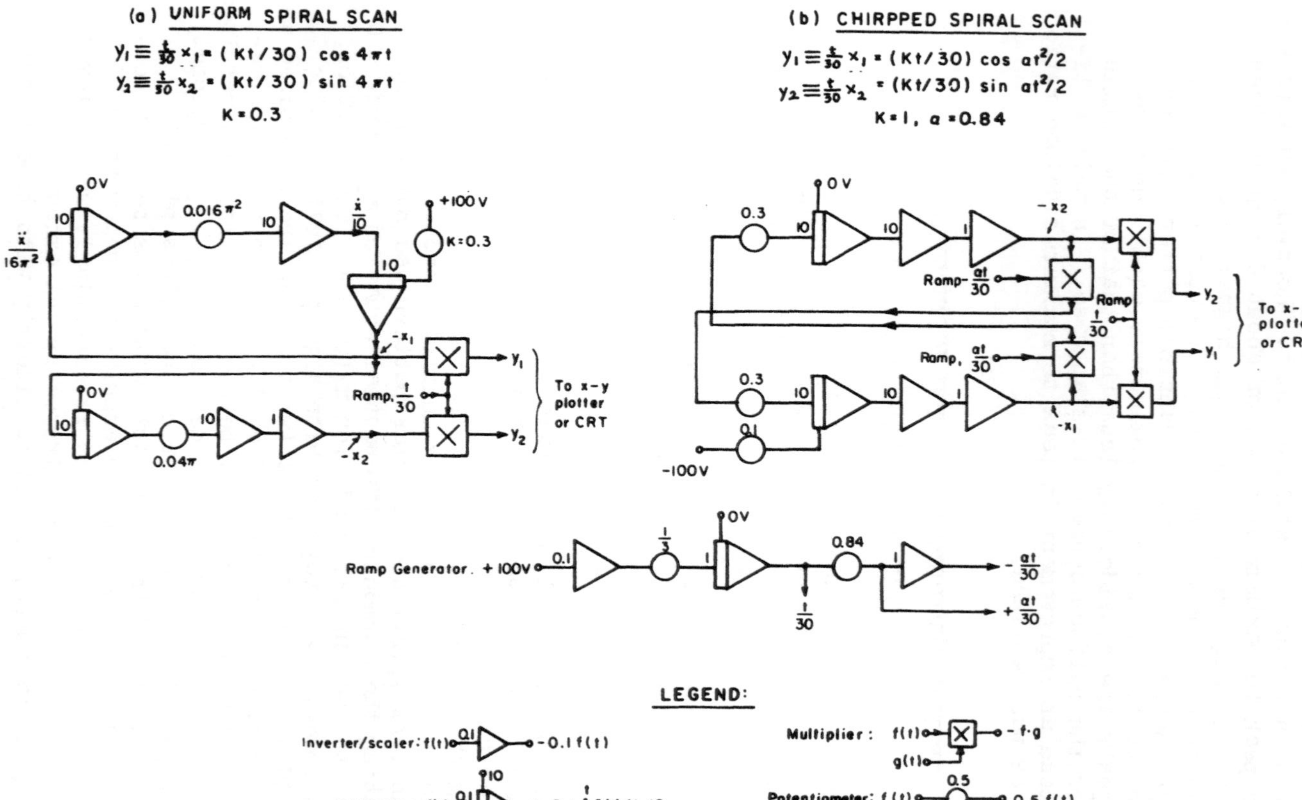

Fig. 5 ANALOG COMPUTER SIMULATION DIAGRAM FOR UNIFORM (a) AND CHIRPPED (b) SPIRAL SCANS

to $\omega = \alpha t$. To realize this, two frequency modulated voltages, $x_1 = K\cos(\alpha t^2/2)$, $x_2 = K\sin(\alpha t^2/2)$ are required. These voltages were generated by simulating the pair of simultaneous equations $\dot{x}_1 = -\alpha t x_2$, $\dot{x}_2 = \alpha t x_1$ subject to the initial conditions $x_1(0) \neq 0$, $x_2(0) = 0$. The constants K and α were chosen conveniently to be ~ 1 and 0.84, respectively, the latter insuring that the average scan line density of the chirpped spiral be equal to that for the uniform spiral.

Photographs of uniform and linearly chirpped spiral scans obtained as described above are shown in Figs. 6 and 7, respectively, together with photographs of their Fourier transforms. The Fourier transforms were obtained optically by illuminating reduced transparency replicas of these spiral scans with a spatially filtered and collimated He-Ne laser beam and observing the Fourier transform patterns in the back focal plane of a convergent lens subjected to the wavefront emerging from the illuminated transparencies.

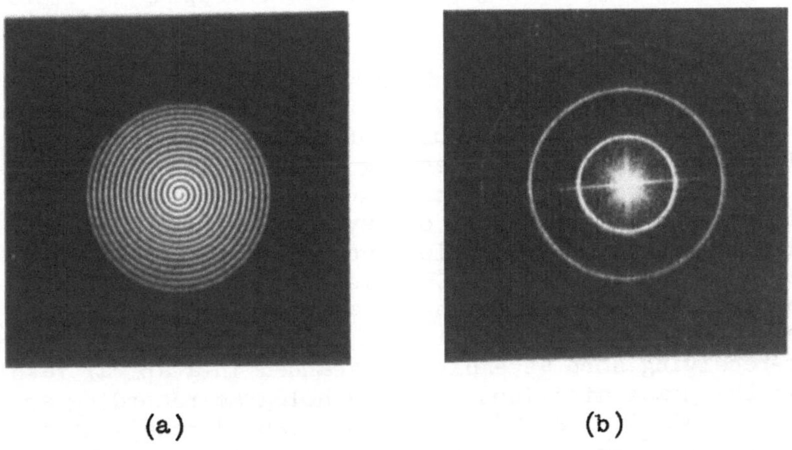

(a) (b)

Fig. 6. Uniform spiral (a) and its Fourier transform (b) (overexposed to show third secondary ring).

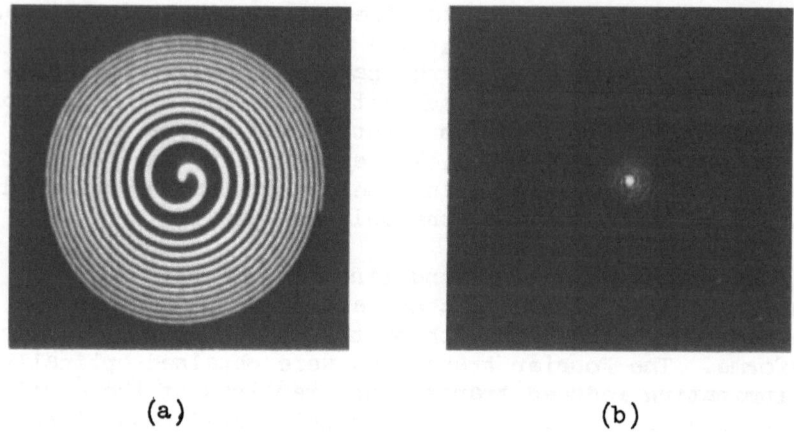

(a) (b)

Fig. 7. Chirpped spiral (a) and its Fourier transform (b).

Inspection of Figs. 6 and 7 shows clearly the disappearance of the secondary peak rings when linear chirpping is employed.

SPIRAL SCANNING IN MICROWAVE HOLOGRAPHY

In this section the results of microwave holographic imaging of a variety of reflecting objects when spiral scanning was employed in the recording process are described. Millimeter wave illumination of wavelength $\lambda = 4.3$ mm was utilized and coherent detection techniques and electronic processing used to obtain a CRT display representing the phase distribution over the hologram recording aperture caused by the object wavefield. A millimeter wave harmonic mixer-receiving horn assembly was scanned in a spiral fashion to map the phase distribution. The hologram recording aperture had a diameter of 75 cm and the scanned receiving horn diameter was .6 cm. A simplified block diagram of the recording system is shown in Fig. 8(a). Detailed description of the system has been given elsewhere [15].

A sketch of the spiral scanner used to scan the millimeter wave harmonic mixer-receiving horn assembly is shown in Fig. 9. The spiral scan is realized by combining radial and circular motions. The mm wave harmonic mixer-horn assembly is mounted on a carriage driven by a small radial

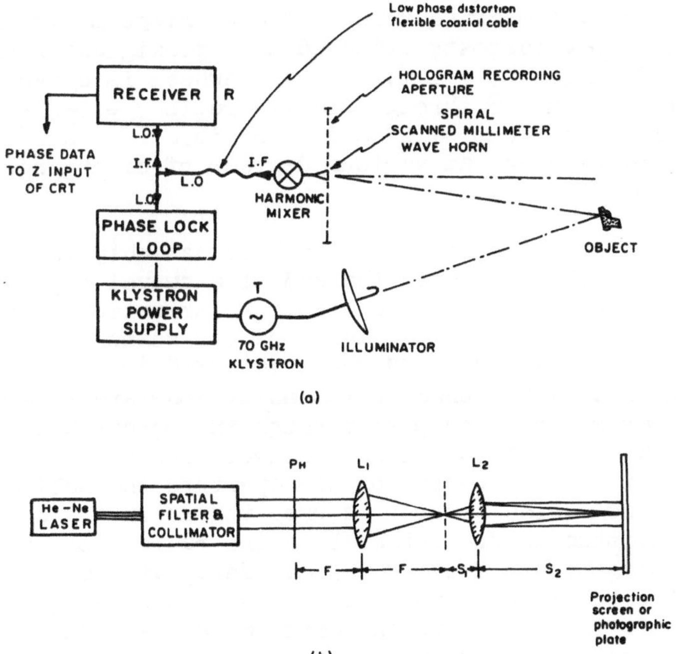

Fig. 8. Recording geometry of millimeter wave hologram employing spiral scanning (a) and image reconstruction arrangement (b).

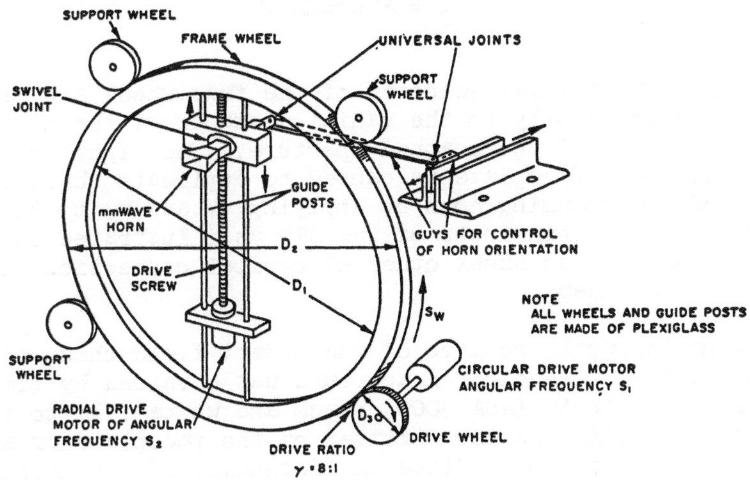

Fig. 9. Spiral scanner.

motion motor and a drive screw. The carriage motion is guided by two guideposts attached to a plexiglass framewheel of inner diameter $D_1 \approx 1$ m. The framewheel is driven by a circular drive motor through a drive ratio $\gamma = D_2/D_3$. Both motors are speed controlled and reversible in order to achieve spiral scans at various degrees of divergence or convergence.

The orientation of the mixer-horn assembly is maintained fixed during the scan with the aid of a swivel joint in order to accommodate for the linear polarization of the field. To display the i.f. output signal of the scanned mixer, it was amplified, limited, and phase detected in the receiver shown in Fig. 8a and used to intensity modulate a CRT. The radial scan motor is coupled through an appropriate gear ratio to a high quality linear helipot such that when the mm wave horn is at the center of the aperture the helipot is at zero position. The helipot is used to generate a dc voltage whose amplitude is proportional to the radial position of the scanning horn. This dc voltage designated by V_{in} is applied to the input terminals of a sine-cosine potentiometer whose shaft is coupled to the circular drive motor through a reduction ratio γ. In this fashion synchronous rotation of the framewheel and the sine-cosine potentiometer are insured. The potentiometer will generate two voltages,

$$\begin{aligned} v_1 &= A \sin\Omega t \\ v_2 &= A \cos\Omega t \ , \end{aligned} \tag{27}$$

where Ω is the angular velocity of the framewheel and $A = V_{in}$ varies proportionally to the radial distance of the horn from the center of the scanned aperture. The light spot on the display CRT faceplate is caused to duplicate the spiral motion of the scanning horn by applying v_1 and v_2 to the x and y deflection terminals of the CRT as illustrated in Fig. 10, which also shows other electrical connections of the spiral scanner.

A photographic replica of the object field phase distribution over the hologram aperture was obtained by time exposing a polaroid (ASA 3000) black and white film to the image of the light pattern written on the faceplate of a CRT by the light spot. Since the intensity of the moving light spot is proportional to the phase signal provided at

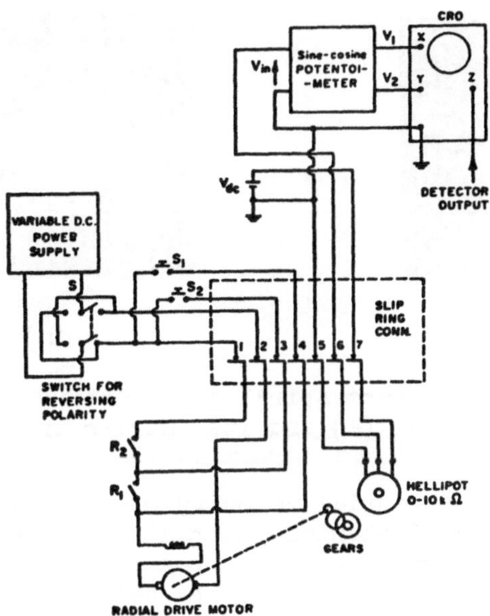

Fig. 10. Diagram of electrical connections and arrangement for display of output of scanner. To limit the radial motion of the horn-mixer carriage, relays R_1 and R_2 break when horn reaches central or outer positions of the scan, respectively. To reactivate in each instance, the polarity switch S is reversed and either of the "push-to-make switches S_1 or S_2 is activated depending on whether reactivation from a central or outer position is required.

the output of the receiver, a phase only hologram or "phasigram" whose optical density is everywhere proportional to the phase of the object field is thus obtained. Because the relative phase between the klystron output and the L.O. does not change during the scan, the phasigram obtained in this fashion is equivalent to one recorded by employing a temporally offset coherent background reference plane wave normally incident on the hologram recording aperture.

A typical constant intensity scan pattern obtained with the spiral scanner described by fixing the Z-axis voltage to a constant value is shown in Fig. 11 together with its Fourier transform. The similarity to the Fourier transform of the uniform concentric circle sampling format discussed in the previous section is evident.

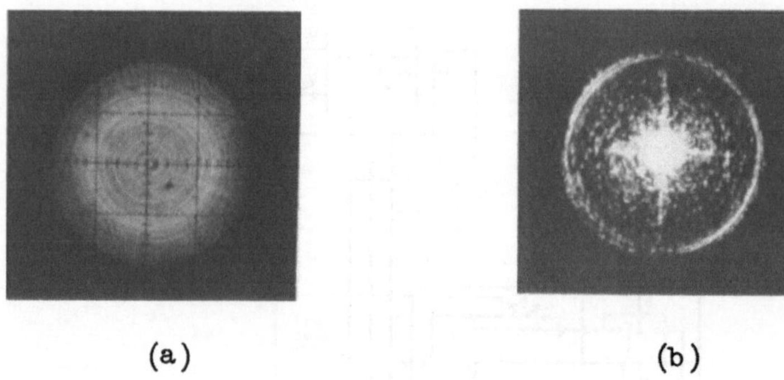

(a) (b)

Fig. 11. Typical uniform spiral scan pattern of the spiral scanner used (a) and its Fourier transform(b).

Because the tangential velocity of the spiral scan is proportional to the instantaneous radius of the scan, and because of the integrating properties of the photographic emulsion, the scan lines in Fig. 11a become fainter and thinner with increased radius. In the case of the chirpped spiral scans this property serves to make the scan resemble more closely the Fresnel zone pattern as is evident in the CRT displayed analog computer generated chirpped spiral scan shown previously in Fig. 7a.

Demagnified transparency replicas of the phasigrams displayed on the CRT were obtained by photographic means. The diameter of the transparency record was roughly 4.7 mm. Visible images were retrieved by placing the resultant transparencies in plane P_H of the image retrieval arrangement shown in Fig. 8b, where they are illuminated with a spatially filtered and collimated He-Ne laser beam. In the arrangement shown, lenses L_1 and L_2 were adjusted to focus a reconstructed real image of the object on the projection screen or photographic plate used to record this image.

Phasigrams and the images retrieved thereof are shown in Fig. 12 for three reflecting objects; a toy gun, a rectangular copper plate, and a metallic wire.

In Fig. 12a the phasigram of the object toy gun was recorded with the gun positioned a distance 1.25 m directly in front of the spiral scanner. The image retrieved is seen to bear good resemblance to the original object. It is worthwhile to mention that the recording of this hologram

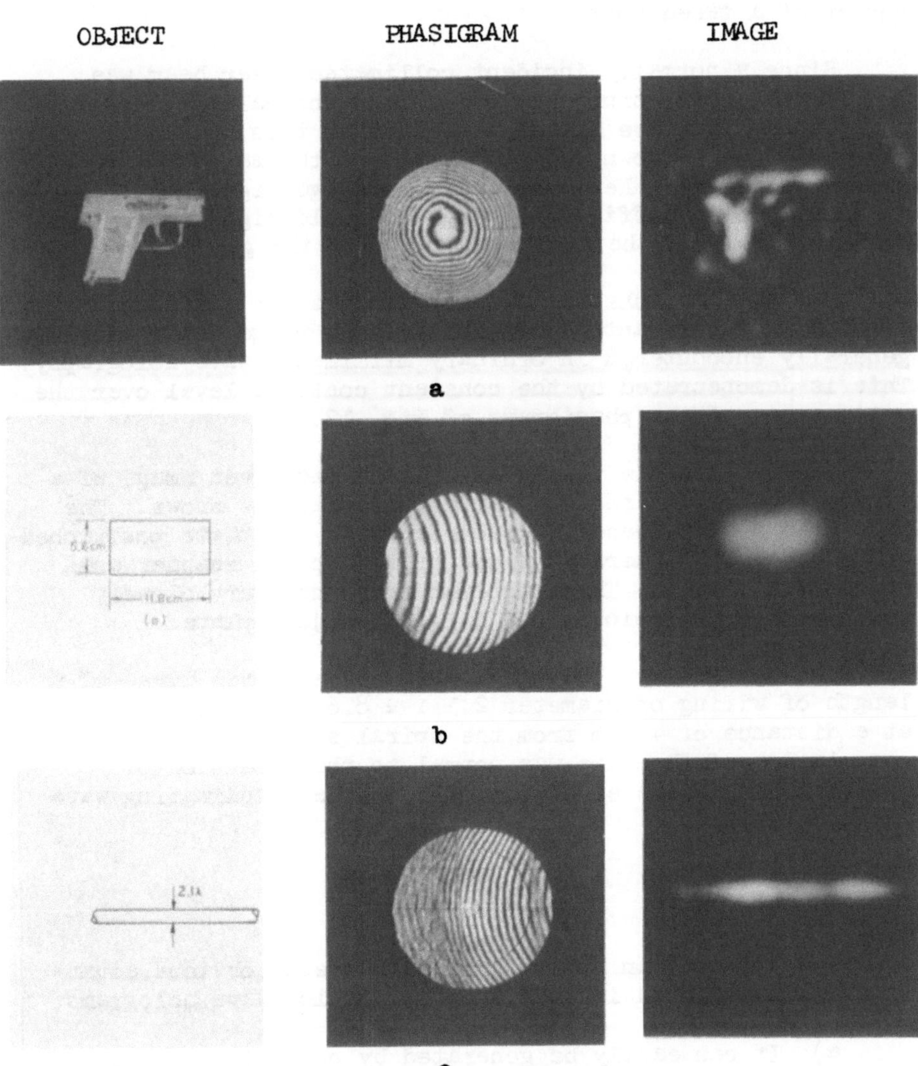

Fig. 12. Objects, phasigrams, and retrieved images of (a) a toy gun, (b) a rectangular copper plate, and (c) a wire, obtained employing uniform spiral scanning in the recording step.

was accomplished when the toy gun was concealed in the pocket of a tweed coat.

Since a normally incident collimated laser beam was used in the reconstruction and an electronically synthesized reference plane wave in the recording of this phasigram, the retrieved image shown was superposed on the zero order undiffracted light. Nevertheless, the image signal to background ratio is sufficiently high to yield high quality visualizations of the microwave image of the gun.

Working with the phase information alone was found to alleviate significantly the wide dynamic range difficulties generally encountered in ordinary millimeter wave holography. This is demonstrated by the constant contrast level over the entire area of the phasigrams of Fig. 12.

In Fig. 12b the phasigram and the retrieved image of a (5.6 cm x 11.8 cm) rectangular copper plate is shown. The phasigram in this case was recorded with the plate positioned a distance of 7 meters in front of the spiral scanner and slightly off axis. The retrieved image has very closely the same aspect ratio of the original object plate.

In Fig. 12c the phasigram and the retrieved image of a length of wiring of diameter 2.1 λ = 8.6 mm located off axis at a distance of 4.5 m from the spiral scanner is shown. The orientation of the wire was normal to the direction of polarization of the electric field in the illuminating wavefield.

CONCLUSIONS

The spiral scan format provides several obvious advantages when employed in the recording of longwave holograms:

a) It can easily be generated by combining a radial and circular motion to obtain a flyback free continuous scanning motion. By reversing the sense of the radial motion only, consecutive identical diverging and converging scans can be generated, thus eliminating flyback completely when repetitive scans are desired.

b) The resultant signal components generated by the scanned sensor are not related as in the case of the linear raster scan to the spatial frequencies of the irradiance distribution along one direction only, but because of the circular nature of the motion, the signal is related to both components of the spatial frequency vector associated with the intensity distribution being mapped.

With these attractive properties in mind a diffraction theory treatment of the effect of sampling on retrieved image quality in holography was undertaken to investigate further the properties of the spiral scan. It was found that the reconstructed image from a sampled hologram is equivalent to the convolution of the image obtained from an ideal nonsampled recording (nonsampled image) with the Fourier transform of the sampling function which represents the scan format and the aperture properties of the scanned sensor.

The Fourier transform of a uniform concentric circle sampling format approximating the uniform spiral scan was determined analytically and experimentally and found to consist of a sharp central peak surrounded by secondary radial peaks in the form of concentric rings. The secondary rings are spaced proportionately to the radial spatial frequency of the concentric scan lines. Good images can be obtained using this scan format as a result of the convolution of the nonsampled image with the sharp central peak. This image will be superposed on a faint smeared background caused by convolution of the nonsampled image with the rings. Aliasing is thus not as serious as in the linear raster scan. Obviously here also, as in the case of the raster scan, aliasing can be eliminated completely by more closely spacing the concentric scan lines. The rate of divergence or convergence of the spiral scan can be changed in a linear fashion to realize a linear chirpped spiral scan that possesses focusing properties similar to the Fresnel zone pattern.

The Fourier transform of the linearly chirpped spiral scan inferred by the well known properties of the Fresnel zone plate was determined using a digital computer and Fourier optics techniques and was found to be void of strong secondary rings and to consist of a single sharp peak surrounded by a broad low level fluctuating background. Thus, images obtained with this scan format are expected to be of good signal to background ratio.

The high quality of millimeter microwave holographic images shown were obtained making use of a spiral scanner and to demonstrate the practicality of the spiral scan. The results obtained suggest that high quality images can be obtained by circular scanning of a linear array of receivers about one of its ends or about a central point to realize a circularly scanned circular aperture. The distribution of elements along the length of the linear array can be linear or chirpped. The geometry of such an array lends itself to high revolution rates. Then, assuming that proper provision could be made to convey the outputs of sensors used to the stationary processing gear and display unit, an economical method for real time recording of longwave holograms presents itself.

A disadvantage of the spiral and circular scan formats is the **inconvenience** of interfacing their output with a computer when digital image retrieval is desired.

REFERENCES

[1] H. E. Stockman and B. Zarwin, "Optical film sensors for R.F. holography," Proc. IEEE (Letters), vol. 56, p. 763, April 1968.

[2] C. F. Augestine, et al, "Microwave holography using liquid crystal area detectors," Proc. IEEE (Letters), vol. 57, p. 1333, July 1969.

[3] L. G. Gregoris and K. Iizuka, "Visualization of internal structure by microwave holography," Proc. IEEE (Letters), vol. 58, pp. 791-792, May 1970.

[4] K. Iizuka, "Mapping of E.M. fields by photochemical reaction," Elec. Letters, vol. 9, no. 4, p. 68, February 1968.

[5] K. Iizuka, "Microwave hologram by photoengraving," Proc. IEEE (Letters), vol. 57, p. 813, May 1969.

[6] K. Iizuka, "Microwave holograms and microwave reconstruction," Elect. Letters, vol. 5, pp. 26-28, January 1969.

[7] K. Iizuka, "Mapping of electromagnetic fields by photochromics and their application in microwave holography," J. App. Phys. vol. 42, pp. 5553-5555, December 1971.

[8] A. F. Metherell, H. M. El Sum, and L. Larmore (eds), Acoustical Holography, Vol. 1, Plenum Press, New York, 1969.

[9] A. F. Metherell and L. Larmore (eds), Acoustical Holography, Vol. 2, Plenum Press, New York, 1970.

[10] A. F. Metherell (ed.), Acoustical Holography, Vol. 3, Plenum Press, New York, 1971.

[11] A. Papoulis, Systems and Transforms with Applications in Optics, p. 337, McGraw-Hill, New York, 1968.

[12] C. R. Wylie, Advanced Engineering Mathematics, McGraw-Hill, 3rd edition, New York, 1966.

[13] E. A. Chistova, Tables of Bessel Functions of the True Argument and of Integrals Derived from Them, Pergamon Press, New York, 1959.

[14] A. Gray, et al, A Treatise on Bessel Functions, 2nd edition, MacMillan, London, 1952.

[15] N. H. Farhat and W. R. Guard, "Millimeter wave holographic imaging of concealed weapons," Proc. IEEE (Letters), vol. 59, pp. 1383-1384, September 1971.

GRAPHIC DISPLAY FOR ULTRASONIC NONDESTRUCTIVE TESTING

Morio Onoe, Mikio Takagi, Taketoshi Masumoto and Nobuo Hamano
Institute of Industrial Science,
University of Tokyo, Roppongi, Tokyo,
Japan

INTRODUCTION

Acoustical holography can present a "complete" image of three-dimensional ultrasonic field. Because of this capability, its application to nondestructive testing, biomedical examination and especially underwater acoustics seems promising. Although there are several two-dimensional sensors, such as liquid surface or Polhman cell, which are able to directly make a hologram, their sensitivity is rather low. An ultimate sensitivity has been so far obtained by the scanning transducer method. In this method, a transducer is scanned over the aperture of the hologram and its output is recorded in a two-dimensional form. The reconstruction of image is done by either analog processing by coherent optics or digital processing by computer. The latter processing becomes increasingly popular because of its flexibility. The cost of digital processing is also decreasing due to the recent development of integrated circuits. This combination of scanning transducer and digital processing in acoustical holography has undoubtedly a great future. In the present state of art, however, similar capability presenting a complete image is more simply and more easily obtained by a conventional pulse echo method combined with graphic display controlled by a computer, which will be a subject of this paper.

The capability of nondestructive testing techniques heavily depends on the way of collecting, analyzing and presenting data for defect evaluation. The more dependable evaluation calls for the more thorough information. It is often desirable to present data in graphical form, so that the location and the shape of defects within in the sample under test can be seen at a glance.

The most widely used data presentation in pulsed ultrasonic testing is the A-scan on an oscilloscope screen. The vertical deflection shows the amplitude of returned signals from defects. Whereas the horizontal displacement of the trace is proportional to time and hence can be calibrated in terms of the depth of penetration from the surface. No two-dimensional information is available.

In order to obtain graphical output, the B- and C-scans are sometimes used. In the B-scan, the horizontal sweep is the same time-base as in the A-scan, but the signal modulates the intensity of a light spot on the oscilloscope screen. The vertical deflection is proportional to the movement of an ultrasonic probe. Hence, a cross-section view is obtained. In order to build up a complete picture, either a photographic film or a long-persistence screen must be used. The accuracy of signal amplitude information is rather poor because of limited linearity and dynamic range of characteristics of film or screen.

In the C-scan, the vertical and the horizontal deflections are proportional to the movement of the probe along the X and Y direction, respectively. The light intensity is also modulated by the signal as in the B-scan, but this time the signal must be first time-gated in order to avoid the overlap with such unwanted signals as transmitting pulse, back surface reflection, etc. Hence the information outside of the gate as well as the depth information are lost.

These conventional scans are often unsatisfactory for the evaluation of complex structures of defects, because the information available is limited only to one-dimensional line or prefixed two-dimensional plane.

Fig. 1 shows a proposed system of pulsed ultrasonic testing, in which the collection and the processing of data are separately handled in different places. Data are collected at fields and sent to a central processing station. Thus an emphasis at fields will be in the mechanization of data collection with minimum man power. Whereas at the center more elaborate data processing becomes possible than those available at fields. Although there are many pattern recognition techniques, a completely automatic evaluation of defects is still far from practical reality, unless the processing is limited to very simple defects. Hence human judgments will be inevitable at various stages of processing. The use of graphic display controlled by a computer will be a very convenient tool for such a man-machine interaction.

In the proposed system, original A-scan data are sampled at a considerably slower rate than the repetition frequency, digitized and stored in a computer memory together with the positional information of a scanning probe. Then the computer rearranges the data and presents the results on a graphic display. The following are main features of the present system.

(1) B-scan or C-scan display at any desirable cross-section can be obtained.

(2) Perspective or stereographic display can be obtained.

(3) In the angle beam method for testing welds, true location of defects in a structure can be displayed.

(4) Several images with such different parameters as frequency, diameter of a probe, beam angle etc., can be displayed.

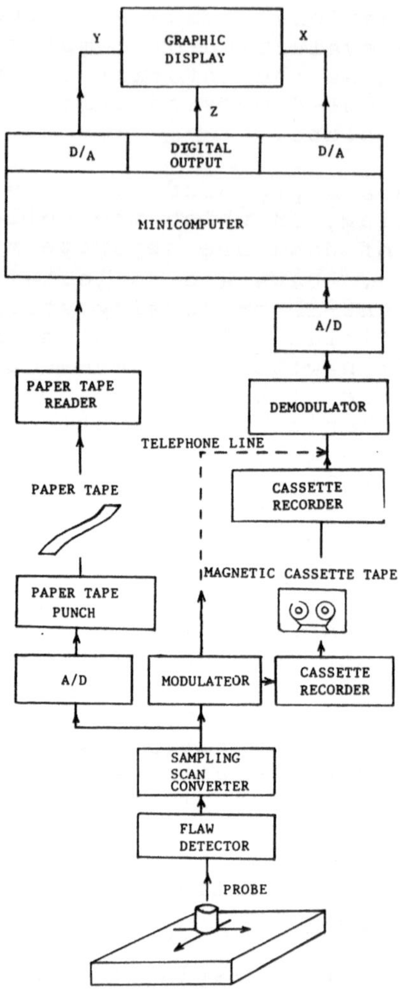

Fig. 1. Schematic diagram of the proposed system.

(5) S/N improvement by signal averaging or contrast enhancement can be easily applied.

COLLECTION OF DATA

The collection of data at fields should be economically done. The signal bandwidth of the A-scan is several hundred kHz which yields the distance resolution of a few mm in steel. A direct acquisition or recording of data with such a wide bandwidth is not feasible with low-cost instruments. Fortunately the scanning speed of an ultrasonic probe is much slower than the repetition rate of the signal. Hence the signal waveform is essentially repetitive and can be read out at a slow speed after the manner of a sampling oscilloscope. The bandwidth of the output is less than one handread Hz when the read out time is one second. With a suitable modulation it can be directly sent to the center through a telephone line. When many portable flaw detectors are used, intermediate recordings are preferred to the on-line transmission. At first a combination of an A/D converter and a paper tape punch was used, but later a small audio cassette tape recorder is used. The signal is recorded in analog AM form with a carrier of 800 Hz. The time base signal is recorded in the same channel using another carrier of 1200 Hz. It has been found that the time base signal may be omitted in most applications, since the tape speed is fairly constant. The positional information and data identifications are coded and recorded in sequence. Aural comments may be recorded if necessary.

GRAPHIC DISPLAY

At the center the data are first stored in a memory of a minicomputer. If the data are on a paper tape, they can be directly read into the computer. If the data are sent over a telephone line, the signal is digitized after the demodulation. If the data are on a cassette tape, aural comments are discarded. The signals are first demultiplexed by filters corresponding to each carrier and then digitized after the demodulation. Fig. 2 shows a comparison between the original input to the cassette recorder and the demodulated

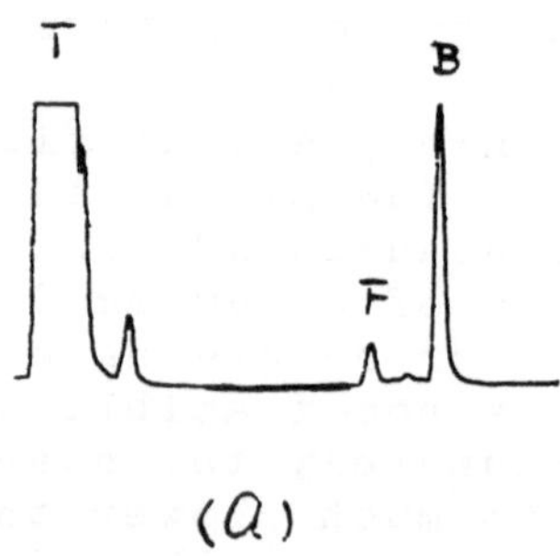

(a)

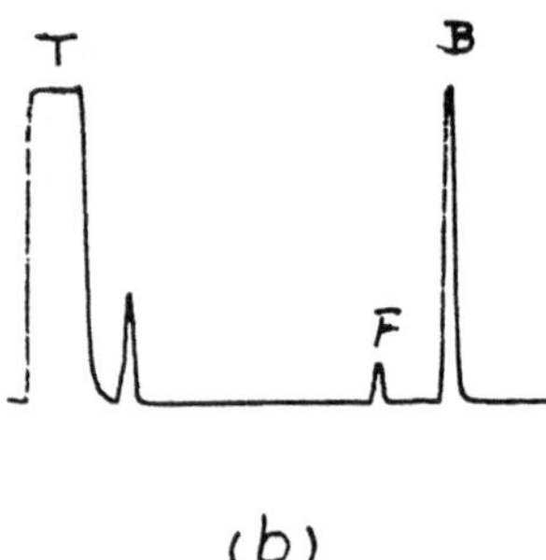

(b)

Fig. 2. Comparison between the original input to the cassett recorder and the demodulated output.

output. There is little distortion in wave form. The positional information and data identifications are read into the computer in a similar manner.

Since the depth (Z) information of defects is obtained from the A-scan signal, only the two-dimensional scanning of the probe is necessary to acquire the whole three-dimensional information. The computer rearranges the data and presents it on a graphic display in a specified format.

The graphic display used here is a modified conventional oscilloscope with a capability of light intensity modulation. The X- and the Y-axis are driven by analog positional signal supplied by D/A converters attached to the computer. Whereas the light spot is turned on and off in binary fashion directly by digital pulse output of the computer. The number of pulses at a specified point is proportional to signal amplitude. In this way better linearity of total light intensity is obtained without using a D/A converter than that of conventional light intensity modulations. Presently 256 x 256 picture elements with 32 gray levels can be displayed.

EXAMPLE

A steel plate of 100x100x35 mm is examined by a vertical probe of 2.25 MHz as shown in Fig. 3. The A-scan signal is recorded at every 0.5 cm in both X and Y directions. The A-scan signal can be sampled at up to 2048 points, but the present case, is sampled at 512 points, which is good enough for faithful reproduction of waveform. This number is too large in comparison with the sampling density

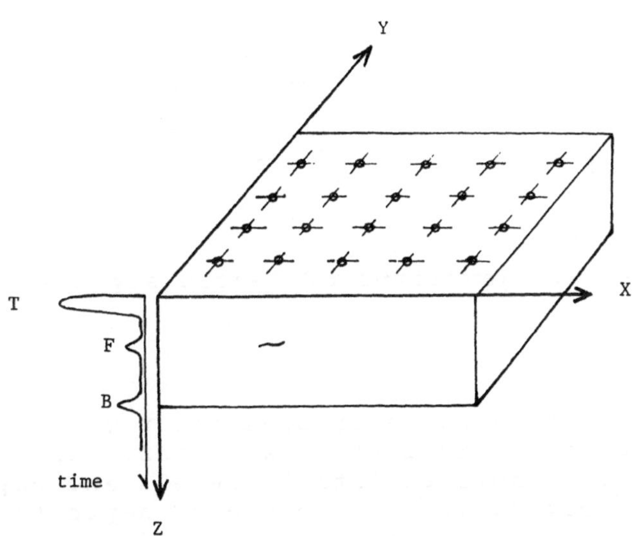

Fig. 3. Steel plate with its coordinate system.

of the X-Y scanning. Hence the data are divided into blocks of four points and the peak value within each block is taken as a new sample. All these parameters are under the control of the computer and can be programmed at will.

Fig. 4 shows changes of waveforms of the A-scan signal plotted by the X-Y recorder when the probe is scanned along a line. The large peaks in the front and the end represent the transmitting pulse and the back surface reflection, respectively. In between there can be seen flaw reflections in some of the middle traces.

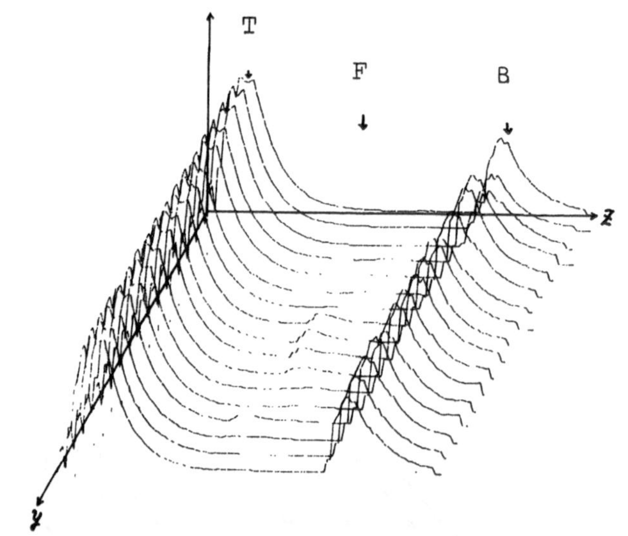

Fig. 4. Waveforms of A-scan signals.

Fig. 5 (a) shows a cross-section view in X-Y plane which is equivalent to a C-scan. The signal amplitude is quantized into 8 levels and represented on gray scale. The range of depth (Z) is limited from the 60th block to the 80th block, which is equivalent to a time gate. Fig. 5 (b) shows the same view, but the binary black and white representation is used in order to distinguish large signals.

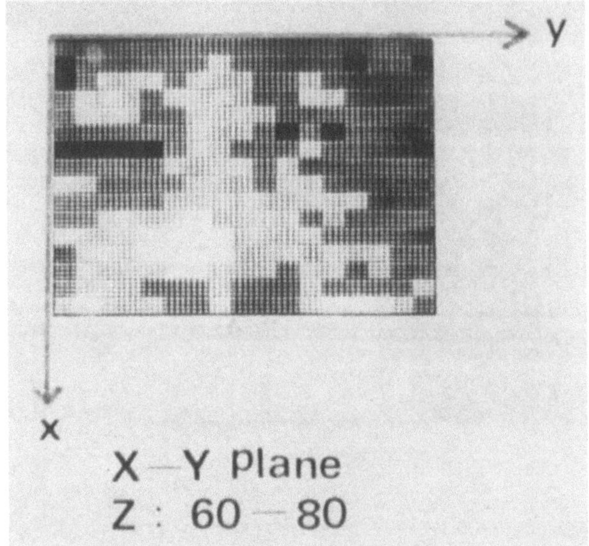

Fig. 5 (a)

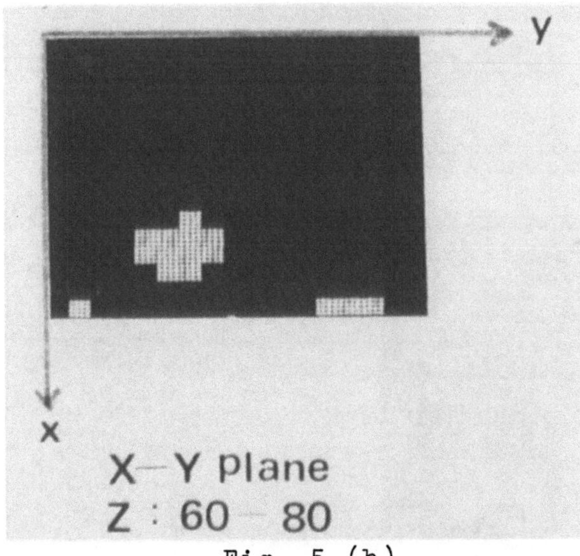

Fig. 5 (b)

Fig. 5. Cross-section view in X-Y plane.

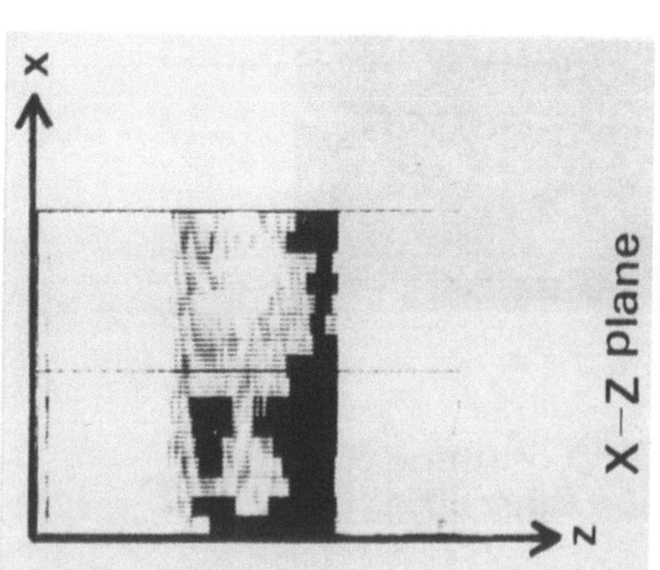

Fig. 6 (a)

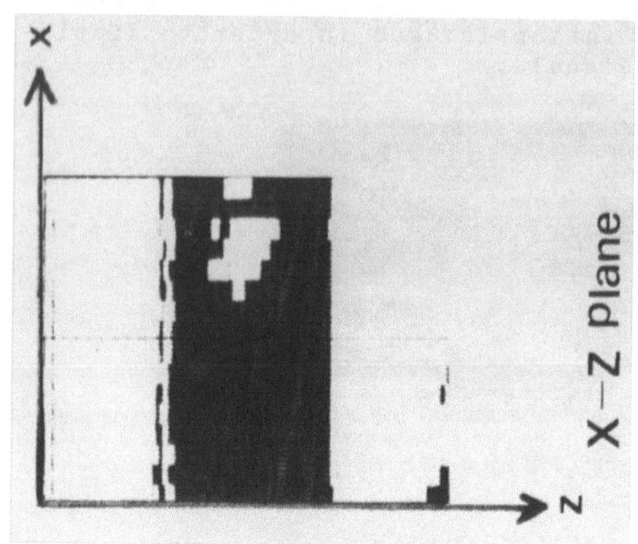

Fig. 6 (b)

Fig. 6. Cross-section view in the X-Z plane.

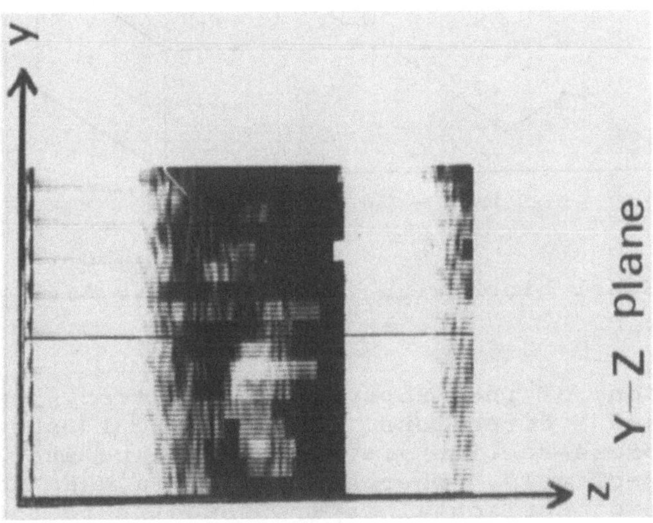

Fig. 7 (a)

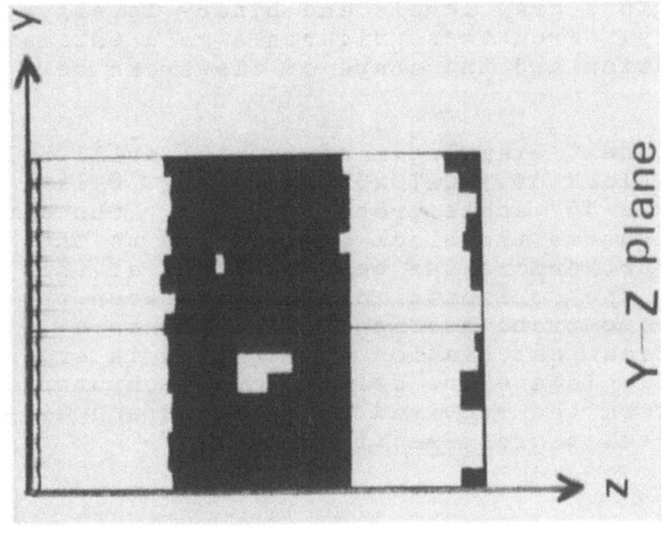

Fig. 7 (b)

Fig. 7. Cross-section view in Y-Z plane.

Fig. 6 shows similar cross-section views in X-Z plane, which are equivalent B-scans. Whereas Fig. 7 shows the views in Y-Z plane. They are also shown in 8 gray levels and binary levels, respectively. From these figures a good estimate of the location and the shape of flaws can be obtained.

As the next example, a horizontal drill hole in a steel block of 300x70x50 mm in Fig. 8 is examined by a 45° angle probe of 5 MHz. The ultrasonic beam enters the block at an angle to the surface. Furthermore the beam reflects at the bottom when the horizontal distance between the defect and the probe becomes large. Hence a trigonometrical calculation is required in order to accurately locate the defect. The computer can easily perform the required calculation and display a corrected cross-section view.

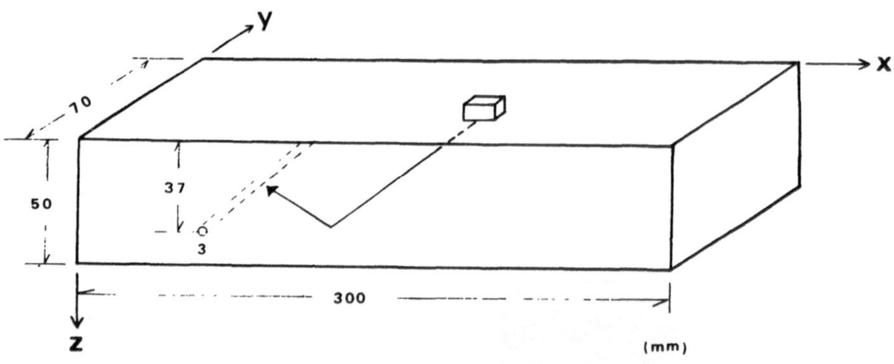

Fig. 8. Steel block with its coordinate system.

Positions of the probe are set at every 5 mm in both X and Y directions. Figs. 9 and 10 show corrected cross-section views in Y-Z plane and X-Z plane, respectively. Whereas Fig. 11 (a) and (c) show views at horizontal layers touching to the top and the bottom, respectively, of the hole. (b) shows a view of a layer in between. Thin lines in Figs. 10 and 11 show coordinate lines which are ± 1 cm apart from the axis of drill hole.

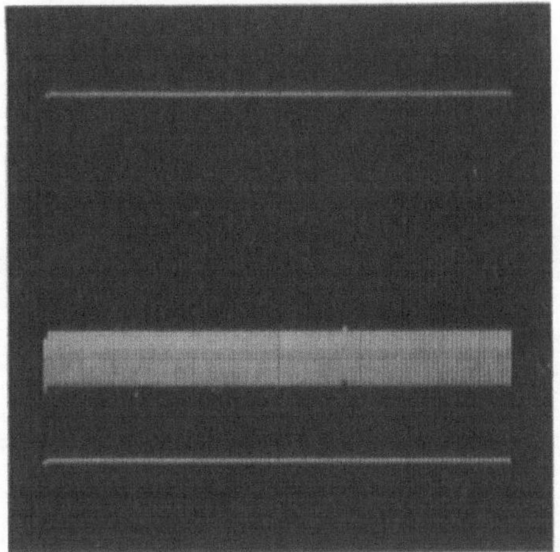

Fig. 9. Cross-section view in Y-Z plane.

Fig. 10. Cross-section view in X-Z plane.

The existence of a hollow space are clearly seen in Figs. 10 and 11 (b).

Perspective views of the drill hole are obtained. Fig. 12 shows the position of eyes in relation to the specimen. Both sides of the hole are scanned this time at every 1 mm interval.

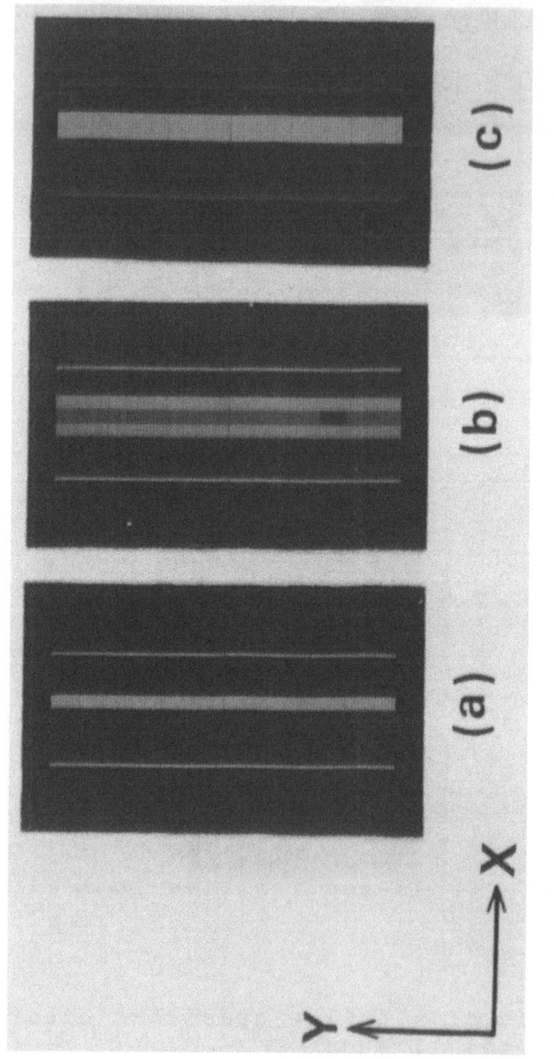

Fig. 11. Cross-section views in horizontal X-Y planes.

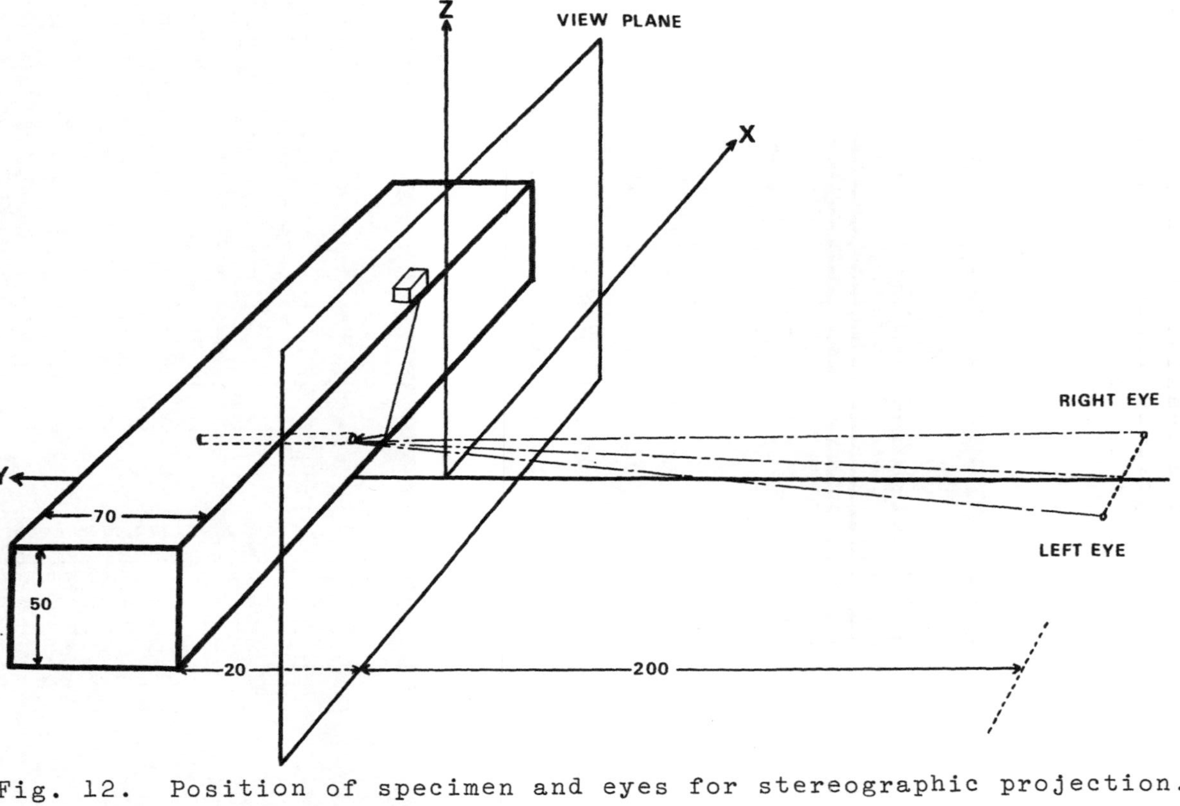

Fig. 12. Position of specimen and eyes for stereographic projection.

Fig. 13 shows the front view of the hole. Fig. 14 shows a stereographic pair when the specimen is rotated 54° around the Z-axis. When one looks at it through a stereographic glass, a three-dimensional view is obtained.

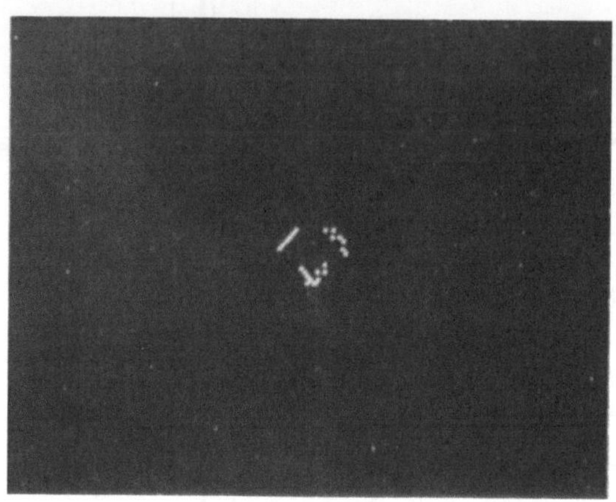

Fig. 13. Front view of the hole.

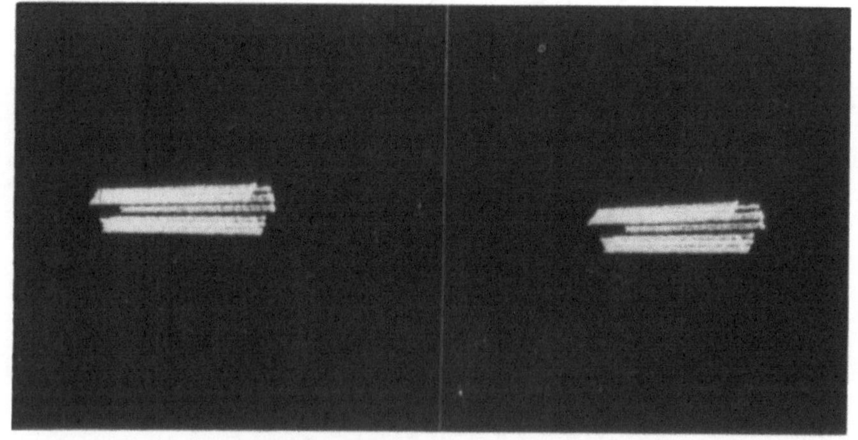

Fig. 14. Stereographic pair for the rotated specimen.

CONCLUSIONS

In conclusion, it has been shown the pulsed ultrasonic method combined with graphic display controlled by a computer can present a complete image of three-dimensional ultrasonic field. In the present state of art, this system has the following advantages over the acoustical holography using a scanning transducer and digital data processing. (This type of acoustical holography will find a usefulness at a low-frequency range, here transducers with good directivity are not available.)

A. A transducer with higher directivity can be used, which yields higher sensitivity and higher S/N.

B. In solid, there are both extensional and shear waves. Reflection and transmission at a boundary are usually accompanied with mode conversion between two types of waves. Hence, ultrasonic field to be recorded by acoustical holography may become complicated, when there are boundaries in the specimen. Testing weldment with an angle beam probe is an example. Whereas the pulsed ultrasonic method can avoid this trouble thanks to its high temporal resolution.

C. The data processing in the present system is mostly simple transformation of coordinates and much faster than Fourier transform required in the reconstruction of image from a hologram.

D. Pulsed flow detectors with the capability of C-scan have been widely used in the fields. They can be expanded into the present system without much modification.

Acknowledgement: The authors thank Mr. H. Yamada who conducted most of ultrasonic work.

ACOUSTICAL RECONSTRUCTION OF HOLOGRAMS, AND THEIR POTENTIAL USE

J. L. Pfeifer

Bendix Research Laboratories

Southfield, Michigan 48076

Extensive work has been performed to date in coherent optical processing. Coherent acoustic processing, on the other hand, has not progressed as far as its coherent optics counterpart, although it has identical operational capabilities. Some progress had been reported earlier[1] in the development of acoustical holographic matched filters, but the published results were not definitely positive. Before now, one of the major factors hindering the development of acoustic processing techniques was the lack of a good acoustic hologram meant to operate directly in an acoustic field. In the past, the detected acoustic holographic information has invariably been used to generate a photographic transparency (of reduced size) which reconstructed a visible image upon irradiation with laser light. Usually, in the conversion process, the image quality suffered because of the orders of magnitude difference between the acoustical and optical wavelengths.

In this article, a technique will be described for operating entirely in the acoustic domain using a new type of acoustic hologram which can be irradiated with acoustic waves to yield an image as an acoustic intensity pattern. Furthermore, this new acoustic hologram will allow processing techniques to be carried out entirely in the acoustic realm. The basis for these acoustic holograms has been borrowed from optical holography - the development of computer-generated binary holograms. These binary holograms give acoustic holography new potential by presenting the holographic information in a form that can be etched into a metal plate. This plate then acts on acoustic waves just as a photographic negative hologram operates on optical waves.

BINARY HOLOGRAMS

Over the recent years, research has been carried out in generating binary masks for spatial filtering[2] and binary holograms[3,4] for optical reconstruction with the aid of computers and plotters. These computer-generated holograms are binary in that they consist of an array of transparent dots or rectangular apertures on an opaque background; the transmittance of the hologram assumes values of either zero or one. Figure 1 shows one of the earlier binary Fraunhofer holograms[3] of the letters ICO; an optical reconstruction of the image obtained in the lab from the hologram (as copied from the published paper) is shown in Fig. 2. Generally, the images from the binary holograms are about as good as those from ordinary holograms of the same aperture.

The amplitude and phase information in the binary holograms is contained in the relationship of the size and spacing of the small dots with respect to each other. In ordinary holograms the amplitude or grey scale is controlled by the density of the photographic film, while in binary holograms it is controlled by varying the size of the apertures. This aperture size variation to achieve amplitude is similar to the technique of achieving grey scale in halftone printing.

The binary and ordinary grey holograms are similar in the manner in which they operate on the incoming wave front and produce the desired phase in the outgoing image wave fronts. The method which achieves the phase shift is often referred to as detour phase. This is often explained in terms of slight dislocations in the lines of a diffraction grating giving rise to "ghosts" in the diffracted wave front. This difference or deviation from an integral wavelength is called detour phase.

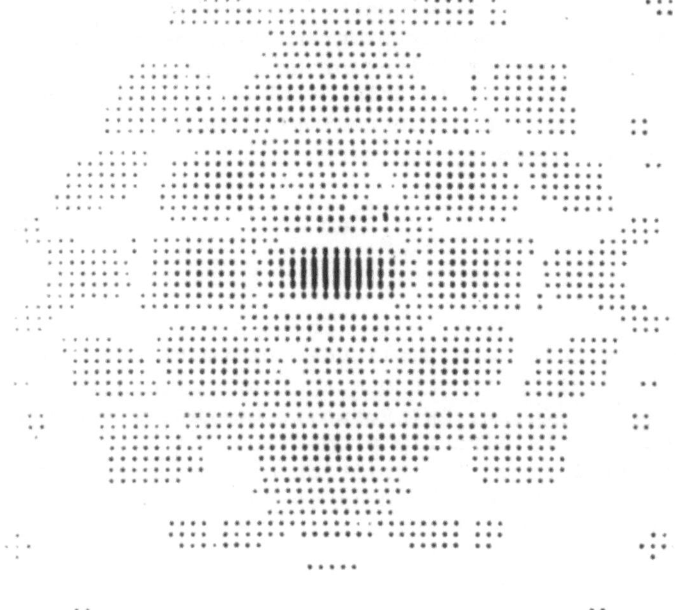

Fig. 1 - Computer-Generated Binary Fraunhofer Hologram[3] of ICO

Fig. 2 - Optical Reconstruction Obtained From the Binary Hologram

By using these techniques of detour phase and grey scale simulation, computer programs have been devised to synthesize the complex filter or hologram corresponding to the desired image. In this synthesis, it is assumed that the hologram can be approximated by a sampled function with values

$$H_{nm} = A_{nm} e^{i\alpha_{nm}}$$

where nm specifies each cell. There are many ways of assigning the phase and amplitude information to the cell; the most typical is to use one aperture within the cell with a fixed width, c, but variable height, w. Mathematically then, A_{nm} is directly proportional to w. The phase then is controlled by the position of the aperture within the cell. The size of the cell is such that it can accommodate movement of the aperture to take care of a $\pm\pi$ shift in the detour phase.

In addition, the value assigned to the phase and amplitude of each cell is often broken down into many discrete levels, that is quantized. As shown in one of the preceeding papers given at the Fourth Symposium,[5] the number of discrete levels can be increasingly reduced to a small number of steps before the reconstructed image is seriously degraded.

This presents just a rough treatment of the binary holographic concept; more precise detail can be obtained elsewhere.[2,3,4,5] What is important, however, is that the binary hologram concept can also be directly applied to acoustic holography. The acoustic holographic information may be converted to a binary form and etched into a metal plate; the binary format contributes to the structural stability of the metal plate. The aperture openings will then have a transmittance of one while the metal will have a transmittance of nearly zero (i.e., small enough in comparison to the transmittance of the aperture to be considered zero). Hence this etched metal plate hologram should act on acoustic waves in the same manner that the binary photographic hologram acts on optical waves.

EXPERIMENTAL ACOUSTICAL RECONSTRUCTION

In order to test whether these binary etched metal holograms would indeed form an image as an acoustic intensity pattern, the following experiment was carried out. The computer-generated binary Fraunhofer hologram of the letters ICO by Lohmann and Paris,[3] see Fig. 1, was selected because of its ready availability. (Its structure was based on a 64 x 64 array of cells.) This hologram, reduced to a square area of 2.5 cm side length, was reproduced as a hole pattern in 50 μm thick stainless steel shim stock by a chemical etching process. This new type of acoustic hologram was immersed in a tank of water and irradiated by a 7 MHz ultrasonic beam generated by a 2.5 cm diameter air-backed barium titanate crystal. The image information was diffracted at an angle of $\sim 30^{\circ}$ from the hologram because of the particular hologram hole spacing and acoustic wavelength used. A plexiglas acoustic lens, with a focal length of 4.8 cm, was used to form an image of the letters ICO in the form of an acoustic intensity pattern at its focal plane. A small air-backed 7 MHz quartz crystal, whose sensitive area was limited by placing a thin acoustically opaque disc with a 1 mm diameter aperture over the top of the crystal, was used as the image detector. This detector was scanned across the image plane using a motor-driven micropositioner. The signal from the detector was fed to a selectively tunable RF microvoltmeter. The I.F. output from the microvoltmeter was demodulated and used to control the intensity of a scope display. Position helipots attached to the detector drive provided the scope scanning voltages. A Polaroid oscilloscope camera was used to record the display. A sketch of this experimental arrangement is shown in Fig. 3.

Figure 4 shows the acoustic reconstruction of the letters. The height of the letters is approximately 3 mm, in good comparison with the height achieved in laser light reconstruction when the hologram size scaling factor is taken into consideration. The streaked appearance is from the individual raster scan lines; these are spaced approximately 0.25 mm apart. Several raster lines show up darker because the microvoltmeter became detuned during the scan.

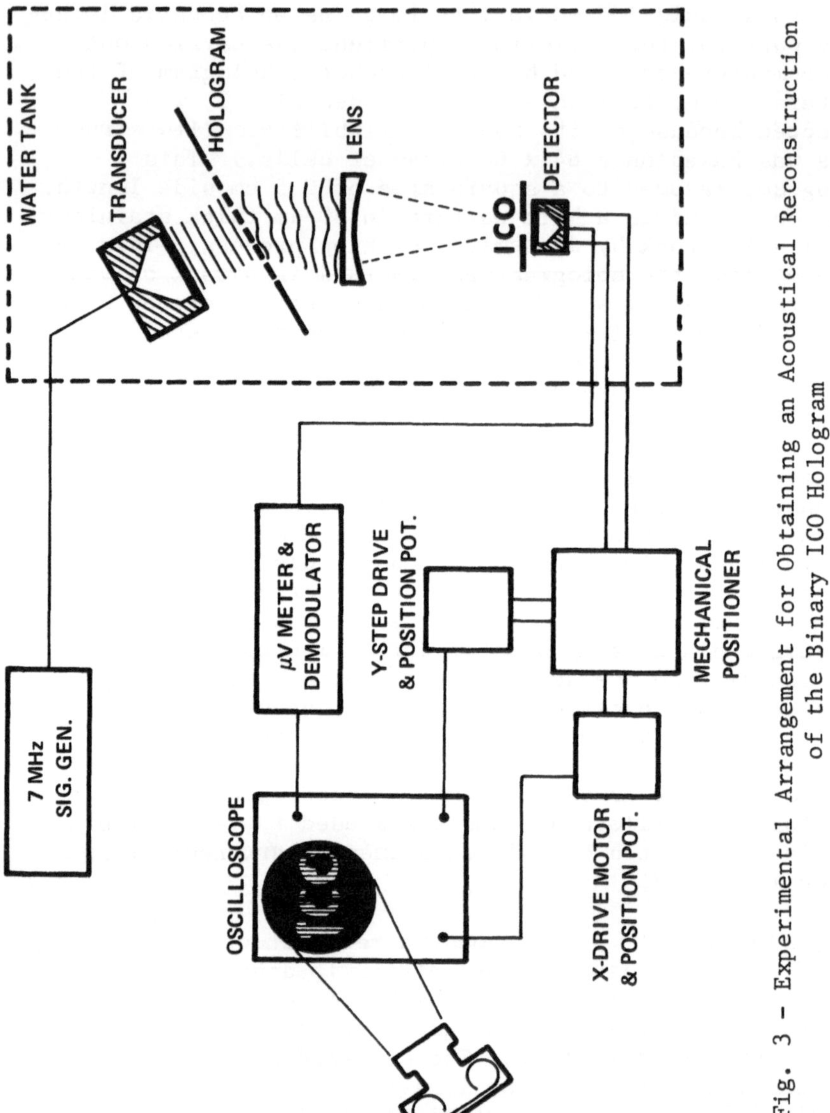

Fig. 3 - Experimental Arrangement for Obtaining an Acoustical Reconstruction of the Binary ICO Hologram

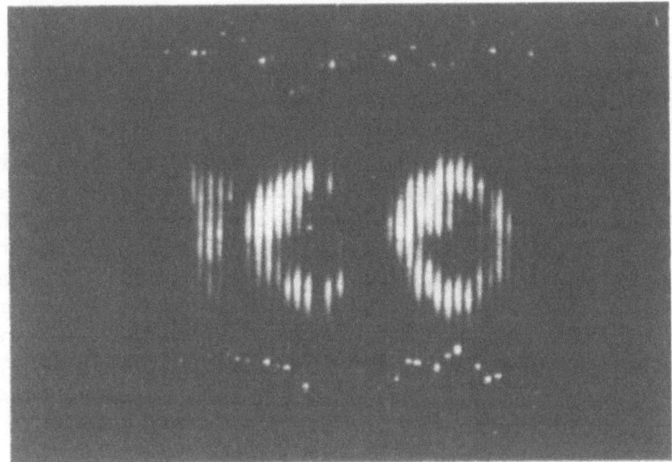

Fig. 4 - Acoustical Reconstruction of the Binary Hologram

The extra thickness in the C and O letters in one particular spot is thought to be caused by the image being slightly out of focus. When out of focus, some of the spurious signals within the C and O letters seen in the optical reconstruction (Fig. 5) become stronger. The wide detector aperture (as compared to the image size) caused this spurious noise to blend together with the letters. In general, however, the quality of the acoustic reconstruction is good.

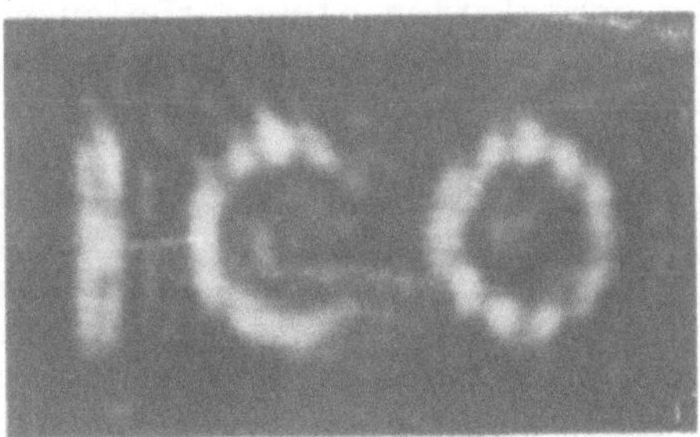

Fig. 5 - Optical Reconstruction of the Binary Hologram

CORRELATION EXPERIMENTS

After demonstrating that the etched plate ICO hologram would reconstruct acoustically, the next logical step was to see if an acoustic correlation could be achieved between this Fraunhofer hologram and an ICO letter object. Since originally the computer-generated hologram was developed from a nonexistent target, it was necessary that the ICO target geometry be estimated from past holographic reconstructions of the ICO image in both the acoustic and laser light domain. For the particular experimental arrangement and acoustic lenses used, the dimensions of the lettering needed was determined to be approximately 0.58 cm high, by 1.5 cm long, and 0.13 cm in letter width. The lettering was etched through metal shim stock to form a positive type target (i.e., acoustic energy passing through the letters only).

A sketch of the experimental arrangement is shown in Fig. 6. A 3.8 cm diameter 7 MHz air-backed quartz crystal was used as the transducer. The binary Fraunhofer hologram was actually attached to a micropositioner to assist in proper alignment of the hologram in front of the transformation lens. Considerable care in the alignment of the transducer, focusing lens, and detector, as well as rotational alignment of the ICO target had to be undertaken. A small quartz crystal with a special electrode configuration to give an effective detecting area of 0.12 mm^2 was used as a detector; a preamplifier unit was mounted behind the crystal in a water-tight housing. This unit was scanned across the correlation plane; the output was fed into a RF microvoltmeter, demodulated, and displayed on a scope or X-Y recorder.

An idea of what the correlation function should look like was obtained by experimentally determining the convolution between two positive ICO targets using an image-casting method. The correlation distribution to be expected when one or more of the ICO letters in the target object is blocked out was also determined. Figure 7 shows the expected correlation functions for the selected target geometry, where, for simplicity, only the amplitude through the center of the correlations is shown.

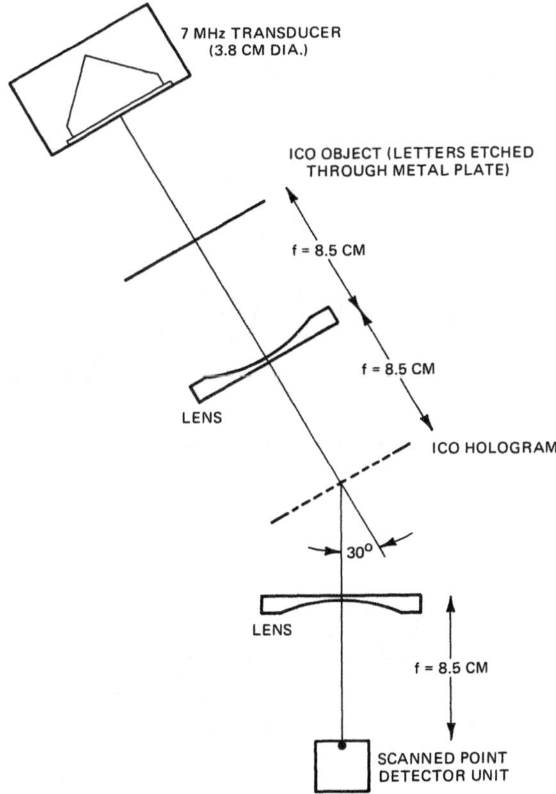

Fig. 6 - Experimental Arrangement for ICO Correlation

The actual experimental correlation distribution obtained is shown in Fig. 8; this was achieved with the target object located on-axis. When the ICO target was shifted around in the object plane, then the correlation distribution function would shift in a corresponding manner in the correlation plane. In comparing the predicted with the actual results, one can see that some similarity was achieved in the overall shape. The central correlation peak was not as high as it should be, indicating that maximum or good correlation was not achieved. However, the distribution did change in the manner expected when one of the letters was blocked in the target. For instance in Fig. 8, when the letter O was blocked from the

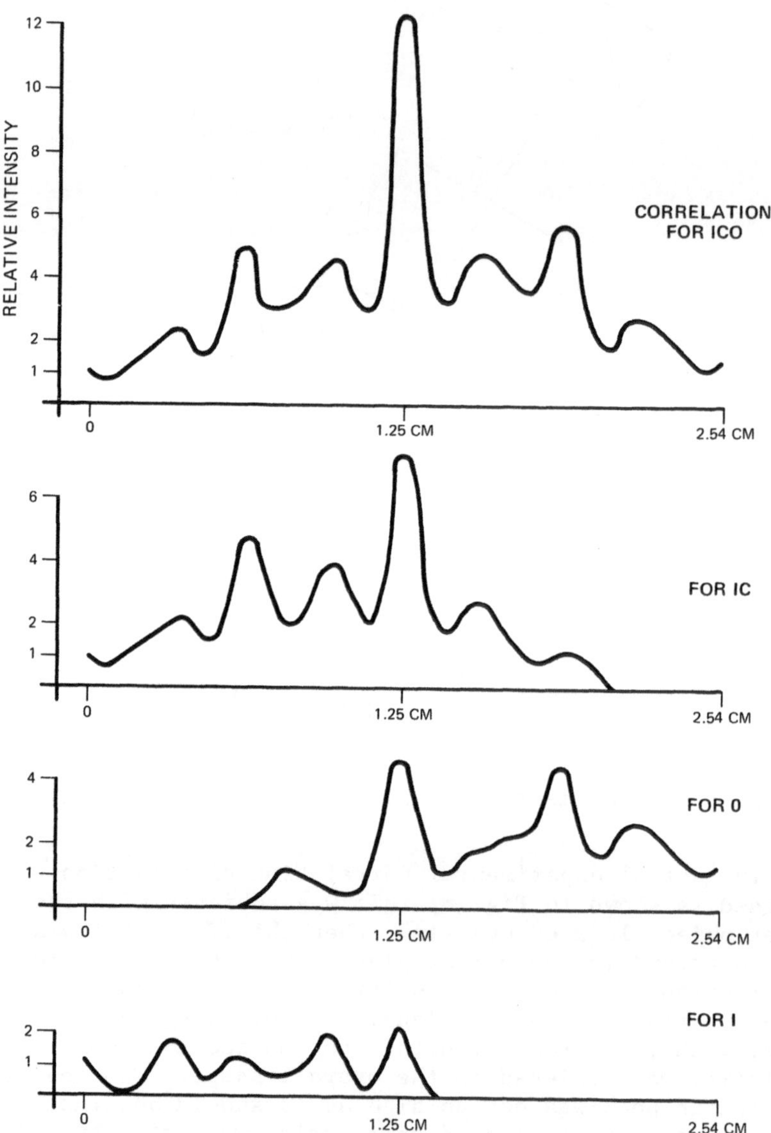

Fig. 7 - The Expected Correlation Function - Center Cut

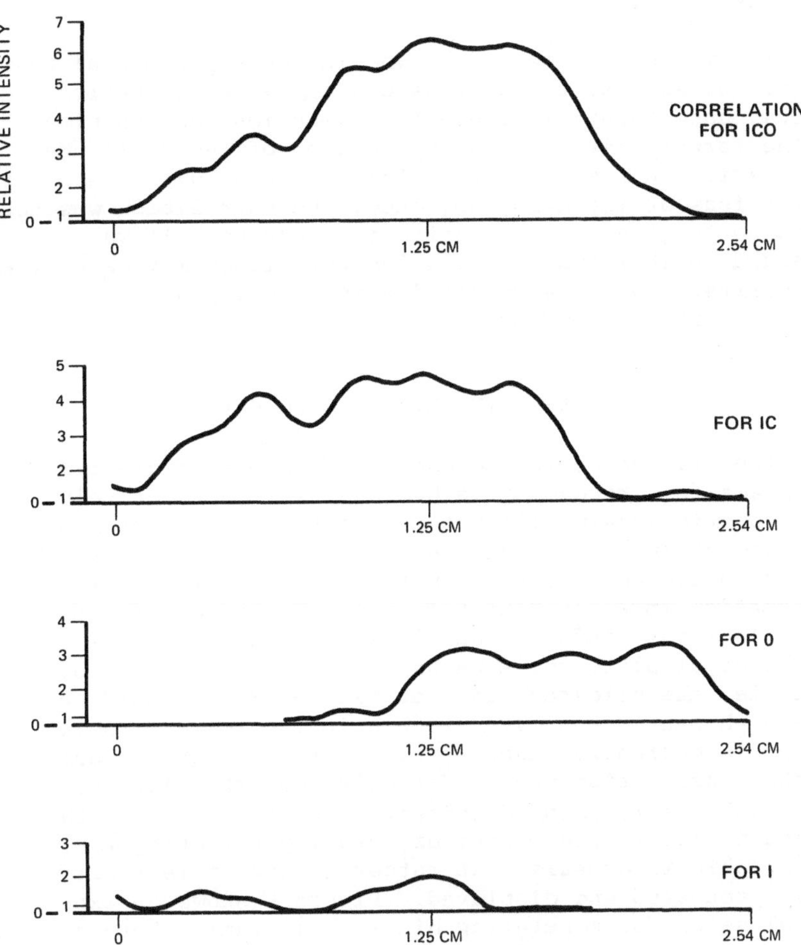

Fig. 8 - The Experimental Correlation Results - Center Cut

target (correlation for IC), the amplitude in the right half of the correlation was reduced in a manner predicted in Fig. 7. Please note that the intensity scale in Fig. 8 is nonlinear due to the demodulator circuit and that the intensity scale in Fig. 7 does not have the same absolute value as in Fig. 8.

In view of the many unknowns in the experimental tests performed, such as distortions caused by the acoustic lenses, uncertainty in the exact dimensions and spacing in the target object, some suppression of the amplitude information in the central region of the hologram, aperture limitations of the hologram, etc., the correlation was not expected to give exact quantitative results. It was intended to show that correlation techniques may be carried out entirely in the acoustic domain by using binary etched plate acoustic holograms.

GENERATING BINARY HOLOGRAMS

The true value in the use of the acoustic binary type holograms for processing techniques performed entirely in the acoustic domain will not lie in the synthesizing of a hologram of an imaginary object in a computer program, but rather in the ability to transform actual holographic information into the binary format. The binary format of the holographic information allows the pattern to be etched into a metal plate while mechanical integrity is maintained (that is, the hologram does not fall apart as would be the case with many normal grey-tone holographic patterns). A method is currently under development by which the holographic information may be directly converted into binary form while it is being detected. The uniqueness of this method is not in the manner of actually detecting the holographic information but rather in how it is electronically processed and displayed. Figure 9 shows an experimental setup for generating binary holograms. Here a point detector scans the hologram plane in a raster fashion; the holographic fringe information is then converted into the binary format and displayed on the CRT. A long time exposure with a camera records the complete hologram. The reference wave is electronically simulated and impinges normally onto the hologram plane, while the signal beam strikes the plane at a 15° angle from the normal. This helps to simplify the electronics; however,

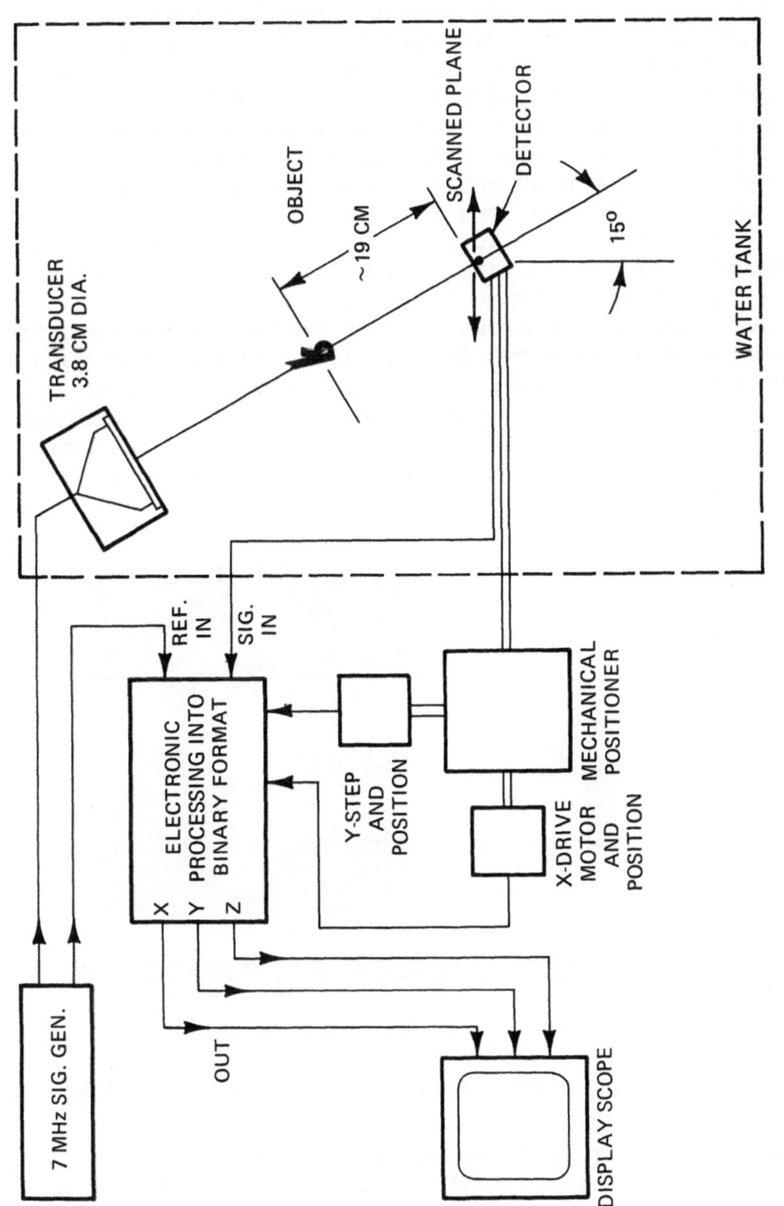

Fig. 9 - Generation of a Binary Hologram

the signal and reference waves can be switched by introducing the proper phase-shifting electronics to the reference signal. The detector is a small quartz disc with a special electrode configuration that gives an effective detecting area of approximately 0.12 mm^2; this is tilted normal to the signal wave to enhance reception. Some preliminary testing of the method has been done using simple negative-type letters as objects. Figure 10a shows the target object R with approximate outside dimensions of 1.2 cm x 1.3 cm; the long tail was used to hold the target in position. The dotted outline shows the approximate holographic scanned area (2.5 cm x 2.5 cm) in relation to the target size. Figure 11 shows the preliminary binary hologram of the R object. An optical reconstruction from the hologram is shown in Fig. 10b. In this hologram, the parameters controlling the binary format were not yet fully optimized. The reconstruction is not to be construed as showing any true resolution capabilities, but rather, only that binary acoustical holograms of actual objects can be generated.

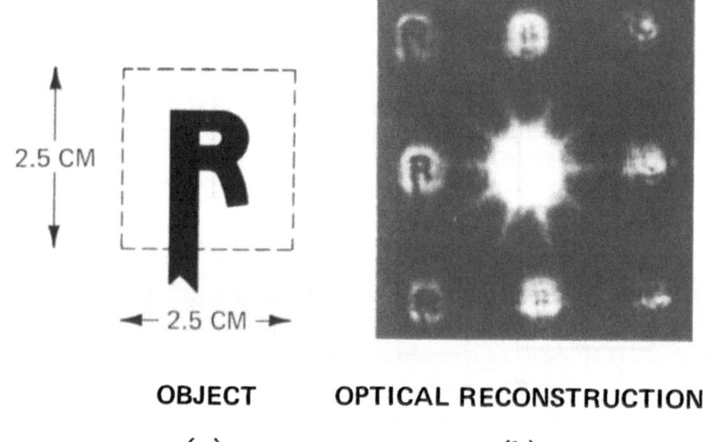

Fig. 10 - Image from (Preliminary) Binary Hologram

(a) Object target in relation to scanned area.
(b) Optical Reconstruction from the binary hologram.

Fig. 11 - Preliminary Binary Hologram of Letter R

POTENTIAL APPLICATIONS

An area of particular interest in acoustic holography is acoustic correlation. Etched metal binary holograms will allow correlations to be made entirely in the acoustic realm. This, in turn, should permit the construction of instruments for pattern recognition, material testing, etc. that will be simplified in form, complexity, and operation. This will stem from the elimination of converting the acoustic information for either complex computer or optical processing. One potential application is shown in Fig. 12 where rapid nondestructive testing of assembly line parts could be performed. Once a correlation hologram of a good part has been produced, succeeding parts could be quickly tested for internal flaws, mis-drilled holes, etc. This method should be excellent for testing on an accept-or-reject basis and where specific examination of a flaw is not necessary.

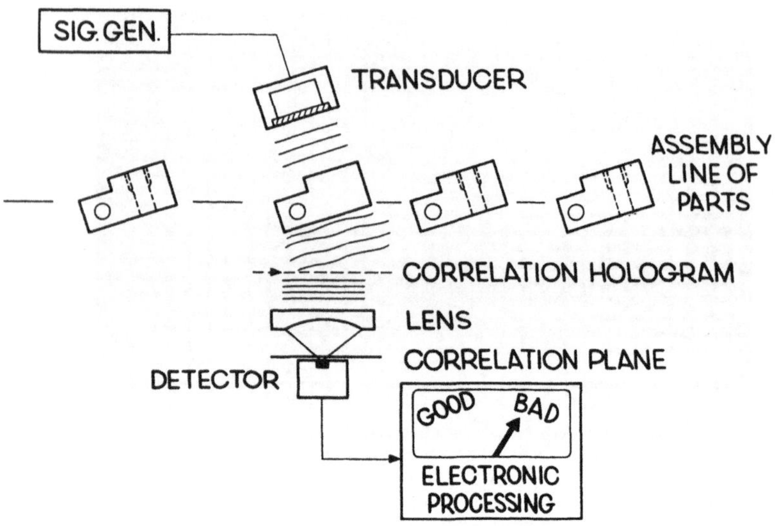

Fig. 12 - NDT of Parts Using Acoustic Holographic Correlation Method

In summary, the demonstrated ability to make reasonable holograms which can operate directly on acoustic waves should greatly expand the realm of coherent acoustic processing.

ACKNOWLEDGEMENTS

I would like to thank Drs. R. K. Mueller and E. Marom for their suggestions during the early stages of this work, and also to Dr. P. N. Keating.

REFERENCES

1. P. Greguss, Acoustical Holography (Metherell et al., editors), Vol. 1, 1969, p. 259.

2. B. R. Brown and A. W. Lohmann, "Complex Spatial Filtering with Binary Masks," Appl. Opt. 5, 6, 967 (1966).

3. A. W. Lohmann and D. P. Paris, "Binary Fraunhofer Holograms, Generated by Computer," Appl. Opt. 6, 10, 1739 (1967).

4. B. R. Brown and A. W. Lohmann, "Computer-Generated Binary Holograms," IBM J. Res. Develop. 13, 2, 160 (1969).

5. W. J. Dallas and A. W. Lohmann, "Quantization of the Hologram Transmittance," Presented at Fourth International Symposium on Acoustical Holography.

CYLINDRICAL SCAN ACOUSTICAL HOLOGRAPHY

F. G. Geil*
U. S. Department of the Navy, N.U.R.D.C.
San Diego, California

G. Mott
Westinghouse Research Laboratories
Pittsburgh, Pennsylvania

METHOD

In this system, a cylindrical target space is ensonified with both the transmitting and receiving transducers being moved, or scanning, together. Cylindrical scan has two basic forms—what we have called inside-looking-out (I.L.O.) and outside-looking-in (O.L.I.). In the former, the two transducers are fixed to a short metal arm that is attached and perpendicular to a vertical screw thread. See Figure 1. The screw itself is inside a slotted hollow metal tube and the transducer supporting arm protrudes from the narrow slot. The screw and tube are caused to rotate and may have a relative rotatory motion. The transducers both rotate in a horizontal plane and move vertically and thus trace out a helical path. Hence, a cylindrical space is ensonified having the helix as center.

In the latter, or (O.L.I.) method, we could have used the previous system and simply reversed the transducers causing them to 'look' inwards instead of outwards. See Figure 2. While this is quite feasible, in practice it is not done since it is simpler and more convenient to fix the transducers and to rotate the target space. Sample targets could be affixed to the same screw-slot-arm combination of Figure 1 and rotated past the fixed, inward looking, transducers, as shown in Figure 2.

*Interchange Executive, Westinghouse Research Laboratories

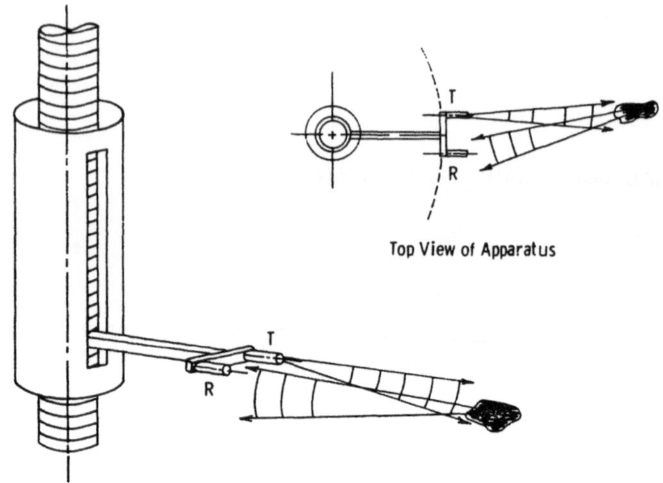

Fig. 1 — Inside—looking—out cylindrical scanning system with both transmitter (T) and receiver (R) being moved

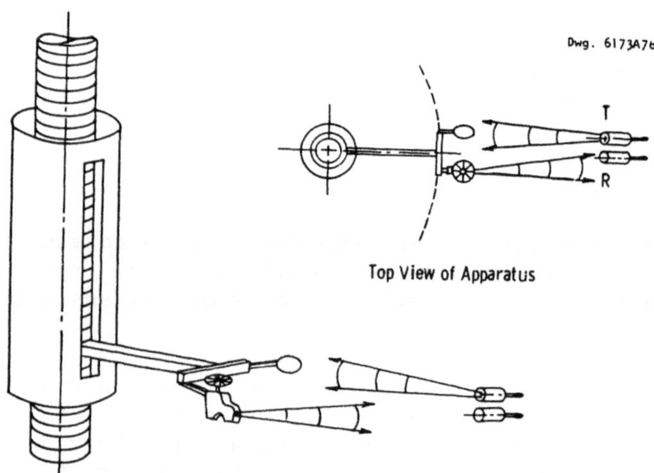

Fig. 2 — Outside—looking—in system

Obvious practical and useful applications of these two methods are 1) a tripod supported I.L.O. system for ocean use could be placed on the bottom and caused to scan a cylindrical volume and 2) non-destructive testing of cylindrical samples by means of an O.L.I. system. See Figure 3. The most obvious advantage of such cylindrical scan systems over planar scans is the easy elimination of backlash in the traverse mechanism.

The hologram is formed by means of a companion scanning cylindrical camera shown in Figure 4. The drum on which a strip of film is wound is, by means of a non-slip drive, coupled to and synchronized with the transducer screw-slot drive mechanism. The cylindrical film is rotated and moved axially past the light spot from a glow tube. The received sonic signals are processed electronically and the resulting signals are made to modulate the intensity of the light from the glow tube and this light is projected onto the film.

Our first holograms were of the I.L.O. type but this method was given over to the O.L.I. method, thus eliminating the need for mercury pool and slip-ring electrical contacts that were used originally.

APPARATUS

A 1 MHz ovened crystal oscillator is the core of the system depicted in Figure 5. Part of the oscillator signal is passed through an amplifier and its output is gated and passed to the transmitting transducer. The signals received from the target space are pre-amplified and taken through a signal processing chain of amplifiers and filters and finally mixed in a synchronous detector with a fiduciary signal from the master oscillator that has been passed through a phase changing network (when off-axis reference signals are used). The output from the synchronous detector is placed on a sample-and-hold I.C. and the desired signal is selected and is then further processed and made to control the glow tube light output. The light from the glow tube is focussed to a fine spot, about a third of a wavelength square, on the film drum.

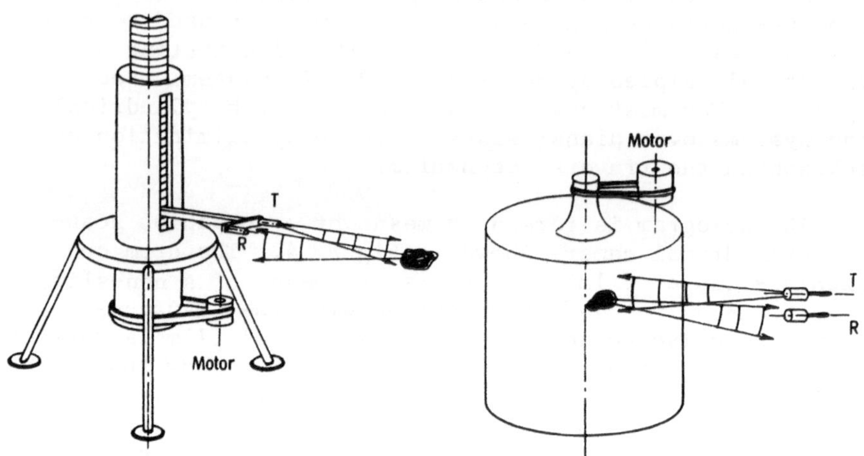

Fig. 3 — I-L-O system for ocean—space application and O-L-I system for N.D.T. use

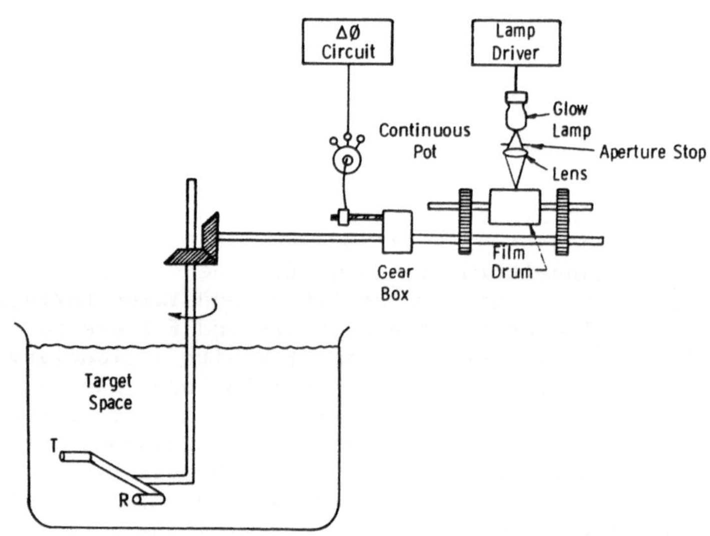

Fig. 4 — The cylindrical camera and its mechanical connection

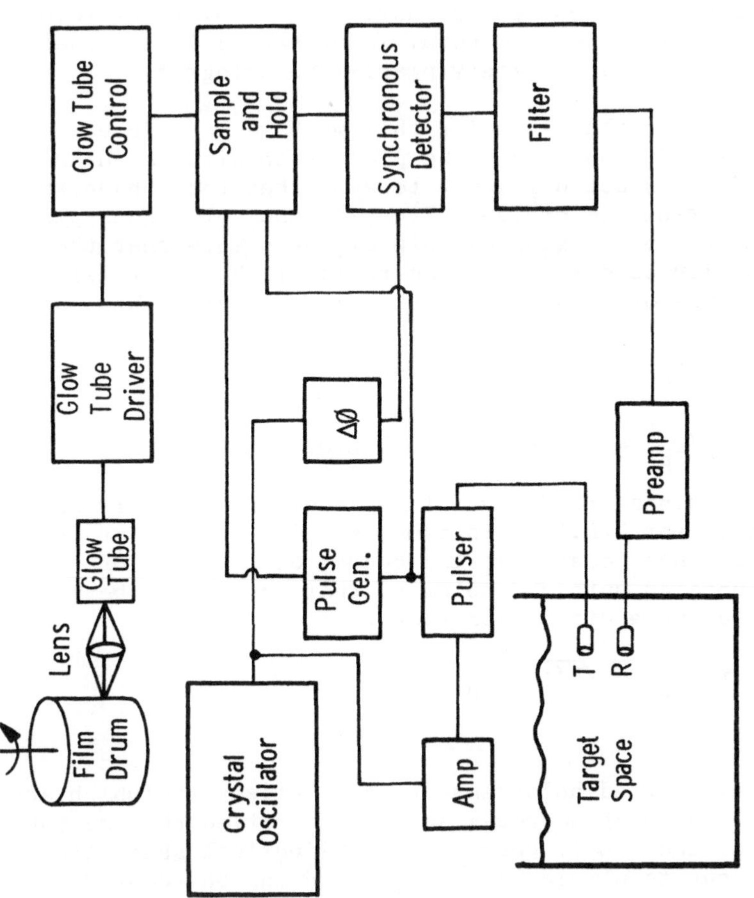

Fig. 5 – Block diagram of the hologram apparatus

HOLOGRAM FILM PROCESSING AND RECONSTRUCTION

The cylindrical film is unwound to a flat strip, developed, decreased in size using photo-reduction techniques by approximately a factor of 10 in length and width and is then formed on a glass plate. Thus, our hologram, which was a cylinder, is now made planar, and in obtaining reconstructions by means of the usual laser set-up it is treated as though it were an ordinary plane-scan hologram.

It is clear that in doing this we have introduced aberrations additional to those present in plane scan holography and it is not difficult to show that these have the effect of making the ordinary Fresnel zones of a zone-plate (hologram of a point source) into ellipses such that the distortion may be defined as the ratio of the major axis to the minor axis of such an ellipse. This is given by

$$r_c/r_p = \sqrt{R/a}, \text{ for O.L.I.}$$

and

$$r_c/r_p = \sqrt{R/(R+a)}, \text{ for I.L.O.}$$

These expressions are derived in Appendix A and it is easy to see that, for O.L.I. holograms, a point source (or its equivalent) when located at the center of rotation of the scan produces an unfolded hologram that appears to be that of a rod having a distortion factor of

$$r_c/r_p = \sqrt{R/a}\ \Big|_{a\to o}$$

which is infinite.

We have formed holograms of ball reflectors (which are equivalent to point sources) and have verified the previous formula for the O.L.I. case. This cylindrical aberration is equivalent to simple astigmatism and can be removed by a cylindrical lens for each point target. However, the amount of astigmatism is clearly a function of the target range (R-a) and constitutes the major difficulty in reconstructing all images by means of simple optics.

We have also been able to show that the apparent angular size of a large object is increased and this is akin to the astigmatism just mentioned. The exact calculation of this distortion of image size is difficult but it can be seen

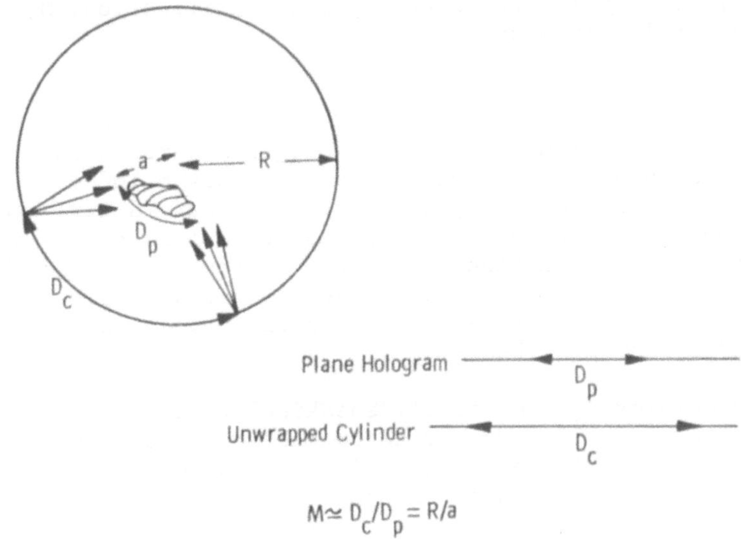

Fig. 6 – Approximate graphical construction for finding the angular magnification, M

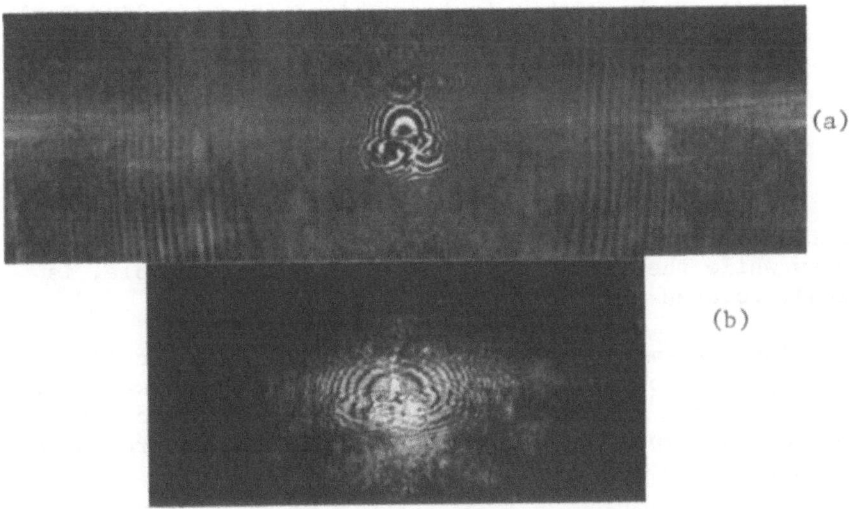

Figure 7. (a) O.L.I. cylindrical hologram of a tap valve and (b) reconstruction of flattened hologram

from Figure 6 that the apparent angular magnification, M, is approximately given by

$$M = R/a$$

Our first experimental O.L.I. holograms were taken from ball targets, a metal valve approximately 6 inches in height and a reflecting plate with slots cut in it. The optical quality of these holograms was quite good but we did not get good reconstructions from them in general, although in the special case of the ball target it was possible to focus to a good image by means of properly chosen and positioned cylindrical lenses.

More complex (geometrically speaking) targets, for example, the tap-valve, gave rise to multiple focussing of variably distorted images. Some of these holograms and reconstructions are shown in Figures 7 and 8.

After these initial experiments we directed our attention to the inspection of a cylindrical steel spin-test sample. This is a piece of steel that has been spin-tested (spun in a centrifuge apparatus until large cracks have appeared or the specimen disintegrates). A piece of steel with large cracks within it is machined into a cylindrical form and holograms of its internal flaws are obtained using the apparatus depicted in Figure 9 and described next.

STEEL SPIN-TEST SAMPLE APPARATUS

This is an O.L.I. apparatus with the slight difference that, here, the transducers are moved vertically on a worm screw while the target space, i.e., the steel sample, is simply rotated in a horizontal plane.

The transmitting transducer is mounted in a cylindrical housing that contains a simple plastic lens that was designed to concentrate the beam formed by the ceramic disc vibrator into a small spot, approximately a half-inch square, on the surface of the steel cylinder.

Although the beam is concentrated on a small spot—the beam width for the major lobe in the water is about 20°, or so—due to the sound velocity ratio of a steel/water interface there is a fanning out of the beam that is refracted into the steel and this is shown in Figure 10. This fanning

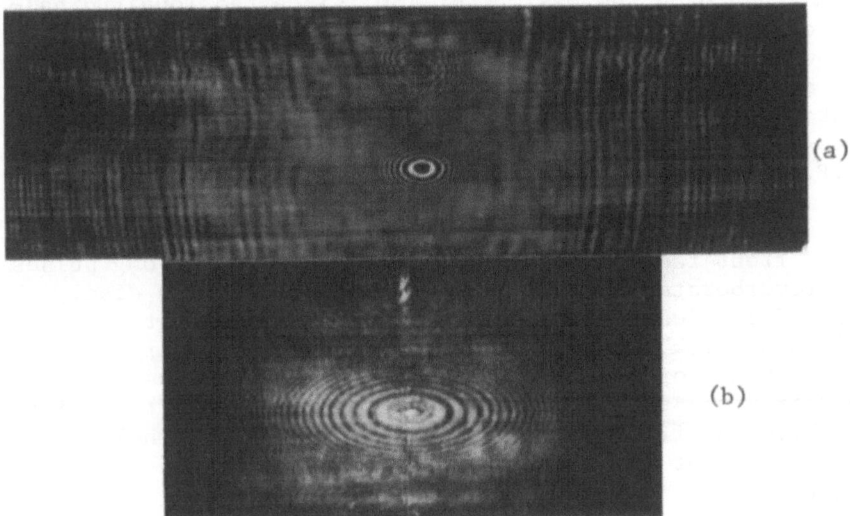

Figure 8. (a) O.L.I. cylindrical hologram of a ball bearing (point source equivalent) and (b) reconstruction of flattened hologram. Note the vertical image in (b) which is due to the slot in the apparatus shown in Figures 1 and 2.

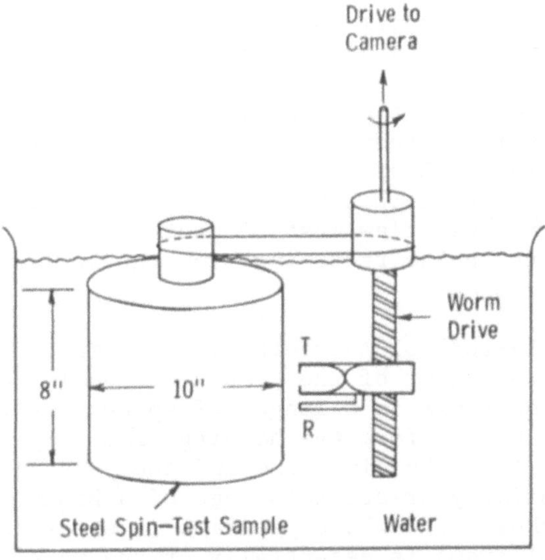

Fig. 9 – Diagrammatic layout of the spin–test sample holography apparatus

out may approach the maximum of 180° since the longitudinal-wave critical angle for steel/water is only about 14° and the outer rays of the incident beam are incident at near to that angle. There is also a large reflection, and a consequent reverberation in the lens housing, that constitutes an 8 - 9 dB loss of signal. This reverberation limits the P.R.F. of the system to around about (10 → 100)/sec.

Another troubling effect occurred due to reflections at the front face of the lens setting up a few strong pulses that reverberated between the lens and the ceramic disc. We lined the lens with absorbing rubber, approximately a half-inch thick and we placed a small disc of the rubber in the center of the lens. This prevented the lens-ceramic reverberation just mentioned but it also made the transmitted beam have an effectively wider beam angle. Nevertheless, this turned out to be the most satisfactory method of operation to date.

PULSE-LENGTH AND SAMPLE-HOLD POINT

The physical length of the pulse in target space (in this instance, steel) determines the maximum length, L_{max}, of the region from which signals returning from flaws can add together at the receiving transducer to produce a simultaneously recorded joint holographic disturbance. This is shown graphically in Figure 11. It is not difficult to see that if the pulse time duration is t seconds and the sound velocity is c, then the length $L_{max} = ct/2$ is half the physical length L of the pulse in space. For simultaneous recording of flaws in this way, the tail of the pulse returning from the nearest flaw must be just able to combine with the front of the pulse returning from the farthest flaw which is L/2 distant.

The sample-and-hold circuit, in this instance, holds the detected resultant of the combined signals returning from all of the flaws in a length of space that was ensonified $t = L/2c$ secs before the holding time. By selecting the position of the holding time and the pulse length, L, one can essentially create holograms of a bounded region of variable thickness L/2 whose position in space is variable. The set of holograms is essentially of the range-gated type and, as shown in Figure 11, the first position of the holding time permits simultaneous detection of signals from flaws A, B and C while, in the second position, flaws B, C and D will be jointly detected.

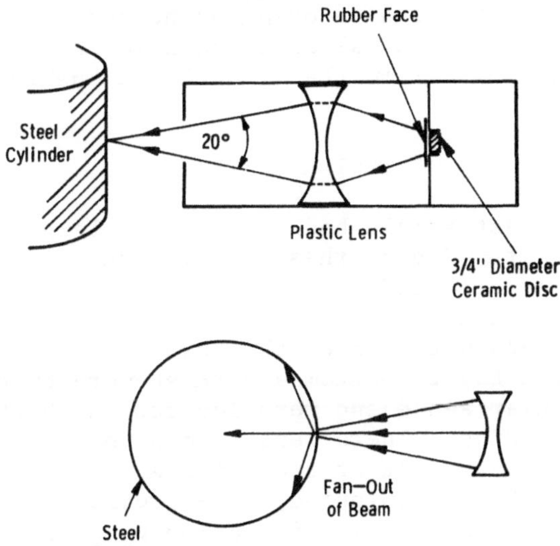

Fig. 10 – Acoustic lens and housing showing beam formation together with sketch of fan–out due to refraction

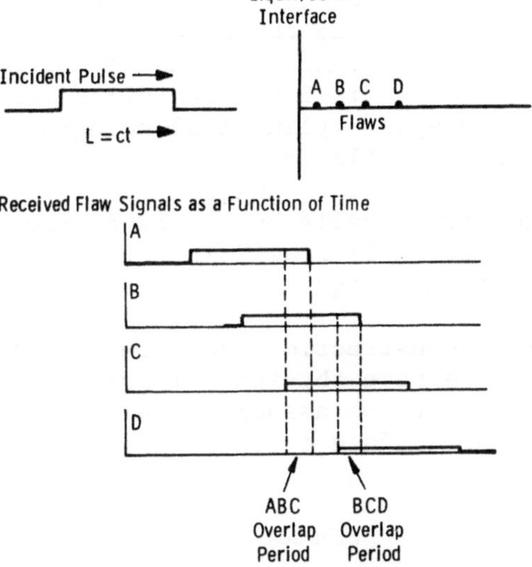

Fig. 11 – Sketch showing importance of pulse length and sample-hold time. Joint hologram of several flaws can only be formed during pulse overlap periods

In this way, by letting L_{max} exceed the diameter of the steel test cylinder we have obtained holograms of flaws on the far side of its central axis and have thus looked at a number of flaws both from the "front" and "behind".

TEST RESULTS

The spin-test sample had a number of large cracks within it. One of these lay in the lower plane surface and was oriented with its length (about a half-inch) more or less radially. It was not quite visible to the naked eye. There were three other cracks that ran almost vertically from top to bottom and these were evenly distributed in an angular sense and were located in about the central third of the radius. There were also several other smaller manufacturing flaws. All of these flaws, both "natural" and those produced by the spinning, were previously detected by conventional ultrasonic flaw detection pulse-echo techniques at 1 → 2.5 MHz.

In Figure 12 is a three-dimensional representation of the flaws in the steel cylinder together with a sketch of the pulse-echo flaw detection test results. These results were obtained by moving detecting transducers over the curved surface and recording the strength, position and radial distance of all flaws by means of radially directed pulses. The flaw record is then displayed two-dimensionally in the same way as the unwrapped cylindrical hologram. Below this conventional pulse-echo flaw record is a hologram that was obtained at 1 MHz. All of the major flaws appear to have been detected holographically but it is difficult to comment unequivocally on the small flaws since the optical quality of our holograms is not high.

No reliable reconstructions have been made to date, but the holograms we have obtained clearly show patterns that are undoubtedly to be associated with the axially oriented and bottom surface cracks.

SUMMARY

Our system should be considered to be in the early stages of development. Our off-axis scan function, $\Delta\phi$, has been used but, unfortunately, there has been no proper way

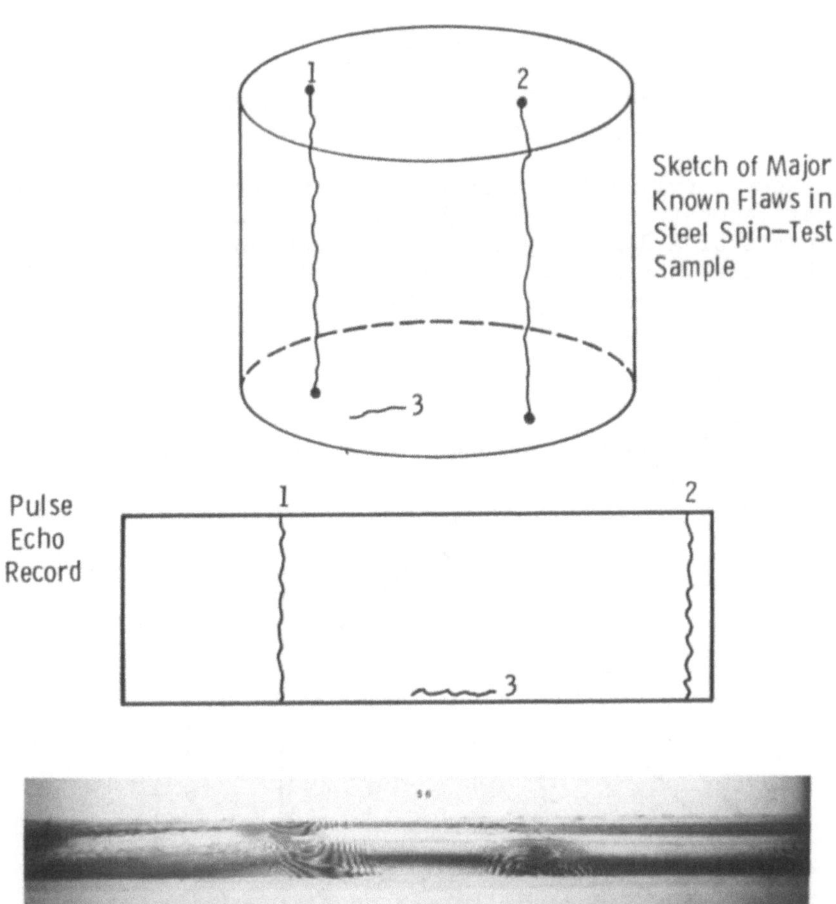

Fig. 12 — Comparison of pulse echo record and cylindrical hologram of flaws in a steel cylinder taken at 1.0 MHz

to assess the level of improvement in the quality of reconstructions since these have been generally poor except in the case of very simple objects. We have, however, shown that some reconstructions, with distortions due to astigmatism, have been found.

APPENDIX A

Cylindrical Scan Aberration Produced by Unwrapping the Hologram

When a cylindrical scan hologram is unwrapped and made planar, reconstructions from this plane hologram will obviously be images of a distorted field. This distortion occurs due to the fact that all wave vectors incident on the hologram have, by the unwrapping, equivalently been rotated by amounts that depend on their original direction and source point.

Some idea of the magnitude of this unusual aberration can be obtained by comparing the flattened cylindrical holograms of point sources with a plane hologram counterpart. Both O.L.I. and I.L.O. cases are considered in the following discussion.

O.L.I.

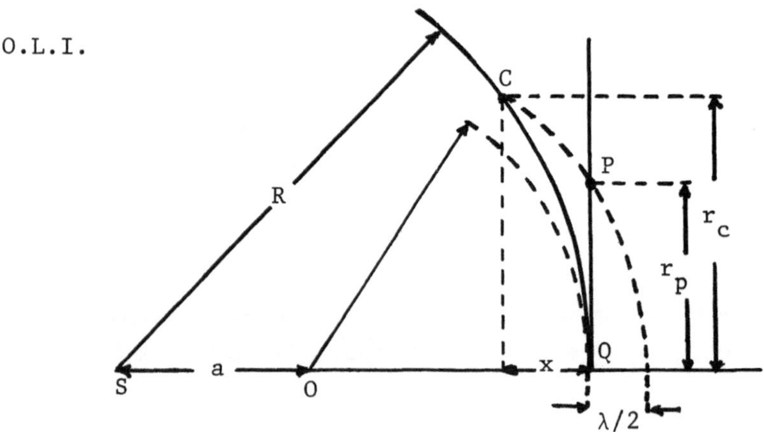

$$r_c/r_p = \sqrt{R/a}$$

Figure A1

Let R be the radius of a cylindrical scan centered on S and let O be a point object at "a" from S. Let P be a point on the plane hologram and C a corresponding point on the cylindrical hologram where the phase of the spherical wave from O is 180° out of phase with the wave at Q. Let r_c and r_p be the vertical heights of C and P above the axis SOQ. Let the wavelength be λ. Then, by the sagittal theorem

$$r_c^2 = 2Rx = 2(R - a + \lambda/2)(x + \lambda/2)$$

Also,

$$x = (R - a)\lambda/2a$$

By eliminating x from these two equations, one obtains

$$r_c = \sqrt{\lambda(R - a)R/a}$$

The value of r_p is easily found to be

$$r_p = \sqrt{\lambda(R - a)}$$

and, thus, the equivalent Fresnel zone which would normally have been circular has been changed to an ellipse whose major and minor axes will be r_c and r_p when the cylindrical 'zone plate' is flattened to become a plane. Hence the distortion may be described in terms of the ratio of the two axes as

$$r_c/r_p = \sqrt{R/a}$$

Clearly, when a = R, there is no distortion, but when a → o, the distortion approaches infinity.

I.L.O.

Using the same meanings for R, a, S, C, P, etc. as previously, then

$$r_c^2 = 2Rx = 2(a + \lambda/2)(\lambda/2 - x)$$

and $$x = a\lambda/2\,(R + a)$$

Elimination of x yields

$$r_c = \sqrt{a\lambda\, R/(R + a)}$$

and r_p can easily be found as

$$r_p = \sqrt{a\lambda}$$

Hence, the distortion is

$$r_c/r_p = \sqrt{R/(R + a)}$$

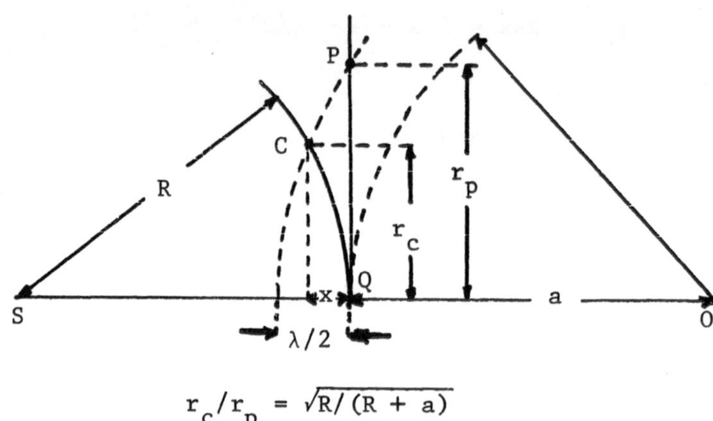

$$r_c/r_p = \sqrt{R/(R + a)}$$

Figure A2

Both of these distortions clearly depend on the range (R - a) of the point source from the cylindrical scanning surface. This unfortunate result represents a major difficulty in the path of reconstruction of all images using simple optical components.

AN EXPERIMENTAL FOCUSED ACOUSTIC IMAGING SYSTEM

Jerry L. Sutton

Naval Undersea Research and Development Center

San Diego, California 92132

ABSTRACT

A high-frequency experimental underwater acoustic imaging system was tested by the Naval Undersea Research and Development Center (NUC) at its Morris Dam Test Facility.

The system consisted of a high-frequency acoustic projector, an acoustic lens, a scanned line array of 100 hydrophones, and a cathode ray tube display. The system exhibited resolutions on the order of 0.3 and 0.4 degrees at ranges up to 29 feet. Results show that useful acoustic images can be obtained by a focused acoustic imaging technique.

BACKGROUND

The experimental underwater acoustic imaging system discussed in this paper was developed for the Navy at the Lockheed Palo Alto Research Lab (LPARL). The unit was evaluated by NUC at its Morris Dam test site. Although this system is not a holographic system it is an acoustic imaging system. It's performance is comparable, in some respects, to that predicted for certain holographic acoustic imaging systems currently under development. Also, it is a real-time system. Thus, the results obtained from this system indicate the quality of images that might be expected from future real-time acoustic imaging systems designed for viewing in the ocean environment.

OBJECTIVES

There were three major objectives of the test and evaluation of the LPARL experimental acoustic imaging system. The primary objective was to test and evaluate the imaging capability of the system. The equipment was assembled by LPARL personnel according to their design and run by NUC personnel in accordance with LPARL procedures. Qualitative results, similar to those obtained by LPARL, as well as quantitative results on range and resolution were obtained in order to evaluate the performance of the system. During these tests much qualitative information on lensed acoustic imaging and acoustic target properties was also obtained.

A second objective of the tests was to determine the performance characteristics of various components of the system and to determine which components most limited performance. Special attention was paid to the 100-element line array of hydrophones, the acoustic projector (insonifier), the liquid acoustic lens, the solid acoustic lens, and the cathode ray tube display.

The third objective of the tests was to determine how a lensed acoustic imaging system would compare with an underwater optical imaging system.

DESCRIPTION OF THE LPARL EXPERIMENTAL UNDERWATER ACOUSTIC IMAGING SYSTEM

The LPARL experimental underwater acoustic imaging system is essentially a physically-toughened laboratory apparatus designed for testing acoustic imaging techniques in a more hostile, but still controlled, environment than a laboratory. The system is not designed for use in the ocean. Furthermore, the system is not intended to have any practical use beyond the test and evaluation of the acoustic imaging techniques involved. The results of those tests, however, have advanced our knowledge of acoustic imaging considerably.

The three major components of the LPARL experimental acoustic imaging system are:

1. Insonifier

2. Underwater acoustic imaging camera

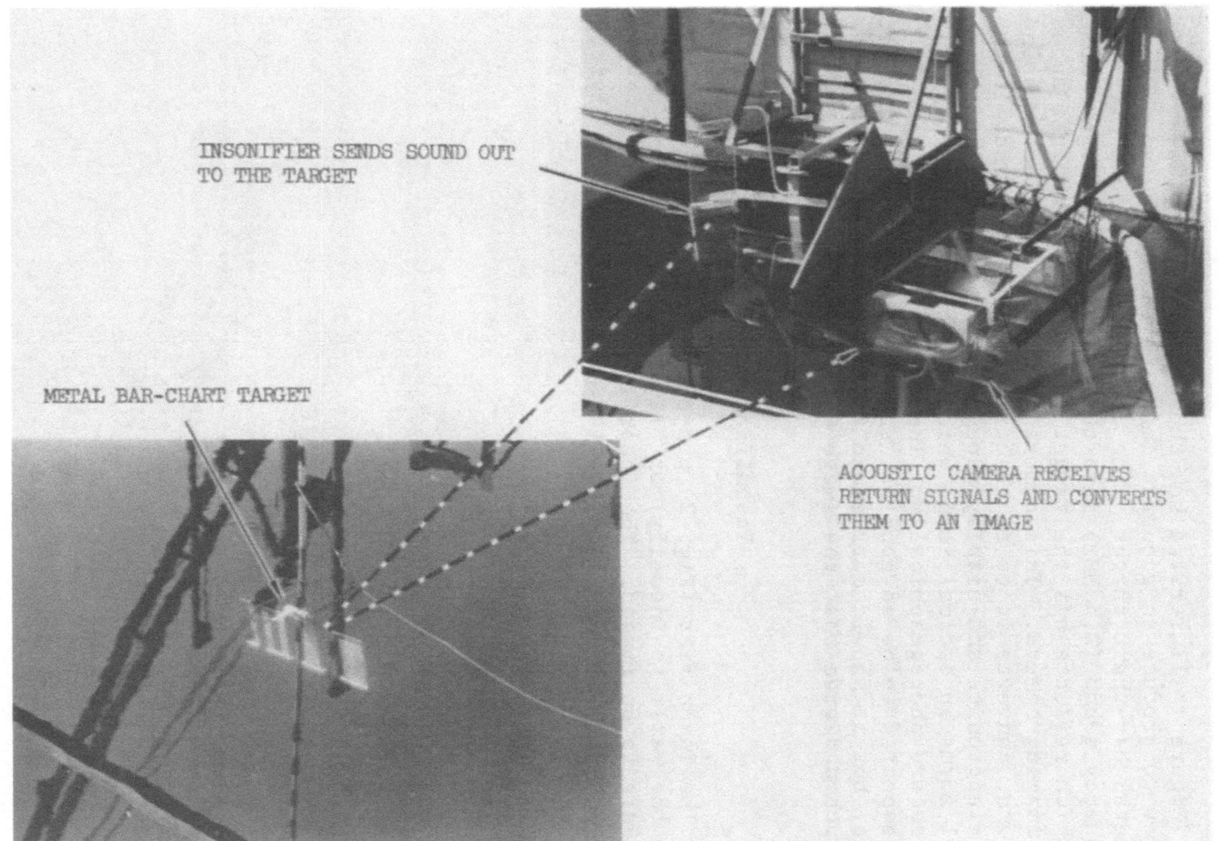

FIG. 1. Setup for the NUC tests of the LPARL experimental acoustic system at Morris Dam.

3. Electronics and display panel

Figure 1 illustrates how the components typically interact to produce an acoustic image in this case, an image of metal bar-chart target. The insonifier projects high-frequency sound (2.5 MHz) into the water at the target, some of which reflects off the target toward the underwater acoustic imaging camera. This camera focuses the reflected sound pattern, converts it to an electric signal, and sends it to the electronics and display panel. Finally, the display panel forms an optical image which the operator can observe and record photographically. The quality of that image is not expected to be as good as, say, a conventional television image, but it should correspond to the target in the water to such a degree that the image can be useful.

Insonifier

The insonifier (Fig. 2) projects high-frequency sound into the water to insonify, or "illuminate", objects (targets) that are to be imaged. The insonifier is made up of a frequency source, two power amplifiers, a piezoelectric crystal,

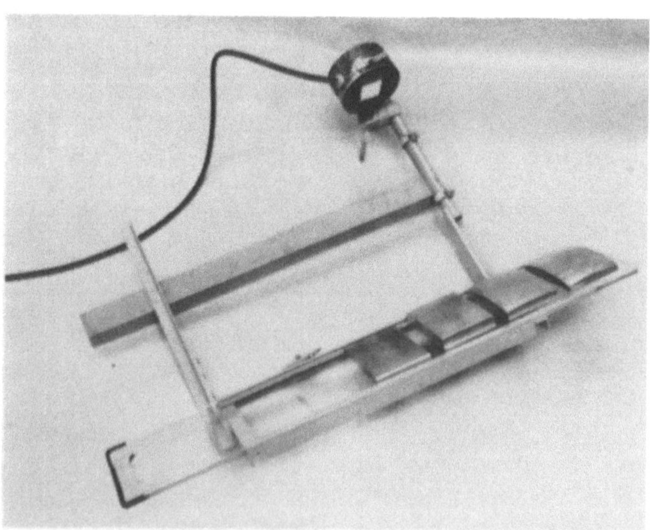

FIG. 2. Insonifier and spherical surface beam spreaders as they are mounted on their frame.

and a beam spreader. The piezoelectric material is a 2-inch square piece of PZT-4 ceramic that is capable of projecting up to 240 acoustic watts into the water. This PZT-4 ceramic has an electrical-to-acoustical conversion efficiency of about 60 percent at resonance (2.5 MHz). Since an acoustic projector that has a large surface area (in acoustic wavelenths) produces a very narrow beam, one of three aluminum spherical surfaces is used to reflect and spread the beam to one of three corresponding beam widths (10, 20, or 30 deg.).

Underwater Acoustic Imaging Camera

Similar to a photographic camera, the underwater acoustic imaging camera (Fig. 3) consists of a lens and a recording plane. The lens, however, is an acoustic lens, which focuses sound waves instead of light waves. And the recording plane is an array of hydrophones instead of a piece of film. In order to avoid the cost and complexity of using a full rectangular array of hydrophones, a line array of 100 hydrophones, in conjunction with a mechanical scanning subsystem, sweeps out a rectangular area in the image plane.

Two lenses can be used with the LPARL system. One lens, a 14-inch diameter plano-concave lens made of Lucite, is shown mounted on the lens carriage in Fig. 3. Focus of this lens is varied by moving the lens carriage by means of a remotely controlled motor and cable arrangement called the "remote focusing subsystem." The other lens, a 12.5-inch diameter liquid lens, is shown in Fig. 4. It consists of two tightly stretched membranes of natural rubber enclosing a volume of Dow Corning 510 fluid between them.[1] Focus is achieved by adding or removing fluid from the lens (i.e., controlling the degree of engorgement) by means of a remote pump and reservoir, also shown in Fig. 4. A 7-inch aperture stop may be used with either lens to reduce degradations caused by spherical aberration. The aperture stop is simply a sheet of aluminum with a 7-inch hole in the center, which allows sound to pass only through the center portion of the lens.

The mechanical scanning of the line array is accomplished by a rocking acoustic mirror (made of stainless steel), which scans the image across the array rather than by moving the hydrophones themselves through the water.

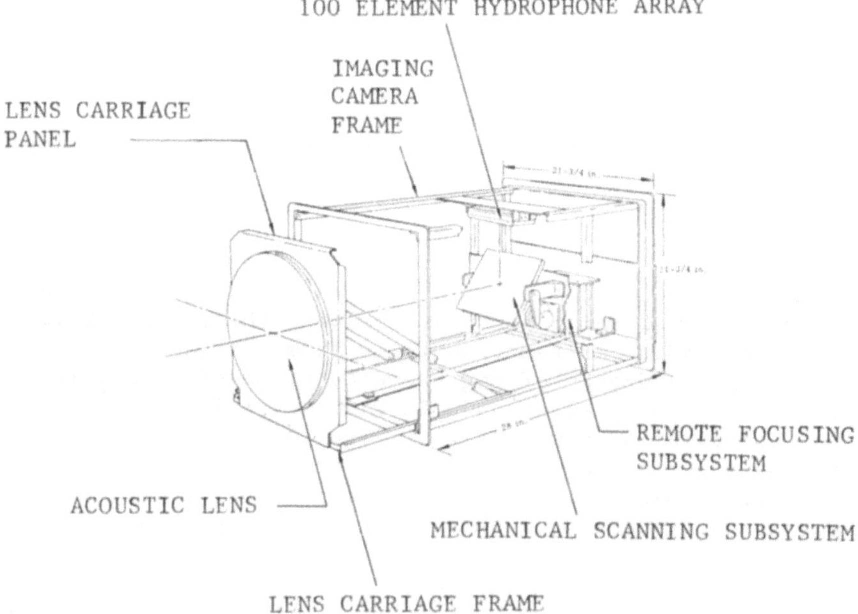

FIG. 3. Underwater acoustic imaging camera with the solid lens in place.

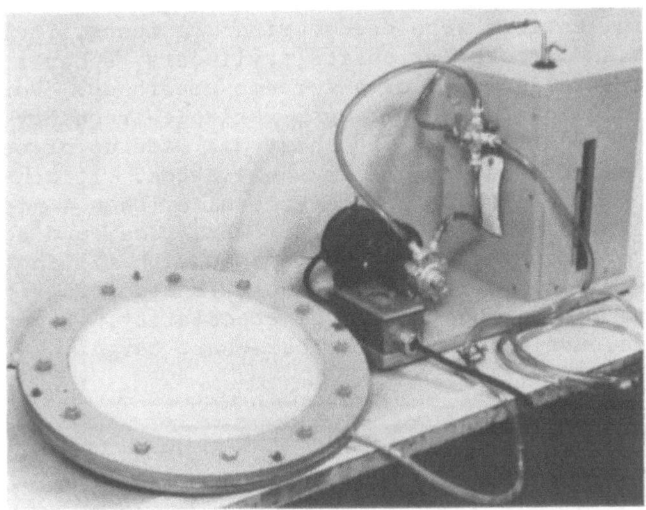

FIG. 4. The liquid lens for the LPARL experimental acoustic imaging system together with the pump and oil reservoirs which change the focal length of the liquid lens.

The hydrophone array consists of a line of 100 piezoelectric elements, together with some processing and multiplexing electronics, all of which are encased in an underwater housing. All of the hydrophones are cut in groups of ten from a single block of PZT-4 such that each of the resulting elements is resonant at about 2.6 MHz. All hydrophones are mounted behind a one-quarter-wavelength-thick "window" cut in the face of the aluminum underwater housing.

Electronics and Display Panel

Information received from the hydrophone array is processed electronically and displayed on a cathode ray tube at the electronics and display panel. The signal coming from the hydrophone array is processed to remove switching transients and unwanted frequencies before it is displayed. The display portion of the panel consists of a Z-axis-modulated oscilloscope with a high-persistence-phosphor display tube. Focus and certain other control functions are also located at this panel.

Several targets were used during the tests, including both rough and smooth bar charts, cylinders, spheres, plates, chain, and a scale model of a torpedo afterbody. For most of the resolution tests, a rough, variable-frequency bar chart was used (Fig. 5). The chart was made up of several strips of quarter-inch aluminum sheet stock. It consisted of a strip, then a space of the same size, then a narrower strip, followed by a similarly narrower space, and so forth. The surface of the bar chart was coated with milling chips of various sizes to produce roughness to acoustic signals in order to reduce the problem of specularity. Figure 5 is a typical acoustic image of the bar-chart target.

PERFORMANCE CONSIDERATIONS

Specularity

A much talked about problem in underwater acoustic imaging is that of target specularity, or shininess. In most instances, "rough" means having randomly distributed reflecting surfaces whose dimensions approximate the illuminating wavelength. The same surface could appear either rough or specular depending only on the wavelength of the illuminating energy. A rough surface has two big advantages over a specular surface:

1. A rough surface reflects an incident beam in many directions, while a smooth, mirror-like surface reflects it in only one direction. The rough surface can, therefore, be seen from many angles.

2. A rough surface is easier to recognize than a smooth one because the image represents only the surface itself, not other objects or sound sources mirrored in the surface. Thus, rough targets were used as much as possible in testing the performance of the LPARL acoustic imaging system, except when the higher reflectivity in one direction or the actual focusing properties of a mirror-like target were desirable.

a. Metal bar chart roughened with milling chips. Overall dimensions approximately 10 x 15 in.

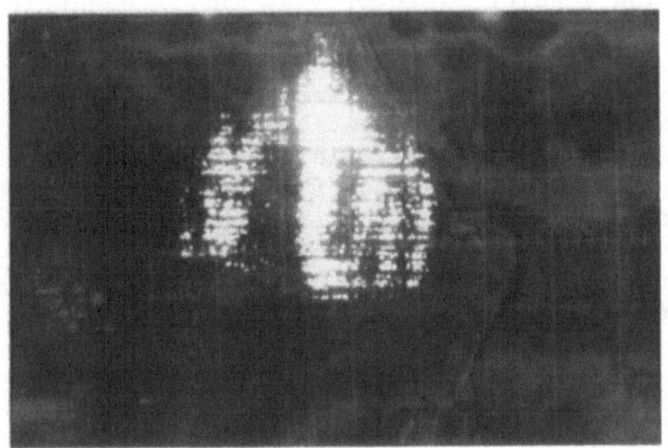

b. Acoustic image of bar chart taken at 10 ft. from the LPARL acoustic camera.

FIG. 5. Rough metal variable frequency bar chart target.

Resolution

An important characteristic of any imaging system is its resolution. Hence, most of the tests were designed to measure system resolution for each of its various configurations.

Angular Resolution. Just as for television and photography, maximum resolution can be determined by finding the minimum resolvable bar width of a variable-frequency bar chart, such as the one in Fig. 5. The minimum resolvable bar is the narrowest one such that it can just be distinguished from the spaces adjacent to it. The angle that this bar subtends at the acoustic camera is the angular resolution of the system.

TV Line Resolution. The angular resolution of a system is not the best indicator of the quality of that system's imaging ability, because angular field of view is also involved. A measurement of resolution that takes both angular resolution and angular field of view into account is the TV line. In television technology, this is the number of just-resolvable lines, alternating black and white, that just fit into the height of a square screen or three-fourths of those that fill the width of a conventional screen with a 3:4 height-to-width aspect ratio. This would correspond to three-fourths the number of minimum resolvable bars that would just fill the field of view, or three-fourths the number of minimum resolvable angles (angular resolution) that would just fill the angular field of view. Thus, the system resolution can be calculated from

$$L = \frac{3}{4}\ \frac{\theta}{a} \qquad (1)$$

where

L = resolution in number of TV lines

= angular field of view

= angular resolution

Alternately, if any one or two of these three variables can be found theoretically, an expectation can be projected for the other variable(s). Since, for the LPARL system, the angular field of view depends on the lens focal length and its position (just as in optics), it is reasonable to use the field of view values measured during the tests.

A geometrical argument can give some indication of what the maximum theoretically possible TV line resolution would be. If there are N resolution elements per horizontal line in the receiver, the maximum number of lines this could display under all conditions would be somewhere between N and N/2. An accepted value from television theory says the number is $N/\sqrt{2}$. Thus the theoretical limit of TV line resolution for a system having 100 resolution elements per horizontal line, as the LPARL system does, is

$$L_t = \frac{3}{4}\frac{N}{\sqrt{2}} = \frac{3}{4}\frac{100}{\sqrt{2}} = 53 \text{ TV lines.} \tag{2}$$

This gives the maximum resolution of the system when the sampling of the image is limiting factor.

When the system is aperture limited, then the usual formula

$$a_t = \frac{1.22\,\lambda}{D} \tag{3}$$

gives the maximum resolution, where λ is the wavelength of sound at 2.5 MHz, and D is the diameter of the aperture.

RESULTS

Resolution

The results of the resolution measurements taken of the LPARL equipment are shown in Table 1. Both the solid (Lucite) lens and the liquid (DOW-510) lens were used with the LPARL system. In addition, an aperture plate with a seven-inch hole in the center was used with each lens, and the liquid lens was used at three lens carriage positions representing three focal length ranges for the lens. Thus seven configurations — those listed in Table 1 — were tested, and the observed angular resolutions and the observed fields of view were listed for each configuration. From these, the TV line resolution and theoretical angular resolution were calculated by means of Eq. 1.

The values presented in Table 1 are average values in all cases. In the case of the solid lens, the receiver field of view varied with the range at which the lens was focused, increasing as the range increased. Focus of the solid lens was achieved by changing the position of the lens with respect

TABLE 1. Resolution data for LPARL experimental focused acoustic imaging system.

Lens and Position	Observed Receiver Field of View	Observed Angular Resolution		Theoretical Angular Resolution (Sampling Limit)		Theoretical Angular Resolution (Aperture Limit)	
	θ, deg.	α, deg.	TV line	α_t, deg.	TV line	α_t, deg.	TV line
Solid lens without aperture stop	13.8	0.38	27	0.19	53	0.12	86
Solid lens with aperture stop	12.7	0.33	29	0.18	53	0.24	40
Liquid lens, position B without aperture stop	11.8	0.55	16	0.17	53	0.13	68
Liquid lens, position C without aperture stop	9.1	--	--	0.13	53	0.13	53
Liquid lens, position A with aperture stop	14.4	0.52	20	0.20	53	0.24	45
Liquid lens, position B with aperture stop	11.4	0.34	25	0.16	53	0.24	36
Liquid lens, position C with aperture stop	8.7	0.23	28	0.12	53	0.24	27

Where:

Position A is lens near detection plane
Position B is lens between positions A and B
Position C is lens far from detection plane

to the sensing array of hydrophones. It was merely a geometrical optical effect. For the ranges of interest (10 to 30 feet) this variation in the field of view was generally on the order of ±10 percent of the value listed in Table 1. In the case of the liquid lens, the variations in field of view were considerably smaller. Focus was achieved only by engorging or draining the lens, while the position of the lens remained fixed. There was also considerable variation (on the order of ±20 percent) in the observed angular resolution. In the case of the solid lens without the aperture stop, for which considerable data was taken, there was a tendency for angular resolution to worsen with range, as would be expected from the increasing field of view; but the effect was on the order of the accuracy of measurement. That is, within the accuracy of measurement, angular resolution was essentially constant with range out to the absorption limit of the system. Hence that effect has been ignored, and only the average values of resolution have been used. In spite of this simplification, the qualitative results were as expected.

Conclusions Relative to Resolution

The test results of Table 1 indicate that the LPARL system behaved as expected from a geometrical optics treatment of the system. The acoustic lenses even had spherical aberrations and edge effects, just as optical lenses do, and an aperture stop improved system resolution performance. Also, the less engorged the liquid lens was, the better its performance, as would also be expected of a thinner optical lens. On the whole, both lenses performed about equally well – the higher acoustic loss through the solid lens being offset by the more severe aberrations of the liquid lens. In all cases, the actual resolution was approximately half of the theoretically possible resolution.

Maximum range achieved by the system (obtained when the solid lens without the aperture was used) for detection and some limited imaging of a specular target was about 30 feet from the hydrophone array. Figure 6 shows some acoustic images of the variable-frequency bar chart at four different ranges.

Most of the data shown in Table 1 were taken at ranges between 9.5 and 23.5 feet from the hydrophone array, because

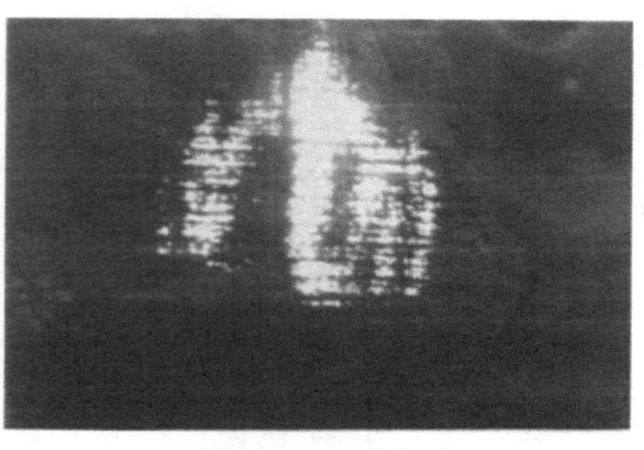

a. Range - 11 ft.

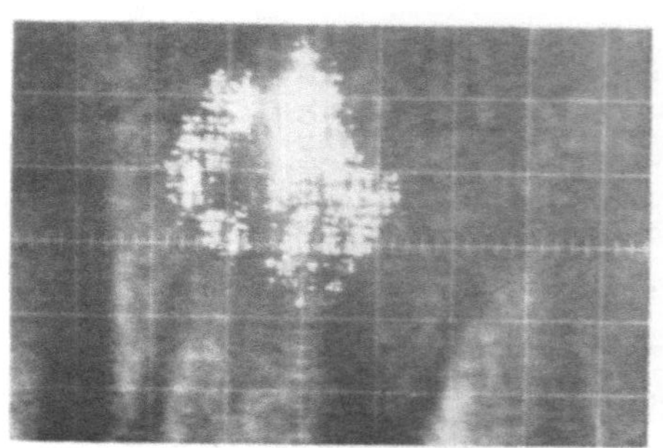

b. Range - 13 ft.

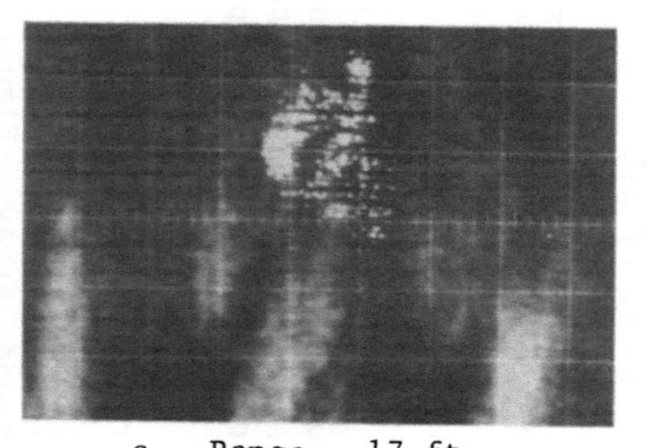

c. Range - 17 ft.

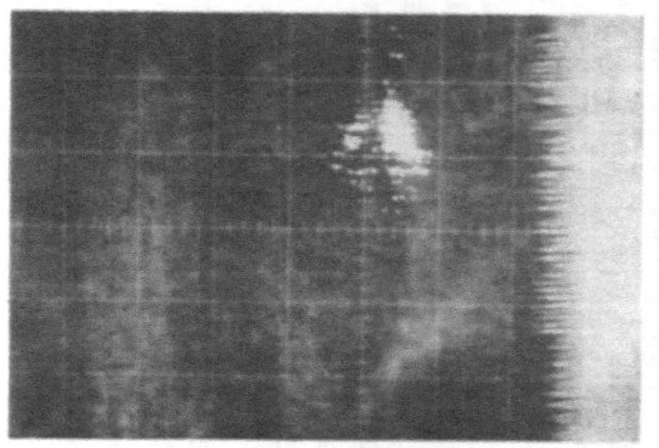

d. Range - 27 ft.

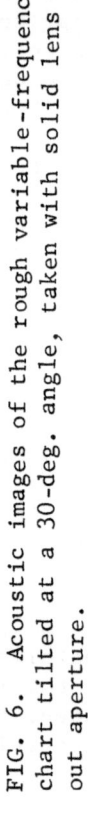

FIG. 6. Acoustic images of the rough variable-frequency bar chart tilted at a 30-deg. angle, taken with solid lens without aperture.

these were the ranges within which imaging could be easily accomplished. The lower limit was established by the relative sizes of receiver field of view and the variable-frequency bar chart, while the upper limit was primarily determined by the absorption of sound energy by the water.

The seven-inch diameter aperture stop limited range to even shorter distances in the case of the solid lens. In this case, maximum range was approximately 18 feet. For the liquid lens at position C (see Table 1) and with the aperture stop, maximum range was about 24 feet.

Acoustic Images of Targets Other Than Bar Charts

Acoustic images were taken of several different kinds of objects in addition to the variable-frequency bar chart (Fig. 5). Among these were: a rough plate (Fig. 7), an array of pipes (Fig. 8), a length of chain (Fig. 9), and a scale model of a torpedo afterbody (Fig. 10). In nearly all cases, the resulting images appeared just as expected. A cylindrical mirror should appear as a line (if oriented perpendicular to the line of sight), and the acoustic image of the pipes indeed appeared as a series of lines. The reflection of a shiny curved surface should appear as a broad,

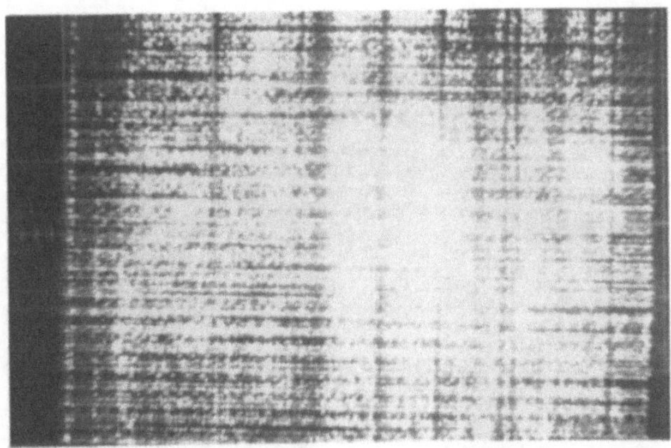

FIG. 7. Nonuniformity of hydrophone response as seen from the acoustic image of a rough plate. Note the diffuse reflection from the rough plate.

FIG. 8a. An array of pipes used as a target.

FIG. 9a. Length of chain about 4 ft. long used as target.

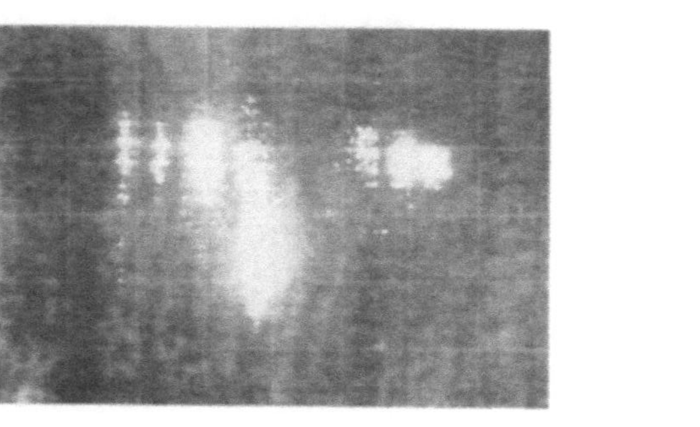

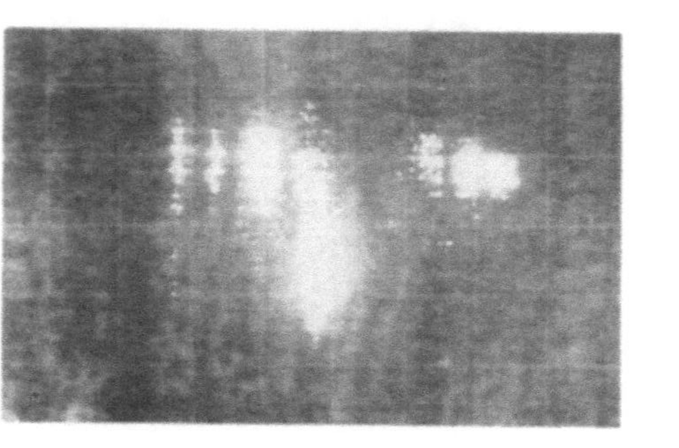

FIG. 8b. The acoustic image of the pipes in 8a taken at a range of 10 ft. with the solid lens without the aperture stop.

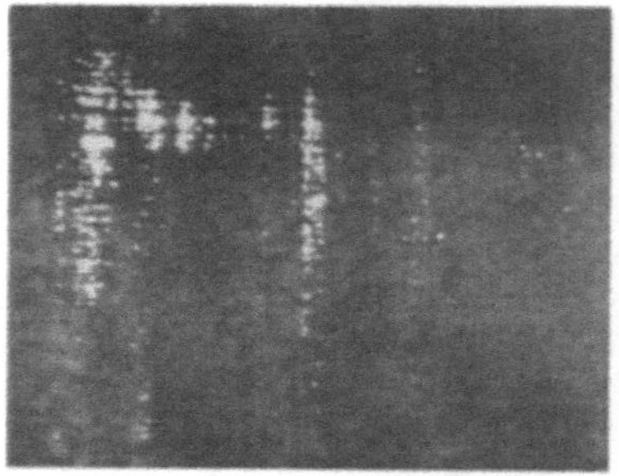

FIG. 9b. The acoustic image of 9a taken at a range of 13 ft. with the solid lens and no aperture.

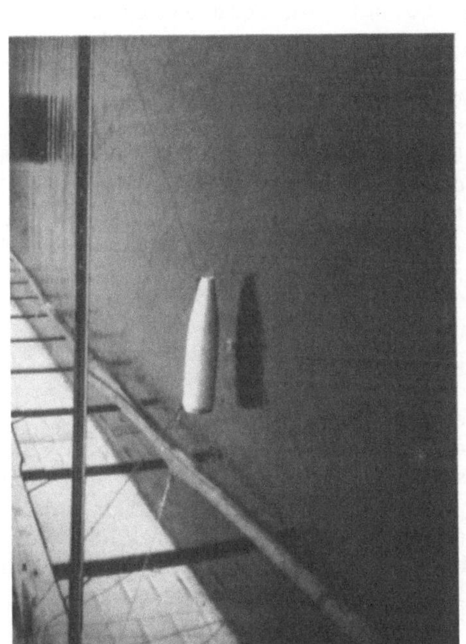

FIG. 10a. Scale model of a torpedo afterbody.

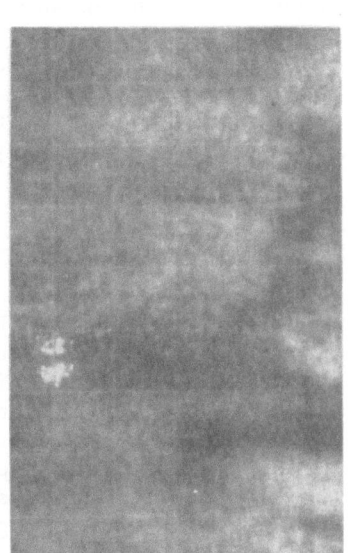

FIG. 10b. Acoustic image of (a) at low signal gain taken with the solid lens without aperture stop at a range of 14 ft.

FIG. 10c. Acoustic image of (a) at maximum signal gain taken with the solid lens without aperture stop at a range of 14 ft. Note that more of object is visible to the left of the specular rear portion of the afterbody, as well as some in the form of background noise that appears below the target.

amorphous spot, and that is exactly how the torpedo afterbody appeared. Finally, a rough plate should appear as a diffusely illuminated surface, and this is more or less how the acoustically rough plate appeared.

COMPARISON OF THE LPARL SYSTEM WITH AN UNDERWATER OPTICAL IMAGING SYSTEM

The LPARL system was operating at a frequency of 2.5 MHz, which, in the lake water at NUC's Morris Dam test site, had an attenuation length of about 1 m. An attentuation length is the distance a collimated beam of energy travels such that the intensity at the end of that distance is 1/e times the intensity at the start, where e is approximately equal to 2.718. Hence, the LPARL system had maximum viewing ranges of from eight to ten meters. This corresponds to maximum viewing ranges of eight to ten attenuation lengths, or about 2.5 times as far as conventional underwater television systems which can see only three or four attentuation lengths under comparable optical-attenuation conditions. This would indicate either that acoustical backscatter (reverberation) is not as detrimental to viewing as optical backscatter underwater, or that it is possible to put out relatively more power or to receive with more sensitivity with acoustics than with optics. In either case, the acoustic imaging system would appear to have an advantage over a conventional optical television system in relatively dirty water, because the acoustic attenuation length is a function primarily of frequency. The optical attenuation length, on the other hand, depends on the optical properties of the water in which it is being used, especially to turbidity (i.e., the number of particles in the water).

CONCLUSIONS

The tests of the LPARL experimental focused acoustic imaging system have shown that useful images of objects can be obtained by means of acoustic imaging techniques and that such systems have advantages under conditions of high optical turbidity. In general, the acoustic images produced by the LPARL system correspond in shape to the targets in the water, and, except for the restrictions imposed by specularity (shininess), those images are recognizable. Where typical underwater television systems are capable of seeing about three or four optical attenuation lengths, this acoustic imaging system is capable of seeing up to ten acoustical attenuation lengths. Thus, when optical attenuation lengths are about the same size as the acoustic attenuation length of a system, the acoustic imaging system can probably see further and/or better than the optical system.

The tests of the LPARL acoustic imaging system have demonstrated specifically that focused acoustic imaging, utilizing both solid and liquid lenses, is capable of producing satisfactory images. It has also indicated, however, that a focused system using simple thin lenses may be limited to a resolution which is a factor of two worse than that predicted by theory; more complicated lenses may alter this result.

These tests have not shown any conclusive evidence as to whether a thin solid lens or a thin liquid lens is more desirable. Each has good and bad attributes that tend to cancel each other in practice. The LPARL system exhibited a best system resolution of approximately 28 TV lines, having angular resolutions of from 0.23 to 0.55 degrees, and fields of view (full angle) of from 8.7 to 14.4 degrees. It also demonstrated imaging capability at ranges up to 30 feet when 240 acoustic watts of 2.5 MHz sound were projected at the target.

REFERENCES

1. Knollman, G. C., Bellin, J. L. S., Brown, A. E., Weaver, J., "Variable-Focus Liquid-Filled Hydroacoustic Lens," Journal of the Acoustical Society of America, Vol. 20, No. 1, 1971, pp. 253-261.

AN EXPERIMENTAL HOLOGRAPHIC ACOUSTIC IMAGING SYSTEM

Newell Booth and Ben Saltzer

Naval Undersea Research and Development Center

San Diego, California 92132

ABSTRACT

An experimental holographic acoustic imaging System for Evaluation and Simulation (SES) has been designed and constructed. The system sensor is a 400 element hydrophone array. Holographic amplitude and phase data are recorded digitally and photographically. The system is designed as a flexible research tool used to evaluate several techniques of obtaining and reconstructing acoustic holograms, and to determine the fundamental limitations to the performance of holographic acoustic imaging systems. The predicted performance capabilities are given and the signal processing techniques and performance parameters to be investigated are described.

INTRODUCTION

As part of the Deep Ocean Technology underwater imaging program, we have been designing and specifying a holographic Acoustic Imaging System (AIS). This system provides real-time acoustical images from a deep submersible. Because there are many research and development problems in the design and construction of such a system we are building, testing, and analyzing an AIS research model. This research tool is called the System for Evaluation and Simulation (SES). Experiments with SES will compare its performance with that predicted by theory and will evaluate many signal processing techniques.

SYSTEM FOR EVALUATION AND SIMULATION

The SES is built around a 400 element square hydrophone array similar to that used by Marom, et al. in their experimental work.[1] There are two configurations of SES. The first, called Stepped SES, requires one transmit-receive (T/R) cycle to obtain holographic data for each hydrophone in the array. This configuration then requires 400 T/R cycles to obtain a frame of holographic data and can examine only stationary targets. The second configuration, called Automatic SES, has electronic channel processing for each element of the array. As a result only one T/R cycle is required to obtain a frame of holographic data, and the constraints on target motion are eased. The Automatic SES configuration also gives a much higher data rate which is useful for determining the effects of medium nonuniformities. Construction of the Automatic SES will begin in June 1972. Because of this the emphasis in this paper is on the Stepped SES.

The SES array, control panel, and recording panel are shown in Fig. 1. The control panel contains the scan, transmit, and receive electronics; a digital volt meter for measuring the amplitude of the received signal; and a phase meter. The recording panel contains the digital tape recorder, the XY"Z" display, the digital memory, and the sequencing circuitry.

In the Stepped SES configuration, the system is very sensitive to relative motion between the target and receiver. In order to achieve the required stability, we are constructing a floating acoustical bench. The target and array are suspended below the 25 foot bench, which is isolated from surface motion by spar buoys. Range, elevation, azimuth, and target orientation can be varied. The SES experiments are being conducted at the NUC Transducer Evaluation Center, which has a pool which is about 200 feet in diameter and 38 feet deep.

A block diagram of the SES is presented in Fig. 2. The system output shown at the right is a series of acoustical images of various targets at different ranges using many alternate signal processing techniques. High frequency sound ($80\text{kHz} < f < 300\text{kHz}$) reflected from the target is detected at the array. In the stepped configuration, the received signal is switched through a preamp to the control panel where the phase, ϕ, relative to the transmitted signal and

SES CONTROL PANEL SES RECORDING PANEL SES ARRAY

FIG. 1. SES Control Panel, Recording Panel, and Array.

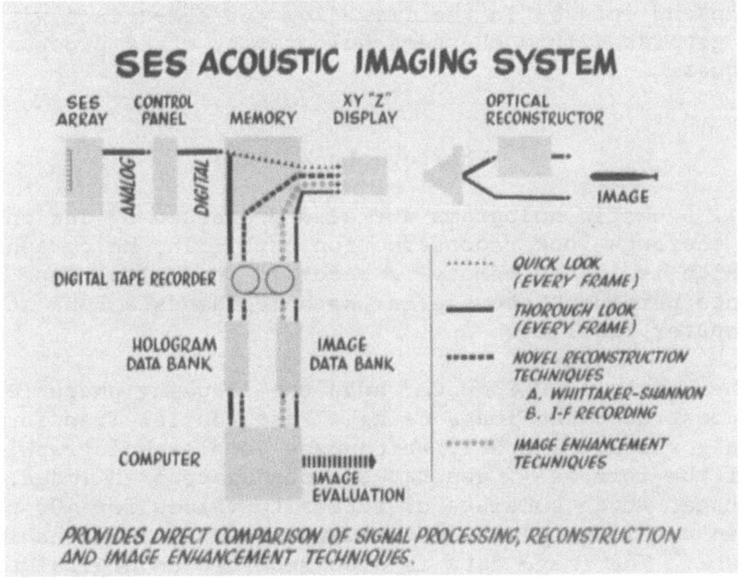

FIG. 2. Data Flow Diagram for SES Acoustical Imaging System.

the amplitude, A, are measured and converted to 18-bit BCD digital data for storage in the digital memory. The memory contains 1,024 digital holographic data words, more than enough to store the amplitude and phase data from each of the 400 hydrophone elements. From the memory the data can take many paths depending upon the signal processing technique used to generate the image.

Quick Look

All acoustical holograms are processed by the quick-look reconstruction path. From the amplitude and phase information in the memory, the holographic term A cos ϕ is generated and displayed as "Z" on the XY"Z" display. A photograph of the display is taken on Polaroid Type 55 P/N film. The negative is an optical hologram equivalent to the acoustical hologram taken at the array.

This negative is inserted into an optical reconstructor which generates the quick-look acoustical image. The optical reconstructor consists of a laser for illuminating the hologram and optics for viewing and photographing the resulting image. The quick-look image is available for viewing within 15 minutes after recording the hologram. A photograph will be taken and entered in the image log for comparison with images generated from the same hologram by other processing techniques.

Thorough Look

All acoustic holograms are also processed by the computer in the thorough-look reconstruction path. The holographic data which is in the memory is recorded on digital tape and read into permanent storage in the hologram data bank at the NUC computer facility.

The Univac 1230 Computer adds the focusing phase factors and reconstructs the image using a Fast Fourier Transform (FFT) algorithm. The FFT, which uses complex holographic data of the form $Ae^{i\phi}$, generates a conjugate free image. This image, which consists of intensity values for 400 positions in the image plane, is permanently stored in the image data bank. The image data is also recorded on digital tape and played into the SES memory. The intensity values from the memory are displayed as "Z" on the XY"Z" display which generates a visual representation of the acoustical image which is then photographed and entered into the image log.

Novel Reconstruction Techniques

Those holograms with computer processed images of sufficient quality and information content to provide meaningful comparison of reconstruction techniques are processed by alternate reconstruction methods.

As was described above, digital processing of complex holographic data generates an image which is free from conjugate interference. Optical processing methods, however, can handle only real holographic data. The alternate reconstruction techniques of Whittaker-Shannon Interpolation[2] and IF Recording[3] are methods of eliminating the conjugate image interference in optical holographic processors. The Whittaker-Shannon hologram is generated from the stored holographic data using the computer, is stored in the SES memory, and is written on the XY"Z" display. The CRT screen is photographed and processed in the optical reconstructor. The IF hologram can be displayed for optical processing directly from the SES memory. The effects of hologram magnification and properties of the film and CRT display on optical reconstruction are also being examined.

Phase-only holograms[4,5] are generated and processed by the computer and optical methods. The quick-look, Whittaker-Shannon, and IF reconstruction techniques are each capable of processing phase-only holograms.

Image Enhancement

Since the spatial frequency spectrum, in the form of the focus-corrected hologram, and the image information are both available in digital form, digital image enhancement processing[6,7] can easily be implemented. The enhanced images are entered in the image log for comparison with images generated by other processing techniques.

Image Evaluation

The numerical evaluation of image quality parameters such as resolution and constrast is usually a time consuming task requiring special targets and subjective evaluation of the images generated. Since the images from SES are available in digital form, computer calculation of image quality is feasible. Image evaluation as a function of system properties, computer reconstruction methods, and image enhancement techniques are done by the computer.

SYSTEM EVALUATION

In addition to evaluating various signal processing techniques, we are comparing the system performance with that predicted by theory. The tests give experimental feedback to improve the approximations made in the analysis. We are paying particular attention to the performance parameters' dependence on system hardware specifications for aid in the design of future systems.

Some of the performance properties which are being studied are presented in Table I. The analytical predictions of some of the important parameters are presented below.

Field of View

The half angle field of view, θ, of a sampled holographic imaging system is given by $\sin\theta = \lambda/2d$ where λ is the wavelength and d is the distance between the sampling hydrophones.[8] For the SES System, θ varies from 16 deg. at 300kHz to 90 deg. at 83kHz.

Resolution

The angular resolution is approximately given by $\alpha = \lambda/D$ where D is the aperture width.[8] α for this system varies from 1.6 deg. at 300kHz to 5.8 deg. at 83kHz.

The resolution is measured from the width of the point spread function. Fine sampling of the point spread function is obtained from the computer reconstruction by inserting the sampled hologram of a point source in a larger matrix of

TABLE I: System Parameters to be Investigated

Performance	Properties
Resolution	Aperture Size, Frequency, Speed of Sound, Position in Field of View.
Dynamic Range in Image	Channel Amplitude and Phase Uniformity, Channel Crosstalk, Dynamic Range of Channel Processor, Array Thinning, Number of Gray Levels in Digital Hologram, Target Strength, Target Motion, Reverberation, Ambient Noise, Medium Nonuniformities.
Maximum Range	Transmitter Power and Directivity, Hydrophone Sensitivity and Directivity, System Noise, Frequency, Pulse Length, Target Strength, Attenuation, Reverberation, Ambient Noise.
Minimum Range	Minimum Range Gate, Reconstructor Focusing, Beam Pattern Overlap, Target Size, Reverberation.
Field of View	Frequency, Sampling Density, Reconstruction Technique, Angle of Reference Wave, Transmitter and Hydrophone Directivity, Speed of Sound.
Depth of Field	Pulse Length, Range, Frequency, Aperture Size, Speed of Sound.

zeros and reconstructing it with FFT techniques. This is also being done with the array focused at different ranges to obtain the depth of field of the system.

Depth of Field

The predicted depth of field has been obtained by calculating the ranges at which the geometrical width of a point source in the image increases to one resolution element. This calculation ignores the effects of range gating. If the system is focused for a range z, point sources at the ranges

$$r_+ = r(1 - \frac{1}{\alpha r - 1})$$

$$r_- = r(1 - \frac{1}{\alpha r + 1})$$

have a width of two resolution elements where r = z/D is the range in aperture diameters and α is the size of a resolution element in radians. Figure 3 is a plot of r_+ and r_- as a function of r for various values of α. If the system is focused at range r, targets are approximately in focus at all ranges between r_- and r_+. Note that at short ranges the depth of field is very narrow. As a result, only portions of targets that have depth are in focus. As in cameras, the depth of field can be increased by reducing the aperture diameter at the expense of resolution.

Dynamic Range

The dynamic range in the image is of particular interest in acoustical imaging systems because of the specularity problem associated with the long wavelength and with the surface properties of most targets of interest. A large dynamic range is required to see the nonspecular returns which carry much information about the target's shape. We are experimentally investigating the dynamic range required for useful imaging and are determining the effects which limit this performance parameter.

The dynamic range in the image, which is directly related to the contrast, is a function of many parameters as shown in Table I. We have paid particular attention to the effects of inaccuracies in the gain and phase measurement of the acoustic signal and to the limitations imposed by temporal noise of the processing electronics. A thorough analysis of these effects is presented in the following paper by Jim Thorn.[9] His results indicate that gain and phase variations in the acoustic wave measurement remove energy from the bright parts of the image and distribute it over the rest of the picture. Thus, dynamic range is also a function of the size of the image highlight.

It may be possible to compensate for the amplitude and phase noise by generating a correction hologram from a known target and its theoretically predicted hologram. These corrections can be applied to other holograms by the computer

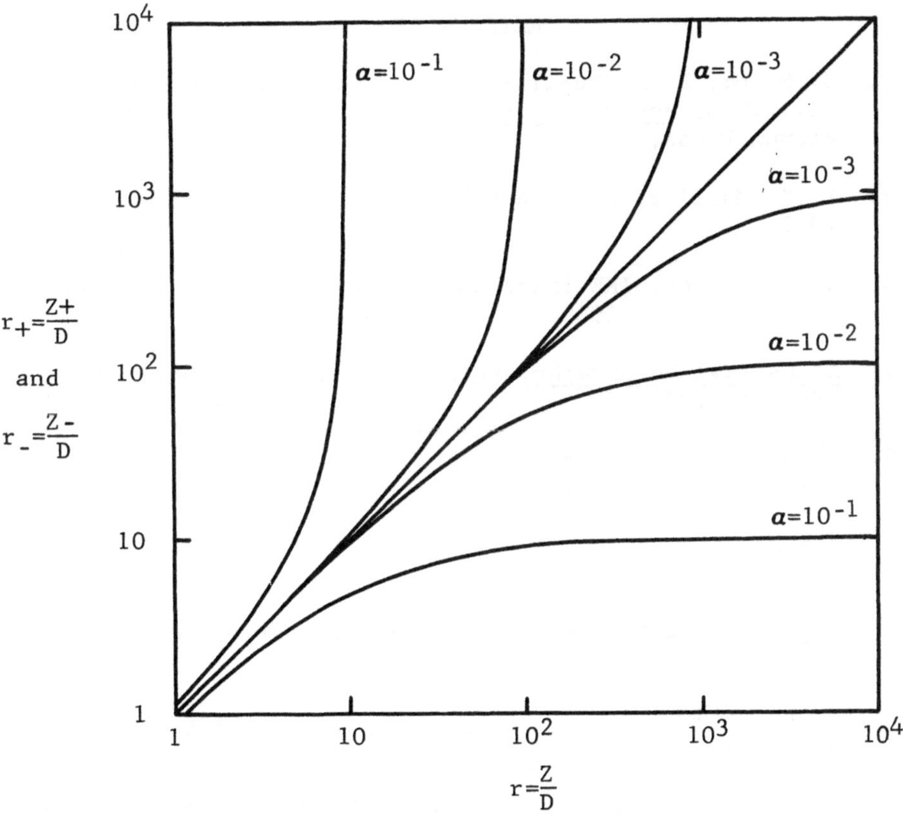

FIG. 3. Depth of Field of Imaging Systems.

to cancel the nonuniformities. This technique of improving the dynamic range is being evaluated with the SES.

SUMMARY

SES is a flexible research tool for the evaluation of the methods and limitations of holographic acoustic imaging systems for underwater use. We are using it to develop design criteria for operational systems and to compare various techniques of obtaining and reconstructing acoustical holograms. The experimental feedback from SES tests is valuable in determining the direction of future analysis as well as the validity of the analysis which has been completed.

REFERENCES

1. E. Marom, R. K. Mueller, R. F. Koppelmann, and G. Z. Zilinskas, Acoustical Holography, Volume 3, p. 191, Plenum Press, 1971.

2. R. K. Mueller, Proceedings of IEEE, Volume 59, p. 1319, 1971.

3. W. G. Hoefer, IRE Trans on Military Electronics, Volume MIL-6, p. 174, 1962.

4. Acoustical Holography, Volumes 1 and 2, Plenum Press, 1969 and 1970.

5. N. H. Farhat, and W. G. Guard, Proceedings of the IEEE, Volume 59, p. 1383, 1971.

6. A Symposium on Sampled Images, Perkin-Elmer Publication, 1971.

7. Image Information Recovery, SPIE Seminar Proceedings, Volume 16, 1968.

8. J. W. Goodman, Introduction to Fourier Optics, McGraw-Hill, 1968.

9. J. Thorn, Acoustical Holography, Volume 4, Plenum Press, p. 569, 1972.

FIRST-ARRIVAL SEISMIC HOLOGRAMS

Gerald L. Fitzpatrick

Physicist, Denver Mining Research Center
U.S. Department of the Interior, BuMines
Bldg. 20, Denver Federal Center
Denver, Colorado 80225

ABSTRACT

Under certain conditions, phase-only impulse seismic holograms of underground objects may be made by using only the source-object-detector travel times to calculate the relative phases of the arriving signals. This circumstance eliminates the necessity of recording and processing entire signals and, in principle, allows one to make seismic holograms by resorting to simple and rapid timing circuitry, rather than the comparatively complex apparatus needed to record and digitize entire signals.

By eliminating such noise contributions as shear and surface waves the technique of first-arrival seismic holography also results in a large improvement in the signal-to-noise ratio of the recorded fringe data.

A comparison of first-arrival seismic holograms made with actual field data and holograms made with entire signals from the same field data shows that the first-arrival holograms (and their reconstructions) are superior to those made with the full seismic records.

INTRODUCTION

Before discussing the first-arrival hologram in full detail, we wish to present a brief description of the original experiment in seismic holography that motivated its development. The interested reader is directed to the original work for a more complete description of that experiment (1)[1]. The method of obtaining the seismic holograms from the field data gathered during the foregoing experiment, though straightforward, was quite involved and time-consuming. During the original experiment, for example, the author and colleagues began by collecting 441 channels of raw seismic data in order to eventually produce a discretely sampled seismic hologram of an underground fracture zone in oil shale illustrated in Fig. 1.

The seismic signals recorded at the ground surface were not continuous waves but rather impulses from a small explosive charge (150 gm pentaerythritetranitrate - PETN) located at a depth of 195 feet below the ground surface. Each time one of these charges was detonated, the output of 11 surface geophones at a particular position in the sampling array was amplified and recorded on magnetic tape for later processing. By moving the 11 phones and firing a sequence of 150-gm charges, the 21 by 21 array in Fig. 1 was eventually sampled. Thus, the raw seismic data collected consisted of 441 channels of wide-band impulse data and not a hologram as such. Further processing of the data had to be undertaken.

Accordingly, the next step in making the holograms involved digitizing these records. Each recorded trace was digitized thereby producing 450 individual amplitude bits per trace, or a total of 198,450 bits for the 441 traces. Having digitized the data, the next step was to choose a reference frequency and a synthetic reference source location so that a reference wave could be combined with the seismic data. Fig. 1 illustrates the position of the synthetic reference source chosen at a depth of 80 feet. The reference frequency chosen was 140 Hz.

---

[1] Underlined numbers in parentheses refer to items in the list of references.

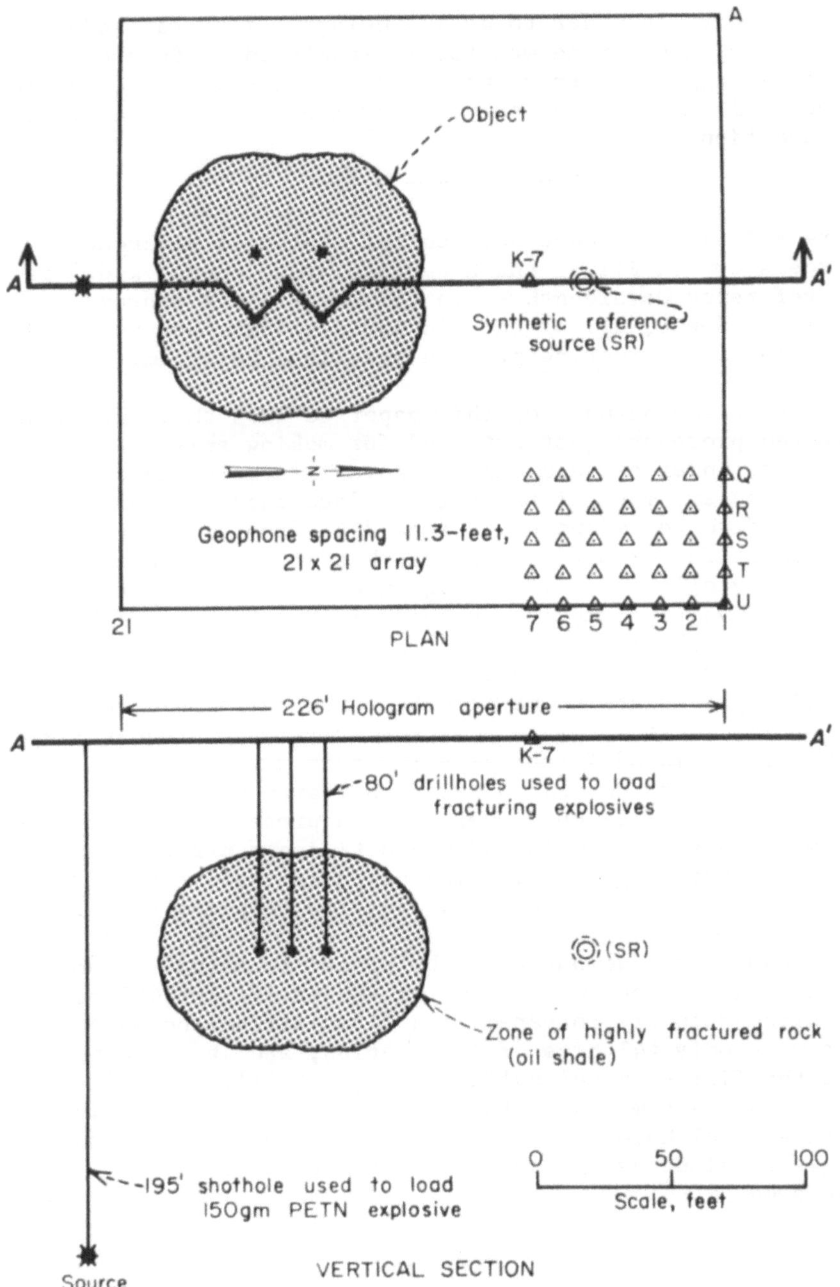

Fig. 1. - Experimental Arrangement.

Finally, in order to make a hologram from this digitized data, each trace was Fourier-analyzed to determine the phase angle Φ corresponding to the chosen reference frequency. A simple phase-only hologram was then computed from the equation

$$H = \cos\left(\frac{2\pi R}{\lambda} - \Phi\right), \tag{1}$$

where R is the distance between the synthetic reference source shown in Fig. 1 and a particular geophone, and λ is the reference wavelength -- in this case λ was approximately 50 feet. Hence, all of the 198,450 data bits had to be manipulated in order to produce the final hologram.

It is the purpose of this paper to show that the rather involved procedure just outlined for making seismic holograms is, under certain circumstances, largely unnecessary. In particular, most of the seismic data that was collected was found to be redundant. Using data from the previous experiment for example one finds that it is possible to make an excellent quality hologram with only 441 bits of first-arrival information instead of 198,450 bits of information, where the first arrival times are simply the time intervals from the source via the object to the detector. The relative phases are to be computed from the simple formula $\Phi = \omega_0 t_{fa}$, where ω_0 is an arbitrary reference frequency and t_{fa} is the time of first arrival. Such a reduction in the quantity of data needed to make a hologram has obvious practical significance. Where such a process is applicable one may resort to simple and rapid timing apparatus rather than the more complicated, time-consuming, and expensive apparatus needed to record and digitize complete signals.

Results are demonstrated by appeal to actual seismic field data. Holograms made using the entire 198,450 bits of information are compared with first-arrival holograms made with only 441 bits of information, and it is shown that the first-arrival holograms are actually less noisy than those made with the full seismic records. One of the first-arrival holograms is reconstructed by a laser and the images obtained using various masks are compared and discussed.

THEORY OF FIRST-ARRIVAL SEISMIC HOLOGRAMS

There are generally two different but equivalent ways to measure the relative phases of two spatially separated detectors in a continuous wave field. Perhaps the most common method is to measure the phase difference between two different wavefronts that intersect the two detectors at the same instant of time. However, an equally valid technique would be to measure the phases at the two detectors by measuring the time it takes for the wavefront to pass from the first detector to the second. The relative phase in this case is then simply the observed time difference times the circular frequency of the radiation. It is the latter approach that we wish to capitalize upon in what follows. Of course it would be difficult in a continuous field to follow individual wavefronts, therefore one would not ordinarily attempt to measure the phase in this way. However, if impulsive signals are used, it is quite easy to follow the wavefronts.

Phase Measurements and Redundant Information

Consider for example the seismic traces illustrated in Fig. 2. These idealized traces are taken to represent an impulsive signal which has either been transmitted through an object or reflected from the surface of a single object. The basic assumption we now make is that the medium through which this signal propagates is linear to a first approximation. It is then reasonable to assume that the first wavefront to arrive at a detector at time t_{fa} is quite comparable in shape and phase to one of the continuous Fourier component waves that makes up the impulse. This approximation is good when the frequency of the component in question is near the highest frequencies contained in the impulse and poor when the frequency is far from this value since in reality higher frequencies are the first to arrive.

Keeping these assumptions and limitations in mind, it is clear that in principle the relative phases needed to make a hologram can simply be determined by noting the times of arrival of the impulsive signal at various detectors. Another way to see that this must be a valid procedure for a transmission experiment (such as that illustrated in Fig. 1) is to note that most of the P-wave information beyond the time of first arrival in Fig. 2a is essentially

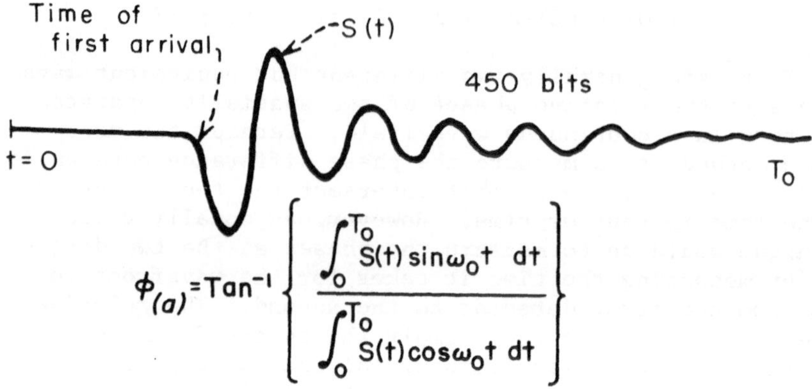

(a). Conventional Calculation of Phase Angle.

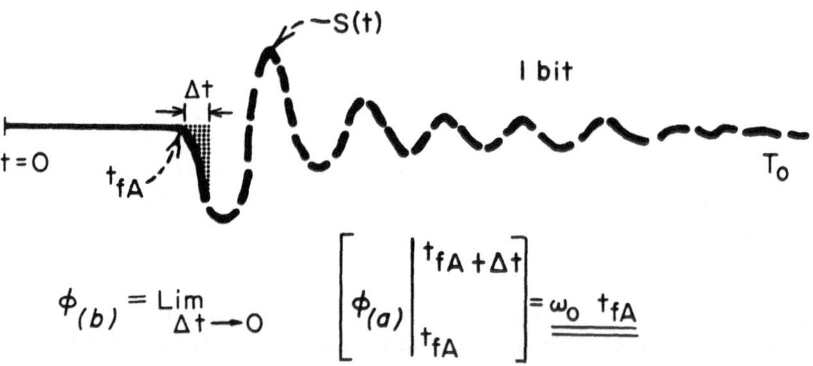

(b). Calculation of Phase Using Only First Arrival Time.

Fig. 2. - Calculation of phase angles of a given Fourier component within an idealized impulse by: (a) Carrying out integrations over full records and (b) Limiting integrations to the interval preceding the arrival of the signal.

of the same type; that is, it involves transmission and scattering information from a single object rather than many objects, and successive P-wavefronts after the first are therefore not expected to carry new information.

This conclusion of course is not true in general in a reflection experiment where many deeper reflectors contribute new information as time goes on. The technique of calculating phases using only first arrivals will obviously not work in this case. However, it will work in the case of a single reflector since all successive wavefronts at a given frequency carry essentially the same information. It would also work in the case where a particular reflector can be "picked" from the records. We are thus led to the idea that in the case of a simple transmission experiment or in a simple reflection experiment, all successive wavefronts after the first are essentially redundant and unnecessary for making holograms.

Computation of First Arrival Holograms and Discussion of Timing Errors

Ordinarily the phase angle used in computing a hologram at reference frequency ω_o from an entire trace using standard Fourier analysis is

$$\Phi = \tan^{-1} \left\{ \frac{\int_0^{T_o} S(\bar{r},t) \sin \omega_o t \, dt}{\int_0^{T_o} S(\bar{r},t) \cos \omega_o t \, dt} \right\} , \qquad (2)$$

where the integration is carried out over the whole trace. If the entire trace in Fig. 2a were used, for example, this calculation would involve all of the 450 digitized points on that record. Since most of this information is redundant and will also contain unwanted shear and surface wave information and since the amplitude is zero previous to t_{fa}, only the interval from t_{fa} to $(t_{fa}+\Delta t)$ of Fig. 2b need be examined. Then the phase angle is

$$\Phi = \tan^{-1}\left\{\lim_{\Delta t\to 0}\left[\frac{\int_{t_o}^{t_o+\Delta t} S(\bar{r},t)\sin\omega_o t\,dt}{\int_{t_o}^{t_o+\Delta t} S(\bar{r},t)\cos\omega_o t\,dt}\right]\right\}, \tag{3}$$

where we have relabeled the time of first arrival $t_{fa} = t_o$. Replacing the signal $S(r,t)$, $\sin\omega_o t$ and $\cos\omega_o t$ by their constant average values over the interval Δt this becomes

$$\Phi = \tan^{-1}\left\{\lim_{\Delta t\to 0}\left[\frac{\sin\omega_o t_o + \sin\omega_o(t_o+\Delta t)}{\cos\omega_o t_o + \cos\omega_o(t_o+\Delta t)}\right]\right\} \tag{4}$$

$$\Phi = \tan^{-1}\left\{\frac{\sin\omega_o t_o}{\cos\omega_o t_o}\right\} = \tan^{-1}\left\{\tan\omega_o t_o\right\} \tag{5}$$

$$\Phi = \omega_o t_o \pm (n-1)\pi \qquad n = 1, 2, 3 \ldots \tag{6}$$

Hence except for the constant reference frequency ω_o and an irrelevant integer n, this phase-angle calculation involves only one bit of information, namely t_o instead of the original 450 bits. One may then compute a phase-modulated first-arrival hologram from the simple formulas[a]

$$H = \cos(2\pi R/\lambda - \Phi) \tag{7}$$

$$\Phi = \omega_o t_o \pm (n-1)\pi \qquad n = 1, 2, 3 \ldots \tag{8}$$

where R is the distance between a point synthetic reference and a particular geophone and where Φ is the first arrival phase $\omega_o t_o \pm (n-1)\pi$ at that geophone.

Of course the first-arrival-time measurements will always be accompanied by errors. The amount of error depends on many different factors such as the noise level of the "quiet" portion of the record prior to the arrival of the signal and the threshold sensitivity of the apparatus

---

[a] It should be noted that since only the relative phases of the arriving signals are needed, the zero time of the reference clock does not have to be the shot time. If the source onset is unknown, the relative arrival times are perfectly adequate.

that might be used to measure it. Such timing errors of course should not exceed about $\lambda/8$, otherwise no hologram could be made owing to an effective loss of coherence. Errors would evidently have to be considerably less than $\lambda/8$ if the time anomalies introduced by the objects to be detected are of this same order.

Reference Frequency

Before closing this section it is worthwhile to point out that the reference frequency ω_0 is arbitrary. One can use any frequency that does not violate the sampling criterion; that is, as long as the final hologram fringes are adequately sampled by the existing geophones in the original sampling array, the particular frequency chosen, even though it might be fictitious, is perfectly acceptable. The use of an arbitrarily high fictitious frequency could improve such things as separation of the image from its conjugate and/or the DC background owing to the smaller spacing of fringes for larger ω_0.

If it is desired to use only realistic reference frequencies one can restrict ω_0 to values within the rms deviation of ω during the first cycle of the signal. If τ is the period of the first cycle the Fourier amplitude is

$$A(\omega) = \frac{1}{\sqrt{2\pi}} \int_{t_0}^{t_0+\tau} S(\bar{r},t)e^{+i\omega t}\, dt \quad , \tag{9}$$

and the average value of ω is

$$\bar{\omega} \stackrel{\Delta}{=} \frac{\int_0^\infty \omega |A(\omega)|^2 d\omega}{\int_0^\infty |A(\omega)|^2 d\omega} \quad . \tag{10}$$

If the root mean square deviation of ω is defined as

$$\Delta\omega = \sqrt{\overline{(\omega-\overline{\omega})^2}} = \sqrt{\frac{\int_0^\infty (\omega-\overline{\omega})^2 \, |A(\omega)|^2 d\omega}{\int_0^\infty |A(\omega)|^2 d\omega}} \, , \tag{11}$$

the real frequencies should be limited to the range,

$$[\overline{\omega} - \Delta\omega] < \omega_o < [\overline{\omega} + \Delta\omega] \; ; \tag{12}$$

whereas frequencies outside this range must be viewed as fictitious and not characteristic of those associated with the first pulse to arrive at the detector. A definition of ω_o, which does not involve the frequency spectra but which is equally valid, is to set $\omega_o = 2\pi/(4t_r)$ where t_r is the typical rise time of the first pulse.

APPLICATION

Holograms

Fig. 3 illustrates the direct application of the foregoing ideas to actual seismic field data collected during the experiment illustrated in Fig. 1.

All holograms were plotted by computer on a cathode-ray-tube screen and photographed on microfilm by first dividing the H data into 12 intensity bins and then employing a 12-gray-level tonal plotting routine. The actual holograms that resulted were approximately 5/16-inch-square microfilm transparencies, which of course correspond to the original hologram aperture in the field that measured 226 feet square.

The holograms in Figs. 3a and 3b were made with the synthetic reference source placed at a distance of ¾ of the hologram aperture from the true source; the holograms of Figs. 3c and 3d were made with the reference source placed at the other side of this aperture. The reference frequency in all cases is 140 Hz, which represents about the largest frequency possible consistent with a geophone sample interval of 11.3 feet since $\lambda/4 = 12.5$ feet. In each case the reference source is located at a mid-object depth of 80 feet, and the true source is at a depth of 195

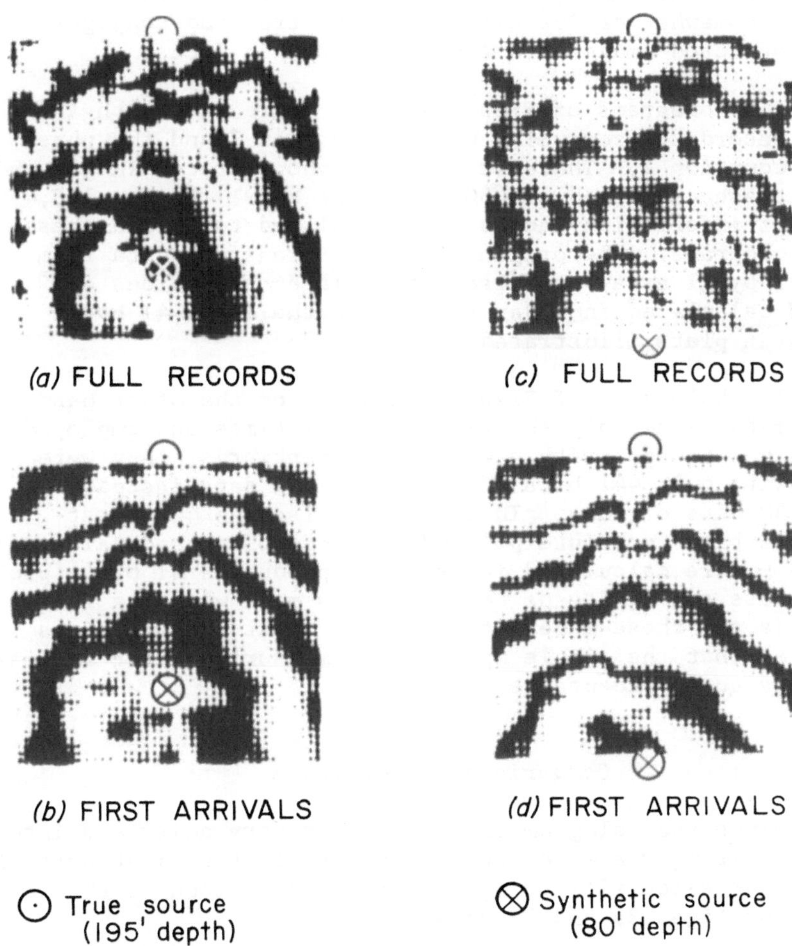

Fig. 3. - Comparison of holograms made using full records and those made using only first arrival times for two different positions of the synthetic reference source.

feet. This type of geometry means that all of these holograms are lensless Fourier-transform holograms, and as we shall see, this makes interpretation of reconstructed images somewhat easier since both the true and conjugate reconstructed images are simultaneously in focus.

The holograms of Figs. 3a and 3c were made with the full records by computing H from equation 7 and the phase from equation 2. There were 441 geophone traces in all and each trace consisted of 450 amplitude bits. Hence, 198,450 bits of information had to be manipulated to obtain these two holograms. The original Fourier transform data from the 21 by 21 array was also interpolated and values of Φ and H calculated in order to produce the final 41 by 41 computer plots illustrated.

The holograms of Figs. 3b and 3d, on the other hand, were made using only the first-arrival times and employed equations 6 and 7 with $n = 1$. In other words, they were made with only 441 bits of first-arrival data instead of 198,450 bits of data. In this case also the 441 first-arrival bits were interpolated to 1,681 bits, and values of Φ and H were calculated in order to produce a 41 by 41 plot. The gross similarity of the holograms made with the full records and those made with only the first-arrival times indicates that the simple approach of measuring first arrivals appears to be acceptable.

Comparison of Holograms

Since the hologram of Fig. 3c is very noisy and interference fringes are not readily apparent, the first part of this discussion will be limited to a comparison of the holograms of Figs. 3a and 3b only.

Both of these holograms have the same general appearance, and both of them show the "hyperbolic" fringes to be expected of two interfering point sources at two different depths from the ground surface. Both holograms show phase modulation (fringes bending toward true seismic source) on the top half of the recording plane, which is an expected characteristic of the low-velocity (high-refractive-index) fracture zone object known to be present at the site (Fig. 1).

Although the holograms of Figs. 3a and 3b are very similar, there are differences worth noting. First, the exact position of the interference fringes appears to be somewhat different in the two cases; secondly, the first-arrival fringes appear to be much more continuous and generally less noisy than those in the original.

Later arrivals representing reflecting horizons at the site are probably responsible for some of the differences in the position of the fringes in the two cases. It is also possible that internal reflections within the object or even resonances are contained in the full records but not in the first arrival information and might also contribute to this effect.

The lack of noise in the first-arrival holograms is perhaps the most interesting and important new feature. This is especially apparent in the comparison of Figs. 3c and 3d where the synthetic reference source is located on the far side of the hologram from the true source. By placing the reference source this far from the true source, the effects of noise in the full-data hologram (Fig. 3c) are highly accentuated; that is, with the reference source farther from the true source, the fringe separation diminishes (approaches $\lambda/2$) to the point that the existing spatial fluctuations in the fringes cause them to overlap and thereby lose their individuality. This effect is so severe in Fig. 3c that one cannot distinguish fringes at all.

The first-arrival hologram of Fig. 3d, however, shows a dramatic improvement in signal-to-noise ratio. The fringes show practically no spatial noise fluctuation at all. The probable explanation for such a dramatic improvement is undoubtedly to be traced to the fact that the first-arrival hologram (Fig. 3d) contains no shear and/or surface wave information or possible later incoherent arrivals from deep reflecting horizons at the site.

The quality of the first-arrival holograms is sufficiently good to suggest that the reconstructions will be better than those obtained from holograms made from the full records. In the case of Fig. 3c, for example, one would not even expect to obtain a reconstruction owing to the extremely noisy nature of this hologram.

Laser Reconstructions

The position of the synthetic reference source for the two holograms shown in Figs. 3a and 3b does not yield a good separation of the reconstructed image from the central focus of the reconstructing lens. This source position was chosen originally because it represented the best possible position that could be achieved, as evidenced by the severe degradation that occurs when this source is moved still further (Fig. 3c). Therefore, no direct comparison between reconstructions of Figs. 3a and 3b will be attempted here since the reconstructions using this reference source position are very poor at best. Neither will a direct comparison be made between reconstructions obtained from Figs. 3c and 3d since it was found experimentally that 3c does not reconstruct an image at all. For these reasons only the hologram of Fig. 3d will be reconstructed here. The interested reader may consult reference (1) for a laser reconstruction of Fig. 3a.

Although the quality of the first-arrival hologram of Fig. 3d is far superior to the holograms made with the full records (Figs. 3a and 3c), the laser reconstruction leaves something to be desired. The hologram reconstructions illustrated in Figs. 6a through 6c (to be discussed shortly) suffer from two basic problems: First, the number of fringes in the hologram aperture is at most about six--a fact that results from the experimental geometry and the fact that the seismic wavelength is 50 feet at a reference frequency of 140 Hz. The obvious result of this situation is that image resolution will be quite low for an object (fracture zone) having an expected maximum diameter of 100 feet.

The second major problem with our data is that the rock at the field test site was both nonuniform and anisotropic. We have not as yet investigated the full significance of the anisotropy and will more or less ignore such effects in what follows. Perhaps a more serious problem is the nonuniform nature of the rock at the field site.

Fig. 4 illustrates a contour map of the times of first arrival of the impulse. There are several important features worth noting. First, the area included within the mask A-B-C includes most of the modulation due to the object; that is, the signal which originates at S is retarded

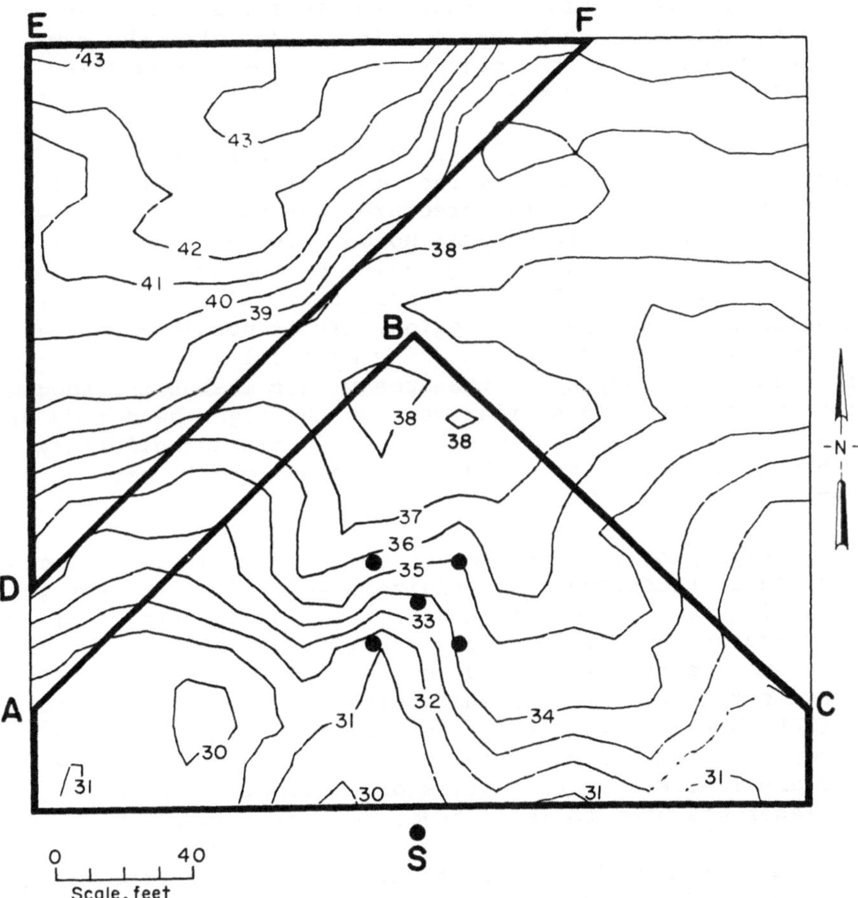

Fig. 4. - First-arrival-time contour map illustrating arrival-time contours in milliseconds. Mask A-B-C covers region of ground surface over which contours bend toward seismic source S owing to low-velocity fracture zone object. Mask D-E-F covers region in which contours are severely distorted by a natural nonuniformity at the site.

in its travel through the zone of fractured rock thereby shifting the travel-time contours toward the source. As mentioned earlier, the hologram fringes also show the same tendency to bend over the object, which constitutes phase modulation of these fringes.

The area within mask D-E-F, on the other hand, constitutes information that we feel has little or nothing to do with the presence of the fracture zone object but rather has to do with a severe natural nonuniformity in the rock at the field site. Note the grouping of the arrival time contours along line D-F. This indicates a slowing down of the seismic P-waves in this region. It is possible that a wedge or fault zone of low-velocity material is causing this result. In any case, the area within mask D-E-F should appear very much as a similar area on the right-hand side of the hologram except for the presence of this nonuniformity, and presumably would have, had this nonuniformity been absent.

By employing these two masks we can examine the effects on the reconstructions of retaining or excluding information originating from the object and/or the nonuniformity. Fig. 5 illustrates the simple reconstruction arrangement used to obtain the images illustrated in Fig. 6. The reconstruction arrangement is a Fourier-transforming arrangement owing to the fact that the synthetic reference source is placed in the central plane of the object so that our seismic holograms are lensless Fourier-transform holograms. A helium neon laser was used to reconstruct the holograms.

The effect of including the nonuniformity in the reconstruction is illustrated by the laser reconstruction of the complete hologram of Fig. 6a, which is identical to the hologram of Fig. 3d. Here we see information that originates from both the object and from the nonuniformity. In Fig. 6b the object information was obscured by the mask A-B-C. The resultant reconstruction shows that the object information has disappeared, but the information from the nonuniformity remains.

If, however, the information from the nonuniformity is masked by the triangle D-E-F, a reasonably good reconstruction of the roughly circular fracture zone object is obtained as shown in Fig. 6c, where the image of the fracture zone object falls inside the large circle and is roughly

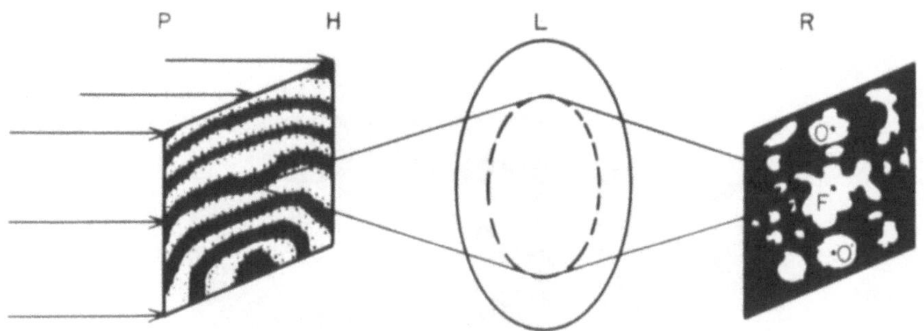

Fig. 5. - Simple Fourier-transforming hologram reconstruction arrangement illustrating incident plane wave of 6328A° laser radiation P, 5/16-inch-square hologram transparency H, reconstructing lens L, and the reconstructed true object image 0, and its conjugate 0′ symmetrically located about lens focus F in reconstruction plane R.

half the diameter of the circle. There also appears to be diffraction ring-like structure surrounding the object that falls both inside and outside the large circle. The quality of the reconstruction is still very low owing to the low resolution of the system; however, by scaling this image, it was ascertained that the object should measure 100 feet in diameter, which is in agreement with known features of the fracture zone.

Of greater significance than the quality of our final image for the purposes of this paper, however, is the simple method by which we obtained the hologram from first-arrival information alone and the fact that the holograms so constructed are actually superior to those made with the full seismic records. Evidently, when better experimental conditions prevail such first-arrival holograms could produce excellent reconstructions.

CONCLUSIONS

Results of this study indicate that when impulsive sources are used and a transmission seismic hologram is being made or when a simple reflection hologram of a single

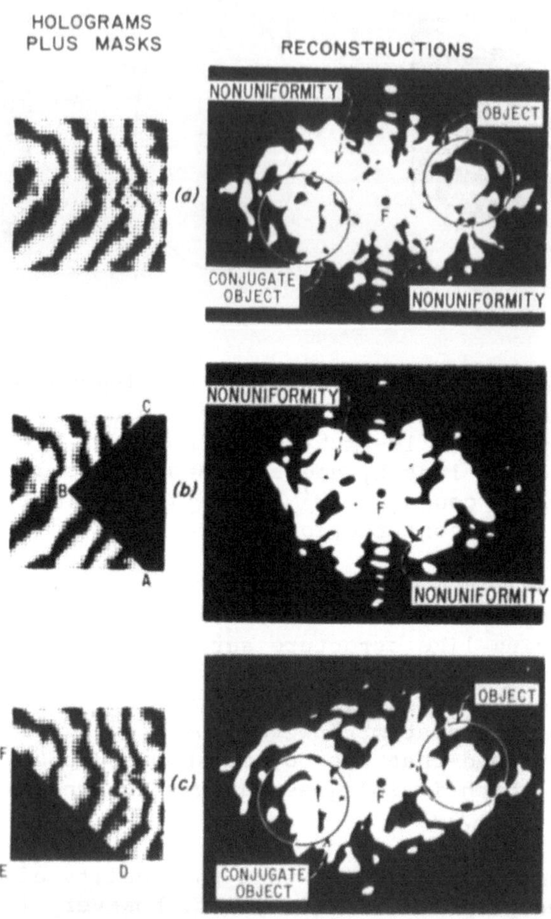

Fig. 6. - Holograms and laser reconstructions of first-arrival hologram of Fig. 3d with: (a) Information from object and natural nonuniformity unobstructed. Both the object and the nonuniformity contribute to final image; (b) Object information masked during reconstruction by mask A-B-C. Object image disappears; (c) Natural nonuniformity obstructed by mask D-E-F. Distortions due to nonuniformity removed and object image improved.

object is being made, the first arrivals are sufficient to make holograms. It is not necessary to examine an entire seismic trace or even record such a trace in making these first-arrival holograms.

One may therefore dispense with equipment needed to record and digitize entire signals and use instead simple rapid and inexpensive timing circuitry. The reference frequency to be used may be obtained by performing a Fourier analysis on a typical signal and choosing a dominant high-frequency component, or it may simply be obtained by measuring the rise time of a typical pulse and defining the reference frequency as being one-quarter of the inverse rise time. Arbitrary frequencies are also possible and might be used to obtain better separation between the object image and the background.

By using only the first arrivals in making the seismic holograms, such things as shear and surface waves are eliminated from the data, thereby resulting in a very great improvement in the signal-to-noise ratio of the stored fringe data.

Although no detailed discussion of seismic sources was presented, it should be clear from the foregoing that the technique of first-arrival seismic holography is not limited to a single point source. In fact, many spatially distributed sources could be used to "illuminate" the object, and a first-arrival hologram could be constructed from each source. One could then add the reconstructions of these various holograms or combine the holograms and reconstruct them simultaneously to produce images with better definition due to the "diffuse illumination."

It should also be clear from the foregoing that any type of impulsive source--be it artificial or natural--that has a spatially coherent component could be used in making such first-arrival holograms. Microseismic or even larger natural seismic events could therefore be used to make these holograms.

REFERENCES

1. Fitzpatrick, G. L., H. R. Nicholls, and R. D. Munson, "An Experiment in Seismic Holography," Bureau of Mines Report of Investigations, No. 7607, 1972, 20 pp.

ANALYSIS OF VARIOUS ULTRASONIC HOLOGRAPHIC IMAGING METHODS FOR MEDICAL DIAGNOSIS

David Vilkomerson

RCA Laboratories

Princeton, New Jersey 08540

I. INTRODUCTION

Acoustical imaging for medical diagnosis is becoming a reality. This paper is meant to be an engineering analysis of the problem of imaging internal tissue masses and of the systems that might solve it. Because experimental data and experience are meager, it is a rough sort of analysis. However, rough analysis is better than no analysis; it is hoped that the results, though approximate, will provide some insight to the problem and to the solutions.

We simplify the problem by assuming it is to image a small disc of "different" tissue immersed in a homogeneous tissue mass. The disc fills one resolution element of the image. The signal power available from this "target" is calculated as a function of the depth of the target and the characteristics of the tissue mass in which it is immersed. Then the minimum signal power required to resolve such a target is calculated for several different imaging systems. Comparing the signal power available to the signal power needed, we can conclude which systems are suitable for what targets. We then extend the results of the analysis of the simple target case to the general imaging problem by recognizing the image as the sum of targets.

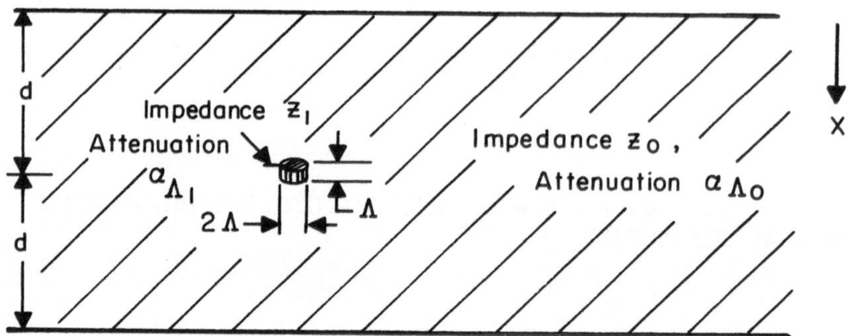

Fig. 1. The model used for calculating the available signal power for diagnostic acoustic imaging. The "target", located d centimeters from the surface and of different acoustical properties from the surrounding "body", fills one resolution element of the output image.

II. DETECTION MODEL

All models are compromises between reality and simplicity. The human body being an extremely complex structure, we avoid any attempt at reality and assume this simple model: the target feature to be imaged is a disc of acoustic impedance Z_1 and attenuation $\alpha_{\Lambda 1}$ immersed d centimeters below the surface in an otherwise homogeneous medium ("the body") of acoustic impedance Z_o and attenuation coefficient $\alpha_{\Lambda o}$. The disc's dimensions are normalized to the insonifying acoustic wavelength Λ, the radius and thickness being set equal to Λ. This model is pictured in Fig. 1. (The biological situation closest to the model is probably a small tumor or cyst in a breast). We insonify with a plane wave of intensity I_o. The plane wave is assumed to pass the interface of the body with no attenuation, and proceed, without distortion, through the body, the intensity varying with the penetration, x, as

$$I(x) = I_o e^{-\alpha_{\Lambda o} x} \tag{1}$$

$\alpha_{\Lambda o}$ being the wavelength-dependent attenuation coefficient of the medium.

When the plane wave encounters the target, a portion of it is reflected and a portion transmitted. The model is two branched, one appropriate to detection by reflection ("reflection-mode"), one appropriate to detection by transmission ("transmission-mode"). We will discuss later where each of the modes might be appropriate.

The power striking the target is the intensity multiplied by the area of the target. The model assumes the target is 2 wavelengths across; this is the resolution capability of a good imaging system.

The intensity at the distance d into the medium where the target lies is (from eq. 1)

$$I_{INC} = I_o e^{-\alpha_{\Lambda o} d} \tag{2}$$

and the power incident on the target is the intensity times the area:

$$P_{INC} = \pi\Lambda^2 \cdot I_o e^{-\alpha_{\Lambda o} d} \tag{3}$$

If the signal comes from reflection, its magnitude depends upon the change in acoustic impedance between the surrounding medium and the target. The reflection coefficient is

$$R = \left| \frac{Z_1 - Z_o}{Z_1 + Z_o} \right|^2 \tag{4}$$

The transmitted wave emerging from the target (which is Λ thick) differs from the surrounding wavefront by

$$\begin{aligned} \Delta I \triangleq I_{TRANS} &= (I_{THRU\ TARGET} - I_{THRU\ BODY}) \\ &\triangleq T \cdot I_{INC} \end{aligned} \tag{5}$$

where

$$T = e^{-(\alpha_{\Lambda 1} - \alpha_{\Lambda o})\Lambda} \tag{6}$$

Because α_Λ in inversely proportional to wavelength [1], and because we have chosen our target to be one wavelength thick, we can express T as a constant:

$$T = \text{ANTILOG}\ [.01(\alpha'_{\Lambda 1} - \alpha'_{\Lambda o})] \tag{7}$$

where α' is the attenuation constant expressed in db/cm at Λ equal to 1 mm. (The .01 accounts for the conversion from logarithms to db and from centimeter units to 1 mm wavelength.)

The coefficients R (eq. 4) and T (eq. 7) measure the interaction of the target with the insonifying wave. Before evaluating these coefficients, let us examine what happens to the altered wavefronts.

According to the model shown in Fig. 1, either the transmitted or reflected wave travels through another d of surrounding medium before it can be detected. Both the reflected and transmitted waves diffract as from an aperture 2Λ in extent. If at the surface we have an energy collection system characterized by an f-number of f/1, i.e. the diameter of the aperture is equal to the distance d, almost all the diffracted power will be collected. We will assume, for simplicity, all is collected.

We can now determine the signal power, defined as the change in detected power due to the presence of the target. If the detection is linear with power, the detectability of the signal is the same whether it appears with no background, as in the case of the reflected signal, or with a constant background, as for the transmitted signal [2].

From equation 2, the power incident on the target is

$$P_{INC} = I_o \cdot \pi\Lambda^2 \cdot e^{-\alpha_{\Lambda o} d}$$

The signal power emerging from the target is given by the coupling, C,

$$P_{SIG} = C \cdot P_{INC} \qquad \begin{array}{l} C = R \text{ for reflection-mode} \\ C = T \text{ for transmission-mode} \end{array} \tag{8}$$

This signal is attenuated by a factor of $e^{-\alpha_{\Lambda o} d}$ before it reaches the detection aperture. The total signal power received is

$$P_s = I_o \cdot C \cdot \pi\Lambda^2 \cdot e^{-\alpha_{\Lambda o} \cdot 2d} \tag{9}$$

It is this signal power, divided by the noise of the detecting system, that determines signal/noise ratio of the detected image.

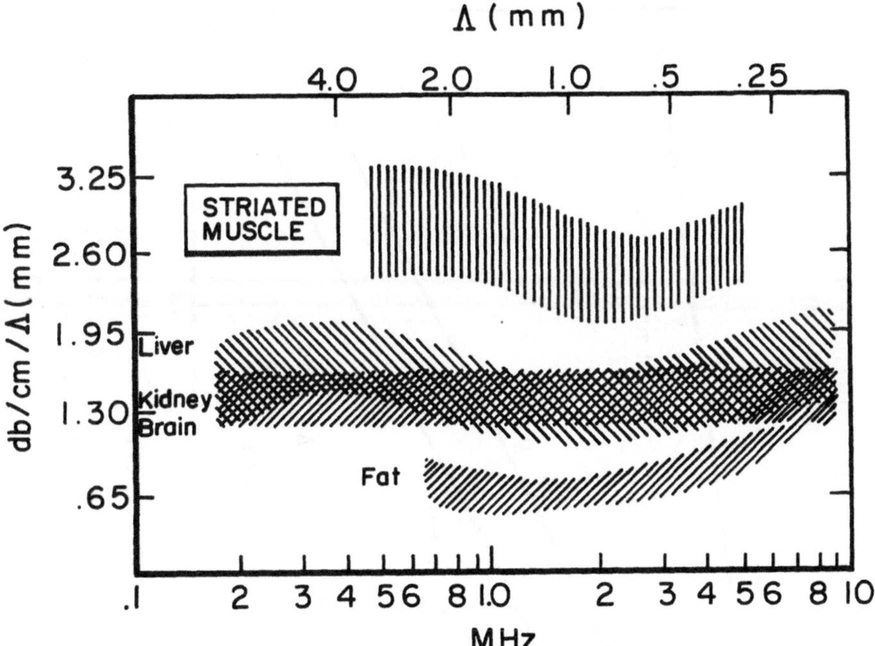

Fig. 2. The attenuation properties of various mammalian tissues as a function of wavelength. The attenuation is approximately inversely proportional to the wavelength; hence, db/cm is divided by the wavelength in mm (after Goldman and Hueter [1], modified).

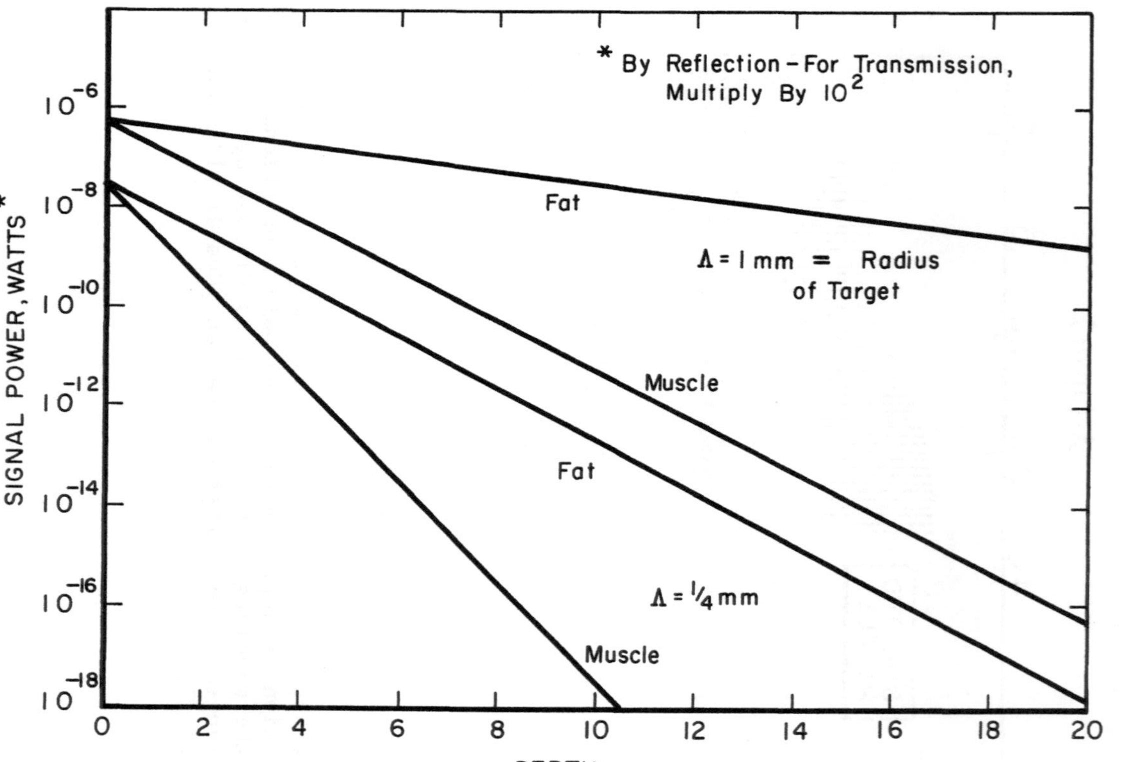

Fig. 3. The signal power received versus target depth in a "fat body" or "muscular body", using the model shown in Fig. 1; a 100 mW/cm^2 insonifying plane wave and an f/1 collection system are assumed. Two wavelengths of insonification, and corresponding target radii, are shown for each body type.

We can evaluate P_s for different media and depths. I_o will be assumed to be 100 mW/cm^2, the recommended maximum diagnostic intensity [3]. The range of possible values for α_Λ for mammalian tissue is shown in Fig. 2 (after Goldman and Hueter [1]). The attenuation ranges from .64 db/cm to 3.0 db/cm for 1 mm wavelength ultrasound. We will calculate the signal power assuming two different body media; $\alpha_{\Lambda o}$ will be assumed to be .65 db/cm/Λ (fat) and 2.50 db/cm/Λ (muscle).

To evaluate the transmission coefficient T of eq. 7, we must know the difference in α_Λ between the target tissue and the body tissue. We will assume a $\Delta\alpha$, the difference in attenuation between the target and the surrounding medium, equal to one-tenth of the total range of α_Λ in mammals. This is arbitrary, but does give some "feel" for the magnitudes involved. The total spread is $\sim$2.4 db/cm, so $\Delta\alpha$ is .24 db/cm/Λ. Inserting this value in eq. 7 gives

$$T \approx 5 \times 10^{-3} \tag{10}$$

The range of impedance values of mammalian tissue is 1.5 to 1.7 x 10^6 acoustic ohms [4] (water is 1.49 x 10^6 acoustic ohms). Using the same (arbitrary) criterion we used for the attenuation, i.e. the difference between the target and the body is one-tenth of the maximum possible difference, eq. 4 yields

$$R \approx 4 \times 10^{-5} \tag{11}$$

We can examine now how the depth of target, wavelength of ultrasound, and attenuation constants of the tissues affects the signal power. Figure 3 presents the reflected signal power plotted against depth of target, using different insonifying wavelengths and the assumption of either "fatty" body tissue or "muscular" body tissue. For transmission-mode, the signal power is 100 times that of the reflection-mode signal power plotted. These signal powers, when compared to the minimum required signal powers of the various systems, will enable us to conclude what is detectable under which conditions using a given system.

III. HOLOGRAPHIC VERSUS DIRECT IMAGE DETECTION

We have calculated in the previous section how much signal power is collected by an f/1 system for targets at different depths and wavelengths. We may use this signal power

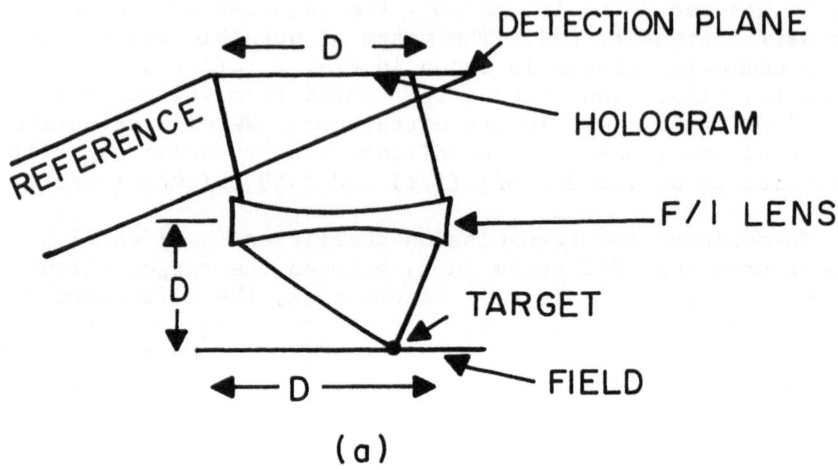

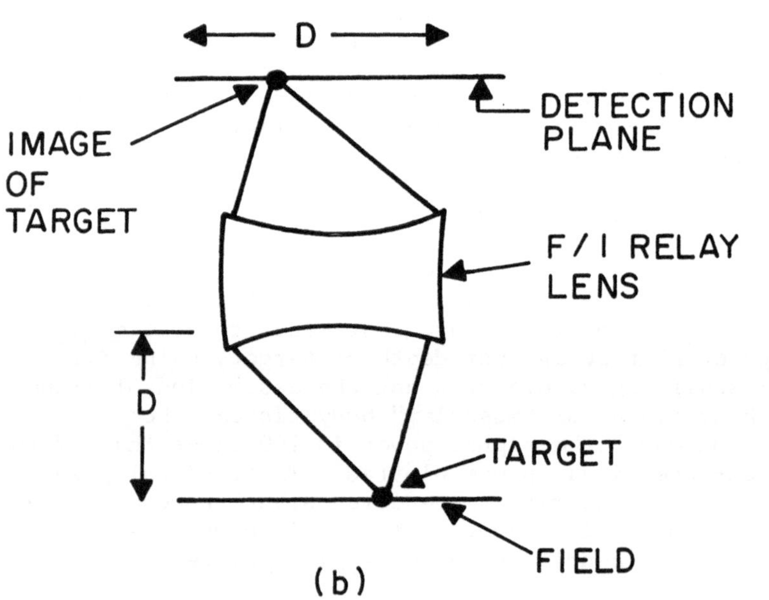

Fig. 4. Two methods of using the signal power: (a) to form a hologram, and (b), to form a direct image.

either to form a hologram or to form an image. Will there be a difference in detectability?

Communication theory can be used to answer this question. One result of communication theory is that the information content in a temporal signal is proportional to the product of the frequency bandwidth and the time-duration of the signal. When extended to the spatial domain [5], the information content should be proportional to the product of the square of the spatial frequency bandwidth and the spatial extent. (The squaring arises from the two-dimensionality of the signal channel). Therefore the limit to the information content is common to both methods.

To demonstrate this equivalence, we will evaluate the number of resolvable points using holography or direct imaging. The two methods are shown in Fig. 4. Note we assume the same f/1 "optics", the same (unobtainably large) image field size for both methods. It will be seen later that the equivalence in results is independent of the field and f-number assumed.

The number of independent positions in each dimension of the output image plane of the hologram is the ratio of the total range in signal frequencies on the hologram plane to the resolution in frequency, Δf, set by the finite aperture of the hologram plane. The range of frequencies, shown in Fig. 5, is equal to $2(\sin \theta_{max}/\Lambda)$ [6], where θ_{max} is the maximum angle of the object beam. The frequency resolution is given by the fourier transform of the aperture; for a single dimensional "slit" of width D, this is the sinc function, shown in Fig. 6, of width 2/D. We say we can resolve two points in the output image if their frequencies are separated by 1/D, as shown by the second function, in dotted line, in Fig. 7. (This is the Rayleigh resolution criterion of optics.)

The number of resolvable points in one dimension, which is the square root of the total number of resolvable points N, is

$$\begin{aligned} \sqrt{N} &= \frac{f}{\Delta f} = \frac{2\ \mathrm{Sin}\theta_{MAX}/\Lambda}{1/D} \\ &= 2\ \mathrm{Sin}\theta_{MAX}\cdot D/\Lambda \end{aligned} \qquad (12)$$

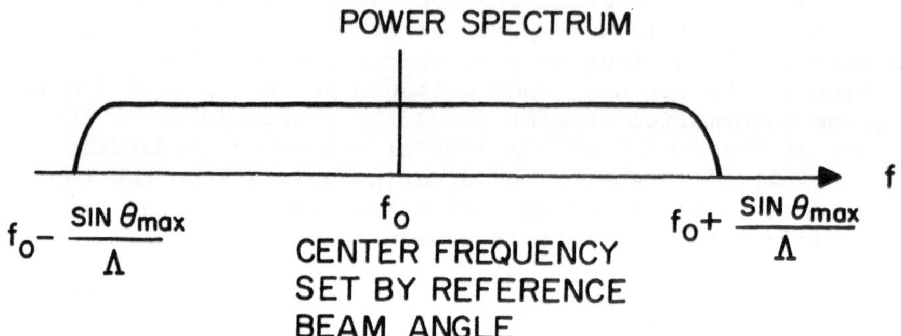

Fig. 5. The power spectrum of a hologram formed from wavefronts within an acceptance cone of $\pm\ \theta_{MAX}$ and a wavelength of Λ.

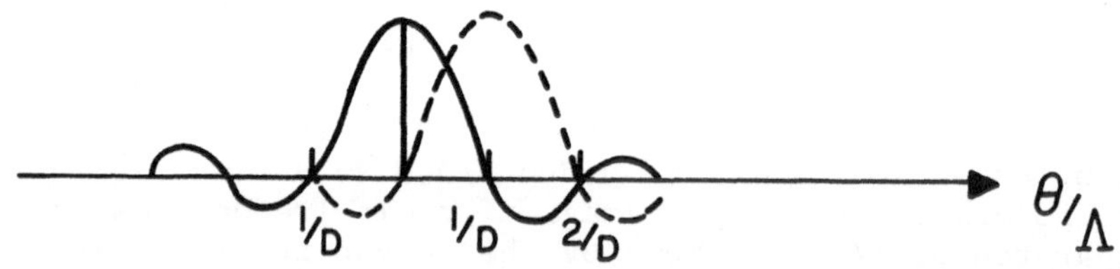

Fig. 6. The fourier transform of a one dimensional aperture of diameter D; this function describes the distribution of amplitude in the spectrum of a plane wave falling on the aperture. It also represents the amplitude in the focal plane of a lens of aperture D focussing a plane wave.

(The fourier transform we performed assumed the same wavelength for playout. If, as is the usual case in acoustical holography, we use a different wavelength, both the width of signal spectrum and aperture spectrum will be multiplied by a factor proportional to the ratio of the playout wavelength to be the sound wavelength; since it is the ratio of these spectra that determines the number of resolvable points, the result is not affected.)

For the direct imaging situation, an f/1 system produces a spot size of twice the f-number times the wavelength, Λ [7]. If we use the same resolution criterion as before (shown in Fig. 7 if the distance between zeroes is $2 \cdot \Lambda \cdot f\#$) the size of each resolution element is half the spot size, or $\Lambda \cdot f\#$. The number of spots in one dimension that we can resolve is the total diameter D divided by the space per spot:

$$\sqrt{N} = D/\Lambda \cdot f\# \tag{13}$$

By the definition of f#, we can relate the f# to the maximum cone angle of acceptance of the field:

$$\frac{1}{f\#} = 2\ \mathrm{TAN}\ \theta_{MAX} \approx 2\ \mathrm{Sin}\theta_{MAX} \tag{14}$$

Therefore

$$\sqrt{N} = 2\ \mathrm{Sin}\theta_{MAX} \cdot D/\Lambda \tag{15}$$

which is the same as (12). We see the same number of resolvable elements for the holographic and the imaging detection method. Examination of this treatment shows that the equivalence is independent of the assumed working field and f-number. (However, the number of resolvable elements is proportional to the field area.)

For an f/1 lens, $\sin\theta_{max} \approx 1/2$, so

$$\sqrt{N} = D/\Lambda \tag{16}$$

and the total number of resolution elements is

$$N = (D/\Lambda)^2. \tag{17}$$

The number of resolvable elements, eq. 17, is proportional to the square of the product of the highest spatial frequency (sin θ/Λ) and the extent of the signal aperture (D), as expected from communication theory.

The spectrum of a single frequency observed over a certain aperture and the image of a point when focussed by a lens of that aperture is the same. The reason lies with the fourier-transforming properties of lenses. The signal power in the image spot and the signal power in the spectrum of the fourier hologram are the same, by Parseval's theorem. In the imaged spot, the power is directly available at the detecting surface; in the holographic case, some sort of signal processing is usually necessary to observe the signal spectrum and find the power concentrated in one small area of the spectrum of the signal (as opposed to one area of the detecting surface for direct imaging).

The conclusion is that either direct imaging or holographic detection produces the same signal power and, with the same detection mechanism, the same signal/noise ratio. This conclusion is in agreement with the results of communication-theoretic studies of radar and other such systems that show the energy of the signal is the critical quantity for detection, e.g. independently of the way the energy is distributed into many high intensity short bursts or one long lower-power signal [8]. We can conclude that the energy of the signal is the critical parameter for imaging, independent of the way the energy is distributed in both space and time. It is for this reason that we have formulated the system analysis in terms of signal power without regard to signal intensity.

In the following section we will discuss the minimum detectable energy requirements of various detection systems. To give practical results, we will assume a standard time of detection of 1/30 second and calculate minimum power. We will indicate parenthetically how the direct imaging and holographic consideration of the requirement can be seen to yield equivalent results; this is to make concrete the somewhat abstract argument of this section.

IV SENSITIVITY OF DETECTION SYSTEMS

A. Method of Calculation

The previous section established the equivalence of

sensitivity for detection by holographic or direct imaging. This equivalence allows us to calculate the minimum detectable power required of a detection system used in either mode.

We will take advantage of previously published analyses that calculated the minimum needed intensity (watts/cm^2) for imaging. This minimum intensity will be transformed into minimum power by assuming that an f/1 imaging system is focussing onto the detection aperture the signal power from a disc of Λ radius. (This is consistent with the target detection model of Section II.) The f/1 imaging system has an impulse response [9] of a Bessel function of radius to first zero of about Λ; when convolved with the object disc of radius Λ, the output image has a radius of about 2Λ. As discussed in Section III, the resolution criterion is equivalent to resolvable points placed one image radius away from each other. We replace the Bessel-function image by a flat-topped one of half the radius, i.e. one Λ, but of equal peak amplitude; the energy content of such a spot is about the same as the true Bessel-function spot [10]. While this approximation leads to somewhat better sensitivities than can be expected, the simplicity gained by its use more than makes up for its lack of accuracy; we may approximate the intensity as being equal to the signal power divided by the area of the spot, $(2\Lambda)^2$.

$$I_{SIG} = P_s/(2\Lambda)^2 \tag{18}$$

In the three sections that follow, three of the most discussed methods of acoustical imaging for medical diagnosis are analyzed to evaluate the minimum signal power required for imaging. The frame time, i.e. the time to assemble one picture, is set as 1/30 of a second (as is used in commercial television). The minimum required signal/noise ratio is set at 10 db (according to commercial television standards, producing a "just viewable" picture).

B. Surface Deformation at a Liquid-Air Interface ("Static-Ripple")

The theory of a surface deformation at a liquid-gas interface has been extensively treated theoretically [11], and various impressive experiments have been performed with this technique, particularly by Brenden et al. [12] and Sikov et al. [13]. The surface deformation at a liquid-air interface will be in the form of a "rippled" surface, where the displacement, b, of the surface is

$$b = K\cdot\Lambda^2\cdot(I_R I_s)^{1/2} \quad \text{(cm)} \tag{19}$$

where I_R is the reference beam intensity, I_s the signal beam intensity, Λ the acoustic wavelength, and K a factor determined by the physical properties of the liquid used and the angles of the two beams to the surface. For a water-air interface with a 1 W/cm^2 reference beam at an angle of 45° and a normally incident signal beam,

$$b \sim 10^{-2}\cdot\Lambda^2\cdot(I_s)^{1/2} \tag{20}$$

where Λ is in centimeters and I_s is in W/cm^2. We can think of the rippled surface of the interface as a reflection phase hologram; the ratio of the image light reflected from this hologram, $P_{\ell s}$, to the incident light, $P_{\ell i}$, is the holographic efficiency, η.

$$\begin{aligned} \eta &= R\cdot(2\pi/\lambda)^2\cdot b^2 \\ &= R\cdot(2\pi/\lambda)^2\cdot 10^{-4}\cdot\Lambda^4\cdot I_s \end{aligned} \tag{21}$$

where R is the intensity reflection coefficient of the interface, and we substituted for b from eq. (20). We will assume R = 1 for simplicity for the rest of the discussion. We consider now the light signal available, $P_{\ell s}$, from the reflection hologram formed on the interface by the reference acoustic beam beating with the focussed spot of signal power. The sound intensity at that spot is

$$I_s = \frac{P_s}{(2\Lambda)^2} \tag{22}$$

The light signal is equal to the product of the holographic efficiency, η, and the light incident on the rippled area, $P_{\ell i}$:

$$\begin{aligned} P_{\ell s} &= P_{\ell i}\cdot\eta \\ &= P_{\ell i}\cdot\frac{4\pi}{\lambda^2}\cdot 10^{-4}\cdot\Lambda^4\cdot\frac{P_s}{4\Lambda^2} \end{aligned} \tag{23}$$

The number of possible image points on the insonified area depends on the aperture field. We will set the number of image points as N, and assume that all the N points on the field are imaged onto the N resolution elements of a television camera tube, such as a vidicon. The ratio of the light power falling on the rippled spot, $P_{\ell i}$, to the total light power, $P_{\ell t}$, is the ratio of the spot size to the total field, or 1/N. The power of the diffracted light (the light signal) is

$$P_{\ell s} = \frac{P_{\ell t}}{N} \cdot \frac{4\pi^2}{\lambda^2} \cdot 2.5 \times 10^{-5} \cdot \Lambda^2 \cdot P_s \tag{24}$$

Because all the signal light is concentrated into 1/N of the total area of the vidicon, we can drop the factor of 1/N if, when calculating the minimum power needed, we take the specification for the total faceplate power.

In actual operation of a liquid-air interface deformation system, the sound must be pulsed to avoid streaming effects [14]. Because it takes a finite time for the deformation to build up, the illuminating light can be on for but a portion of the sound pulse. The duty cycle, r, defined as the ratio of the time the illuminating light power is present to the total frame time, multiplies the frame time to give the effective time. The effective signal power, obtained from eq. 24 and assuming $\Lambda = .5$ mm and $\lambda = .5$ μm, is

$$P_{\ell s} = P_{\ell t} \cdot P_s \cdot r \cdot 10^3 \tag{25}$$

The best, easily available, low-light-level camera tube is a silicon intensifier tube (RCA 4094A). This tube, like all such tubes, is limited by the dark current flow. To achieve a 200 element by 200 element picture, 10^{-9} watts for 1/30 second is required on the faceplate for a 1:1 signal-to-noise ratio [15]. For a 10:1 signal-noise ratio,

$$P_{\ell s} \geq 10^{-8} \text{ W} \tag{26}$$

Therefore, to obtain this minimum light power,

$$P_s \geq 10^{-11}/r \cdot P_{\ell t} \tag{27}$$

The insonification cycle used by Holosonics can be as much as 100 μsec every 2-3 milliseconds [16], equivalent to a duty cycle of about .04. With one watt of illuminating power (neglecting the still shorter portion of the active period that the light is on), this gives a minimum sound power of

$$P_{SMIN} \sim 2.5 \times 10^{-8} \text{ W} \tag{28}$$

It would seem that by increasing the illuminating power the sensitivity could be proportionally increased. However these calculations assume the dominant noise source is electronic noise in the camera tube. This may not be so. The holographic efficiency of the ripple pattern formed by a signal beam of 10^{-8} watt is

$$\eta = \frac{4\pi^2}{\lambda^2} \cdot b^2$$

$$\approx 10^{-5} \tag{29}$$

It is hard to keep the various optical surfaces so clean that the randomly scattered light is less than this value. Therefore, the limiting factor on increasing the light power to reduce on the minimum detectable power may, in practice, be the scattered light. There are additional factors, such as Brillouin scattering, that may also limit operation at high levels [17].

[If we had used the holographic readout with the same static-ripple mechanism, the same signal-to-noise power-ratio would result. While in the holographic readout the intensity of the signal beam, and therefore the diffraction efficiency (eq. 21), would be 1/N as great (the signal being spread over the whole aperture and N being the ratio of the spot area to the field area), this is balanced by the incident light on the signal area being equal to the total light; since for the direct imaged case only 1/N of the total light was incident on the signal beam, the diffracted signal light power would be the same.)

C. Displacement-Amplitude Detection ("Dynamic-Ripple")

To avoid the need of a reference acoustic beam and the difficulties of a liquid-air interface, the displacement component of a propagating acoustic wave can be used to modulate a light beam. For instance, a thin membrane which is transparent to sound but reflective to light can be laser scanned [18]. Because the resulting signal is modulated at the acoustic frequency, the scattered light problems of the surface deformation technique are greatly reduced.

Whitman and Korpel's excellent review of the methods of detecting acoustically induced perturbations [19] shows that all the various methods of detecting the perturbation give approximately the same result for signal-noise ratio as a function of perturbation excursion, b, and frame time, T_{frame};

$$S/N = K \cdot \frac{8\pi^2}{\lambda^2} \cdot b^2 \cdot P_{\ell t} \cdot \frac{T_{frame}}{N} \tag{30}$$

where

$$K = \frac{1}{4}\, q/h\nu \tag{31}$$

q being the detector quantum efficiency and $h\nu$ the energy of the photons being used (in the ideal, shot-noise-limited system).

In water, the displacement, b, is related to intensity by [20]

$$b \sim 3.5 \times 10^{-6} \cdot \Lambda \cdot (I_s)^{1/2} \quad \text{(cm)} \tag{32}$$

We calculate the signal-noise ratio from the signal power detected in the same manner as the previous section, i.e. the power is focussed to a point of area $(2\Lambda)^2$.

We assume: a detector quantum efficiency of .4; 5000Å light; a 200 x 200 element field; and a 1/30 second frame time. The S/N ratio (eq. 30) is

$$S/N \approx 7.5 \times 10^8 \cdot P_s \cdot P_{\ell t} \tag{33}$$

Note that the S/N ratio is independent of Λ; although the displacement goes down with decreasing wavelength, so does the image spot size. We assume a laser scanning beam of .25 watt; to achieve the 10:1 S/N ratio which is our criterion for detection,

$$P_{SMIN} \geq 5 \times 10^{-8}\ \text{W} \tag{34}$$

In the "dynamic ripple" method used by the Zenith group [21], the displacement upon reflection at an interface is utilized. The magnitude of the displacement is increased by a factor of two, increasing the signal power by four; including losses in the required transducer plate, the minimum signal power for such a system is about 2×10^{-8} W, or about the same as for the static-ripple method of part B.

[The same minimum signal power would be required if a holographic readout was used. For direct readout, the focussed spot is scanned by the laser beam; the bandwidth of the signal-processing circuitry must be the inverse of the time to scan one spot diameter. For the holographic readout, the signal is spread over the entire field, so the signal processing for the holographic readout has a bandwidth of the inverse of the entire field scan time. The noise power, which is proportional to the bandwidth, is therefore N times greater for the direct imaging than holographic, since there are N spot diameters to the entire field. This factor of N cancels the N times greater intensity obtained by focussing the sound power, so the signal/noise ratio is the same for both methods.]

D. Piezoelectric Electronic Detection

With piezoelectric transducers, acoustic energy is efficiently converted into electronic signals without the intermediary of a light beam, as in the previously considered detection methods. This removes one of the important limitations on sensitivity, as will be seen.

We will analyze this detection method on the basis of holographic readout, as this will probably be its mode of use for reasons that will be discussed. We consider a mosaic of piezoelectric transducers covering a field of area $N \cdot 4\Lambda^2$, appropriate to N resolution elements. We assume an amplifier for each transducer whose impedance is matched, so that half of the received power is transferred to the amplifier. The signal power into each amplifier is

$$P_e = \frac{I_i A_j}{2} \tag{35}$$

where I_i is the incident intensity and A_j is the area of the transducer. The area of each transducer is assumed to be $(\Lambda)^2/4$, with spacing between transducers of Λ so that for f/1 and higher f-numbers, good sampling of the wavefront is achieved. The transducers occupy 1/4 of the detection plane.

The noise power of the matched amplifier is [22]

$$P_n = KT \cdot \Delta F \cdot N.F. \tag{36}$$

where ΔF is the temporal bandwidth and N.F. is the noise figure of the amplifier. The noise figure is assumed to be 2 (3 db). We read out all the amplifiers in parallel.

The signal we detect corresponds to the sum of the spatial signals from the radiating points in the object space. We can synthesize a matched filter for each possible point in the object space and measure how much of the spatial signal appears at the output for a particular filter; the output tells how "bright" that point in object space is. Communication theory states that such a matched filter is the optimum detection system for a signal embedded in white noise [23]. For our discrete array of n_D (= 4N) detectors, the matched filter processing can be imagined as multiplying the signal of each detector by a complex number of unit magnitude; the phase is such that for a signal from a specified point, all the amplitude products are in phase and, because the signals are coherent, the total power is n_D^2 the power of each detector. Signals from other points are orthogonal to the matched filter, and give no net signal. The amplifier noise, being random, remains random after the multiplication process and the noise power is n_D times the noise power of each detector, the noise signals being incoherent. Therefore, the processing increases the signal-noise ratio by the number of detectors, n_D.

The intensity I_i corresponds to the total signal power, P_s, divided by the area of detection, $n_D\Lambda^2$. The processor output signal/noise ratio is

$$S/N = n_D \cdot \frac{P_e}{P_n}$$

$$= n_D \left(\frac{P_s/2}{n_D\Lambda^2} \cdot \frac{\Lambda^2}{4} \right)/2KT\cdot\Delta F$$

$$S/N = \frac{P_s}{16KT \cdot \Delta F} \tag{37}$$

The bandwidth of the amplifier, ΔF, is the inverse of the frame time; therefore

$$S/N = \frac{P_s \cdot T_{FRAME}}{16\ KT} \tag{38}$$

This is the signal/noise ratio for one point. To be consistent with our model of a target $4\Lambda^2$ in area, the signal power is allocated equally to the four points in the object space to which such a target corresponds. We evaluate the signal-noise ratio for the target at operation at 300°K and a 1/30 second frame-time:

$$S/N = 1.6 \times 10^{17} \cdot P_s \tag{39}$$

The minimum signal power, by our usual 10:1 criterion, is

$$P_{SMIN} = 6 \times 10^{-17}\ W \tag{40}$$

Note that the minimum required signal power is independent of the size of the field; we can supply the same electric power to every element of the detecting aperture independently of the number of elements. In the light-coupled systems, the available light power had to be divided among the resolution elements, and the greater the number of elements, the less the signal power per element that could be detected.

[If a direct imaging system were to be used with the piezoelectric-transducer array, the signal/noise ratio of each of the four energized detectors would be

$$\begin{aligned} S/N &= \frac{1/2 \cdot P_s/4 \cdot 1/4}{P_N} \\ &= \frac{P_s \cdot T_{FRAME}}{32\ KT} \end{aligned} \tag{41}$$

which when evaluated gives the same result as eq. 39.]

POWER FOR DETECTION

FRAME TIME = 1/30 SEC.

PIEZOELECTRIC MOSAIC: $P_{SMIN} \simeq 6 \times 10^{-17}$ W

LIQUID-AIR INTERFACE: $P_{SMIN} \propto \left[\left(\frac{\Lambda}{\lambda}\right)^2 \cdot r \cdot (P_\ell/N)\right]^{-1}$

$P_{SMIN} \sim 2.5 \times 10^{-8}$ W

$\lambda = .5\,\mu$
$\Lambda = .5$ mm
$r = .04$
$P_\ell = 1.0$ W
$N = 200 \times 200$

SURFACE DEFORMATION: $P_{SMIN} \propto \left[\frac{1}{\lambda^2} \cdot (P_\ell/N)\right]^{-1}$

$P_{SMIN} \sim 2.5 \times 10^{-8}$ W

$\lambda = .5\,\mu$
$P_\ell = .1$ W
$N = 200 \times 200$

Table I. The minimum power required to produce an image of a target one resolution element in size, at 10:1 signal/noise ratio. (P_{SMIN} is the minimum required power, λ the light wavelength, Λ the acoustic wavelength, P_ℓ is the light power, r is the duty cycle, and N is the number of resolution elements.)

E. Sensitivities Compared

Table I shows the results of the calculations. The piezoelectric transducer mosaic was found to require a minimum power, for 1/30 second frame time, of 6×10^{-17} watt. The methods using light to couple to the acoustic field, the "static ripple" and "dynamic ripple" methods, required about the same minimum required power for 1/30 of a second frame time: about 3×10^{-8} watts [24]. The problem with the light-coupled detection schemes is that light is not well-coupled to acoustic fields, while the piezoelectric transducers are very well coupled, approaching 80% efficiency in converting acoustic power to electrical power and vice versa.

The light-coupled detection methods have the advantage that they are presently practical, while the piezoelectric transducer mosaic, or similar solid-state device, is not. The liquid-air interface method ("static ripple") in particular is simple, since the scanning is electronic. Moreover, because there is a heterodyne-type gain involved in mixing the signal beam with a powerful reference beam, the displacement, b, is much greater in the static ripple than in the dynamic ripple; at Λ of 1 mm, the static ripple is over 100 times greater (see equations 19 and 32). However, by the unstable nature of the interaction of the sound with the interface, i.e. streaming effects, this ripple is available only for a small portion of the frame time. The dynamic ripple is continuously usable. Other system aspects of dynamic ripple that compensate for the smaller displacement, b, are a more sensitive light detection method and an inherent discrimination against scattered light. It should be noted that while the sensitivity of the dynamic ripple method is independent of wavelength (neglecting material attenuation), the sensitivity of the static ripple method is proportional to the wavelength squared.

It may be advantageous to combine the static ripple method's advantage of greater displacement with the dynamic ripple method's greater signal discrimination and detector sensitivity by laser scanning a static ripple, using the scanning beam velocity to generate a doppler shift to the signal. How much advantage could be gained depends upon the duty cycle that could be achieved and the frequency of operation. It is estimated a factor of 100 might be gained at wavelengths around one millimeter.

Less widely-used imaging methods exist, such as Bragg-Diffraction [25], Sokolov tube [26], and capacitive transducer arrays [27,28]; the same analytical procedure can be used to obtain the minimum required signal power for these systems.

V. DETECTING TARGETS IN THE BODY

We now combine the results for the sensitivity of detection systems, obtained in section III, with the calculations for the signal energy from the target, from Section II. We can estimate the limits of depth at which we can detect a target [29].

In section II we found the signal power by transmission was about 100 times that by reflection. However, the transmission mode often leads to much longer paths through tissue and losses by attenuation much greater than the factor of 100 gained. For example, if the kidneys are to be examined, transmission-mode examination would require propagation through 20 centimeters or more of viscera; besides greatly attenuating the signal, the wavefront would be distorted by passage through the varying tissue layers. Reflection-mode examination, on the other hand, would require only penetration of the muscle layer of the back to reach the kidneys. Generally, to examine most of the structures in the abdomen, reflection-mode will be necessary.

For other, more external structures transmission-mode would be satisfactory. In particular, examination of the breast by transmission is feasible [30]. Not only is it possible to have transmission through just the breast, but in addition the breast is fairly simple in structure and is predominantly adipose tissue that does not attenuate strongly.

Figure 7 shows a plot of target size imaged versus depth for several acoustic wavelengths, using the attenuation coefficient for fat tissue and assuming a minimum required signal of 10^{-7} watts, appropriate to the light-coupled systems. (These curves are somewhat optimistic in that they neglect the effect of the skin.) According to anatomy texts [31], the average mature female breast is about 12-14 centimeters in lateral extent and about 12 cm vertically, extending from the chest wall 3-5 cm. Experience indicates fairly wide variation in size of the breast; however, using the "standard"

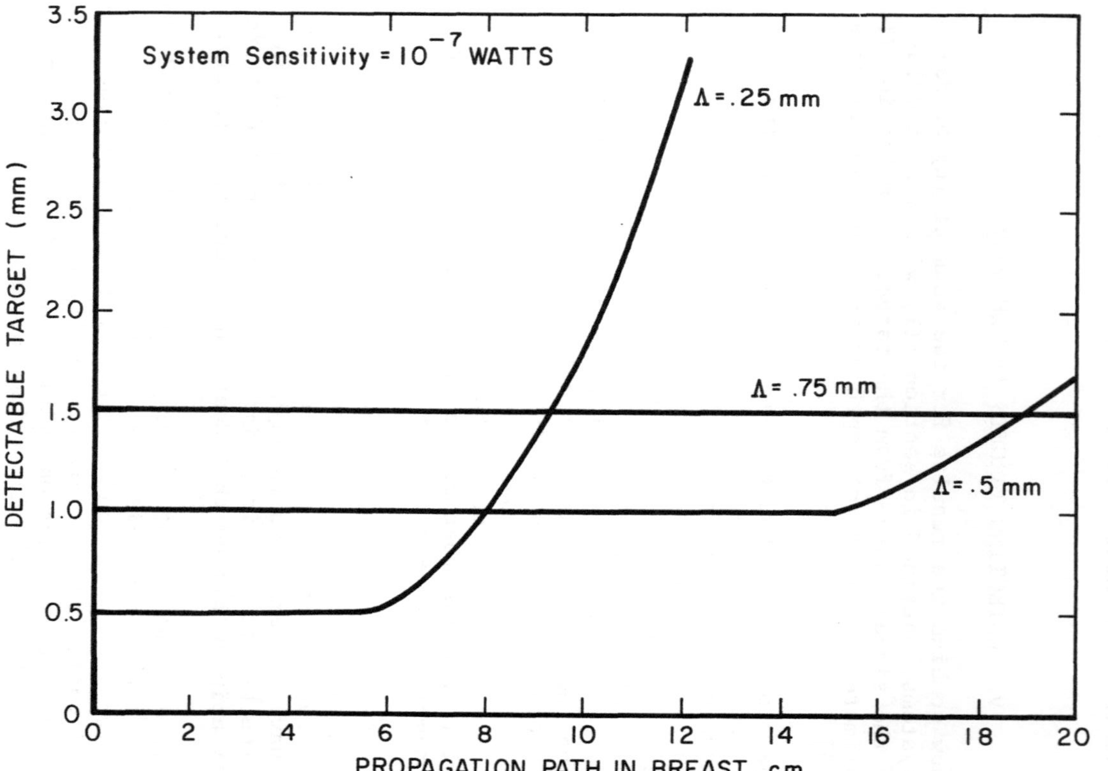

Fig. 7. The size of target required to produce a signal of 10^{-7} W (or more) versus propagation path in the breast for several acoustic wavelengths. A transmission-mode system is assumed. (The average breast is 12 cm laterally [31]).

size in conjunction with Fig. 7 indicates that detection of targets in the breast as small as 1 mm in diameter should be possible using light-coupled systems.

Now we consider the problem of examining the kidney by ultrasonic reflection. Before reaching the kidney, the sound must pass through 3-4 centimeters of highly-attenuating muscle tissue. Then, to see the interior (ventral) surface of the kidney, another 3 cm of kidney tissue, of intermediate attenuation (see Fig. 2) must be traversed. If we examine Fig. 3, the reflected signal power to be expected is between about 10^{-9} watts for 1 mm waves and 10^{-13} watts for .25 mm wave. This is outside the range of light-coupled systems, even if long frame-times can be realized. For piezoelectric transducer arrays, however, such signal levels are quite adequate.

Indeed, examining Fig. 3 indicates that even such deeply lying tissue as the gall bladder and pancreas can give signals large enough to be detected. For such organs, pulsed insonification and time-grated detection will be necessary to eliminate interference from overlying tissue. The minimum required power will be increased by the ratio of the repetition period to the duration of the gated signal. This ratio is approximately equal to the total thickness of the body to the thickness of the region to be examined, or on the order of 10^2. Using gating, a minimum power for detection required by a piezoelectric transducer array is about 10^{-15} watts. Referring to Fig. 4, organs 10-12 cm into the body, which is just about half-way through, should be able to be visualized with .25 mm waves in this manner.

In reflection-mode detection a diffuser will have to be used to fill the detection aperture, since the tissues are smooth compared to acoustic wavelengths. To avoid the specular image caused by such a diffuser used with coherent illumination, a time-varying diffuser will be needed. The sum of the images formed from different diffuser structures will tend to eliminate "speckle", as has been demonstrated for optical speckle [32].

VI. DISCUSSION & CONCLUSIONS

As would be expected from communication theory, the sensitivity of a system using direct imaging or holographic imaging is the same. The advantage of holographic recording is that all depth planes can be reconstructed from the one hologram after it is made; its disadvantage is the increased complexity of signal processing involved and the associated loss of signal-to-noise ratio. For medical diagnostic work, real-time visualization is important; the effect of a different viewing angle or the effect of moving some nearby part can be evaluated.

An important advantage of holographic detection is that a synthetic aperture can be implemented [33]. Rather than building a complete mosaic of piezoelectric transducers, for instance, a small portion of an array, such as the outer ring, can be built; while the sensitivity will be reduced by the ratio of the N of the full array to the N' of the reduced array, the resolution can be maintained and the fabrication greatly simplified. Since piezoelectric transducers require about 10^{-9} of the signal power of the light-coupled detection schemes, such partial arrays would still have a great advantage in sensitivity.

The calculations we have performed indicate that acoustic imaging of such structures as the breast at a resolution of about a millimeter is already practical, using light-coupled detectors such as the static or dynamic ripple methods. For higher resolution or imaging of deep-lying tissues in the body, piezoelectric mosaics or other highly efficient detectors are necessary. With such, however, even the deepest lying tissue in the body (except that encased in bone) should be able to be visualized at a resolution of about a millimeter, using safe powers of insonification.

REFERENCES

1. D. Goldman and T. F. Hueter, Tabular Data of the Velocity and Absorption of High-Frequency Sound in Mammalian Tissues, J. Acoust. Soc. Am. 28, 35 (1956)

2. For an interesting analysis of various detecting modes using square-law detection, see K. Wang and G. Wade, Threshold Contrast for Various Acoustic Imaging Systems, this volume, page 431

3. A. Sokollu, Ultrasound Effects on Animal Tissue, Acoustical Holography Vol. 3, Plenum Press (New York), p. 3 (1971)

4. G. D. Ludwig, The Velocity of Sound Through Tissues and the Acoustic Impedance of Tissues, J. Acoust. Soc. Am., 862 (1950)

5. E. L. O'Neil, Introduction to Statistical Optics, Addison-Wesley (Reading), Chap. 2 (1963)

6. J. W. Goodman, Introduction to Fourier Optics, McGraw-Hill (New York), Chap. 8 (1968)

7. M. Born and E. Wolf, Principles of Optics, 3rd Ed. Pergamon Press (Oxford), Chap. 6 (1965)

8. P. M. Woodward, Probability and Information Theory, with Applications to Radar, McGraw-Hill (New York), (1953)

9. J. W. Goodman, op. cit., Chap. 2

10. The relations between signal-noise ratio, resolution, and display effectiveness are quite complicated; see, for instance, O. H. Shade, Resolving Power Functions and Integrals of High Definition Television and Photographic Cameras, RCA Review 32, 567 (1971)

11. R. K. Mueller and P. N. Keating, The Liquid-Gas Interface as a Recording Medium for Acoustical Holography, Acoustical Holography Vol. I, Plenum Press (New York), p. 49 (1969)

12. B. B. Brenden, Real Time Acoustical Imaging - Interesting Examples and Practical Applications, this volume, p. 1

13. M. R. Sikov et al., Biomedical Studies Using Ultrasonic Holography, this volume, p. 147

14. R. B. Smith and B. B. Brenden, Refinements and Variations in Liquid Surface and Scanned Ultrasound Holography, IEEE Tran. Sonics Ultrasonics SU-16, 29 (1969)

15. RCA Electronic Components, Camera Tube 4804A (1971)

16. A. F. Metherell, Acoustical Holography as a Tool for Biomedical Imaging in Quantitative Imagery in the Bio-Medical Sciences, Vol. 25, SPIE (Redondo Beach) (1971)

17. See, for example, the analysis of noise sources given by R. A. Smith and G. Wade, Noise Characteristics of Bragg Imaging, Acoustical Holography Vol. III, A. F. Metherell, Ed., Plenum Press (New York), Chapter 6 (1971)

18. D. Gabor, Tilted Phase Contrast Method, French Patent 1,479,712, Issued 3/67

19. R. L. Whitman and A. Korpel, Probing of Acoustic Surface Perturbations by Coherent Light, Appl. Optics 8, 1567 (1969)

20. T. F. Hueter and R. H. Bolt, Sonics, John Wiley (New York), Chap. 2 (1955)

21. R. L. Whitman, M. Ahmed, and A. Korpel, A Progress Report on the Laser Scanned Acoustic Camera, this volume, p. 11

22. P. F. Panter, Modulation, Noise, and Spectral Analysis Applied to Information Transmission, McGraw-Hill (New York) (1965)

23. S. J. Mason and H. J. Zimmerman, Electronic Circuits, Signals, and Systems, John Wiley (New York), Sec. 7.11 (1960)

24. When the observed minimum detectable sound intensities reported at this Symposium are converted to the minimum required signal power form of our calculations, the agreement is quite good.

25. J. Landry et al., Bragg-Diffraction Imaging: A Potential Technique for Medical Diagnosis and Material Inspection, this volume, p. 127

26. H. W. Jones, The Ultimate Sensitivity of Sokolov Image Converter Tubes, this volume, p. 599

27. A. K. Nigam et al., Foil-Electret Transducer Array for Real-Time Acoustical Holography, this volume, p. 173

28. P. Alais, Acoustical Imaging by Electrostatic Transducers, this volume, p. 237

29. Some of the ignored effects of distortion caused by overlying tissue layers is discussed by P. Green, Considerations for Diagnostic Ultrasonic Imaging, this volume, p. 97

30. Brenden has demonstrated such imaging of the breast; see ref. 12

31. C. M. Goss, Gray's Anatomy 38th Ed., Lea and Febiger (Philadelphia), p. 1330 (1970)

32. W. Martienssen and S. Spiller, Holographic Reconstruction without Granulation, Phys. Lett. 24A, 126 (1967)

33. J. L. Kreuzer, A Synthetic Aperture Coherent Imaging Technique, Acoustical Holography Vol. III, A. F. Metherell, Ed., Plenum Press (New York), Chap. 16 (1971)

THRESHOLD CONTRAST FOR VARIOUS ACOUSTIC IMAGING SYSTEMS

Keith Wang and Glen Wade

Department of Electrical Engineering
University of California, Santa Barbara
Santa Barbara, California 93106

ABSTRACT

The ability of an imaging system to produce a good picture can be evaluated in terms of resolution and contrast. Because of noise, there exists a threshold in the contrast between adjacent resolution elements below which the diffraction-limited resolution cannot be achieved. Threshold contrast, being a function of input quantities and system characteristics, provides a cogent measure of the system sensitivity. This paper presents an analysis of the ultimate ideal performance of some acoustical imaging systems in terms of the above threshold contrast as limited by thermal noise or quantum noise.

The analytical procedure involved, although simple, reveals crucial information concerning the systems and serves as the basis for a valid competitive evaluation. For example, the minimum insonification in each system that is needed to obtain an effective image under the ideal circumstances hypothesized can be calculated. This kind of information is important, especially for medical applications where the sensitivity of biological tissue to sound intensity is a critical factor in the operation of the system.

The analysis in this paper concentrates on various modes of acoustic scanning. The contrast sensitivities and the significance of the comparisons for different levels of input intensities are discussed.

INTRODUCTION

This paper presents a comparison of the ultimate performance under ideal conditions of certain acoustic-imaging systems as characterized by contrast sensitivity. Such a characterization makes it possible to evaluate consistently a number of different acoustic imaging systems in terms of environmental and system parameters. The analytical procedure to accomplish the evaluation is not complicated and, once established, may be employed to select an optimum scheme for a particular application.

BASIC CONSIDERATIONS

The ability of an imaging system to produce a good picture can be evaluated in terms of resolution and contrast. Because of noise, for each achievable resolution there exists a minimum contrast between adjacent resolution elements that can be detected. Consequently, there is a threshold in contrast between these adjacent elements below which the desired resolution cannot be achieved. In acoustic imaging the quality of the final visual image depends on the acoustic contrast of the object. Therefore, in comparing different acoustic imaging systems, we are interested in finding the one that will produce an observable contrast in the image for the lowest acoustic contrast between adjacent resolution elements in the object, assuming that the different systems have the same resolution capability and use the same amount of power to insonify the adjacent object resolution elements. Such information is available if we calculate the acoustic threshold contrast C_a, defined as the ratio of the smallest difference in the acoustic transmittance of two adjacent resolution elements of the object when the corresponding two adjacent resolution cells of the final visual image can just barely be distinguished from each other (i.e., as having different shades of intensity) to the larger transmittance of the two adjacent object resolution elements.

The system configuration we will consider in this paper is both simple and basic. The object plane is located between a transmitting plane and a receiving plane as shown in Figure 1. The imaging is achieved by means of a transmission mode of operation.*

*The transmission mode of operation instead of the reflection mode of operation is considered because it is the better

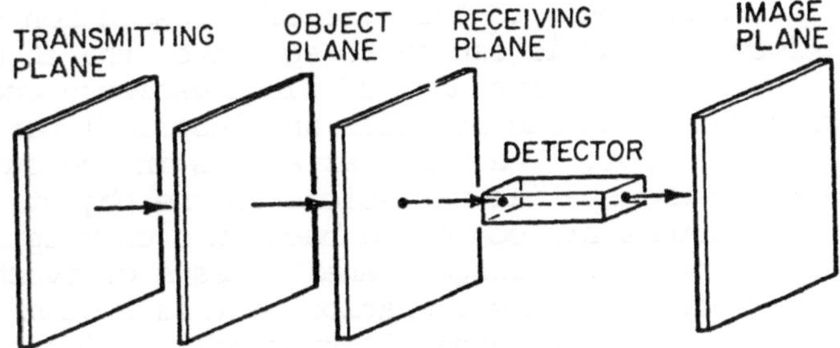

Figure 1 - The basic system configuration for transmission-mode imaging.

Four types of scan are possible in principle:

1. Positively scanning transmitter (PST)

In this system, as indicated in Figure 2, the sound beam from the transmitting plane is focused to a narrow cross section at the object plane. The cross-sectional area

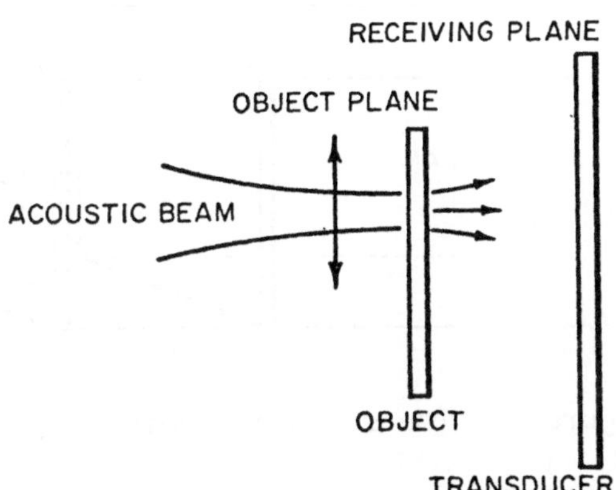

Figure 2 - The positively scanning transmitter system.

---

candidate for biomedical application.[1] However, the same approach can be employed to analyze reflection-mode imaging.

at that plane corresponds to the size of a resolution element. The focused beam is caused to scan over the object plane. A single large transducer of plane geometry occupies the receiver plane and provides an electrical output signal to the detector. The time variations in the output signal stem from the spatial variations in the object. These time variations are converted back into corresponding spatial variations in the image plane by means of synchronous scanning. Note that the operation of this system is analogous to that of a scanning electron-beam microscope.

2. <u>Positively scanned receiver (PSR)</u>

In this case, as shown in Figure 3, a broad collimated sound beam from the transmitting plane insonifies simultaneously and uniformly the entire object plane. The receiving plane contains an array of receptors, each receptor

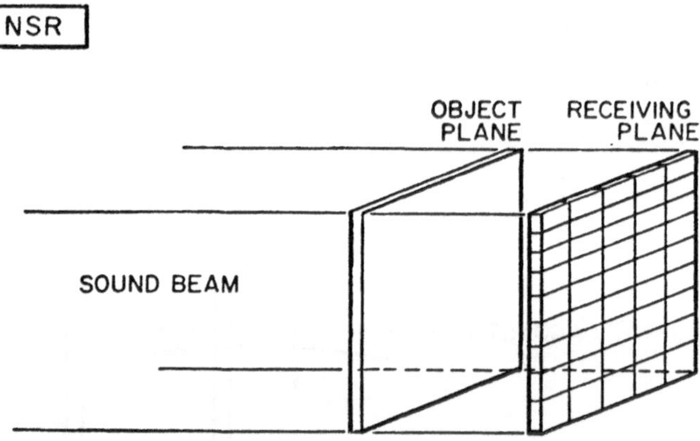

Figure 3 - The scanned receiver systems.

covering an area corresponding to the size of a single resolution cell. The receptors are activated one by one in sequence. The signal output from the receiving plane is processed as in the PST case, the image scanning being synchronous with the scanning of the receptors.

3. Negatively scanned receiver (NSR)

In this system (also see Fig. 3), the sonic beam from the transmitting plane covers the entire object plane as in the case of the PSR. Here also the receiving plane contains an array of receptors which are scanned. The only difference between this and the PSR is that here a "hole" is scanned in the receiving plane. All of the receptors are always kept "on," except for one receptor. The position of the "turned-off" receptor is scanned sequentially in synchronism with the image-plane scanning. Professor B. A. Auld's scanned acoustic microscope[2] is an example of the implementation of such a system.

4. Negatively scanning transmitter (NST)

Here the transmitter produces a broad collimated beam in which a narrow pencil-like "hole" in the beam (whose cross section at the object plane corresponds to a picture element) is scanned over the object plane (see Fig. 4). The rest of the system is the same as in the case of the PST.

NST

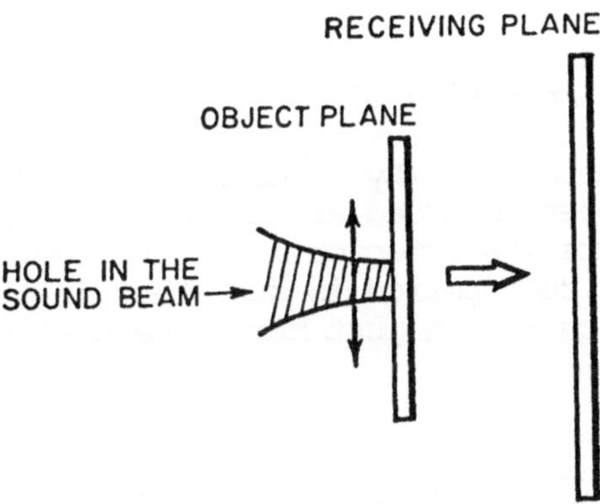

Figure 4 - The negatively scanning transmitter system.

We will start the analysis by making the following assumptions:

1. For any of the systems, the resolution is diffraction limited.*

2. The number of resolution elements per picture frame is m^2.

3. An ideal detector which generates no noise is used to process the signal coming from the receiving plane. The detector supplies an output for the image plane. The no-noise assumption is made for simplicity and to permit treating thermal noise and quantum noise equally. In the calculation that follows a square-law detector preceded by a filter is assumed. The amplitude of the detector output is used to control the intensity in the image plane. At any point in the image the intensity is proportional to the corresponding rms acoustic power incident at the receiving plane. Note that specifying a square-law detector results in very little loss of generality.

4. The detector detects only a fraction η of the incoming phonons.

5. The scanning period, or integration time, for one resolution cell is τ; the frame time is τ_f.

6. The receiver bandwidth is $B = 1/2\tau$.

7. The receiving plane has a total sensitive area of size A. For the PSR and NSR cases, the receiving plane contains $m \times m$ switchable receptors, each with area δA.

Consider any two adjacent resolution cells at the image plane. The intensity of each is proportional to the power absorbed at the receiving plane during the corresponding scanning period τ. (By absorbed power we mean the incident acoustic power absorbed at the receiving plane and delivered to the square-law detector.) Assume that the power absorbed during each such period is such that in the image the two resolution cells can just barely be distinguished.

*We assume here that the system operates on a "diffraction-limited" basis; that is, the system is capable of giving the resolution predicted by the Rayleigh criterion. Our problem, therefore, is to calculate the necessary threshold contrast for the system to achieve Rayleigh resolution.

Let

C_i = the ratio of the smallest difference in the power absorbed at the receiving plane during each of the above two scanning periods when the two corresponding resolution cells of the image plane can just barely be distinguished from each other (i.e., as having different shades of intensity) to the larger amount of power absorbed during the two adjacent periods. C_i is called the input threshold contrast. The subscript i refers to the detector input.

k_i = the ratio of the above absorbed power difference to the noise power which ultimately limits the distinguishability of the two adjacent cells. k_i is called the input threshold signal-to-noise ratio.

From the above definitions

$$k_i = \frac{P_s}{N_i} = \frac{C_i S_i}{N_i} \tag{1}$$

$$\frac{k_i}{C_i} = \frac{S_i}{N_i} \tag{2}$$

where P_s is effectively the signal power for the imaging system, defined as the previously referred to difference in the absorbed power; i.e. the difference between the power absorbed at the receiving plane during the scanning periods corresponding to the above two adjacent resolution cells. S_i and N_i refer to average transmitter power and noise power respectively at the input to the detector during the scanning period corresponding to the brighter image cell.

Similarly, we define image plane quantities using the subscript o to denote output from the detector.

C_o = the ratio of the smallest difference in intensity at the image plane between two adjacent resolution cells which can just be distinguished from each other to the intensity of the brighter of the two cells. C_o is the output threshold contrast.

k_o = the ratio of the above difference in intensity to the intensity variations corresponding to picture noise on the image. k_o is the output threshold signal-to-noise ratio.

Let us assume that the image is viewed by the human eye. Then k_o is associated with the visual process. It has a numerical value of from 3 to 6, as stated by Rose[3].

From the above definitions, we have

$$\frac{S_o}{N_o} = \frac{k_o}{C_o} \quad .$$

Because of assumption 3, we can write

$$C_o = C_i \quad .$$

Therefore

$$\frac{S_o}{N_o} = \frac{k_o}{C_i} \tag{3}$$

The input (receiving plane) and output (image plane) characteristic quantities have the same relationship for all four types of scan because the same detection scheme is assumed to link the two planes.

The most basic noise that limits the ultimate performance of devices is thermal noise and quantum noise (see Appendix I). Under many ordinary conditions of interest as far as studies of this type are concerned, it is commonly found that either thermal noise or quantum noise predominates. There are two definable regimes (depending on signal strength and on a comparison of the magnitudes of $\hbar\Omega$ and kT) for which one or the other of the two kinds of noise (but not both) determines the limitations on performance. Therefore, in the interest of simplicity and clarity, we will confine our considerations to first the one and then the other of these regimes. We will start with a general expression for the case where the system performance is limited by thermal noise. We will then consider the case where the system performance is limited by quantum noise.

Thermal Noise

From Eq. (3) and Eq. (A.4) in Appendix I, we obtain

$$C_{it} = \frac{k_o}{\left(\frac{S_o}{N_o}\right)_t} = 2k_o\sqrt{\frac{KTB}{S_i}} = k_o\sqrt{\frac{2KT}{S_i\tau}} \qquad (4)$$

where the subscript t implies thermal-noise limited performance. Note that k_o does not need a subscript t because the value of k_o is determined by the visual process and it is the same for both thermal and quantum noise. Recall that S_i represents the transmitter power absorbed at the receiving plane corresponding to the brighter of the two image cells being distinguished.

Quantum Noise

Here we consider fluctuations at the image plane stemming from the quantum fluctuations in the absorbed transmitter power at the receiving plane. As shown in Appendix I, the quantum fluctuations at the image plane are given by

$$\overline{(\delta N^2)}_q = \overline{N}$$

where $\overline{(\delta N^2)}_q$

represents the mean square fluctuation of the number of quanta (phonons) associated with the image and N is the average number of these quanta. In the quantum-noise regime, we assume that essentially all of the image-plane quanta stem from the presence of the acoustic transmitter power in the system; only a negligible number of these quanta, therefore, come from other sources (such as, for example, thermal-noise sources). The above equation can also be readily derived by assuming a Poisson distribution associated with the quantum-mechanical absorption process. The quantum noise can be regarded as coming unavoidably from fluctuations in the transmitter power reaching the receiving plane. For simplicity, we assume that there is a one-to-one correspondence between phonons absorbed at the receiver plane and the response produced at the image plane; i.e. each absorbed phonon produces a commensurate output response. Therefore,

$$C_{iq} = \frac{k_o}{\left(\frac{S_o}{N_o}\right)_q} = \frac{k_o}{\left(\frac{S_i}{N_i}\right)_q}$$

where the subscript q denotes quantum-noise limited performance.

Let N_p be the average number of phonons absorbed at the receiving plane during the scanning period corresponding to the brighter of the two cells.

$$N_p = \frac{S_i \tau}{\hbar \Omega}$$

where $\hbar$ is Planck's constant divided by 2π and Ω is the angular frequency of the insonification.

The rms fluctuation in the above phonon number is $\sqrt{N_p}$, therefore the power fluctuation due to quantum noise in the absorbed energy is

$$N_{iq} = \sqrt{N_p}\,\frac{\hbar \Omega}{\tau}$$

The input threshold contrast can be expressed as

$$C_{iq} = k_o \left(\frac{N_i}{S_i}\right)_q = k_o \frac{\sqrt{N_p}\,\frac{\hbar\Omega}{\tau}}{N_p \frac{\hbar\Omega}{\tau}} = k_o \sqrt{\frac{\hbar\Omega}{S_i \tau}} \tag{5}$$

THRESHOLD CONTRAST FOR THE FOUR BASIC SYSTEMS

We will now derive the equations of threshold contrast for the four basic systems described in the preceding section. In each case we will consider first the operation as if it were limited by thermal noise only, and second, the operation as if limited by quantum noise only.

The object plane, in all four systems, can be visualized as being divided into rectangular resolution elements which correspond to the picture resolution cells at the image plane.

(See Figure 5.) Let two immediately adjacent resolution elements that are barely resolved at the image plane as having different intensities be called E_m and E_n.

Positively Scanning Transmitter (PST)

Assume that when the focused beam insonifies the more transparent of the two elements, the power incident at the receiving plane is P. Then when the focused beam moves from element E_m to element E_n, the power incident at the receiving plane changes by the amount C_aP where C_a is the previously defined acoustic threshold contrast. As indicated in assumption 4, η denotes the transducer conversion efficiency of the receiver. The signal power absorbed at the receiving plane is therefore

$$P_s = \eta C_a P \quad .$$

Now S_i, as previously defined, is the power absorbed at the receiving plane when the more transparent element is insonified. Hence

$$S_i = \eta P \quad .$$

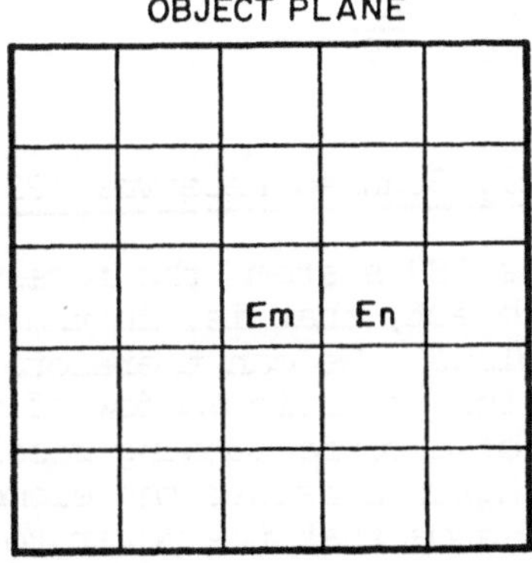

Figure 5 - Division of object plane into resolution elements.

The input threshold contrast is given by

$$C_i = \frac{P_s}{S_i} = \frac{\eta C_a P}{\eta P} = C_a \tag{6}$$

Thermal-Noise-Limited Performance

From Eqs. (4) and (6)

$$C_{at} = C_{it} = k_o \sqrt{\frac{2KT}{\eta P \tau}} \quad . \tag{7}$$

This is the relationship we seek. As previously stated, we will use this acoustic threshold contrast function for comparisons between systems.

Quantum-Noise-Limited Performance

For this system

$$S_i = \eta P \quad .$$

From Eqs. (5) and (6)

$$C_{aq} = C_{iq} = k_o \sqrt{\frac{\hbar\Omega}{\eta P \tau}} \quad . \tag{8}$$

Positively Scanned Receiver (PSR)

Assume that in the PSR system the receiving plane is located close to the object, that is, in the very near field region of the object plane. We can therefore think of the receptors as being in the acoustic shadow of the object, each covered by the shadow of a corresponding resolution element in the object plane. Again consider the elements E_m and E_n as in the PST case. Assume that the power to the receptor in the shadow of the more transparent of these two elements is P. The difference in the power to the two adjacent receptors is then C_aP. Therefore, when the above two receptors are being scanned, the absorbed power at the receiving plane S_i corresponding to the more transparent element is given as before by ηP and the absorbed signal power is

$$P_s = \eta C_a P \quad .$$

The input threshold contrast is

$$C_i = \frac{P_s}{S_i} = \frac{\eta C_a P}{\eta P} = C_a \tag{9}$$

just as for the PST case.

Thermal-Noise-Limited Performance

From Eqs. (4) and (9)

$$C_{at} = C_{it} = k_o \sqrt{\frac{2KT}{\eta P \tau}} \quad . \tag{10}$$

Quantum-Noise-Limited Performance

From the previous section we have

$$S_i = \eta P \quad .$$

Using Eqs. (5) and (9) we obtain

$$C_{aq} = C_{iq} = k_o \sqrt{\frac{\hbar\Omega}{\eta P \tau}} \quad . \tag{11}$$

Negatively Scanned Receiver (NSR)

As in the PSR case, we assume that the receiver plane is in the very near field region of the object and that each receptor is covered by the shadow of the corresponding object resolution element. The only difference between the operation of the NSR system and that of the PSR system is in the receptor scanning, the NSR scanning being the negative version of that for the PSR. Since the sensitivity of the prescribed systems depends on the entire receiver input during the two relevant adjacent scanning periods, it is reasonable to expect, in general, different sensitivity for the NSR case than for the PSR case.

Consider the two object elements E_m and E_n. As in the PSR system, let the power incident at the receiver plane in the shadow of the more transparent element be P. Then the difference in incident power at the receiver plane in the two shadows is C_aP. Assume that the total incident power at the receiver plane in the shadow of all the other elements (background elements) is P_B. There are m^2-2 of these background elements. When the two receptors are being scanned, the change in the absorbed power and hence the signal power is

$$P_s = \eta C_a P \quad .$$

The absorbed power is $\eta(P+P_B)$ when the receptor in the shadow of the less transparent element is turned off. This amount of power is greater than the power absorbed when the receptor in the shadow of the more transparent element is turned off. Therefore

$$S_i = \eta(P+P_B)$$

and the input threshold contrast is

$$C_i = \frac{\eta C_a P}{\eta(P+P_B)} = C_a \frac{P}{P+P_B}$$

and we have

$$C_a = \frac{P+P_B}{P} C_i \tag{12}$$

Thermal-Noise-Limited Performance

From Eqs. (4) and (12)

$$C_{at} = \frac{P+P_B}{P} C_{it} = \frac{P+P_B}{P} k_o \sqrt{\frac{2KT}{\eta(P+P_B)\tau}}$$

$$= k_o \sqrt{\frac{2KT}{\eta P \tau}} \sqrt{\frac{P+P_B}{P}} \tag{13}$$

Quantum-Noise-Limited Performance

From Eqs. (5) and (12)

$$C_{aq} = \frac{P+P_B}{P} C_{iq} = \frac{P+P_B}{P} k_o \sqrt{\frac{\hbar\Omega}{\eta(P+P_B)\tau}}$$

$$= k_o \sqrt{\frac{\hbar\Omega}{\eta P\tau}} \sqrt{\frac{P+P_B}{P}} \tag{14}$$

Both of the above threshold contrast expressions, Eq. (13) and Eq. (14), contain a background factor

$$\sqrt{\frac{P+P_B}{P}} \quad .$$

To pinpoint the effect of this background factor on threshold contrast, let us consider the following three cases.

Assume first that the two elements of interest E_m and E_n are located in a background completely opaque to the sound waves. In this situation the background factor has the minimum value of unity since $P_B = 0$. Equations (13) and (14) then become, respectively,

$$C_{at} = k_o \sqrt{\frac{2KT}{\eta P\tau}} \tag{15}$$

$$C_{aq} = k_o \sqrt{\frac{\hbar\Omega}{\eta P\tau}} \tag{16}$$

Second, assume that the object plane contains completely transparent resolution elements, except for E_m and E_n. Then the background factor becomes a maximum. Let the insonification power onto each resolution element of the object be P_i and let the acoustic transmittance of the more transparent of the two elements be t. We then have

$$P = P_i t \tag{17}$$

and

$$P_B = (m^2-2)P_i = (m^2-2)\frac{P}{t} \tag{18}$$

By applying Eq. (18) to Eqs. (13) and (14), we have

$$C_{at} = k_o \sqrt{\frac{2KT}{\eta P\tau}} \quad \sqrt{1+\frac{m^2-2}{t}} \tag{19}$$

$$C_{aq} = k_o \sqrt{\frac{\hbar\Omega}{\eta P\tau}} \quad \sqrt{1+\frac{m^2-2}{t}} \tag{20}$$

Third, consider a special case of the above when t = 1 (i.e. the case of an object consisting of one element which is not completely transparent in a completely transparent background). This circumstance leads to simpler expressions.

$$C_{at} \simeq mk_o \sqrt{\frac{2KT}{\eta P\tau}} \tag{21}$$

$$C_{aq} \simeq mk_o \sqrt{\frac{\hbar\Omega}{\eta P\tau}} \tag{22}$$ *

Since, in general, $m^2 >> 1$.

Negatively Scanning Transmitter (NST)

The operation of the NST system differs from that of the PST system only in the nature of the acoustic beam scanning at the object plane, the NST scanning being the negative version of that for the PST. Consider the two relevant scanning periods. Assume that when the scanning pencil-like "hole" in the sound beam of the NST system moves from object element E_m to object element E_n, the power incident at the receiving plane changes by the amount C_aP, where P is the power scattered to the receiver plane from the more transparent of the resolution elements when the "hole" is at the less transparent element. Let P_B denote the portion of the incident power at the receiving plane due to sound transmitted through the background elements during the relevant scanning periods (i.e. the sound from all the object elements except E_m and E_n). The greater amount of incident power during the two periods is then $P+P_B$. We have

$$S_i = (P+P_B)$$

*The equivalent of this equation has also been derived by Prof. B. A. Auld of Stanford (private correspondence).

and the signal power absorbed is

$$P_s = \eta C_a P \quad .$$

The input threshold contrast is

$$C_i = \frac{P_s}{S_i} = \frac{\eta C_a P}{\eta (P+P_B)} = \frac{P}{P+P_B} \; C_a \quad .$$

Thus we have for the acoustic threshold contrast

$$C_a = \frac{P+P_B}{P} \; C_i \quad . \tag{23}$$

Equation (23) is similar to Eq. (12) for the NSR system.

Thermal-Noise-Limited Performance

It follows from Eq. (4) and Eq. (23) that the acoustic threshold contrast is given by

$$C_{at} = \frac{P+P_B}{P} \; C_{it} = \frac{P+P_B}{P} \, k_o \sqrt{\frac{2KT}{\eta (P+P_B)\tau}}$$

$$= k_o \sqrt{\frac{2KT}{\eta P \tau}} \; \sqrt{\frac{P+P_B}{P}} \tag{24}$$

Quantum-Noise-Limited Performance

From Eq. (5) and Eq. (23), we obtain

$$C_{aq} = \frac{P+P_B}{P} \; C_{iq} = \frac{P+P_B}{P} \; k_o \sqrt{\frac{\hbar\Omega}{\eta (P+P_B)\tau}}$$

$$= k_o \sqrt{\frac{\hbar\Omega}{\eta P \tau}} \; \sqrt{\frac{P+P_B}{P}} \tag{25}$$

The general expressions for the threshold contrasts of the NST system are the same as those of the NSR system. Therefore considerations involving various classes of background elements will lead to the same results in the two systems. For a completely opaque background, $P_B = 0$, we have

$$C_{at} = k_o\sqrt{\frac{2KT}{\eta P\tau}} \tag{26}$$

$$C_{aq} = k_o\sqrt{\frac{\hbar\Omega}{\eta P\tau}} \tag{27}$$

For a completely transparent background, we obtain

$$P_B = (m^2-2)\ \frac{P}{t}$$

$$C_{at} = k_o\sqrt{\frac{2KT}{\eta P\tau}}\ \sqrt{1+\frac{m^2-2}{t}} \tag{28}$$

$$C_{aq} = k_o\sqrt{\frac{\hbar\Omega}{\eta P\tau}} \tag{29}$$

For the case of a single opaque element in a completely transparent background $t=1$, $m^2>>1$, and we have

$$C_{at} \simeq mk_o\sqrt{\frac{2KT}{\eta P\tau}} \tag{30}$$

$$C_{aq} \simeq mk_o\sqrt{\frac{\hbar\Omega}{\eta P\tau}} \tag{31}$$

DISCUSSION OF THEORETICAL PERFORMANCE OF THE BASIC SYSTEMS

The derived expressions for the acoustic threshold contrast for the four basic systems are presented in Table I. P has the same significance in each of the expressions and represents the transmitter power passing through the more transparent of the two object elements under consideration when that element is insonified. For medical applications of acoustic imaging, the sensitivity of biological tissue to ultrasound must be considered. Therefore, in trying to obtain an image, we should pay attention to the incident quantity P_i, the amount of acoustic power incident on an object element of arbitrary size. As before, let t be the transmittance of the more transparent element. We have

$P=P_i t$. Substitution of this relationship into the expressions of Table I yields the expressions in Table II.

TABLE I

Contrast for the Four Basic Systems in Terms of the Transmitted Power P

	PST,PSR	
C_{at}	$k_o\sqrt{\frac{2KT}{\eta P\tau}}$	$k_o\sqrt{\frac{2KT}{\eta P\tau}}\sqrt{\frac{P+P_B}{P}}$ (1) $k_o\sqrt{\frac{2KT}{\eta P\tau}}$ (opaque background) (2) $k_o\sqrt{\frac{2KT}{\eta P\tau}}\sqrt{1+\frac{m^2-2}{t}}$ (transparent background) (3) $mk_o\sqrt{\frac{2KT}{\eta P\tau}}$ (transparent background) (t=1)
C_{aq}	$k_o\sqrt{\frac{\hbar\Omega}{\eta P\tau}}$	$k_o\sqrt{\frac{\hbar\Omega}{\eta P\tau}}\sqrt{\frac{P+P_B}{P}}$ (1) $k_o\sqrt{\frac{\hbar\Omega}{\eta P\tau}}$ (2) $k_o\sqrt{\frac{\hbar\Omega}{\eta P\tau}}\sqrt{1+\frac{m^2-2}{t}}$ (3) $mk_o\sqrt{\frac{\hbar\Omega}{\eta P\tau}}$

TABLE II

Acoustic Threshold Contrast for the Four Basic Systems in Terms of the Incident Power P_i

	PST, PSR	NSR, NST
C_{at}	$k_o\sqrt{\frac{2KT}{\eta P_i t\tau}}$	$k_o\sqrt{\frac{2KT}{\eta P_i t\tau}}\sqrt{\frac{P+P_B}{P}}$ (1) $k_o\sqrt{\frac{2KT}{\eta P_i t\tau}}$ (2) $k_o\sqrt{\frac{2KT}{\eta P_i t\tau}}\sqrt{1+\frac{m^2-2}{t}}$ (3) $mk_o\sqrt{\frac{2KT}{\eta P_i t\tau}}$
C_{aq}	$k_o\sqrt{\frac{\hbar\Omega}{\eta P_i t\tau}}$	$k_o\sqrt{\frac{\hbar\Omega}{\eta P_i t\tau}}\sqrt{\frac{P+P_B}{P}}$ (1) $k_o\sqrt{\frac{\hbar\Omega}{\eta P_i t\tau}}$ (2) $k_o\sqrt{\frac{\hbar\Omega}{\eta P_i t\tau}}\sqrt{1+\frac{m^2-2}{t}}$ (3) $mk_o\sqrt{\frac{\hbar\Omega}{\eta P_i t\tau}}$

The tables illustrate the fact that the two positive systems (PST and PSR) have the same expressions, and the two negative systems (NSR and NST) also have the same expressions. The acoustic threshold contrast for the two negative systems depends on the degree of transparency of the background elements in the object. This is not true of the acoustic threshold contrast for the two positive systems. Because of this, for the same set of parameters, the acoustic threshold contrast for the negative systems has, in general, a larger value due to the presence of the background factor.

For each system, by choosing a set of parameters (P_i, τ, η, Ω, T) for operation, we can readily calculate C_{at} and C_{aq} as functions of t from Table II. A comparison of their numerical values (or, more simply, a comparison of the magnitudes of KT and $\hbar\Omega$) shows the predominance of one or the other of the two prescribed noise phenomena in limiting the system performance.

Also, for each system, when we fix all the parameters but one, we can study the dependence of C_a on that particular parameter. For instance, with the values for τ, η, t, Ω, T chosen, the dependence of C_{at} and C_{aq} on P_i can be investigated. The equations for the two positive systems involving a general object are the same as those for the two negative systems involving an object in which the two elements of interest are located in a completely opaque background. Under these circumstances, we have (by taking the logarithm):

$$\log C_{at} = \log\left(k_o\sqrt{\frac{2KT}{\eta t\tau}}\right) - \frac{1}{2}\log P_i \tag{32}$$

$$\log C_{aq} = \log\left(k_o\sqrt{\frac{\hbar\Omega}{\eta t\tau}}\right) - \frac{1}{2}\log P_i \tag{33}$$

Eq. (32) and Eq. (33) indicate that the log-log plot of C_{at} vs. P_i and of C_{aq} vs. P_i have the same slope. A typical superposed plot of the two is shown in Fig. 6.

Note that Fig. 6 is also a typical depiction for the two negative systems even when the object has an arbitrary background. The slope of the C_{at} line and the slope of the C_{aq} line must always have the value $-\frac{1}{2}$. However, the positions of intersections with the coordinate axes depend

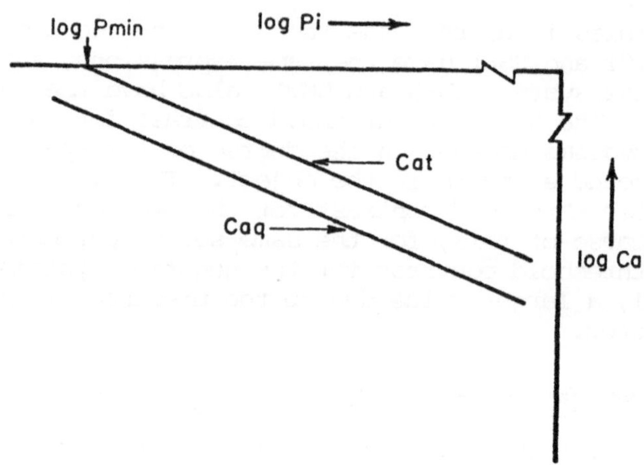

Figure 6 - A Typical Log-Log Plot of C_a vs. P_i.

upon all the parameters in the expressions of Table II. A change in one or more of these parameters will result in translations of both lines.

Because the intensity of sound for imaging biological tissue is recommended not to exceed 0.1 w/cm ^{4} and we are usually interested in a resolution better than 1 cm, P_i should be less than 1 watt, making log P_i a negative quantity. Also, the actual acoustic contrast (defined as the ratio of the difference in the acoustic transmittance to the larger transmittance of the two adjacent elements) cannot exceed unity, making log C_a a negative quantity. Therefore, we can expect the third quadrant to be the region of interest.

Consider two specific adjacent object elements, one completely opaque, the other completely transparent. The actual acoustic contrast is then unity, and the value for t is also unity. The minimum insonification power P_{min} needed to resolve this contrast is then the minimum detectable power onto an object element for the system and, as such, is an important figure of merit. P_{min} can be easily calculated by substituting unity for t and for the threshold acoustic contrast in the appropriate expression. It also corresponds to the intersection of the predominating line with the horizontal axis in Fig. 6. For operating conditions of interest, we have $KT \gg \hbar\Omega$ and the C_{at} line is always above the C_{aq} line. The intersection of the C_{at} line with the horizontal axis therefore gives P_{min}.

Consider a numerical example. A PSR system operates at room temperature in water. The object insonification intensity is 10^{-8} watt/cm^2 at 3 MHz. Resolution is five-wavelengths.

$$\Omega = 6\pi \times 10^6 \text{ Hz} \quad , \quad \eta = 0.1$$

$$I_i = 10^{-8} \text{ watt/cm}^2 \quad , \quad \text{res.} = 5\,\Lambda$$

$$M = 300 \quad , \quad k_o = 5$$

$$\tau_f = 0.1 \text{ sec} \quad , \quad t = 1$$

Since $KT \gg \hbar\Omega$, C_a is given by C_{at}.

$$\tau = \tau_f/m^2 = 1.1 \times 10^{-6} \text{ sec}$$

$$P_i = I_i(5\,\Lambda)^2 = 6.25 \times 10^{-10} \text{ watt}$$

$$C_a = k_o\sqrt{\frac{2KT}{\eta P_i t\tau}} = 5.4 \times 10^{-2}$$

Therefore, for an ideal PSR system operating under the above conditions we may conclude that the sensitivity of the system is essentially limited by thermal noise. For the given insonification level only acoustic contrasts greater than 5.4×10^{-2} can be resolved between adjacent resolution elements.*

If the same set of figures can be used for the NSR system to image an element in a completely transparent background (this corresponds to case 3 in the preceding discussion), the acoustic threshold contrast is, from Table II,

---

* This means simply that the actual acoustic contrast must be greater than C_a, the threshold acoustic contrast, in order for the 2 elements to be observed.

$$C_a = C_{at} = mk_o\sqrt{\frac{2KT}{\eta P_i t\tau}}$$

and is m times higher than the calculated value for the PSR system. For m=300, its value exceeds unity. Therefore, under the specified conditions for the NSR system, 10^{-8} watt/cm^2 is not a detectable insonification intensity.

Let us consider a second numerical example. This time we will calculate the minimum detectable insonification intensity for the positive systems. Again we assume the same values for the parameters as in the preceding example.

$$T = 300° \text{ K}, \ \Omega = 6\pi \times 10^6 \text{Hz}, \ \eta = 0.1, \text{in water}$$

$$m = 300, \ \tau_f = 0.1 \text{ sec., res.} = 5\Lambda, \ k_o = 5$$

Setting $C_a = 1$, $t = 1$ we have

$$C_a = C_{at} = k_o\sqrt{\frac{2KT}{\eta P_{min}\tau}} = 1$$

$$P_{min} = k_o^2 \frac{2KT}{\eta\tau} = 1.8 \times 10^{-12} \text{ watt}$$

$$I_{min} = P_{min}/(5\Lambda)^2 = 2. \times 10^{-11} \text{ watt/cm}^2 \ *$$

If ideal target storage can be achieved, τ should be replaced by τ_f in the expressions and both P_{min} and I_{min} can be reduced by a factor of m^2.**

As can be seen from the tables, with the same set of operating parameters, the positive systems have the better theoretical performance. For the same level of insonification onto each object element, lower acoustic contrasts are detectable with the positive system than with the negative systems. Also to make any actual acoustic contrast observ-

---

* This is essentially in agreement with Eq. (8) of Korpel and Kessler[5].

** In this case, P_{min} becomes 2×10^{-17} watts which corresponds to the figure quoted by Vilkomerson[1].

able, the positive systems require lower levels of insonification onto each of the object elements than do the negative systems.

In the above discussion, we have postulated the same transducer conversion efficiency for the different scanning systems. However, often such a postulate cannot be justified.* It can be shown that the effective conversion effieiency depends very much on the fraction of the total sensitive area of the matched independent transducer that is used. Therefore, it depends critically on the realization of (or design behind) the receiving plane. For each conceivable implementation, the conversion effieiency is a sensitive function of the acoustic field distribution at the receiving plane.

Assume that each receptor, large (the big single transducer for the PST system and the NST system) or small (elemental receptors for the PSR system and the NSR system), can achieve the same efficiency η. Consider for the moment an implementation of the NSR system such as the scanned acoustic microscope of Dr. Auld.[2] It may appear that for Case (3) of the tables (an object with a transparent background) the system is always less effective than for Case (1) (an object with an opaque background) because of the factor m. But such a conclusion is not correct. In Case (1), most of the receptors are not insonified and it can be shown that there will be an inherent reduction in the conversion efficiency by a factor of m^2. For Case (3), there is no such reduction.

Theoretically, in general, the PSR system has better performance than the NSR system. The degradation in the NSR performance comes partly from the effect of the background, and partly from the above cited reduction in the conversion efficiency.

Let us compare the two positive systems. When the same set of parameters (including the conversion efficiency) are involved in the operation, both systems require the same level of P_i to make a desirable acoustic contrast observable. In this case, the PST system has the advantage of using less total insonification energy into any single element during

---

* This fact was first called to our attention by Professor B. A. Auld (private conversation).

a single scanning period τ for each frame time τ_f. This means that the average incident power on any one element is only

$$P_i \frac{\tau}{\tau_f} = \frac{P_i}{m^2} .$$

However, for the PSR system, each element is essentially insonified continuously throughout the frame time, and therefore, P_i is the average incident power to each. From the standpoint of average incident power, the PST system is therefore the better candidate for biomedical application.

One additional point should be brought out. Because of difficulty in efficiently using the sensitive area of a single large transducer, a PST system may not achieve the conversion efficiency that each elemental recptor of a PSR system can achieve. In principle, perhaps the best conceivable system would be one incorporating both positive scanning transmission and positive scanned reception into one system to avoid the above possible reduction in conversion efficiency. This would involve a simultaneous source and receiver scan.

The acoustic threshold contrast is a measure of the sensitivity of a system. The smaller the acoustic threshold contrast for a given set of parameters, the more sensitive is the system in acoustic imaging. One figure of merit for the sensitivity of a system would be the reciprocal of the acoustic threshold contrast.

As indicated in Appendix I, the thermal noise expressions derived in the preceding sections apply in the case of a strong signal-related power (S_i) at the input to the detector. For the configuration considered above, this is always satisfied; S_i being several times greater than KTB for a detectable power.

The above ideal sensitivity of the systems can be enhanced if further improvement of the signal-to-noise ratio can be achieved, for instance, by the proper insertion of a noiseless optimum filter.*

---

* It could be a matched spatial filter for examination of a special known pattern in the object.

Consider such an insertion behind the square-law detector such that the minimum detectable signal-related power at the input to the detector becomes small compared to KTB. At low insonification near the minimum detectable level, C_{at} takes on a different form than at high insonification (See Appendix II) and the log-log plot of C_a vs. P_i shows different slopes for C_{at} and C_{aq}.

SUMMARY OF CONCLUSIONS

In summary, there are certain circumstances under which all four systems have the same theoretical performance as a function of P_i. The two positive systems have better sensitivity than the two negative systems with an object not having a completely opaque background. In medical diagnosis when small acoustic contrasts in living biological tissue are to be recongized by imaging, the required insonifiction intensity (P_i divided by the area of a resolution element) may approach the maximum level for safe operation. In this case, the PST system would be preferred because of the need for less total insonification.

Under operating conditions where the theoretical performance shows little difference, the decision of employing one particular system will have to depend upon practical implementation and its deviation from ideal behavior.

ACKNOWLEDGEMENTS

This work was supported by the Public Health Service (Grant No. GM-16474-04).

APPENDIX I

Thermal Noise and Quantum Noise

Gabor[6] has given the following simple equation which includes both the quantum effect and the thermal effect of detection

$$\overline{\delta N^2} = \bar{N} + 2\bar{N}\bar{N}_T - \bar{N}_T^2 \qquad (A1)$$

Here $\overline{\delta N^2}$ represents the mean square fluctuation of phonon number per mode, $\bar{N}$ the mean number of total phonons per mode (including signal and thermal phonons), $\bar{N}_T$ the mean number of thermal phonons. Now

$$\bar{N}_T = \frac{1}{e^{h\nu/KT}-1}$$

For $h\nu >> KT$, $N_T << 1$ and only the first term prevails. Thus

$$\overline{(\delta N^2)}_q \simeq \bar{N} \qquad (A2)$$

where the subscript q denotes that the fluctuation is predominatly "quantum noise."

On the other extreme when $KT >> h\nu$, two different situations are involved. In the absence of signal, or in the presence of a very weak signal,

$$\bar{N} \simeq \bar{N}_T$$

and we have

$$\overline{(\delta N^2)}_t \simeq \bar{N}_T^2$$

This equation is basic in deriving the well-known sign-independent thermal noise power expression KTB. The subscript t above denotes "thermal noise" predominance. In the presence of a strong signal, $\bar{N} >> \bar{N}_T$, the middle term $2\bar{N}\bar{N}_T$ then predominates and we have

$$\overline{(N^2)}_t \cong 2\bar{N}\bar{N}_T \qquad (A3)$$

The effect represented by the above expression can be interpreted in terms of beating between signal and noise quantities. By relating $\bar{N}$ to S_o and δN^2 to N_o We obtain

$$\left(\frac{S_o}{N_o}\right)_t = \frac{\bar{N}}{\sqrt{\overline{\delta N^2}}} = \sqrt{\frac{1}{2}\frac{S_i\tau}{\hbar\Omega}\frac{\hbar\Omega}{KT}} = \frac{1}{2}\sqrt{\frac{S_i}{KTB}} \qquad \text{(A4)}$$

The same result can be obtained from a classical spectral analysis of a square-law detector output. In such an analysis we assume a sinusoidal input with a zero-mean random amplitude modulation embedded in narrow band Gaussian noise with spectral density KT.

APPENDIX II

Threshold Contrast Under Various Different Circumstances

The noise at the output of a noisless square-law detector results from the interaction of the input signal with the input noise and the interaction of the input noise with itself. The relative importance of the two interactions depends on the input signal-to-noise power ratio. Consider sine-wave signal with zero-mean random amplitude modulation embedded in narrow-band Gaussian noise. For a large input signal-to-noise power ratio, we can show by classical spectral analysis that

$$\left(\frac{S_o}{N_o}\right)_t = \sqrt{\frac{1}{4}\frac{S_i}{KTB}} \qquad \text{(A5)}$$

where the symbols have the same meaning as in the text. This is the same equation as Eq. (A4) in Appendix I and the C_{at} expressions in Table I and Table II were derived using the above equation.

For a very small input signal-to-noise power ratio, the relationship is as follows:

$$\left(\frac{S_o}{N_o}\right)_t = \frac{1}{\sqrt{2}}\frac{S_i}{KTB} \qquad \text{(A6)}$$

Equations (A5) and (A6) are obviously quite different and lead to different expressions for C_{it} and C_{at}. However, for the ordinary operation of the systems discussed in the text, the assumption upon which (A6) is based corresponds to a power level lower than the minimum detectable level of insonification and hence is not of interest. Therefore, Table I and Table II hold under ordinary conditions of operation.

Now consider a more sensitive system obtained by inserting an optimum filter between the detector and the image plane of our previous configuration. For simplicity assume that the only function of the optimum filter is to introduce an improvement factor R of the signal-to-noise power ratio. Let the image plane quantities be S_o, N_o, and k_o, C_o and the receiving plane quantities be k_i, C_i, S_i, N_i. Let the corresponding quantities at the output of the square-law detector and hence at the input of the optimum filter be k, C, S, N.

$$\frac{k_o}{C_o} = \frac{S_o}{N_o} = R\,\frac{S}{N} = R\,\frac{k}{C}\,, \quad C_o = C_i$$

If R is large enough, the minimum detectable S_i can be very small compared to KTB. For low insonification near the minimum detectable level, we would then have for S/N the expression of (A6). Thus

$$\left(\frac{S}{N}\right)_t = \frac{1}{\sqrt{2}}\,\frac{S_i}{KTB} \tag{A7}$$

Under these circumstances

$$C_{it} = k_o\Big/\left(\frac{S_o}{N_o}\right)_t = \frac{k_o}{R\left(\frac{S}{N}\right)_t} = \frac{k_o}{\sqrt{2}R}\,\frac{KT}{S_i\tau} \tag{A8}$$

Following similar procedures to those described in the text, we obtain the following equation for the four systems

$$C_{at} = \frac{k_o}{\sqrt{2}R}\,\frac{KT}{\eta P_i t\tau} \tag{A9}$$

For a sufficiently high level of transmitter power even

with the optimum filter we obtain for S/N an expression similar to that in (A5). Hence

$$\left(\frac{S}{N}\right)_t = \sqrt{\frac{1}{4}\frac{S_i}{KTB}} \tag{A10}$$

Under these circumstances, we have

$$C_{it} = \frac{k_o}{\left(\frac{S_o}{N_o}\right)_t} = \frac{k_o}{R\left(\frac{S}{N}\right)_t} = \frac{k_o}{R}\sqrt{\frac{2KT}{S_i\tau}} \tag{A11}$$

From Eq. (A11), the threshold contrast for the two positive systems becomes

$$C_{at} = \frac{k_o}{R}\sqrt{\frac{2KT}{\eta P_i t\tau}} \tag{A12}$$

For the two negative systems, we have

$$C_{at} = \frac{k_o}{R}\sqrt{\frac{2KT}{\eta P_i t\tau}}\sqrt{\frac{P+P_B}{P}} \tag{A13}$$

REFERENCES

1. D. Vilkomerson, "Analysis of Various Ultrasonic Holographic Imaging Methods for Medical Diagnosis," Acoustical Holography, vol. 4, Plenum Press, New York, pp. 401-430 (1972).

2. B. A. Auld, et. al., "A 1.1 GHz Scanned Acoustic Microscope," Acoustical Holography, vol. 4, Plenum Press, New York, pp. 73-96 (1972).

3. A. Rose, "Television Pickup Tubes and the Problem of Vision," Advances in Electronics, vol. 1, pp. 131-166 (1943).

4. A. Sokollu, "Irreversible Effects of High Frequency Ultrasound on Animal Tissue and the Related Threshold Intensities," Acoustical Holography, vol. 3, Plenum Press, New York, Chapter 1 (1971).

5. A. Korpel and L. Kessler, "Comparison of Methods of Acoustic Microscopy," Acoustical Holography, vol. 3, Plenum Press, New York, Chapter 3 (1971).

6. D. Gabor, "Communication Theory and Physics," Philosophical Magazine, vol. 41, pp. 1161-1187 (1950).

INFLUENCE OF QUANTIZATION AND OF OTHER NONLINEAR DISTORTIONS OF THE HOLOGRAPHIC SIGNAL

W. J. Dallas and A. W. Lohmann

Applied Physics and Information Science
University of California, San Diego
La Jolla, California 92037

First we will describe why quantization and other nonlinearities occur in holography. Then we will show how the influence of those nonlinear distortions upon the quality of the reconstructed image can be understood.

A $10 \times 10\ cm^2$ photographic plate might contain the information of up to 10^{10} image points. On the other hand an acoustical hologram is worth only 10^3 to 10^5 image points. Microwave holograms and computer holograms fall into the same category. The reasons for this scarcity of data are not fundamental, only economical. One of the consequences is that in acoustical holography one has to squeeze the utmost of information out of a system. When approaching the performance limits so closely deteriorations are sometimes unavoidable. For example many receivers become nonlinear when used with ultimate sensitivity. In the language of photographic holography the amplitude transmittance T of the hologram is no longer linearly proportional to the exposure E, but is described for example by a third order polynomial. Equation 1 describes fairly well the performance of high resolution photographic emulsions.

$$T_A(E) = \sum T_m E^m; \quad m = 0,1,2,3. \tag{1}$$

From a small hologram one gets only a dim image. Hence one bleaches the hologram in order to improve the light efficiency. Bleaching is a nonlinear process even if the exposure E occurs linearly in the exponent (equ. 2).

$$T_B(E) = \exp(ibE) \tag{2}$$

$$T_{NB}(E) = \exp[ibD(E)] \tag{3}$$

The nonlinearity becomes even more involved if the bleach phase is assumed to be proportional to the density D which in turn is a nonlinear function of E when very wide exposure ranges occur (equ. 3).

Yet another nonlinearity might afflict the hologram if the measured exposure distribution E(x) in the hologram plane is used for synthesizing an optical hologram whose amplitude transmittance $T_Q(E)$ is quantized (equ. 4, fig. 1).

$$T_Q(E) = n/N \quad \text{if } n/N \leq E < (n+1)/N;$$
$$n = 0,1,\ldots N-1; \quad 0 \leq E \leq 1. \tag{4}$$

Most automatic plotters will introduce a certain amount of quantization error.

Now we want to present a unifying theory of all these nonlinear effects. We will find that any nonlinearly distorted hologram can be thought of as a superposition of bleached holograms (equ. 2). We assume that the exposure has a maximum value which we arbitrarily call one. Hence the nonlinear curve T(E) is defined on the E-axis only between zero and one. Now we try a formal trick which will turn out to be useful. We extend the T(E) curve periodically over the whole length of the E-axis. This periodically extended T(E) curve describes the same nonlinearity as the original T(E) curve, since only E-values between zero and one occur anyway. But this periodic continuation allows us to describe the nonlinear curve as a Fourier series (equ. 5).

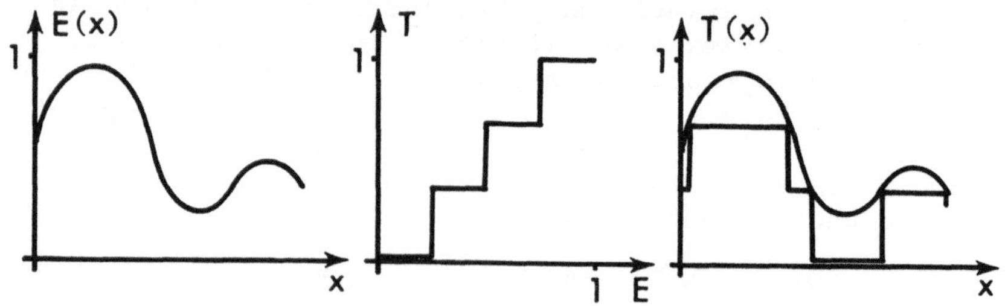

Figure 1. Quantization of the exposure distribution E(x) via the nonlinear curve T(E) into the amplitude transmittance T(x) of the hologram.

$$T(E) = \sum_{(m)} A_m \exp(2\pi imE);$$

$$A_m = \int_0^1 T(E) \exp(-2\pi imE)\, dE. \qquad (5)$$

In the case of N equally spaced quantization levels (equ. 4, fig. 1) we find the coefficients as presented in equ. 6.

$$A_0 = 1/2; \qquad A_m = i/2\pi m;$$
$$\text{but} \quad A_N = 0 = A_{-N} = A_{2N} = 0 \ldots \qquad (6)$$

The essence of equ. 5 becomes more evident after rewriting equ. 2 and equ. 5 in a slightly modified form (equ. 7).

$$T_B(E(x);b) = \exp[ibE(x)];$$
$$T(E(x)) = \sum_{(m)} A_m\, T_B(E(x);2\pi m). \qquad (7)$$

We may interpret equ. 7 as saying that a hologram which has been distorted by a general nonlinear curve T(E) is equivalent to the superposition of several bleached holograms $T_B(E;2\pi m)$. This result is valuable since the impact of bleaching is experimentally and theoretically well understood. The most common theoretical treatment of bleached holograms makes use of the fact that the hologram exposure E(x) is a carrier frequency signal (equ. 8). Applying the Jacobi-Bessel formula yields a series with (n=+1) as the true image term and (n>1) as higher diffraction orders which deviate far off to the side.

$$E(x) = |u_0(x) + \exp(2\pi i\nu_0 x)|^2$$
$$\approx 1 + 2|u_0(x)| \cos[2\pi\nu_0 x - \varphi_0(x)];$$

$$\exp(iA \sin t) = \sum_{(m)} J_n(A) \exp(int);$$
$$\exp[ibE(x)] \approx \sum J_n(2b|u_0(x)|)$$
$$\exp[in(2\pi\nu_0 x - \varphi_0 + \pi/2) + ib]. \qquad (8)$$

The special case of two-level quantization (N=2) is of particular interest since it leads to a black-white hologram, which is very easy to produce. Two-level quantization, which appears to be a very severe case of nonlinear distortion, turns out to be surprisingly harmless. The (m=0) term describes only an uninteresting average transmittance. The (m=+1) term corresponds to a bleached hologram with a phase $2\pi E$. The (m=-1) term is the complex conjugate of the (m=+1) term. Hence the corresponding reconstructions will relate to each other like true and conjugate holographic images. By designing the setup properly these two images (m=+1, m=-1) will not overlap [1,2,3]. The (m=+2, -2) terms do not exist for N=2 and the (m=3) term is nine times weaker in intensity as compared to the (m=1) term since $(A_3/A_1)^2 = 1/9$. Hence we may disregard these higher order terms under many circumstances. It helps also that these higher order terms will often send their light into areas far away from the regular image. More details are published elsewhere [4].

In conclusion we can say that nonlinearities occur in holography for a variety of reasons, many of them due to economical constraints. In general a nonlinear process will distort the reconstructed image. But it is not necessary to avoid nonlinearities at any price since image distortions can be spatially divorced from the image if the system is designed accordingly.

References

[1] J.W. Goodman, A.M. Silvestri, IBM J.R.&D. 14, 478 (1969).
[2] W.J. Dallas, Appl. Opt. 10, 673 (1971).
[3] W.J. Dallas, A.W. Lohmann, Appl. Opt. 11, 192 (1972).
[4] W.J. Dallas, A.W. Lohmann, Opt. Comm. 5, (1972).

THE EFFECTS OF SCANNING POSITION AND MOTION ERRORS ON HOLOGRAM RESOLUTION

H. D. Collins and B. P. Hildebrand

Holosonics, Inc. and Battelle-Northwest

2950 George Washington Way, Richland, WA. and
P. O. Box 999, Richland, WA. 99352

ABSTRACT

This paper presents the results of an evaluation of image defects caused by transducer and object position and motion errors in generating an acoustical hologram. The image parameters we studied were lateral and longitudinal resolution. An error analysis, assuming slowly varying gaussian statistics, was used to estimate the mean and standard deviation. The results show that two types of errors must be considered; i.e., position errors may approach an appreciable fraction of the hologram resolution limits; however, motion errors (especially longitudinal motion) are more restrictive and appear to be limited by coherence between object and reference signals. Daubin[1] has previously noted longitudinal motion restrictions.

A limited experimental program was followed to validate the theoretical conclusions. The procedure used was to increment transducer and object errors as the hologram was generated. We found that the theoretical predictions were borne out, at least on a subjective basis. Incremental longitudinal position errors as large as 75 wavelengths

---

1. S. C. Daubin, "System Requirements for Underwater Holographic Mapping", IEEE Trans. Geosci. Electron Vol. GE-8, Oct., 1970, pp. 313-320.

caused little visible image defects while longitudinal motion errors of one-half wavelength per scan line completely destroyed the image.

INTRODUCTION

Scanned acoustical holography has many potential applications such as under-sea imaging, nondestructive testing or seismic imaging. In all of these applications one or more transducers are physically moved in order to synthesize a two-dimensional aperture. Whenever a physical element is moved it is quite likely that random or systematic position errors will arise. These errors can be three-dimensional; i.e., in-plane and out-of-plane. The errors can arise in laboratory instruments because of imperfect mechanical drive mechanisms, in ocean imaging systems because of wave action and in seismic imaging systems because of errors in laying out the geophone strings. Even when filled two-dimensional arrarys are used, it is of interest to know what deviations from flatness can be tolerated, since this has a bearing on the cost of such an array. It is important, therefore, to know what effects such errors have on the final image.

In this paper, we make an attempt to define the effect of these kinds of errors on the image resolution, since this is the primary concern of all of these applications. We approach the problem in a way easily related to practice. That is, we assume that we have separate scanning and recording systems. The scanning system is designed to operate in a certain programmed manner. The signals from the transducers are transmitted to the recording system which is another scanning system, usually electronic. Although both systems are subject to position error, we lump them all together into the scanning system, since this is where the major errors will occur.

The statistics of the random error process are assumed to be independent in the three directions (x,y,z) and Gaussian with small variance. By small, we mean that the variance of the error is less than one-tenth of the error itself. This insures that the scanning deviations are slowly varying across the aperture and allows us to expand the resolution equations in a Taylor's series and retain only the linear terms.

The analysis shows that surprisingly large scanning errors can be tolerated without serious degradation of the hologram resolution. A series of experiments was performed to verify the analytical conclusions. It is, of course, difficult to prove a statistical result in the laboratory since many trials must be made. However, we feel that the simple experiments we performed verified the theory satisfactorily.

ANALYSIS

The formal mathematical analysis begins with a short review of the hologram resolution and the law of propagation of errors.

Lateral and Longitudinal Hologram Resolutions

The lateral resolution (Δx_1) in the object space is defined as the distance through which the object point can be displaced before the total number of interference fringes in the hologram aperture changes by one.[2] This means the phase changes by 2π between the aperture extremes. The phase at the receiving transducer in the hologram plane during the recording process is:

$$\Phi(x,y,z) = \frac{2\pi}{\lambda_s} [r_o{}^1 + r_1{}^1 - r_2{}^1]. \tag{1}$$

where λ_s is the sound wavelength.

The hologram geometry and symbols are shown in Figure 1. If the object distances are large compared with the aperture dimensions then the distance terms can be expanded in a binomial series with retention up to the second order terms. Equation No. 1 can then be expressed as:

$$\Phi(x,y,z) \cong \frac{2\pi}{\lambda_s} [r_o + r_1 - r_2 + \frac{x^2}{2} (\frac{1}{r_1} + \frac{1}{r_o} - \frac{1}{r_2}) -$$

$$- x (\frac{x_o}{r_o} + \frac{x_1}{r_1} - \frac{x_2}{r_2}) \tag{2}$$

where we consider only the x-z plane.

The phases at the aperture extremes are:

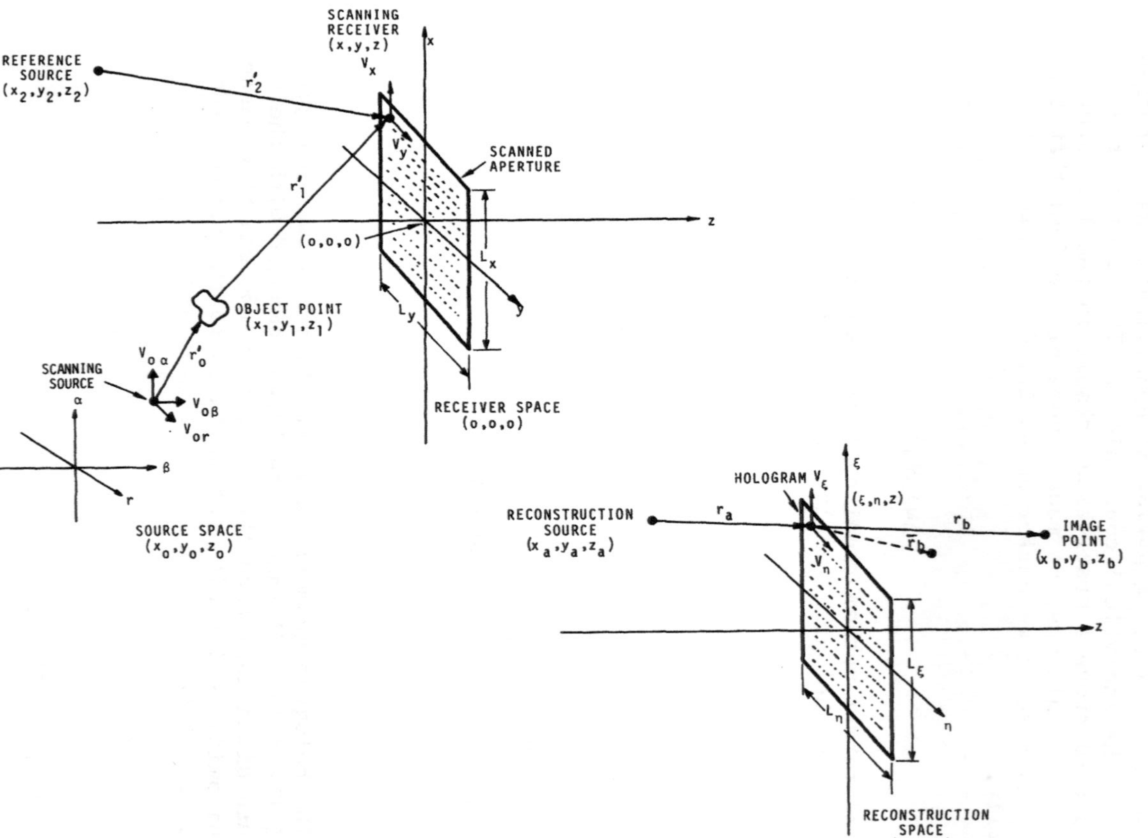

Figure 1. Geometry for Scanned Acoustic Holography

$$\Phi(x = +L/2) \cong \frac{2\pi}{\lambda_s} [r_o + r_1 - r_2 + \frac{L^2}{4} (\frac{1}{r_1} + \frac{1}{r_o} - \frac{1}{r_2}) - $$
$$- \frac{L}{2} (\frac{x_o}{r_o} + \frac{x_1}{r_1} - \frac{x_2}{r_2})] \quad (3)$$

and

$$\Phi(x = -\frac{L}{2}) \cong \frac{2\pi}{\lambda_s} [r_o + r_1 - r_2 + \frac{L^2}{4} (\frac{1}{r_1} + \frac{1}{r_o} - \frac{1}{r_2}) + $$
$$+ \frac{L}{2} (\frac{x_o}{r_o} + \frac{x_1}{r_1} - \frac{x_2}{r_2})] \quad (4)$$

The phase difference between the aperture extremes is:

$$\Phi = \frac{2\pi}{\lambda_s} L (\frac{x_o}{r_o} + \frac{x_1}{r_1} - \frac{x_2}{r_2}) \quad (5)$$

The incremental phase change as a function of the incremental change in the lateral object position is given by the following expression:

$$\Delta\Phi = \frac{2\pi}{\lambda_s} \frac{L}{r_1} \Delta X_1 . \quad (6)$$

The final result for the lateral hologram resolution (stationary source) is then:

$$\Delta X_1 \cong \lambda_s (\frac{r_1}{L}) . \quad (7)$$

and for simultaneous source-receiver scanned holograms.[2]

$$\Delta X_{1_s} = \lambda_s (\frac{r_1}{2L}) . \quad (8)$$

The longitudinal or depth resolution is more difficult to define mathematically than the lateral resolution. The longitudinal resolution as stated by Hildebrand[3] can be related to a decrease in intensity of the image as a result of an incremental change in r_b. The object point can then be moved an incremental distance Δr_1 to achieve the same decrease in intensity of the image. The longitudinal resolution can then be expressed as:

$$\Delta r_1 \cong 2\lambda_s \left(\frac{r_1}{L}\right)^2. \tag{9}$$

Analysis of Hologram Resolution Errors

The analysis consists of finding the approximate variance and mean value of the hologram lateral and longitudinal resolution. It is assumed that the resolution functions vary slowly in the region where the values of the independent variables remain within one or two standard deviations of their mean or expected value. The functions can then be adequately represented by the linear terms of their Taylor series expansions (see Appendix A and B).

Let v_x, v_y and v_z be the actual acoustic receiver velocity components in the recording aperture (i.e., [x,y,z] coordinate plane) and v_ξ, v_η, v_γ the simulated velocity components in the hologram plane. The ξ, η, γ coordinates are related to the x, y, z coordinates by the following equations:

$$x = m_x \xi \tag{10}$$

$$y = m_y \eta \tag{11}$$

$$z = m_z \gamma \tag{12}$$

We now substitute Equations 10, 11 and 12 into Equations 8 and 9 and assume that random errors exist in the simulated position of the acoustic receiver during the hologram construction. The following linear statistical models are used:

$$\xi = \bar{\xi} + e_\xi \tag{13}$$

$$\eta = \bar{\eta} + e_\eta \tag{14}$$

$$\gamma = \bar{\gamma} + e_\gamma \tag{15}$$

where $\bar{\xi}$, $\bar{\eta}$, and $\bar{\gamma}$ are the mean values and e_ξ, e_η, and e_γ are statistically independent random errors. It is assumed that the errors are normally distributed with zero means and their respective variances (σ_ξ^2, σ_η^2, σ_γ^2). In order to derive the approximate variances and expected values of the hologram resolution, we then expand the functions Δx_1 and Δr_1 in a Taylor's series about the point $(\bar{\xi}, \bar{\eta}, \bar{\gamma})$ and retain only the linear terms of the expansions. This approximation is usually quite adequate for obtaining the approximate variance and expected value of an arbitrary function[4]. The final expressions for the linear terms are given by the following equations:

$$\text{Lateral resolution } (\Delta x_1) \cong \left.\frac{\lambda_s r_1}{L}\right|_{\bar{\xi},\bar{\eta},\bar{\gamma}} +$$

$$+ \frac{\lambda_s m_x}{L r_1}(m_x \xi - x_1)\left|\frac{(\xi-\bar{\xi})}{\bar{\xi},\bar{\eta},\bar{\gamma}}\right. + \frac{\lambda_s m_y}{L r_1}(m_y \eta - y_1)\left|\frac{(\eta-\bar{\eta})}{\bar{\xi},\bar{\eta},\bar{\gamma}}\right. +$$

$$+ \frac{\lambda_s}{L r_1} m_z (m_z \gamma - z_1)\left|\frac{(\gamma-\bar{\gamma})}{\bar{\xi},\bar{\eta},\bar{\gamma}}\right. + \ldots \tag{16}$$

$$\text{Longitudinal resolution } (\Delta r_1) \cong \lambda_s \left(\frac{r_1}{L}\right)^2\Big|_{\bar{\xi},\bar{\eta},\bar{\gamma}} +$$

$$+ \frac{2\lambda_s}{L^2} m_x (m_x \xi - x_1)\left|\frac{(\gamma-\bar{\gamma})}{\bar{\xi},\bar{\eta},\bar{\gamma}}\right. + \frac{2\lambda_s}{L^2} m_y (m_y \eta - y_1)\left|\frac{(\eta-\bar{\eta})}{\bar{\xi},\bar{\eta},\bar{\gamma}}\right. +$$

$$+ \frac{2\lambda_s}{L^2} m_z (m_z \gamma - z_1)\left|\frac{(\gamma-\bar{\gamma})}{\bar{\xi},\bar{\eta},\bar{\gamma}}\right. + \ldots \tag{17}$$

where the partial derivatives are evaluated at the point $(\bar{\xi},\bar{\eta},\bar{\gamma})$.

The expected or mean values of the lateral and longitudinal resolutions are given by the following expressions:

$$E[\Delta x_1] \cong \left.\frac{\lambda_s r_1}{L}\right|_{\bar{\xi},\bar{\eta},\bar{\gamma}} + \text{bias} \tag{18}$$

$$E[\Delta r_1] \cong \lambda_s \left(\frac{r_1}{L}\right)^2\Big|_{\bar{\xi},\bar{\eta},\bar{\gamma}} + \text{bias} \tag{19}$$

The variance of the stationary source hologram lateral resolution (Δx_1) when statistically independent random errors are introduced is:

$$V[\Delta x_1] \cong \frac{\lambda_s^2}{L^2 r_1^2}\left\{m_x^2(m_x\bar{\xi}-x_1)^2\ \sigma_\xi^2 + m_y^2(m_y\bar{\eta}-y_1)^2\ \sigma_\eta^2 + \right.$$

$$\left. + m_z^2(m_z\gamma-z_1)^2\ \sigma_\gamma^2\right\} \qquad (20)$$

and the variance of the longitudinal resolution (Δr_1)

$$V[\Delta r_1] \cong \frac{4\lambda_s^2}{L^4}\left\{m_x^2(m_x\bar{\xi}-x_1)^2\ \sigma_\xi^2 + m_y^2(m_y\bar{\eta}-y_1)^2\ \sigma_\eta^2 + \right.$$

$$\left. + m_z^2(m_z\gamma-z_1)\sigma_\gamma^2\right\} \qquad (21)$$

If the deviations in the simulated receiver positions are assumed never to exceed three standard deviations, then estimators of the variances can be defined as:

$$\hat{\sigma}_\xi^2 = \frac{\Delta\xi^2}{9} \qquad (22)$$

$$\hat{\sigma}_\eta^2 = \frac{\Delta\eta^2}{9} \qquad (23)$$

$$\hat{\sigma}_\gamma^2 = \frac{\Delta\gamma^2}{9} \qquad (24)$$

This assumes the maximum deviation of errors in the ξ, η and γ scanning receiver positions will essentially never exceed these values. The probability of errors greater than $\Delta\xi$, $\Delta\eta$ or $\Delta\gamma$ is approximately 0.003.

If the holograms are sampled for deviations in scanning receiver positions, then estimates can be calculated using the maximum likelihood method[5]. The estimation of the means $\bar{\xi}$, $\bar{\eta}$ or $\bar{\gamma}$ assuming a normal distribution is:

$$\hat{\xi} = \frac{1}{n}\sum_{i=1}^{n} \xi_i = \bar{\xi} \qquad (24)$$

$$\hat{\eta} = \frac{1}{n}\sum_{i=1}^{n} \eta_1 = \bar{\eta} \qquad (25)$$

$$\hat{\gamma} = \frac{1}{n}\sum_{i=1}^{n} \gamma_i = \bar{\gamma} \qquad (26)$$

It is important to note that the estimates of $\bar{\xi}$, $\bar{\eta}$ and $\bar{\gamma}$ have not involved estimates of σ_ξ^2, σ_η^2, or σ_γ^2. If $\bar{\xi}$, $\bar{\eta}$ and $\bar{\gamma}$ are unknown, then unbiased estimates of σ_ξ^2, σ_η^2, and σ_γ^2 can be expressed as:

$$\hat{\sigma}_\xi^2 = \frac{1}{n-1} \sum_{i=1}^{n} (\xi_i - \bar{\xi})^2 \tag{27}$$

$$\hat{\sigma}_\eta^2 = \frac{1}{n-1} \sum_{i=1}^{n} (\eta_i - \bar{\eta})^2 \tag{28}$$

$$\hat{\sigma}_\gamma^2 = \frac{1}{n-1} \sum_{i=1}^{n} (\gamma_i - \bar{\gamma})^2 \tag{29}$$

If we assume errors only in the γ or z direction (i.e., aperture flatness) and unity magnification, the variances simplify to the following expressions:

$$V[\Delta x_1]_z \cong (\frac{\lambda_s}{L} \sigma_z)^2 \tag{30}$$

$$V[\Delta r_1]_z \cong (\frac{2\lambda_s}{L^2} z\sigma_z)^2 \tag{31}$$

where $z_1 = 0$ and $\sigma_\gamma = \sigma_z$.

The standard deviations of the lateral and longitudinal resolutions can now be expressed in terms of the position errors in the "z" direction.

$$\sigma[\Delta x_1]_z = \frac{\lambda_s}{9L} \Delta z \tag{32}$$

$$\sigma[\Delta r_1]_z = \frac{2\lambda_s}{9L^2} z\Delta z \tag{33}$$

The effects of a non-flat or "lumpy" aperture on the hologram resolutions can now be investigated by plotting the coefficient of variations as a function of the longitudinal position errors for discrete object distances.

$$[CV\Delta x_1]_z = \frac{\sigma_z}{r_1} \tag{34}$$

$$[CV\Delta r_1]_z = \frac{2\sigma_z}{r_1} \tag{35}$$

Figures 2 and 3 show the resolution errors as a function of the receiver's longitudinal position errors with range as a parameter. The effects of relatively large receiver errors on the lateral or longitudinal resolutions appear to be minimal. A $50\lambda_s$ position error induces only a 5% lateral resolution error at a range of $1000\lambda_s$. The longitudinal resolution error is approximately twice the lateral or 10% for the same position error. The curves show that at ranges greater than $500\lambda_s$ the resolution errors are less than 4% for a $10\lambda_s$ error in the receiver's longitudinal position. Even if the position error is $50\lambda_s$, the induced error in the lateral or longitudinal resolution is less than 10%. This means systems with scanning receivers or towed line arrays can have large excursions in the depth position without serious degradation of the resolution. This naturally assumes the line sampling is sufficient to freeze motion during each line or series of line scans.

The curves indicate or show that if the induced errors are less than the resolution limit of the system then they should be undetectable in the reconstructed image. In other words, if one calculates the system's lateral or longitudinal resolution and then allows position errors never to exceed these limits, the degradation in the reconstructed image should be undetectable. Consider a typical example when $r_1 = 500\lambda_s$ and $L = 100\lambda_s$. The lateral and longitudinal resolutions are approximately $5\lambda_s$ and $50\lambda_s$ respectively. This means lateral position errors of $5\lambda_s$ or less and longitudinal position errors of $50\lambda_s$ or less are essentially undetectable in the reconstructed image. Much larger errors are tolerable, but with the associated loss in resolution.

If the errors in the scanning position are only in the ξ or η direction (i.e., lateral position errors) then the variances can be expressed as:

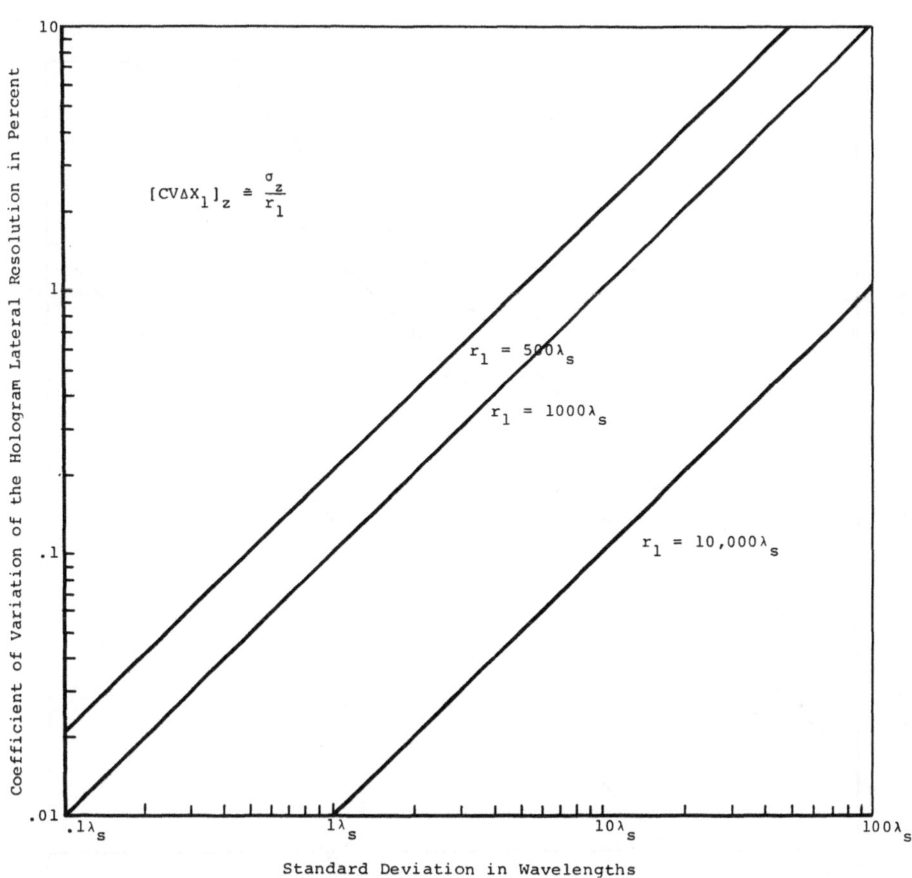

Figure 2. Percent error in the hologram lateral resolution for longitudinal position errors.

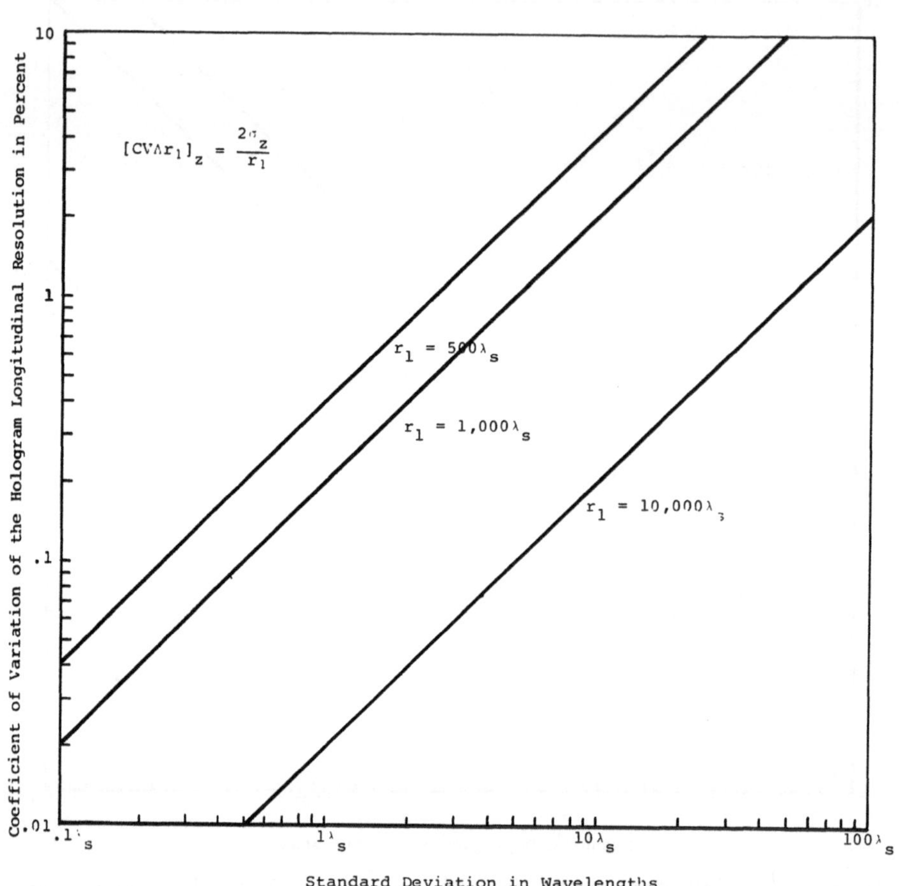

Figure 3. Percent error in the hologram longitudinal resolution for longitudinal position errors.

$$V[\Delta x_1]_{\xi,\eta} = \frac{\lambda_s^{\,2}}{L^2 r_1^{\,2}} \Big\{ m_x^{\,2}(m_x\bar{\xi}-x_1)^2\sigma_\xi^{\,2} + $$

$$+ m_y^{\,2}(m_y\bar{\eta}-y_1)^2\sigma_\eta^{\,2} \Big\} \tag{36}$$

$$V[\Delta r_1]_{\xi,\eta} = \frac{4\lambda_s^{\,2}}{L^{\,4}} \Big\{ m_x^{\,2}(m_x\bar{\xi}-x_1)^2\sigma_\xi^{\,2} + $$

$$+ m_y^{\,2}(m_y\bar{\eta}-y_1)^2\sigma_\eta^{\,2} \Big\} \tag{37}$$

If the same conditions are assumed valid as in the "γ" direction errors, then the standard deviations for the lateral and longitudinal resolutions can be expressed as:

$$\sigma[\Delta x_1]_{\xi,\eta} = \frac{\lambda_s}{L r_1} \sqrt{m_x^{\,2}(m_x\bar{\xi}-x_1)^2\sigma_\xi^{\,2} + m_y^{\,2}(m_y\bar{\xi}-y_1)^2\sigma_\eta^{\,2}} \tag{38}$$

$$\sigma[\Delta r_1]_{\xi,\eta} = \frac{2\lambda_s}{L^2} \sqrt{m_x^{\,2}(m_x\bar{\xi}-x_1)^2\sigma_\xi^{\,2} + m_y^{\,2}(m_y\bar{\xi}-y_1)^2\sigma_\eta^{\,2}} \tag{39}$$

Now we can investigate the effects of lateral or horizontal position errors on the hologram's lateral and longitudinal resolutions by plotting the coefficient of variation as a function of the horizontal position deviation σ_ξ or σ_η. If we now assume errors only in ξ direction, the coefficients of variations are:

$$[CV\Delta x_1]_\xi = \frac{m_x^{\,2}\bar{\xi}\sigma_\xi}{r_1^2} \tag{40}$$

$$[CV\Delta r_1]_\xi = \frac{2m_x^{\,2}\bar{\xi}\sigma_\xi}{r_1^{\,2}} \tag{41}$$

where $x_1 \gg m_x\bar{\xi}$

These equations for lateral position errors are only valid away from the center of the aperture where the first partial derivative in the expansion dominates the second derivative (i.e., $\frac{\partial f}{\partial \xi}\Big|_{\bar{\xi},\bar{\eta}} >> \frac{\partial^2 f}{\partial \xi}\Big|_{\bar{\xi},\bar{\eta}}$).

The region around the (ξ,η,γ) origin can now be investigated with respect to scanning errors if in the Taylor's series expansion of ξ we retain the terms containing the second derivatives. This will provide information on the rate of change of the tangent. The variance can then be expressed by the following equation:

$$V[f(\xi)]_{\xi} = E[(\frac{\partial f}{\partial \xi})^2 (\xi-\bar{\xi})^2 + (\frac{\partial^2 f}{\partial \xi^2})^2 (\xi-\bar{\xi})^4 + + 2(\frac{\partial f}{\partial \xi}\frac{\partial^2 f}{\partial \xi^2}) (\xi-\bar{\xi})^3] \quad (42)$$

The third term in the equation is zero if we assume symetrical errors (i.e., odd moments are zero). The variance can then be reduced to the following expression:

$$V[f(\xi)]_{\xi} = (\frac{\partial f}{\partial \xi})^2 \sigma_{\xi}^{2} + 3(\frac{\partial^2 f}{\partial \xi^2})^2 \sigma_{\xi}^{4} \quad (43)$$

The coefficient of variations for the lateral and longitudinal holographic resolutions can now be expressed in the following form:

$$[CV\Delta x_1]_{\xi} \cong \frac{m_x^{2}\sigma_{\xi}}{r_1^{2}} \sqrt{\bar{\xi}^2 + 3\sigma_{\xi}^{2}} \quad (44)$$

and

$$[CV\Delta r_1]_{\xi} \cong \frac{2m_x^{2}\sigma_{\xi}}{r_1^{2}} \sqrt{\bar{\xi}^2 + 3\sigma_{\xi}^{2}} \; . \quad (45)$$

Now we can investigate the effects of lateral and longitudinal receiver position errors on the holographic resolutions by plotting the coefficient of variations as shown in Figures 4 and 5. Figure 4 shows the error in percent of the lateral resolution from the expected value. The average or expected value of receiver's position is zero. The curves are valid for a small region around the center of the scanning aperture. The resolution errors are shown to decrease very rapidly as the range increases. The resolution

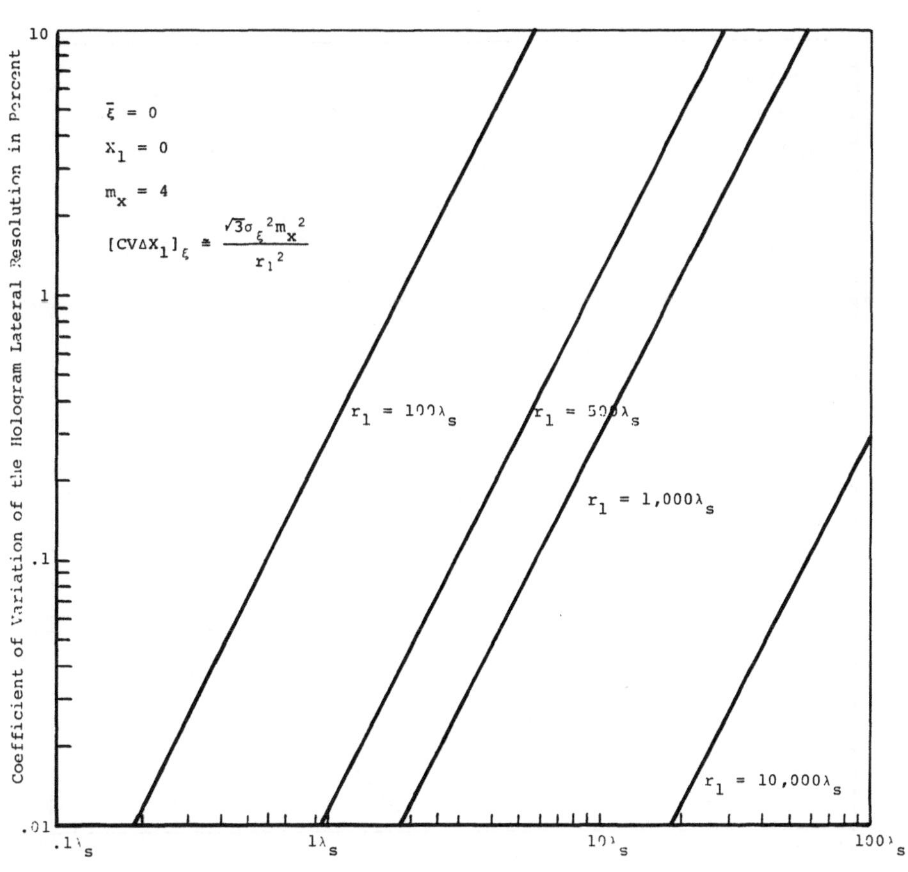

Figure 4. Percent error in the hologram lateral resolution for lateral position errors.

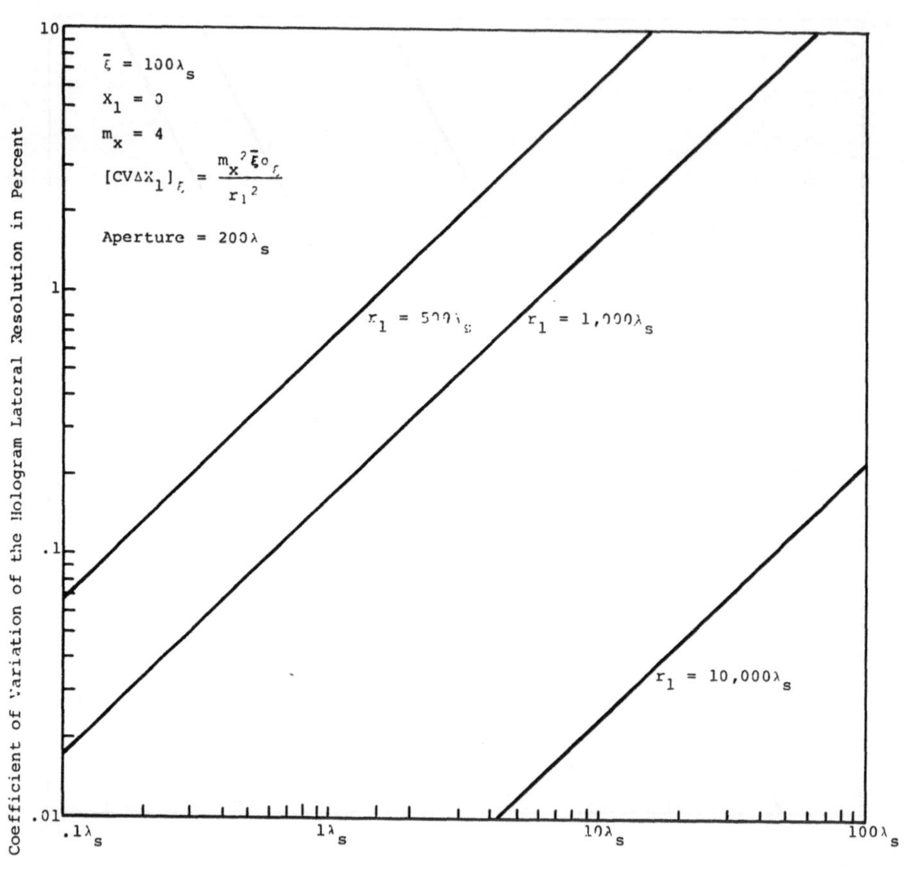

Figure 5. Percent error in the hologram lateral resolution at the aperture extremes for lateral position errors.

errors are extremely small for receiver position errors up to ten wavelengths. This means that horizontal or lateral position errors greater than a quarter of a wavelength at ranges greater than $100\lambda_s$ can be tolerated without serious degradation in resolution. This is extremely important when considering holographic imaging in the oceans with towed line arrays, etc.

Figure 5 shows the lateral resolution errors for regions away from the center of the aperture with the object located on axis. Naturally, if the object is on axis, the errors at the aperture extremes will be greater than around the center. If we consider a point source located on axis, the hologram fringe density increases away from the center. Thus, near the aperture extremes, the finge density is very high and position errors have a much greater effect on the resolution. For example, the resolution error near the center of aperture is 1% with a $10\lambda_s$ position error assuming the range is $500\lambda_s$. The error increases to greater than 6% at a distance of a $100\lambda_s$ from the aperture center.

Experimental Results

A series of holograms was constructed with simulated motion and position errors. Objects were moved in the lateral and longitudinal directions when the receiver was scanning (i.e., motion errors) and also when the objects were essentially stationary during a single or series of line scans (i.e., position errors). Position errors of this type would be encountered in a rapid sampling system moving in both the depth and lateral directions where motion is essentially frozen with respect to the object's position while scanning a single line or series of lines. If object motion is greater than $\lambda_s/4$ in the longitudinal direction during the line scan or series of line scans, the reconstructed image is severely distorted. In other words, longitudinal position errors within the resolution of the system are acceptable if motion errors are less than $\lambda_s/4$ during each line scan or series of lines. Position errors in depth as great as $75\lambda_s$ during construction of the hologram have resulted in a good reconstructed image of the object. (See Figure 9F)

Lateral motion errors, as a rule of thumb, are tolerable if they are less than the lateral resolution of the system. Slowly varying lateral position errors were simulated up to

$10\lambda_s$ without serious image degradation at a range of $500\lambda_s$. Experiments indicated that lateral motion and position errors within the resolution of the system are acceptable provided the effective motion variation in the longitudinal object distance is less than $\lambda_s/4$.

Longitudinal Errors

Figure 6 shows the various simulated transducer longitudinal scanning profiles for two point objects located 2.5 mm apart and approximately $500\lambda_s$ from the scanning aperture. Profiles 1 through 5 correspond to the reconstructions (7A through 7E). Holograms were constructed at 2.5 MHz with the object in motion for Profiles 1, 2 and 3. Figure 7A shows the reconstruction, for Profile No. 1, of the two point objects (9.4 mm in diameter). The reconstructed image represents the error free example. Figure 7B (Profile 2) shows the results of longitudinal (motion) errors approaching $\lambda_s/4$. The image is starting to show deterioration but easily recognized as the two points. Figure 7C (Profile 3) shows the image has been completely destroyed by motions approaching $\lambda s/2$. The experimental results verify conclusively that longitudinal motion errors less than $\lambda_s/4$ are acceptable in holographic imaging without serious loss of image resolution.

Figure 7D (Profile 4) and Figure 7E (Profile 5) represent simulated longitudinal transducer position errors without motion during a single line or series of line scans. The object was moved only when the receiver was motionless. The reconstructions of the two points show very conclusively that longitudinal position errors greater than $\lambda_s/4$ and up to the depth resolution of the system are acceptable without serious degradation of the image. The only restriction is simply that the sampling per line must be sufficient to freeze motion over a suitable aperture distance to provide the required resolution (i.e., Rayleigh criterion). For example, if the required lateral resolution is $5\lambda_s$ and the range $500\lambda_s$, the sampling must essentially freeze motion (i.e., $\Delta z \leq \lambda_s/4$) with respect to the object over at least a $100\lambda_s$ aperture. This means that a line array could be sampled electronically very fast and move between line samples in the depth direction and construct reasonably good holographic images. It also means matrix arrays could be constructed without severe flatness requirements between individual lines or sections.

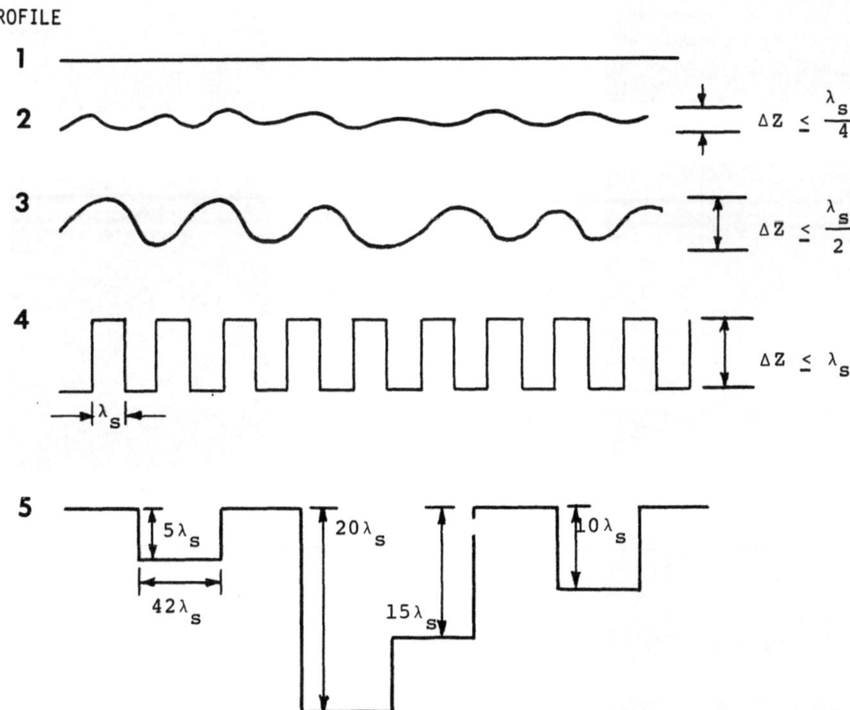

Figure 6. Transducer scanning profile simulating longitudinal position errors -- imaging two point objects.

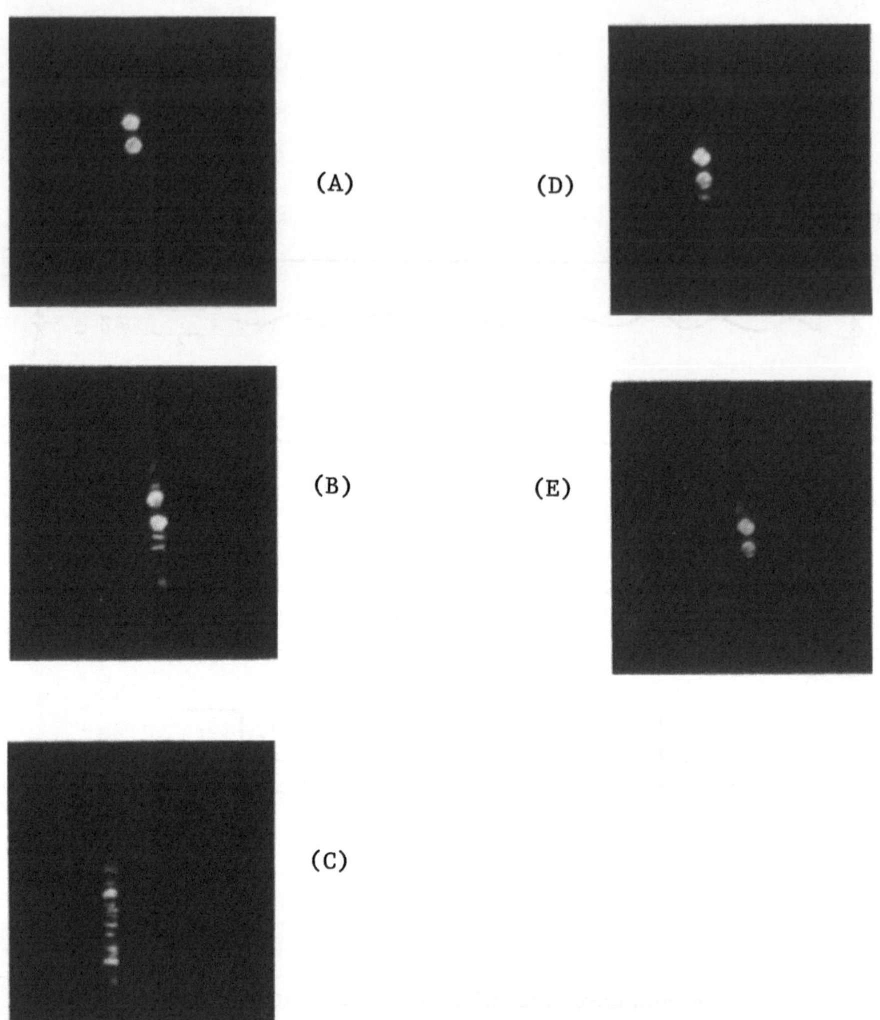

Figure 7. (A) Reconstruction of two points without position errors. (B) Reconstruction with longitudinal motion errors less than $\lambda_s/4$. (C) Reconstruction with longitudinal motion errors less than $\lambda_s/2$. (D) Reconstruction with longitudinal position errors (no motion) of approximately λ_s. (E) Reconstruction with severe discrete step errors less than or equal to $20\lambda_s$.

Figure 8 shows the simulated transducer longitudinal scanning profiles for the letter "F". Holograms were constructed at 5.1 MHz with the object stationary (i.e., the object was moved only when the receiver was motionless). Figure 9A is the reconstruction for Profile #1 of the letter "F" free of position errors. Figures 9B, 9C, 9D, 9E and 9F represent the reconstructions associated with the longitudinal position error Profiles 2, 3, 4, 5 and 6 respectively. The reconstructed images show very slight deterioration and excellent quality considering the extremely non-flat scanning aperture.

Lateral Errors

Figure 10 shows the simulated lateral scanning profiles for the two point objects located 2.5 mm apart and $500\lambda_s$ from the scanning aperture. Profiles 1 through 6 correspond to the reconstructions (11A through 11F). Holograms were constructed at 2.5 MHz with the object in motion while the receiver was scanning. Figure 11A shows the reconstruction of the two points associated with Profile 1 (i.e., error-free example). Figure 11B is the reconstruction of the two points where the object's lateral motion excursions were less than or equal to $\lambda_s/4$. The two points are still easily resolvable in the reconstruction. The lateral motion was then increased to have maximum excursions of λ_s and the reconstruction is shown in Figure 11C. The two points are still resolvable indicating the resolution is acceptable. The lateral motions were then increased to have $5\lambda_s$ excursions and the two points are again completely resolvable in the reconstruction (Figure 11D). Finally, lateral motions of $10\lambda_s$ were imposed and still the two points (see Figure 11E) were resolvable but there is fading present. Figure 11F is the reconstruction showing the two points are unresolvable with lateral motion exceeding $10\lambda_s$ but not greater than $20\lambda_s$.

Now, if one calculates the change in the longitudinal distance (i.e., distance from the receiver to the object) with respect to the lateral change ($20\lambda_s$), the result is approximately $\lambda_s/2$. This indicates longitudinal motions (i.e., Doppler shift) must be restricted to less than a half wavelength in the hologram construction. The $10\lambda_s$ variation exceed the lateral theoretical resolution by a factor of approximately three, but not the longitudinal motion restriction. Thus, if the lateral motion excursions are kept

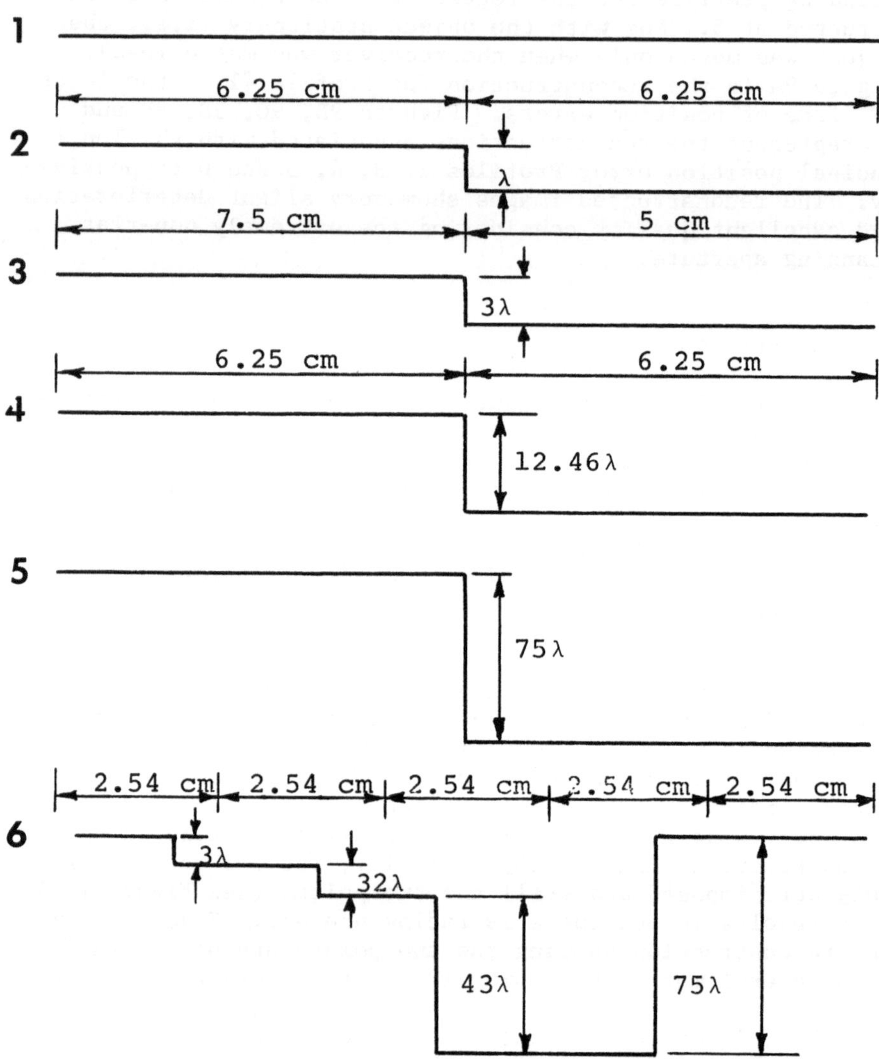

Figure 8. Scanning transducer profiles simulating longitudinal position errors imaging the letter "F."

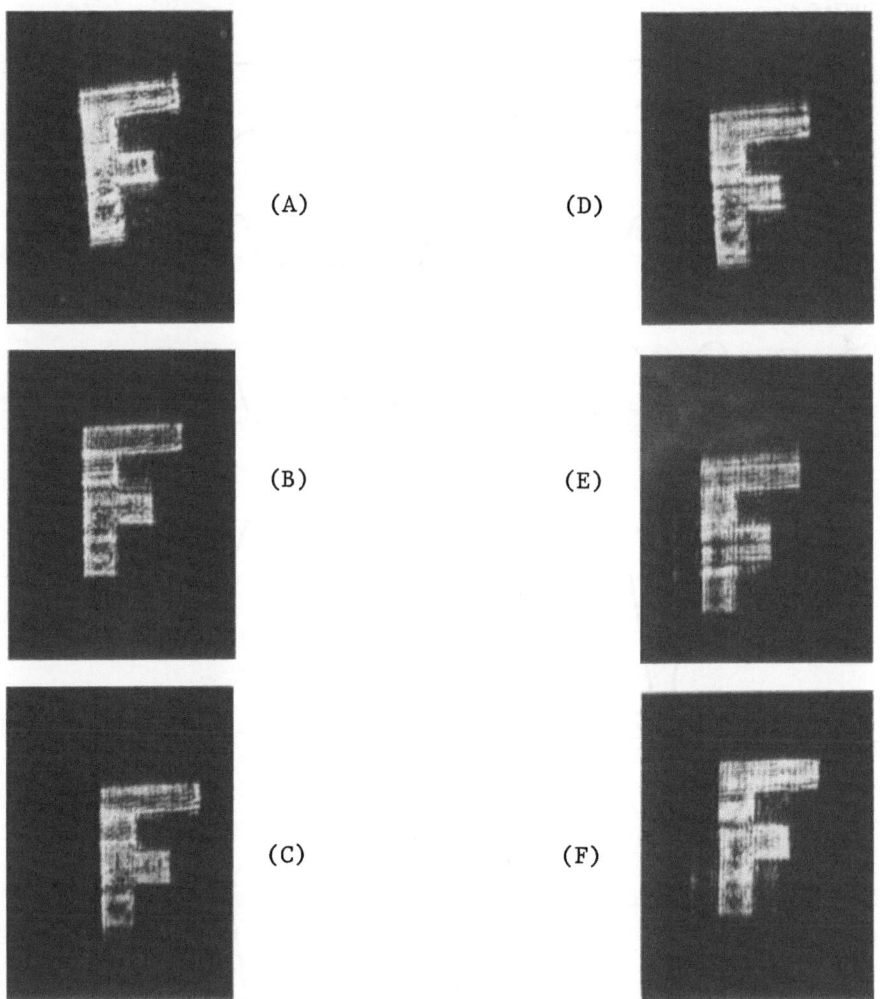

Figure 9. (A) Reconstruction of the letter "F" without longitudinal position errors. (B) Reconstruction with a step longitudinal position error of λ_s. (C) Reconstruction with a step longitudinal position error of $3\lambda_s$. (D) Reconstruction with a step longitudinal position error of $12.46\lambda_s$. (E) Reconstruction with a step longitudinal position error of $75\lambda_s$. (F) Reconstruction with three discrete step longitudinal position errors varying from $3\lambda_s$ to $75\lambda_s$.

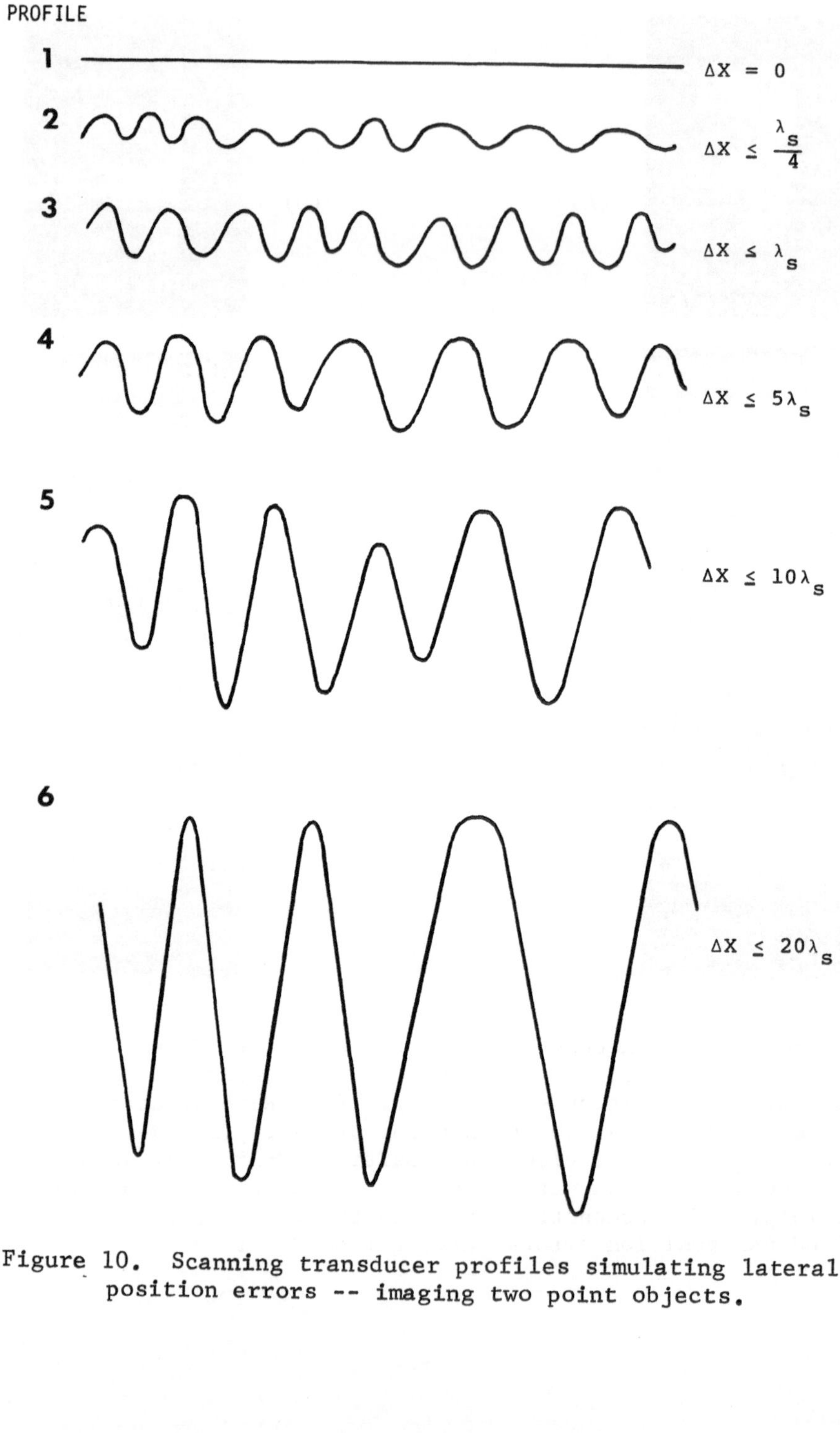

Figure 10. Scanning transducer profiles simulating lateral position errors -- imaging two point objects.

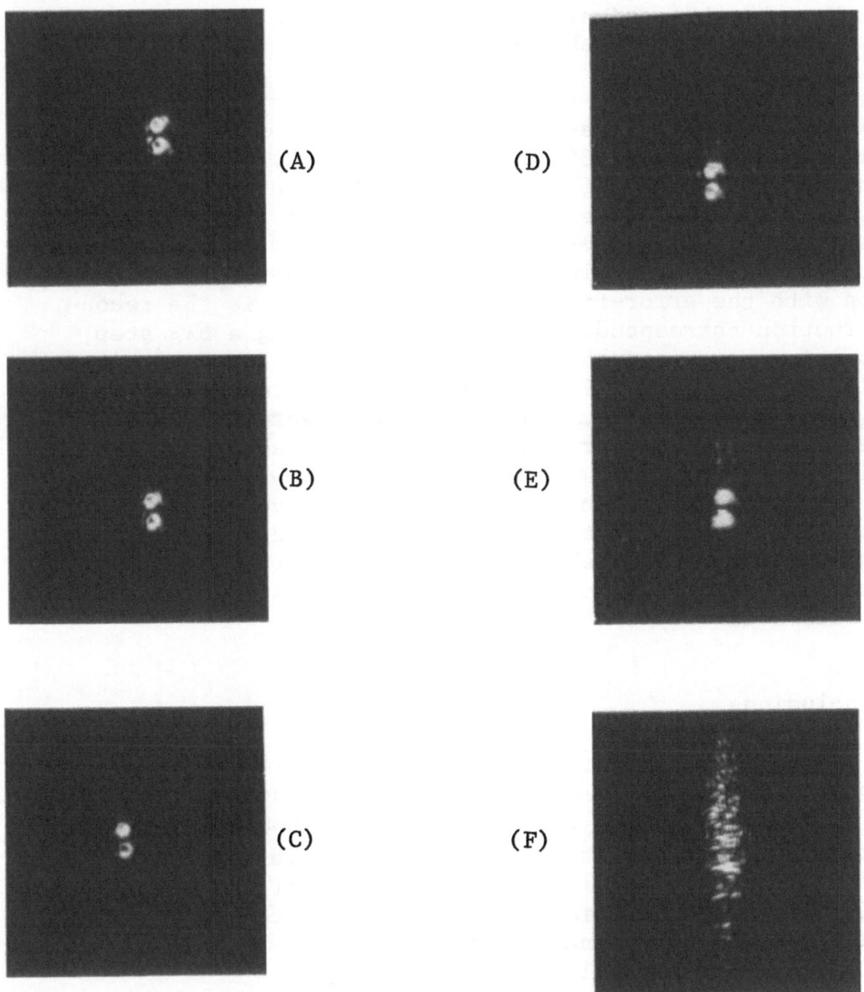

Figure 11. (A) Reconstruction of the two points without lateral position errors. (B) Reconstruction with lateral motion errors less than $\lambda_s/4$. (C) Reconstruction with lateral motion errors less than λ_s. (D) Reconstruction with lateral motion errors less than $5\lambda_s$. (E) Reconstruction with lateral motion errors less than $10\lambda_s$. (F) Reconstruction with lateral motion errors less than $10\lambda_s$. (F) Reconstruction with lateral motion errors less than $20\lambda_s$.

within the theoretical lateral holographic resolution, the reconstructed image should be acceptable.

Figure 12 shows the transducer's lateral scanning profiles (1 through 7) associated with the hologram reconstructions (13A through 13G) of the letter "H". Figure 13A is the reconstruction of the letter "H" associated with Profile 1. Figure 13B is the reconstruction with a discrete step lateral position error of $\lambda_s/2$. The image is of excellent quality and the resolution has not been degraded in comparison with the error-free image. Figure 13C is the reconstruction corresponding to Profile 3 having a $6\lambda_s$ step position error and the image shows slight deterioration. Figures 13D, 13E, 13F and 13G correspond to Profiles 4, 5, 6 and 7 respectively. The images show varying degrees of deterioration, but are easily identified as the letter "H". Figure 13G represents the reconstruction with lateral motion errors up to $9\lambda_s$. The image is still very acceptable in quality compared with the error-free image. The following images compare favorably in resolution with the two point examples having typically the same excursions or lateral position errors.

Conclusions

Longitudinal motions greater than $\lambda_s/4$ (i.e., Doppler) between the receiver and the object void holographic imaging. If the sampling rate is sufficient to essentially freeze motions over a series of line scans, deviations approaching the holographic longitudinal resolution are tolerable without serious degradation of the reconstructed image. The minimum effective line scan width must be long enough to insure adequate lateral resolution.

Lateral motion errors are tolerable if they remain less than or equal to the lateral holographic resolution.

Using this criteria, holograms can be constructed of objects in the projected scanning aperture without serious degradation of resolution in the reconstruction.

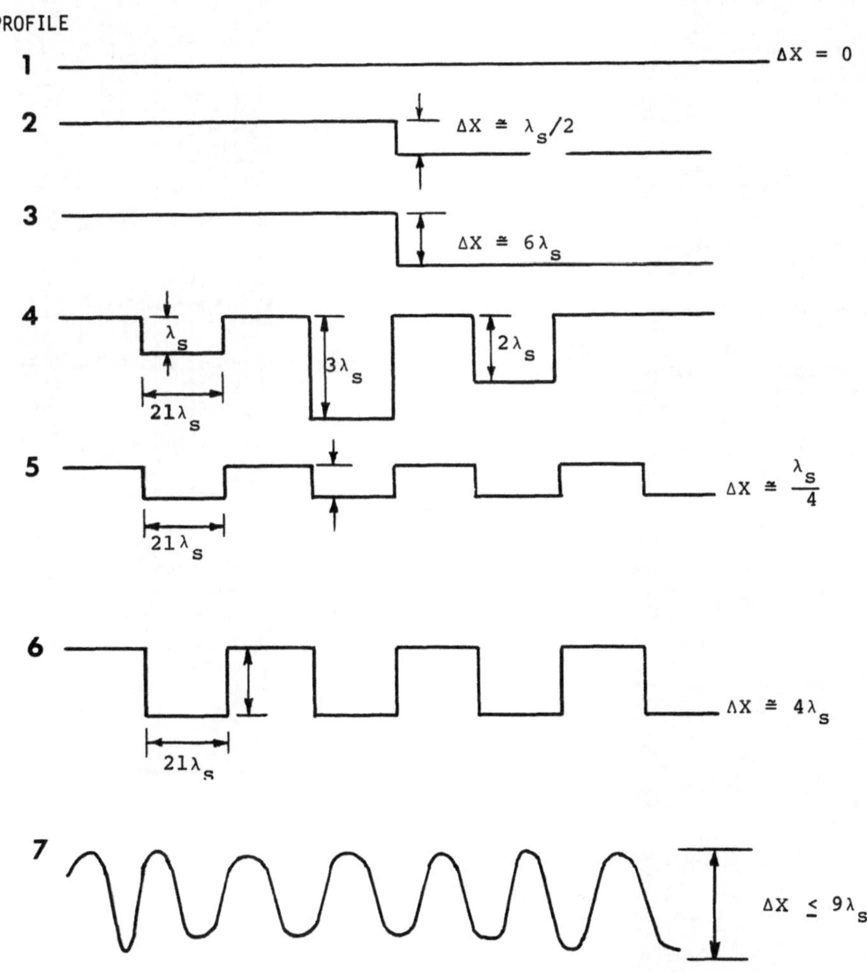

Figure 12. Scanning transducer profiles simulating lateral position errors -- imaging the letter "H."

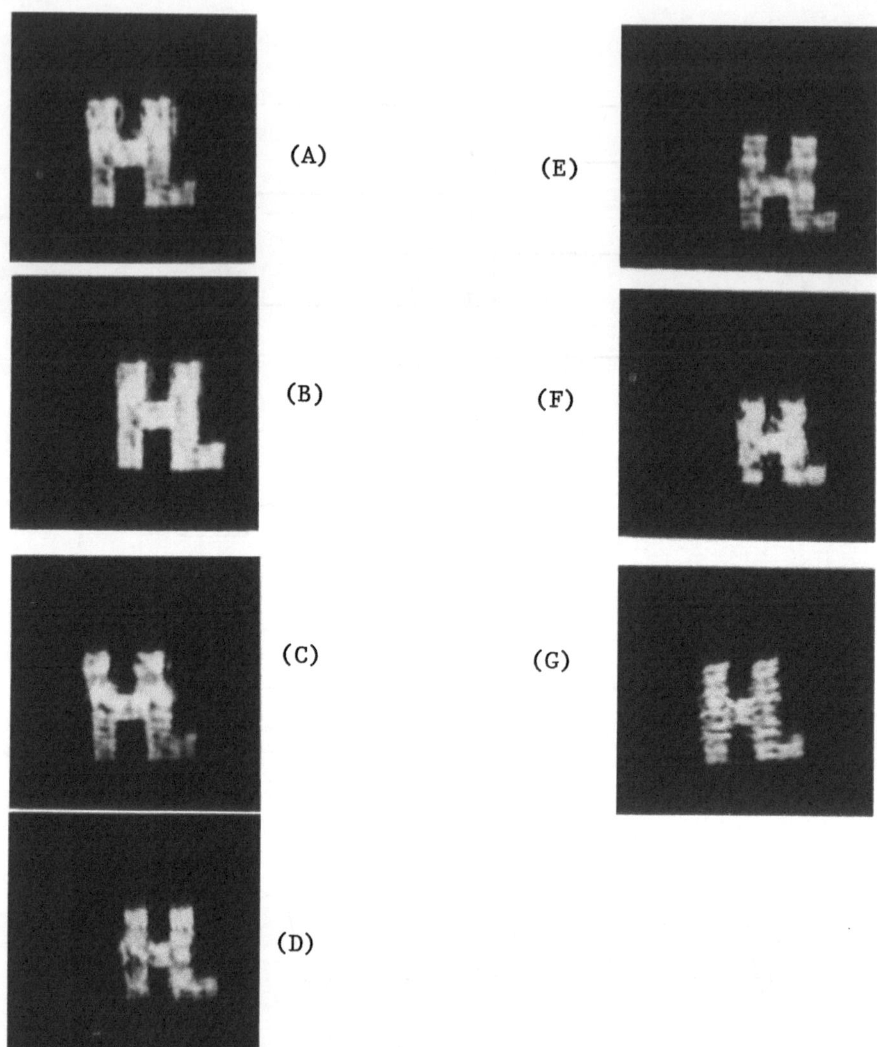

Figure 13. (A) Reconstruction of the letter "H" without lateral position errors. (B) Reconstruction with a step lateral position error of $\lambda_s/2$. (C) Reconstruction with a step lateral position error of $6\lambda_s$. (D) Reconstruction with three step lateral position errors less than $3\lambda_s$. (E) Reconstruction with fourteen step lateral position errors approximately $\lambda_s/4$. (F) Reconstruction with fourteen step lateral position errors approximately $4\lambda_s$. (G) Reconstruction with motion errors of less than or equal to $9\lambda_s$.

References

1. S. C. Daubin, "System Requirements for Underwater Holographic Mapping", IEEE Trans. GeoSci. Electron Vol. GE-8, Oct., 1970, pp. 313-320.

2. B. P. Hildebrand and K. Haines, "Holography by Scanning" Journal of the Optical Society of America, Vol. 59, 1969, pp. 1-6.

3. B. P. Hildebrand and B. B. Brenden, "Introduction to Acoustical Holography", BNWL-SA-3467, Battelle, Richland, Washington, 1971 (book in press).

4. Carl Bennett and Norman Franklin, "Statistical Analysis in Chemistry and the Chemical Industry", Second Ed., New York, Wiley, 1960, p. 724.

5. K. A. Brownlee, "Statistical Theory and Methodology", Second Ed., New York, Wiley, 1960, p. 590.

APPENDIX A

The Law of Propagation of Errors

The exact calculation of the variance of a nonlinear function containing several variables that are subject to error is generally a problem of considerable mathematical complexity. Generally, approximations are used and it is not necessary to solve these difficult problems exactly. There exists a process called linearization that allows the replacement of any nonlinear function with a linear one, for the purpose of obtaining approximate estimates of the variances. The approximation is usually quite adequate for most applications. Linearization is based on the Taylor's series expansion of the nonlinear function with retention of only the linear portion of the expansion.

Let us consider first a function of a single random variable:

$$M = f(X)$$

For example, X might represent the simulated scanning velocity of the receiver and M its magnification. We are interested in the random error of M as a result of random errors in X.

To an error ε in X, corresponds an error δ in M, given by:

$$\delta = f(X + \varepsilon) - f(X) \ . \tag{A-1}$$

If we assume ε to be small with respect to X*, it can be treated as a differential increment. Then the following approximation is assumed valid.

$$\frac{f(X + \varepsilon) - f(X)}{\varepsilon} \approx \frac{df(X)}{dX} \tag{A-2}$$

and

---

* In practice, if $\sigma\varepsilon$ is of the order of 10% of X or smaller, the law can be reliably used.

$$\delta = \frac{df(X)}{dX}\,\varepsilon = \frac{dM}{dX}\,\varepsilon \qquad \text{(A-3)}$$

where the derivative of M with respect to X is evaluated at the measured or average value of X. The variance is given by the following expression:

$$V(\delta) = \left(\frac{dM}{dX}\right)^2 V(\varepsilon) \qquad \text{(A-4)}$$

where $\frac{dM}{dX}$ is a constant.

For example, consider the lateral hologram magnification:

$$M_L(x)ss = \pm \frac{\lambda_L}{\lambda_S}\frac{V_x}{V_\xi}\frac{r_b}{r_1} \quad . \qquad \text{(A-5)}$$

The derivative with respect to the velocity V_ξ is:

$$M_L(x)ss = \pm \frac{\lambda_L}{\lambda_S}\frac{V_x}{V_\xi^2}\frac{r_b}{r_1} \qquad \text{(A-6)}$$

and the variance

$$V(\delta) = \left(\frac{\lambda_L}{\lambda_S}\frac{V_x}{\bar{V}_\xi^2}\frac{r_b}{r_1}\right)^2 V(\varepsilon_\xi) \qquad \text{(A-7)}$$

where ε is the random error in the velocity simulation, and δ the random error, induced by ε, in the magnification. If $\varepsilon_V \sim N(0,\sigma_\varepsilon^2)$, then the standard deviation is:

$$\sigma\delta = \frac{\lambda_L}{\lambda_S}\frac{V_x}{V_\xi^2}\frac{r_b}{r_1}\,\sigma_\varepsilon \quad . \qquad \text{(A-8)}$$

Expressing the error as a coefficient of variation, we have

$$\%\ C\ V_{M_L} = \frac{\sigma_\varepsilon}{\bar{V}_\xi} \times 100 \quad . \qquad \text{(A-9)}$$

Equation (A-4) expresses the law of propagation of errors for the case of a single independent variable. Its proof is based on the assumption that the error ε is small with respect to the measured value of x (i.e., $\sigma_\varepsilon < 0.1$ of x).

The law of propagation of errors for the case of several random variables is just an extension of Eq. (A-4). Assume:

$$u = f(x,y,z, \ldots) \tag{A-10}$$

where x,y,z . . . represent random variables. Let ε_1, ε_2, ε_3, . . . represent statistically independent errors of x, y,z . . ., respectively. The error induced in u as a result of errors ε_1, ε_2, ε_3, . . . has a variance of

$$V(\delta) = \left(\frac{\partial f}{\partial x}\right)^2 V(\varepsilon_1) + \left(\frac{\partial f}{\partial y}\right)^2 V(\varepsilon_2) + \ldots \tag{A-11}$$

The partial derivations are evaluated at values equal or close to the measured values of (x,y,z, . . .). If $\varepsilon_1 \sim N(0,\sigma_x^2)$, $\varepsilon_2 \sim N(0, \sigma_y^2)$, etc., then Eq. (A-11) can be

$$V(\delta) = \left(\frac{\partial f}{\partial x}\right)^2 \sigma_x^2 + \left(\frac{\partial f}{\partial y}\right)^2 \sigma_y^2 \ldots \tag{A-12}$$

APPENDIX B

Approximate Average Value and Variance of an Arbitrary Function

To derive the exact mean (i.e., average value) and variance of an arbitrary function containing several random variables that are subject to error is generally a problem of considerable mathematical complexity. Generally, approximations are used and it is not necessary to solve these difficult problems exactly. Usually the functions involved vary slowly in the region where values of the independent variables remain within one or two standard deviations of their mean and the function can adequately be represented by the linear terms of its Taylor series expansion. This analysis extends the conventional law of propagation of errors theory which assumes the errors are small so they can be treated as a differential. If the functions vary slowly in the assumed region or if the errors are small, for the approximation to be valid either the second partial derivatives evaluated at the mean values or the deviation of the random variables from their respective means can be small.

Let $U = f(\xi,\eta)$, where ξ and η are random variables with mean values $\bar{\xi}$ and $\bar{\eta}$ and variances $\sigma_\xi{}^2$, and $\sigma_\eta{}^2$ and a convariance $\sigma_{\xi\eta}$. If we use the following statistical models

$$\xi = \bar{\xi} + e_\xi \text{ and}$$

$$\eta = \bar{\eta} + e_\eta$$

where the random induced errors e_ξ and e_η are assumed distributed, $N(0,\sigma_\xi{}^2)$ and $N(0,\sigma_\eta{}^2)$. If we expand U about the point $\bar{\xi}$ and $\bar{\eta}$ in a Taylor series, then

$$U \equiv f(\bar{\xi},\bar{\eta}) + \left|\frac{\partial f}{\partial \xi}\right|_{\bar{\xi},\bar{\eta}} (\xi-\bar{\xi}) + \left|\frac{\partial f}{\partial \eta}\right|_{\bar{\xi},\bar{\eta}} (\eta-\bar{\eta}) + \frac{1}{2}\left|\frac{\partial^2 f}{\partial \xi^2}\right|_{\bar{\xi},\bar{\eta}} (\xi-\bar{\xi})^2 +$$

$$+ \frac{1}{2}\left|\frac{\partial^2 f}{\partial \eta^2}\right|_{\bar{\xi},\bar{\eta}} (\eta-\bar{\eta})^2 + \left.\frac{\partial^2 f}{\partial \varepsilon \partial \eta}\right|_{\bar{\xi},\bar{\eta}} (\varepsilon-\bar{\varepsilon})(\eta-\bar{\eta}) + \ldots \qquad \text{(B-1)}$$

Where the partial derivatives with respect to ξ and η are evaluated at the point $(\bar{\xi},\bar{\eta})$.

The approximate expected or mean value of U is given by:

$$E[U] \cong f(\bar{\xi},\bar{\eta}) + \frac{1}{2}\left|\frac{\partial^2 f}{\partial \xi^2}\bigg|_{\bar{\xi}} \sigma_\xi^{\ 2} + \frac{\partial^2 f}{\partial \eta^2}\bigg|_{\bar{\eta}} \sigma_\eta^{\ 2}\right| +$$

$$+ \frac{2\partial^2 f}{\partial \varepsilon \partial \eta}\bigg|_{\bar{\xi},\bar{\eta}} \sigma_{\varepsilon\eta}^{\ 2} \Bigg| \quad \text{(B-2)}$$

where only the first few terms of the expansion are retained. The second term of Eq. (B-2) is referred to as the bias. The approximate variance of U is given by the following expression where only the linear terms of the expansion have been retained and since the variance of a constant $f(\bar{\xi},\bar{\eta})$ is zero:

$$V[U] \cong \left|\left(\frac{\partial f}{\partial \xi}\right)^2\right|_{\bar{\eta},\bar{\xi}} \sigma_\xi^{\ 2} + \left|\left(\frac{\partial f}{\partial \eta}\right)^2\right|_{\bar{\xi},\bar{\eta}} \sigma_\eta^{\ 2} + 2\left|\frac{\partial^2 f}{\partial \xi \partial \eta}\right|_{\bar{\xi},\bar{\eta}} \sigma^2_{\xi\eta} \quad \text{(B-3)}$$

If ξ and η are independent random variables, then the variance reduces to the following expression:

$$V[U] \cong \left|\left(\frac{\partial f}{\partial \eta}\right)^2\right|_{\bar{\xi},\bar{\eta}} \sigma_\xi^{\ 2} + \left|\left(\frac{\partial f}{\partial \eta}\right)^2\right|_{\bar{\xi},\bar{\eta}} \sigma_\eta^{\ 2} \quad . \quad \text{(B-4)}$$

The above results may be extended to a function of n independent random variables (i.e., $u = f(\xi_1,\xi_2 \ldots \xi_n)$. If $E[\xi i] = \bar{\xi}_i$, $V[\xi_i] = \sigma_i^2$, we have the following approximation, assuming that all the derivatives exist:

$$E[U] \cong f(\bar{\xi}_1,\bar{\xi}_2 \ldots \bar{\xi}_n) + \frac{1}{2} \sum_{i=1}^{n} \frac{\partial^2 f}{\partial \xi_i} \sigma_i^{\ 2} \quad \text{(B-5)}$$

$$V[U] \cong \sum_{i=1}^{n} \left(\frac{\partial f}{\partial \xi_i}\right)^2 \sigma_i^{\ 2} \quad . \quad \text{(B-6)}$$

where all the partial derivatives are evaluated at the point $(\bar{\xi}_1,\bar{\xi}_2 \ldots \bar{\xi}_n)$.

SPHERICAL ABERRATION IN LONG WAVELENGTH HOLOGRAPHY

M. D. Fox, F.L. Thurstone, O.T. von Ramm

Department of Biomedical Engineering

Duke University, Durham, North Carolina

1. INTRODUCTION

Images reconstructed from long wavelength holograms have aberrations as a result of large construction to reconstruction wavelength ratios. For many practical recording geometries, these aberrations of which spherical aberration is often the largest may seriously degrade image resolution.

A number of investigators[1-3] have analyzed aberrations in holograms using a classical wave theory approach. Since these analyses employ a binomial expansion for the phase factor, they become invalid as object distances approach hologram aperture diameters. Others[4-6] have used a ray optics model to determine the effects of aberration in both the near and far fields for conventional optical holograms.

Previous efforts to develop an aberration correction scheme have been based on the use of appropriate geometry to cause the spherical aberration to balance astigmatism, thus allowing total aberrations to be minimized.[7,8] These techniques are valid only for off axis holograms with an appropriate wavelength shift. Furthermore, they generally reduce rather than eliminate aberrations, require a complicated analysis of each new hologram configuration to determine the optimum viewing geometry, and are applicable only for restricted values of hologram size and object position.

In the present study, a ray optics model was used to determine the effect of aberrations on hologram resolution for a number of practical geometries. This investigation showed that spherical aberration becomes the limiting factor in resolution for small object distance to aperture diameter ratios. The analysis is carried further to ascertain the possibility of reducing or eliminating aberrations. Two possible methods of correcting aberrations in acoustic holograms are computer generation of a correction hologram and direct insertion of the correction in the hologram formation process. In either case, an unaberrated image will be produced at a single point, and spherical aberration will be eliminated over a plane. Using computer synthesized holograms, these techniques were tested and found to produce significantly improved holographic reconstructions.

2. RAY ANALYSIS OF ABERRATIONS

Elements of the ray tracing approach that we will use here were presented by Sherwood.[9] The method is distinct from other ray tracing analyses due to the utilization of simplifications made possible by limiting consideration to long wavelength holography. To avoid the confusion that would result from attempting to show the aberrations in all possible configurations, we have restricted our consideration here to a few representative imaging geometries.

We shall begin our derivation by examining the holographic formation process illustrated in Figure 1. Assuming a plane wave reference beam the fringe spacing will be given by

$$\delta_1 = \lambda_0/\sin\alpha_1 \tag{1}$$

where λ_o is the wavelength of the object signal. Directing our attention to the reconstruction process, Figure 2, it is apparent that if the reconstructing beam is a plane wave we have

$$\delta_2 = \lambda_c/\sin\alpha_2 \tag{2}$$

Now if the hologram was magnified by a factor m between formation and readout, $\delta_2 = m\,\delta_1$ and from (1) and (2)

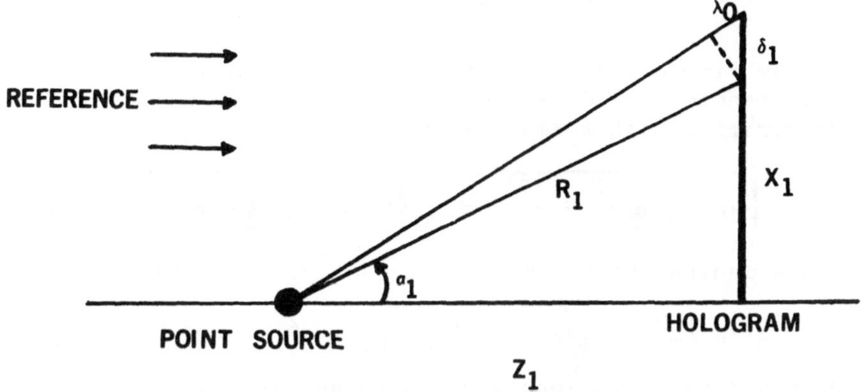

Fig. 1. Hologram formation geometry.

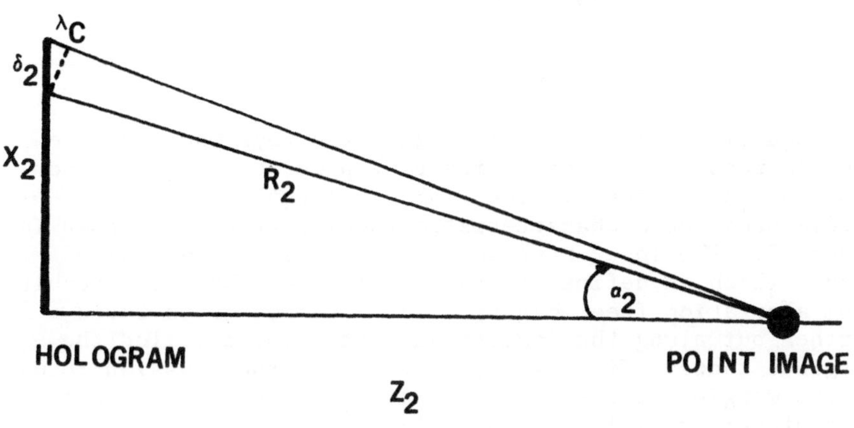

Fig. 2. Hologram reconstruction geometry.

$$\lambda_0/\sin\alpha_1 = \lambda_c/m\sin\alpha_2 \tag{3}$$

expressing $\sin\alpha_1$ and $\sin\alpha_2$ in terms of the object and image distances z_1 and z_2 and the distance from the axis, x_1, recognizing that $x_2=mx_1$ we arrive at

$$\left\{\lambda_0/x_1\right\}\sqrt{x_1^2+z_1^2} = \left\{\lambda_c/m^2x_1\right\}\sqrt{m^2x_1^2+z_2^2} \tag{4}$$

With some manipulation, this can be arranged to give

$$z_2 = \left\{m^2/\mu\right\}\sqrt{(1-\mu^2/m^2)x_1^2+z_1^2} \tag{5}$$

where $\mu = \lambda_c/\lambda_o$. In long wavelength holography μ is often 1/500 or less, so the condition $\mu^2/m^2 \ll 1$ is typically satisfied and we have

$$z_2 = \left\{m^2/\mu\right\}\sqrt{x_1^2+z_1^2} \quad . \tag{6}$$

It is apparent from this expression that spherical aberration is inherent in long wavelength holography since the position of the image point, z_2, is dependent on the lateral position in the hologram, x_1.

Spherical Aberration

Equation 6 can be used to generate rays whose deflection corresponds to the diffracting power of different portions of the hologram. The long wavelength hologram ray bundle produces a characteristic focal pattern, as shown in Figure 3. The inner rays come together at the Gaussian focus, which is defined as the focal point that will result from aberration free imagery. The outermost rays focus further out along the imaging axis at a point called the marginal focus. If a card is moved from the Gaussian focus to the marginal focus, at some point in between, a round spot of uniform illumination will be seen. This spot is called the circle of least confusion and its size can be determined graphically by drawing rays and noting the point at which the resulting ray bundle reaches its minimum. In the following studies, the circle of least confusion was determined by a digital computer program which essentially imitated this graphical procedure mathematically.

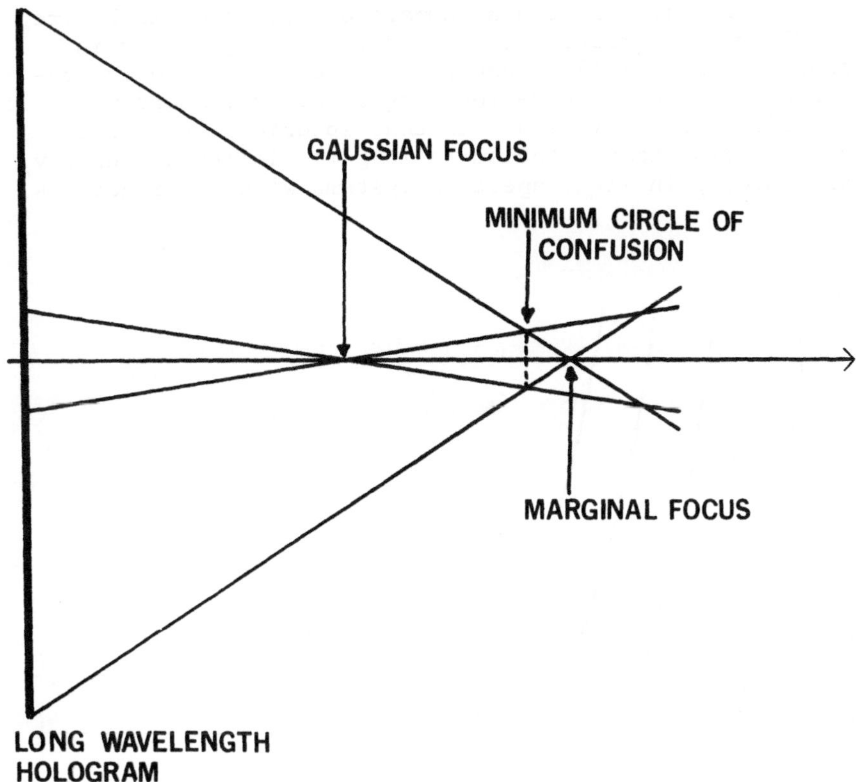

Fig. 3. Spherical aberration in long wavelength holography.

The minimum spot size due to aberrations, once calculated, can be compared to the spot size imposed by diffraction, i.e., the diameter of the first lobe of the Airy disc, to determine which of these factors is the primary limitation on resolution. It must be cautioned that this comparison can only yield general guidelines due to the different origins of these limitations on spot size, but it should clearly delineate the regions in which aberration or diffraction becomes the dominant factor in determining resolution.

The results of such a comparison are shown in Figure 4. Clearly, aberration becomes the limiting factor in resolution for small f numbers. In addition, the aberration becomes more severe for larger hologram apertures. It is evident from this analysis that spherical aberration poses a significant problem in long wavelength holography, particularly in large aperture systems with low f numbers.

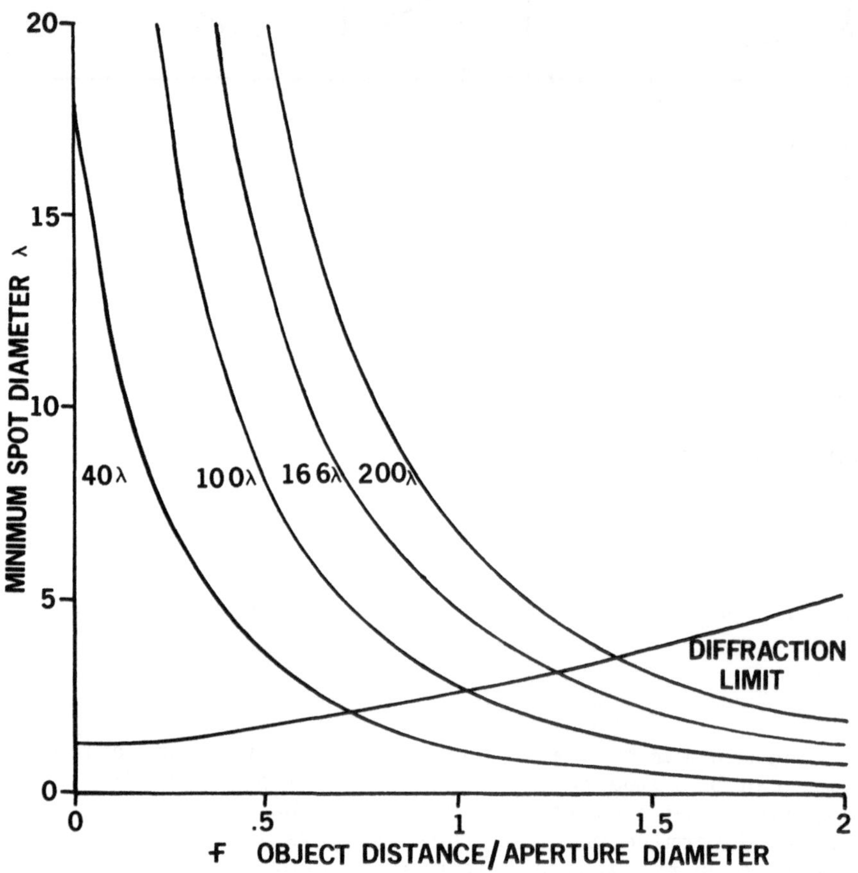

Fig. 4. Diffraction and aberration limited spot diameter versus hologram f for long wavelength holograms. The aberration limited spot diameters are given for various hologram aperture diameters.

Correction Volume

Although correction techniques are usually aimed at eliminating aberrations at a single point, a factor that has considerable bearing on the practical usefulness of a correction scheme is the volume around the corrected point which will also be significantly corrected. It is possible, using ray tracing techniques, to assume that the fringe spacings have been altered to correct the hologram for a particular image point, and to examine the effect of aberrations on neighboring points in the image space.

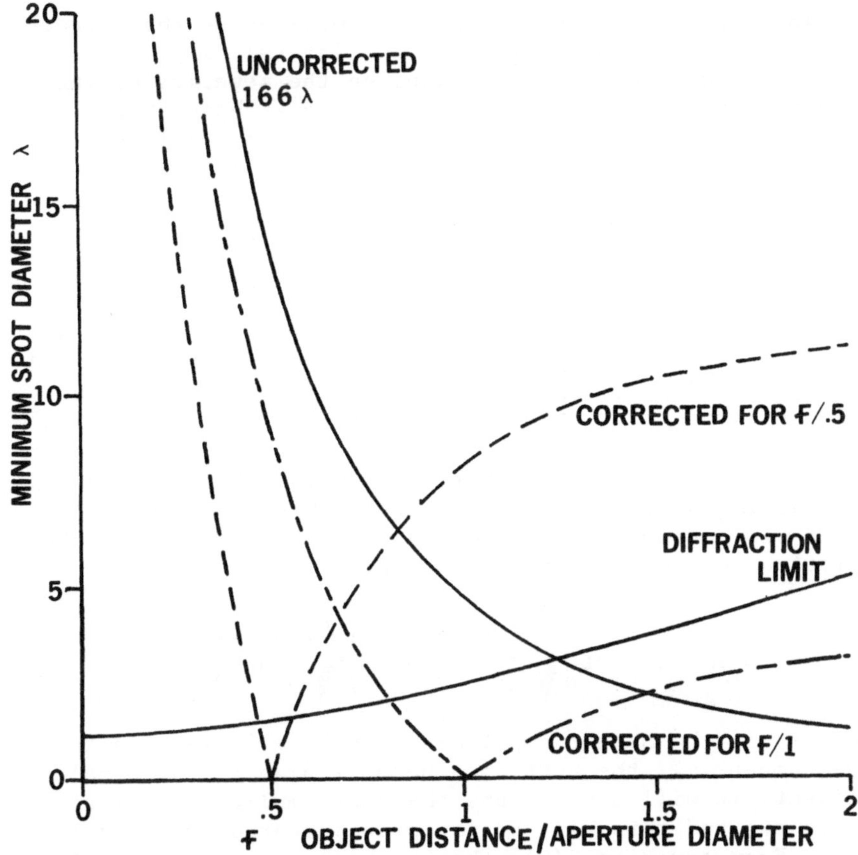

Fig. 5. Effect of correcting long wavelength holograms on the aberration limited minimum spot diameter. All curves are drawn for 166 λ hologram aperture diameter.

Figure 5 shows the results of such a calculation for a typical acoustic hologram geometry. It is apparent that if the aberration is very large at the point to be corrected, the useful correction will be limited to a relatively small volume. If the aberration is small, on the other hand, the effectively corrected volume may be large. For example, the curve in Figure 5 for the hologram corrected for f/1 suggests that the crossover point between diffraction and aberration limited imaging can be moved in from f/1.2 to f/.8 by the application of a correction for a point at f/1, with no resolution loss at higher f numbers.

Another quantity of interest in evaluating the correction volume is the correction in the lateral image dimension. This will once again depend on the severity of the aberration to be corrected, but our studies have shown that the aberrations can be substantially reduced for image points laterally adjacent to the corrected image point.

3. WAVE ANALYSIS OF ABERRATIONS

To implement a correction scheme, it is necessary to determine the phase error at each point in the hologram and to inject a compensatory phase correction. To avoid excessive mathematical complexity, we will show in this section how the phase error can be determined for a point source on axis, with plane wave, normally incident, reference and reconstructing beams. Referring back to Figure 1, the point object will produce a spherical wave which will have spatial variation in the hologram plane given by

$$\overline{E}_o = A_o \exp jk_o \sqrt{x_1^2 + y_1^2 + z_1^2} \Big/ \sqrt{x_1^2 + y_1^2 + z_1^2} \qquad (1)$$

where A_o is the object amplitude, and $k_o = 2\pi/\lambda_o$ is the wave number. If the reference beam is planar and normally incident, it will not add any phase or amplitude variation of its own and this will be the signal stored in the hologram. From equation 2(6), the Gaussian or unaberrated image point will be located at $z_2 = (m^2/\mu)z_1$ assuming magnification m and wavelength ratio μ. Thus to produce a perfect image in the reconstruction process, we would need the following wavefront in the hologram exit plane

$$\overline{E}_i = A_o \exp\left\{ jk_o (m/\mu) \sqrt{x_1^2+y_1^2+(mz_1/\mu)^2} \right\} \Big/ m \sqrt{x_1^2+y_1^2+(mz_1/\mu)^2} \quad (2)$$

To convert (1) into (2) it is necessary to inject a phase correction of

$$\phi_\Delta = (mk_o/\mu) \sqrt{x_1^2+y_1^2+(mz_1/\mu)^2} - k_o \sqrt{x_1^2+y_1^2+z_1^2} \quad (3)$$

and to multiply (1) by an amplitude correction

$$A_\Delta = \sqrt{x_1^2+y_1^2+z_1^2} \Big/ m \sqrt{x_1^2+y_1^2+(mz_1/\mu)^2} \; . \quad (4)$$

For off-axis object points at position (x_o, y_o) the correction factors can be easily modified by replacing x_1 by (x_1-x_o) and y_1 by (y_1-y_o) in (3) and (4). This analysis can be generalized to apply to arbitrary hologram geometries.

By injecting the phase and amplitude corrections suggested, the correction volumes discussed in the previous section can be achieved. In the next section we will investigate the various means by which such corrections might be introduced into the holographic process.

4. STUDY OF ABERRATIONS USING COMPUTER SYNTHESIZED HOLOGRAMS

In order to test aberration correction schemes, it is necessary to have high quality long wavelength holograms in which spurious sources of image degradation are held to a minimum. For this reason, it was decided to use computer simulation to obtain high quality acoustic holograms.

Hologram synthesis techniques have been investigated by Lesem, Hirsh, and Jordan[10] for general geometries and by Farr[11] in long wavelength holography. These studies have established the feasibility of computer generated holograms and their utility in simulating real holograms for the purpose of studying the holographic process.

An exact representation of the object field can be obtained by considering the object to be composed of a superposition of point sources of the form given in equation 3(1). Such a formulation will yield valid results in both the near and far field of the hologram.

Each point can be recorded on the plotter as it is calculated, therefore no large core or memory capacity is necessary. The programming was done in Fortran IV, and carried out using seven place accuracy on a SDS Sigma 5 computer. Hologram synthesis time was limited by the computation and varied from 30 minutes to two hours depending on the complexity of the object and the type of plotting.

The calculated intensities are recorded on an oscilloscopic plotter manufactured by the Dicomed Corporation of Minneapolis, Minnesota. This device can record up to 1024 by 1024 image points with 64 gray levels on a storage screen with an active area approximately 28 cm. square. Analysis using the sampling theorem indicated adequate resolution was available to resolve the holograms under consideration.

Once the synthesized hologram is plotted, the image is photographed using a Nikkormat 35 mm camera with close-up adaptor positioned about 35 cm. from the Dicomed screen. Care must be taken to align the camera so that it is directly centered on the hologram pattern, or aberrations may be introduced due to the skew angle at which the picture is taken.

The objective in the reconstruction process was to use as few conventional optical elements as possible so that minimal aberrations would be introduced. The reconstruction system is illustrated schematically in Figure 6.

5. RESULTS

The results show that the hologram synthesis and reconstruction system is capable of producing clear, sharp images. In addition, they indicate that aberrations can be removed both by direct injection of appropriate phase correction and by computer synthesized correction plates.

To check the hologram synthesis, we developed an unaberrated hologram which would have the same image geometry as the long wavelength holograms. An example of a reconstruction from such an unaberrated hologram of the letter H formed by seven point sources .4 mm apart is shown in Figure 7.

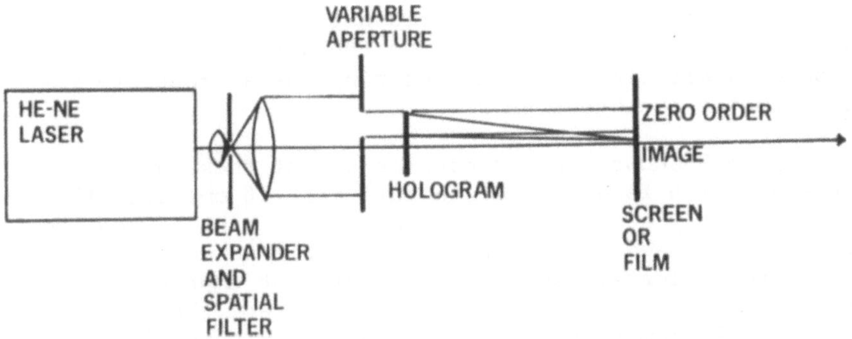

Fig. 6. Hologram reconstruction schema.

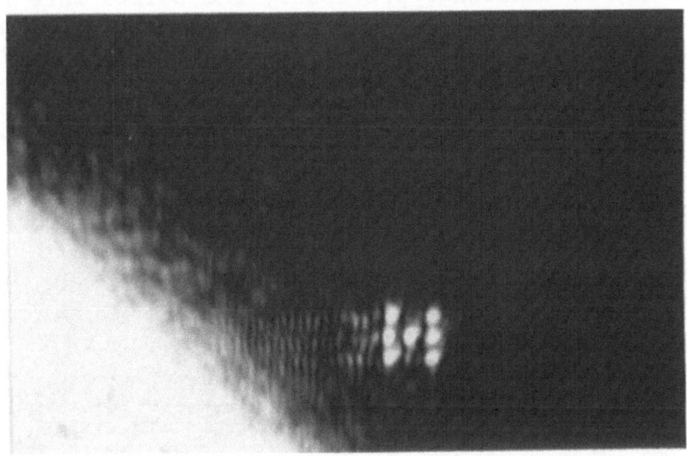

Fig. 7. Reconstruction from unaberrated hologram of seven point letter H. Image formed 75.9 cm from hologram, points .16 mm apart.

Our experiments in long wavelength hologram correction were conducted with the same object as shown in Figure 7, a letter H composed of 7 point sources spaced at .4 mm. The holograms were synthesized and reconstructed using the test geometry of Figure 8. We employed the direct injection method of aberration removal inserting the appropriate phase factor in the synthesis process.

Figure 9 shows the reconstructions of these holograms in the Gaussian focal plane, in this case 76 cm. from the hologram. Only about half the area of the synthesized holograms was used to obtain these reconstructions. The uncorrected hologram reconstruction did not improve materially at other focal distances. As predicted, the corrected acoustic hologram reconstruction shown in Figure 9 (B) forms a letter H composed of seven distinguishable points. Some residual aberrations remain; we feel these are primarily due to limitations in the plotting and reconstruction systems.

The Rayleigh resolution limit for this geometry is about .3 mm; as mentioned previously the points were spaced at .4 mm or just above the Rayleigh limit. Thus we have demonstrated that long wavelength holograms are capable of diffraction limited imagery even in highly aberrated configurations.

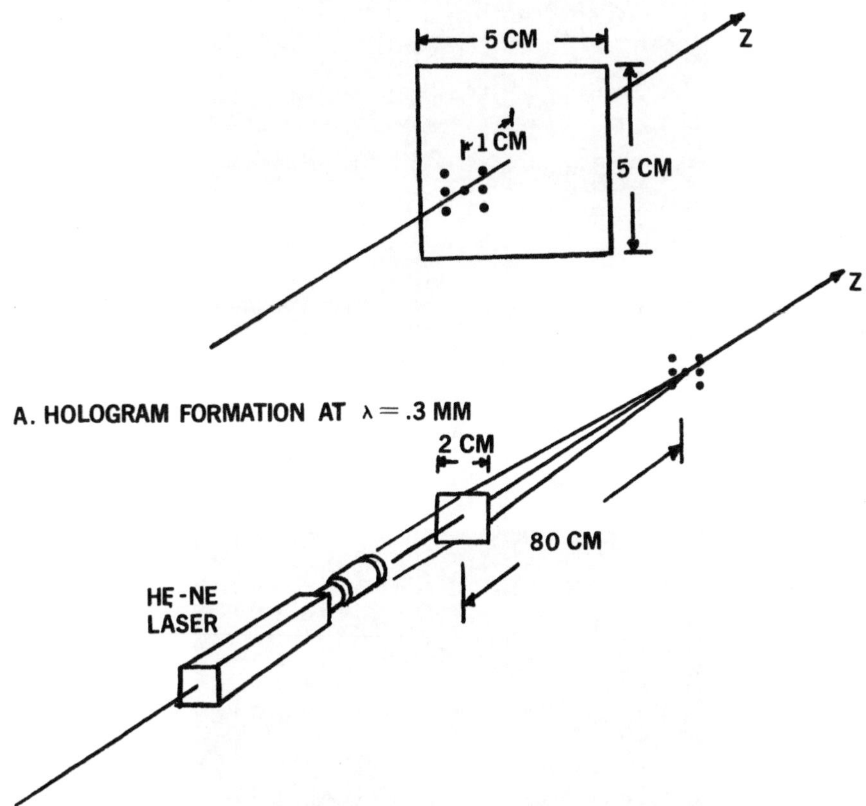

Fig. 8. Test geometry used to synthesize and reconstruct long wavelength holograms. From Figure 4 this can be identified as a highly aberrated configuration.

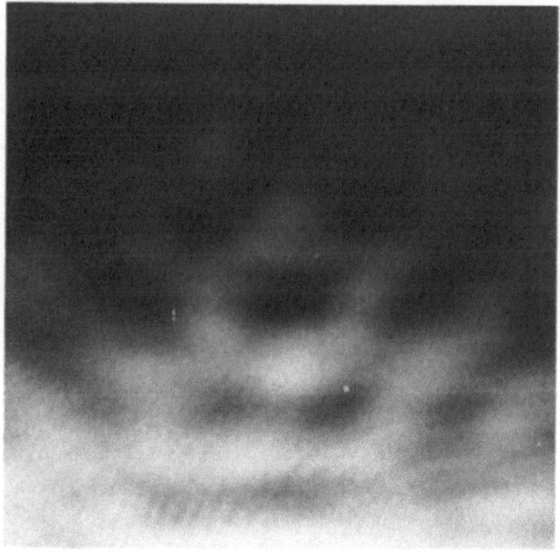

A

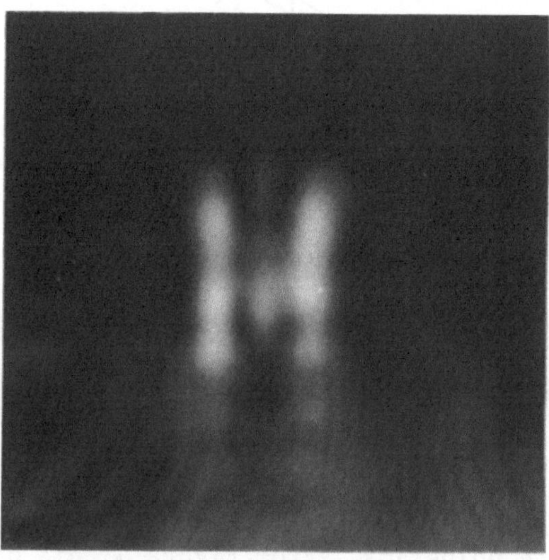

B

Fig. 9. Reconstructions from (A) uncorrected and (B) corrected long wavelength holograms, using the geometry illustrated in Figure 8. The direct injection method of aberration correction was employed.

Correction Plate

Equally effective results can be obtained using a holographic correction plate, but they are considerably more difficult to realize practically due to the low amplitude of the corrected image which follows from the low diffraction efficiency of holographic imaging elements. Another problem is the additive effect of plotting errors in the synthesized hologram and correction plate. The correction plate, which is shown in Figure 10, is placed immediately behind the uncorrected hologram.

The results are shown in Figure 11. Once again both reconstructions are taken in the Gaussian plane, but changing the focal distance did not significantly improve the image from the uncorrected hologram.

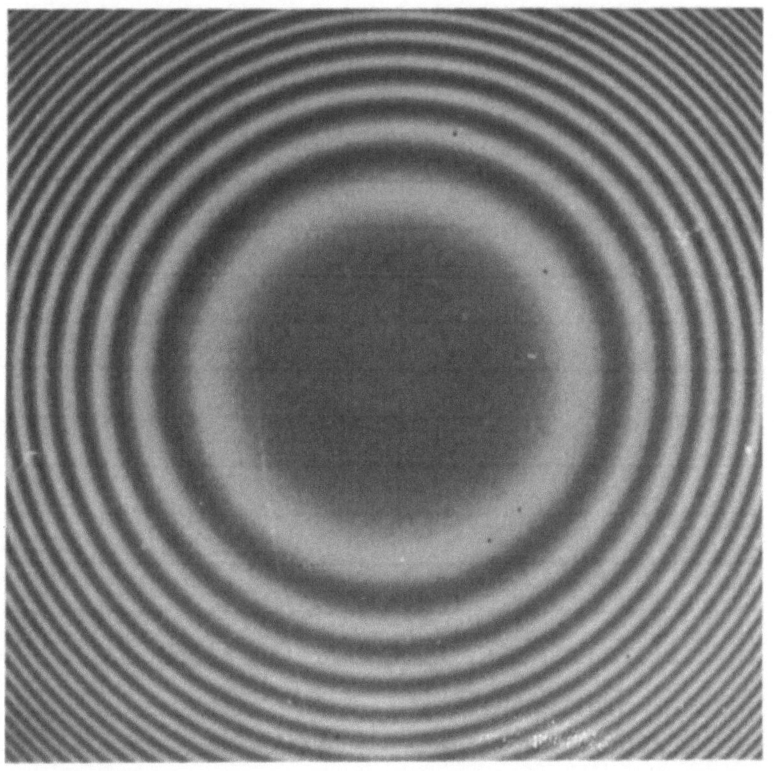

Fig. 10. Central portion of correction plate hologram.

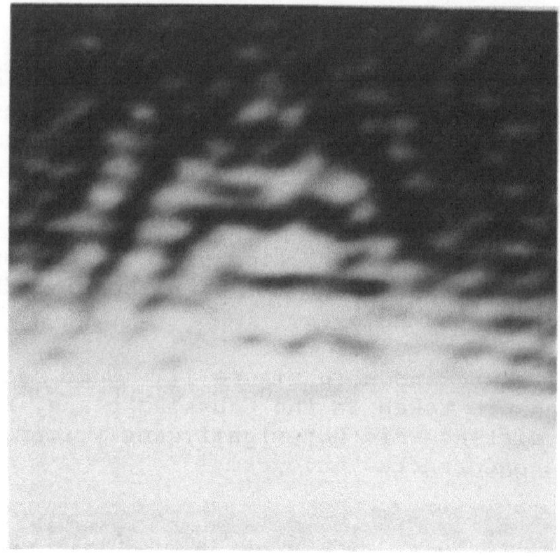

A

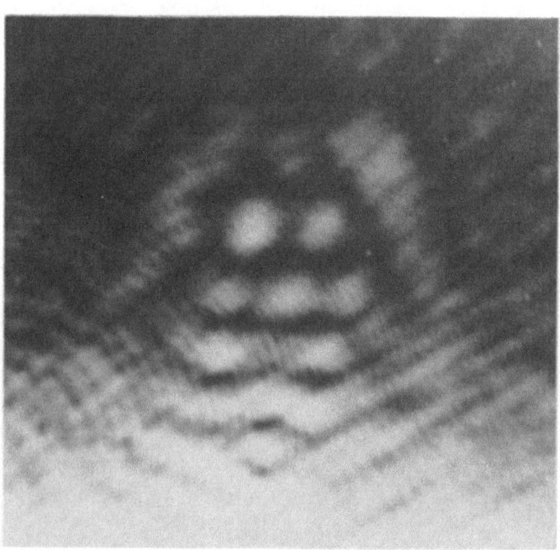

B

Fig. 11. Reconstructions from (A) uncorrected and (B) uncorrected with superimposed correction plate long wavelength holograms. The test geometry of Figure 8 was used.

6. CONCLUSION

A technique has been developed which uses any of several possible methods to correct for the aberrations which are inherent in optical imaging of long wavelength holograms. Although this approach can in general eliminate all aberrations at only a single point, the useful corrected volume around that point may be substantial. Removal of aberrations results in higher resolution and a consequent improvement in image quality.

The primary difficulty in long wavelength holography today is probably the lack of a detector analogous to photographic film in conventional holography. Progress is being made toward solution of this problem through improved acoustic films and better piezoelectric or electret detection arrays. Assuming the development of an adequate acoustic film, which is primarily a technological rather than a theoretical problem, it is interesting to speculate on what role long wavelength holography will play. Since sophisticated direct imaging systems can probably achieve diffraction limited imagery in a single plane more easily than holographic methods, it seems likely that holography will be used in situations where its unique three dimensional recording and/or ability to image through aberrant media are crucial.

It should be pointed out that the three dimensional imaging properties of holography are most pronounced in the near field region, where aberrations become the limiting factor in resolution. In addition, computer synthesized corrections could aid in compensating for the effects of nonrandom aberrant media. Thus aberration correction would appear to be a prerequisite for the practical utilization of long wavelength holography in many representative imaging geometries.

The present study was carried out entirely using computer generated holograms. It should certainly be an objective of future investigations to apply these techniques to real long wavelength holograms, since aberrations could well be the cause of much of the lack of image quality that is present in acoustic holograms today. The application of this work to practical imaging problems could extend the usefulness of long wavelength holography.

Acknowledgments

This work was supported in part by HL 12715, HL 41131 and GM-15892 grants. The authors are grateful to Dr. Frank Starmer for many helpful discussions and for his assistance in the application of the Sigma 5 computer and Dicomed plotter system.

REFERENCES

(1) R. W. Meier, J. Opt. Soc. Am., 55, 987 (1965).

(2) E. N. Leith, J. Upatnieks, and K. A. Haines, J. Opt. Soc. Am., 55, 981 (1965).

(3) J. A. Armstrong, IBM J. Res. Devel., 9, 171 (1965).

(4) C. W. Helstrom, J. Opt. Soc. Am., 56, 433 (1966).

(5) A. Offner, J. Opt. Soc. Am., 56, 1509 (1966).

(6) I. A. Abramowitz and J. M. Ballantyne, J. Opt. Soc. Am., 57, 1522 (1967).

(7) J. N. Latta, Appl. Opt., 10, 609 (1971).

(8) D. C. Winter, Appl. Opt., 10, 1074 (1971).

(9) A. M. Sherwood, Image Anomalies in Reconstructions from Long Wavelength Holograms, Ph.D. Thesis, Duke University, 1970.

(10) L. B. Lesem, P. M. Hirsh, and J. A. Jordan, Jr., Computer Generation and Reconstruction of Holograms, IBM Publication 322-0321 (1967).

(11) J. B. Farr, Acoustical Holography, Volume 2, Edited by A. F. Metherell and L. Larmore, 225 (Plenum Press, 1970).

AMPLITUDE-ONLY AND PHASE-ONLY HOLOGRAMS*

Osman K. Mawardi

Case Western Reserve University
Cleveland, Ohio

Abstract

A theoretical basis for phase-only and amplitude-only holography is developed. A criterion is found for the range of wave lengths within which phase-only holograms yield satisfactorily reconstructed images in the Fraunhoffer diffraction region.

Introduction

A number of investigators have attempted to modify the recorded signal from scanned acoustical holograms. The purpose behind this modification has been varied: Metherell [1] in his "phase-only" holograms argued that such holograms which require only half as much data as conventional holograms will yield brighter reconstructed images because the amplitude transmittance is everywhere at its maximum value, and will eliminate the difficulty of the dynamic range problem; Wade and his collaborators [2] on the other hand pointed out that "phase-only" or "amplitude-only" holograms† have the definite advantage of requiring half the storage and half the number of calculations needed to process the data of such holograms.

It is clear that a modification of the recorded signal from the "unaltered" hologram is bound to introduce distortions in the reconstructed image, and indeed it does. The extent of this distortion is difficult to predict accurately.

Wade and his collaborators have performed a series of numerical experiments on modified holograms for objects of simple geometric shapes (Ref. 2) and have been able to demonstrate the relative superiority of one scheme over the other. They do not give, however, guidelines for the estimation of the distortion produced in the reconstructed image. Although Lohmann [3] did not find the general relations for the distortions from these truncated holograms, he obtained by means of a most ingenious argument a restricted class of objects whose "phase-only holograms" do not seriously distort the reconstructed image. Wade's empirical approach to obtain the holograms from computer simulated objects of diverse shapes has been followed by others. [4] In spite of the large amount of work which has been done on this subject, a generalized treatment for the theory of these modified holograms is still lacking.

The present investigation attempts to develop the basis for the estimate of the distortion referred to above, for scattered fields in the Fraunhoffer region.

Scattered Signal and Imaging Techniques

Several aspects of imaging procedures by holography make use of scattered signals. For this reason it will be helpful to review a few fundamental concepts of scattering theory.

As a consequence of causality, a reciprocal and unique relationship exists between the scattered signal from an obstacle such as O and the illumination of O. To be more specific if the acoustic velocity potential at the surface of O as a result of an illuminating source Q (point source, plane wave, etc.) is denoted by $\psi_o(\underline{r})$, and the scattered signal by $\psi_s(\underline{r})$, then the required relation is equivalent to an exterior boundary value problem.

Indeed, one has

$$\psi_s(\underline{r}) = \int_{S_o} \partial_n G(\underline{r},\underline{r}')\psi_o(\underline{r}')dS \qquad (1)$$

The inverse of this integral equation, would yield the reciprocity between ψ_o and ψ_s. In the above relation $G(\underline{r},\underline{r}')$ is the appropriate Green's function for the space exterior O, and S_o is the surface of O.

Similarly,

$$\psi_o(\underline{r}') = \int_{S_s} \partial_n G^+(r',r'')\psi_s(\underline{r}'')dS \qquad (2)$$

i.e., the illumination on the obstacle can be deduced from the knowledge of the value of ψ_s on a large sphere S_s . Here G^+ is the appropriate Green's function for the interior of S_s .

The reciprocal property of (1) and (2) leads to the observation that (2), together with additional cues (say, direction of incident plane wave, etc. ...) is in principle an image forming procedure.

Of course, in practice one would never follow such an awkward procedure, furthermore the large sphere S_s is substituted by a flat screen. The scattered field ϕ_s is not known everywhere but only on the screen.

The actual expression for (2) now will be evaluated, since it will be needed to discuss the effect of altering the holograms. As indicated in the introduction, we restrict the arguments to the Fraunhoffer regime.

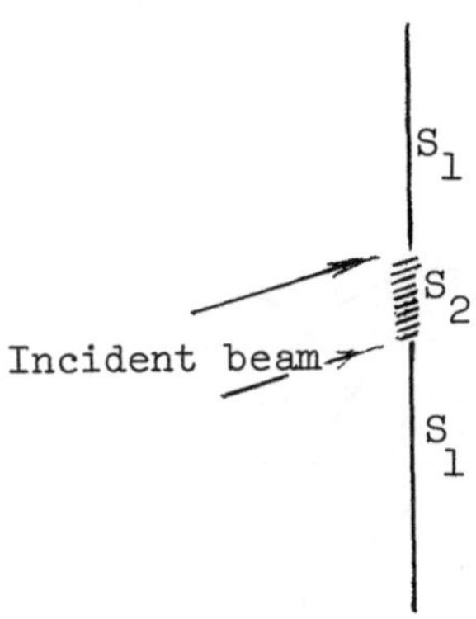

Fig. (1) Geometry of illuminated screen.

Fraunhoffer Scattered Field

It is assumed that the screen is illuminated by the scattered signal ψ_s which is mainly concentrated over a surface S_1. The rest of the screen (area S_2) is imagined to be dark (Fig. 1).

Let it be assumed that the normal derivative of the velocity potential vanish on the screen and that at large distances $\psi(\underline{r})$ asymptotically approaches $f(\theta,\phi)\exp(ikr)/r$.

If $\psi(r)$ and $\chi(r)$ are two functions satisfying the Helmholz wave equation then Green's second identity yields:

$$\int_V [\psi(\nabla^2\chi+k^2\chi)-\chi(\nabla^2\psi+k^2\psi)]d\tau = 0$$
$$= \int_S [\psi(\underline{r})\partial_n\chi(\underline{r})-\chi(\underline{r})\partial_n\psi(\underline{r})]d\sigma \tag{3}$$

Let $\chi(r)$ represent the combination of plane waves

$$\chi(r) = e^{i\underline{K}\cdot\underline{r}}+e^{-i\underline{K}'\cdot\underline{r}} = e^{-ikr\cos\Theta}+e^{-ikr\cos\Theta'} \tag{4}$$

whose direction of propagations is specified by the angles $(\theta',\phi,)$ and (θ'',ϕ'') respectively, as shown in Fig. 2. But from trigonometry

$$\cos\Theta = \cos\theta'\cos\theta + \sin\theta\sin\theta'\cos(\phi-\phi')$$
$$\cos\Theta' = \cos\theta''\cos\theta + \sin\theta\sin\theta''\cos(\phi-\phi'') \tag{5}$$

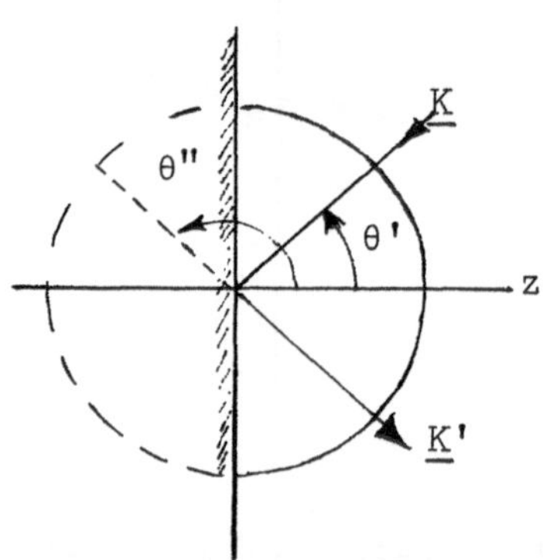

Fig. (2) Geometry describing Eq. 4.

The domain of integration of (3) is bounded by the surfaces S_1, S_2 and the hemisphere S_3. The radius of the hemisphere S_3 is chosen large enough so that the asymptotic value of ψ can be used.

Separating the various parts of the surface integral, one finds by resorting to the boundary conditions.

$$0 = \int_{S_1} \left(\psi(r) \frac{\partial}{\partial z} (e^{-ikr\cos\Theta} + e^{-ikr\cos\Theta'}) - (e^{-ikr\cos\Theta} + e^{-ikr\cos\Theta'}) \frac{\partial\psi}{\partial r}\right)_{z=0} d\sigma + \int_{S_2} \psi(r) \frac{\partial}{\partial z} (e^{-ikr\cos\Theta} + e^{-ikr\cos\Theta'})_{x=0}\, d\sigma +$$

$$\int_{S_3} \{f(\theta,\phi) \frac{e^{ikr}}{r} \frac{\partial}{\partial r} (e^{-ikr\cos\Theta} + e^{-ikr\cos\Theta'}) + (e^{-ikr\cos\Theta} + e^{-ikr\cos\Theta'}) \frac{\partial}{\partial r} f(\theta,\phi) \frac{e^{ikr}}{r}\} d\sigma \qquad (6)$$

Let us choose

$$\theta'' = \pi - \theta' \qquad \phi'' = \phi' \qquad (7)$$

i.e. the combination of an incident and reflected wave.

The first integral now reduces to

$$\int_{S_1} \{\psi(r) \frac{\partial}{\partial z} (e^{-ikz\cos\theta' - ik\rho\sin\theta'\cos(\phi-\phi')} + e^{ikz\cos\theta' - ik\rho\sin\theta'\cos(\phi-\phi')}) + (e^{-ikz\cos\theta' - ik\rho\sin\theta'\cos(\phi-\phi')} + e^{ikz\cos\theta' - ik\rho\sin\theta'\cos(\phi-\phi')}) v_n(\rho,\phi)\}_{z=0}\, d\sigma$$

$$= 2 \int_{S_1} e^{-ik\rho\sin\theta'\cos(\phi-\phi')} v_n(\rho,\phi) d\sigma$$

The second integral vanishes by virtue of the assumptions of S_2 being dark. The third integral is evaluated as follows. The direction (θ',ϕ') is chosen as the polar axis. The process of integration is now carried twice: once for Θ and once for Θ' over all the same hemisphere. Because of the presence of the screen, we can imagine that the function $\psi(\underline{r})$ extends beyond the screen in its mirror image, i.e.,

$$f(\theta,\phi) = f(\pi-\theta,\phi) \ .$$

Consequently, the integral S_3 is to be evaluated as though it were extended over the whole space. Hence,

$$\int_{S_3} = \int_0^{2\pi}\int_0^{\pi} \{f(\theta',\phi') \frac{\partial}{\partial r}(e^{-ikr\cos\Theta}) \frac{e^{ikr}}{r} - e^{-ikr\cos\Theta} \frac{\partial}{\partial r} (f(\theta',\phi') \frac{e^{ikr}}{r} \} r^2 d\Theta \sin\Theta d\Phi = -ik\int_0^{2\pi}\int_0^{\pi} f(\theta',\phi')\cdot (1+\cos\Theta)e^{ikr(1-\cos\Theta)} r\sin\Theta d\Theta d\Phi + \int_0^{2\pi}\int_0^{\pi} f e^{ikr(1-\cos\Theta)} \sin\theta d\Theta d\Phi \tag{8}$$

Integrating by parts the integral over S_3, in the limit $r \to \infty$, reduces to $-4\pi f(\theta',\phi')$. Hence

$$f(\theta',\phi') = \frac{1}{2\pi} \int_{S_1} e^{-ik\rho\sin\theta'\cos(\phi-\phi')} v_n(\rho,\phi)\rho d\rho d\phi \tag{9}$$

The above expression which is quite general will be most helpful in discussing the various alternatives of the truncated holograms.

Now, the function v_n, except for a reference signal is in fact the hologram. It is also evident that v_n bears a definite relationship to the far-zone scattered wave from the obstacle O. Accordingly, in order to evaluate the extent of the modification introduced by any-one scheme (phase-only, amplitude-only, etc. ...) it is important first to obtain the accurate functional dependence of v_n on the scattered wave from O.

Illumination of Screen

The formal expression for the scattered wave from an obstacle of arbitrary shape and for the Neumann boundary conditions (i.e. $\partial_n \psi = 0$ at the obstacle surface) is given by Eq. 1. It is known, however, that the Green's function for the Helmholz equation can be calculated exactly only for those 11 cases [5] for which the wave equation is separable. Consequently, Eq. 1 cannot be obtained exactly in general.

If it is desired to find the characteristics of the "luminous spot" on a screen and which is caused by the scattered wave from an obstacle, then from simple considerations of dimensional analysis [6] it becomes evident that the important dimensions are: the size of the obstacle, the distance of the obstacle from the screen and finally the wave length of the scattered signal. Of course, the detailed shape of the obstacle will dictate the shape of the spot, but the functional relation for the relative "characteristic" dimensions is most certainly independent of shape. For this reason, the calculations performed here, on a sphere can be generalized to obstacles of arbitrary shapes.

Now, the far-zone scattered signal from a sphere of radius "a" and illuminated by a plane wave is expressed by [7]

$$\psi_s \rightarrow \frac{e^{-i\underline{k}\cdot\underline{r}}}{kr} \sum (2n+1) e^{-i\delta_n(ka)} \sin(\delta_n) P_n(\cos\theta) \qquad (10)$$

The arguments $\delta_n(ka)$ are defined in terms of spherical Bessel functions, indeed

$$h_n(z) \equiv -iD_n(z) e^{i\delta_r(z)} \qquad (11)$$

i.e., D_n gives the amplitude and δ_n the phase of the spherical functions. These quantities have been tabulated in a number of places, Ref. [7] for instance.

The evaluation of the value of the potential on the screen is simplified appreciably when $\underline{k}$, the direction of the incident plane wave, is normal to the screen. It can be shown readily that the results for normal incidence can be generalized also to the case of arbitrary angle of incidence. When Eq. (10) is rewritten as

$$\psi_s = \frac{e^{-i\underline{k}\cdot\underline{r}}}{kr} F(\theta) , \qquad (12)$$

the scattered signal distribution on a screen, distant R from the center of the sphere (Fig. 3) can be calculated by a simple Taylor series expansion in ρ , the radial distance

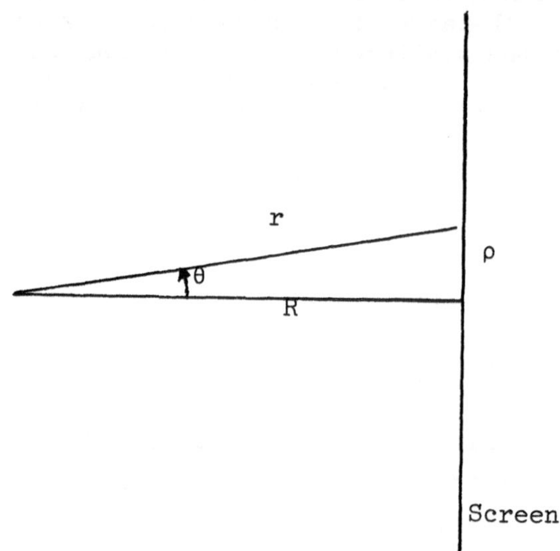

Fig. (3) Geometry for scattered signal from the sphere.

reconed on the screen and from the origin 0 .

Actually,

$$\underline{r} = \underline{R} + \underline{\rho} , \tag{13}$$

hence

$$\underline{k}\cdot\underline{r} = k\sqrt{R^2 + \rho^2} \doteq kR(1 + \frac{\rho^2}{R^2}) \tag{14}$$

Finally,

$$\psi_{S_s} \doteq \frac{e^{ikR}}{kR} e^{ik\rho^2/R}(F(0) + \Delta\theta F'(0) + \frac{\Delta\theta^2}{2!} F''(0) + \ldots)$$

$$= \frac{e^{ikR}}{kR} e^{ik\rho^2/R}(F(0) + \frac{\rho}{R} F'(0) + (\frac{\rho}{R})^2 \frac{F''(0)}{2!} + \ldots) \tag{15}$$

where use has been made of $\Delta\theta \stackrel{\sim}{\sim} \rho/R$, $F^n \equiv \frac{\partial^n F}{\partial\theta^2}$. But, from Eq. (10), it appears that $F(\theta)$ is an even function of θ , hence $F'(0)$ vanishes.

The sequence of functions $F^n(0)$ can be easily evaluated for the case of ka << 1 . For small arguments

$$e^{-i\delta_n} \sin \delta_n = - \frac{i}{2} (1-e^{-i2\delta_n}) \doteq \delta_n \tag{16}$$

But from Ref. [7]

$$\delta_o = ka \quad \text{and} \quad \delta_n = \frac{(ka)^{2n+1}}{1.3.5\ldots(2n+1).1.1.3.5.\ldots(2n-1)} \tag{17}$$

The directivity function $F(\theta)$ and its derivatives are

$$F(\theta) = (ka)+(ka)^3\cos\theta + \frac{1}{12} (ka)^5(3\cos 2\theta+1)+ \ldots$$

$$F(0) = (ka)+(ka)^3+ \frac{(ka)^5}{3} + \ldots$$

$$F'(0) = 0$$

$$F''(0) = -(ka)^3- (ka)^5 \ldots \tag{18}$$

To a first approximation, since (ka) << 1

$$\psi_{S_s} = \frac{e^{ikR}}{kR} e^{ik\rho^2/R}(ka) \tag{19}$$

and

$$v_n = \partial_n \psi_{S_s} \doteq i \frac{e^{ikR}}{R} (ka) e^{ik\rho^2/R} \tag{20}$$

Another quantity which is of importance to the discussion is the "size" of the luminous spot on the screen. This statement needs to be clarified. The expression given above in Eq. (20) leads to the extremely important observation that on the screen and in the vicinity of the origin the scattered signal when (ka) and ρ/R are both much smaller than unity, then the hologram is expressed by a <u>phase only variation</u>. But this is misleading because the directivity function $F(\theta)$ does have side lobes and indeed the illumination falls off at the "edge" of the spot. A rough estimate of the size, is to search for the critical angles which defines $\frac{\partial}{\partial\theta}(\psi_{S_s} \psi^*_{S_s})=0$. The intensity fall off, rather than that of the signal itself is a better indication of the spot size. It can be shown, readily that this critical angle θ_c is found from the transcendental equation

$$\frac{\cos(\delta_o - \delta_1)\sin\delta_o}{3\sin\delta_1} = -\sin\theta_c \tag{21}$$

The size of the spot a_1 is then roughly

$$a_1 = R\theta_c \tag{22}$$

More accurately one should have searched for the angles where the intensity has fallen off to a predetermined value. To fix the ideas the intensity falls off to half value at angle of approximately 45° for ka=3 .

Another graphic way of visualizing the size of a_1 is to examine the dependence of the scattering cross section on the wave number. Indeed, in the limiting case of $k \to \infty$ (i.e. infinite frequency) the cross-section corresponds to the geometric projection on the screen and the size of the spot is the optical shadow. Another limiting case is that for $k \to 0$, in which case the cross section approaches the Rayleigh limit. The behavior of the cross section Q as a function of k is shown in Fig. (4).

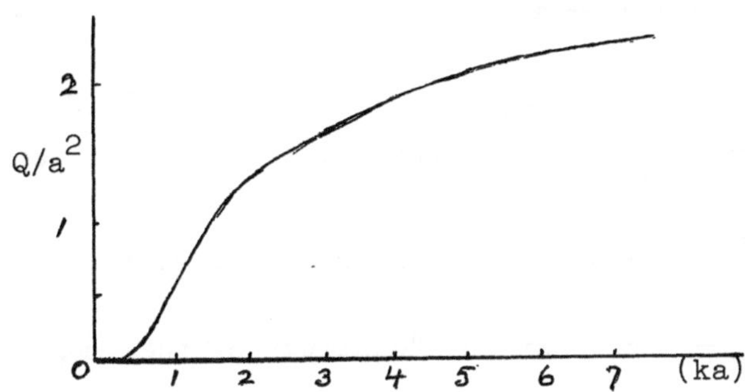

Fig. (4) Scattering cross section for plane wave scattered from a rigid sphere of radius a .

Distortions from Truncated Holograms

When the result of Eq. (20) is substituted in (9) one finds

$$f(\theta',\phi') = \frac{1}{2\pi}\int_0^{a_1}\int_0^{2\pi} e^{-ik\rho\sin\theta'\cos(\phi-\phi')} \left(i\,\frac{e^{ikR}}{R}\,ka\,e^{ik\rho^2/R}\right)\rho\,d\rho\,d\phi \tag{23}$$

In the above equation a_1 is the size of the spot, which will be defined say as the cut-off size of the half intensity fall off point. It is interesting to note, that unless, the scattering becomes violently fluctuating due to unusual resonances on the surface of the obstacle O , then the phase-only approximation is in fact a very reasonable one. This is not true for the amplitude-only case which as pointed out by Lohmann is valid only for a very narrow pencil illuminating the obstacle by scanning it.

The actual amount of distortion produced can be estimated as follows.

$$f(\theta',\phi') = \frac{i(ka)e^{ikR}}{2\pi R} a_1^2 \int_0^1 \int_0^{2\pi} e^{-ika_1\xi\sin\theta'\cos(\phi-\phi')} e^{i\frac{ka_1^2}{R}\xi^2}$$

$$\xi d\xi d\phi = \frac{i(ka)}{2\pi} a_1^2 \frac{e^{ikR}}{R} \int_0^1 J_o(ka_1 \sin\theta'\xi) e^{i\frac{ka_1^2}{R}\xi^2} \xi d\xi \qquad (24)$$

where $\xi = \frac{\rho}{a_1}$. Notice that the reconstructed image is $f(\theta',\phi')\frac{e^{-ikr}}{R}$ so that the "quality" of the image can be inferred from an examination of the behavior of the integral of (24). We first note that because the scattered signal considered is axisymmetric, the azimuthal angle ϕ does not appear. Hence $f(\theta',\phi')$ is really $f(\theta')$.

The expression in (24) has been discussed at length by Lommel [8] in connection with the theory of spherical aberrations. In fact the integral can be written as

$$2\int_0^1 = C(\frac{ka_1^2}{R}, ka_1\sin\theta') = iS(\frac{ka_1^2}{R}, ka_1\sin\theta') \qquad (25)$$

where C and S functions can be expressed in terms of the Lommel functions U_n, V_n which have been tabulated. [9]

Indeed one has

$$C(u,v) = \frac{\cos\frac{1}{2}u}{\frac{1}{2}u} U_1(u,v) + \frac{\sin\frac{1}{2}u}{\frac{1}{2}u} U_2(u,v)$$

and

$$S(u,v) = \frac{\sin\frac{1}{2}u}{\frac{1}{2}u} U_1(u,v) - \frac{\cos\frac{1}{2}u}{\frac{1}{2}u} U_2(u,v) \qquad (26)$$

where $u = \frac{ka_1^2}{R}$ and $v = ka_1\sin\theta'$.

Formally, the distortion can be estimated as the departure of $(C^2+S^2)^{1/2}$ from a constant (as a function of θ'). Stated in a different way, one asks in what manner does the reconstructed image look different from the uniform illumination on the sphere.

Now, for small arguments the dominant terms becomes $C \approx \frac{J_1(ka_1 \sin\theta')}{ka_1 \sin\theta'}$. Consequently, provided that $ka_1 < 3.8$ the first zero of the Bessal function, C is reasonably uniform with θ' . But $a_1 \leq a$, hence the phase-only hologram will be reasonably satisfactory provided the above inequality is satisfied. This means that

$$\frac{a_1}{\lambda} \leq \frac{a}{\lambda} \leq \frac{3.8}{2\pi} \qquad (27)$$

or that the size of the sphere is about 2/3 of incident wave length.

In conclusion, truncated holograms will always lead to distorted images. The distortion from the phase-only holograms is acceptable in the Fraunhoffer region provided $ka < 0.66$. For higher frequencies, the illumination intensity fluctuates too strongly and the amplitude contribution cannot be neglected.

On the other hand, the amplitude-only holograms will lead to acceptable images except for the case where as pointed out by Lohmann the hologram is formed by scanning the object with a very narrow pencil.

Footnotes

*This research was supported in part by the Office of Naval Research.

†D. Vilkomerson brought it to the attention of the author that the terms amplitude-only, phase-only are misnomers since the concept of the hologram is to include both the amplitude and phase. It would have been wiser, perhaps, to call these truncated holograms: phase-only, amplitude-only "signatures."

References

1) A. F. Metherell "Acoustical Holography" Vol. 1 (Plenum Press, New York 1969) Chap. 14 ff.

2) J. Powers, J. Landry and G. Wade "Acoustical Holography

Vol. 2 (Plenum Press, New York 1970) Chap. 13 ff.

3) A. W. Lohmann "Acoustical Holography Vol. 2 (Plenum Press, New York 1970) Chap. 14 ff.

4) An interesting series of papers pertaining to this subject have appeared in: IBM Conference on Holography and the Computers, Houston, Texas, December 1969.

5) L. P. Eisenhart, Phys. Rev. 45, 427, (1934).

6) P. W. Bridgman "Dimensional Analysis" (Yale Univ. Press, New Haven 1931) p. 36.

7) P. M. Morse and H. Feshbach "Methods of Theoretical Physics" (McGraw-Hill, New York, N. Y. 1953) p. 1483.

8) E. Lommel Abh. Bayer. Akad. 15, Abth. 2 (1885), 233; also M. Born and E. Wolf "Principle of Optics" (Pergamon Press, New York, N. Y. 1959) p. 434 ff.

SPATIAL FILTERING CONSIDERATIONS IN BRAGG DIFFRACTION IMAGING

John P. Powers

Electrical Engineering Department
Naval Postgraduate School
Monterey, California 93940

ABSTRACT

Many of the techniques and effects observed in images from systems using Bragg diffraction imaging can be explained using concepts of optical spatial filtering. This is possible because the Bragg diffraction phenomenom can be thought of as an interaction of a plane-wave of light with a plane-wave of sound to produce a plane-wave of diffracted light. Changing the incident light causes a change in the diffracted light. Hence, the input light field of the imaging system plays an analogous role to the filter transparency of an optical spatial filtering system. The plane-wave components of the diffracted light that form the image of the sound field can be selectively changed by modification of the incident light field. When the imaging method is analyzed as a plane-wave: plane-wave interaction, simple explanations can be given for such diverse effects as the observed astigmatic resolution, the dependence of the resolution on the semi-apex angle of the incident light wedge, dark field imaging, and reflection imaging.

INTRODUCTION

Bragg imaging is a method of obtaining a real-time optical image of a cross section of an ultrasonic field. In this method laser light diffracted from a laser beam passing through the sound field forms the image. This acoustic imaging scheme holds promise in the areas of medicine, biology, and nondestructive testing because of its ability to image internal detail in optically opaque material. Figure 1a illustrates the imaging of a biological object - a Silver Dollar fish (*mylossoma argenteum*). Figure 1b is the image of a 7.0-mm thick aluminum plate with two holes (1.5-mm diameter) drilled into the edge with a spacing of 7.5 mm. Internal details are apparent in both of these objects.[1]

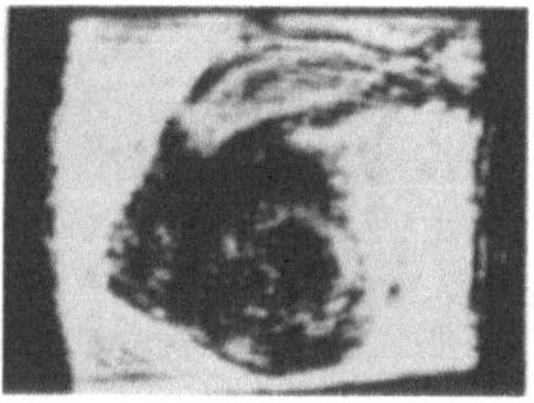

(a)

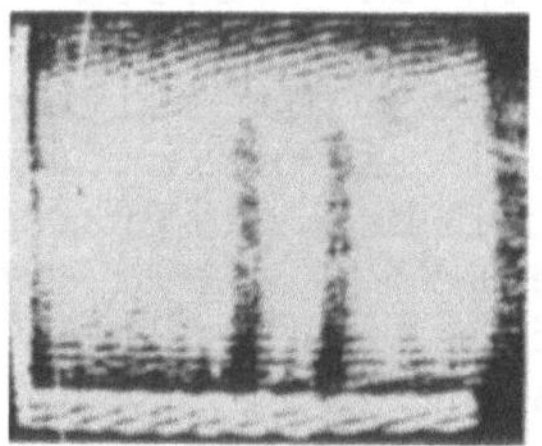

(b)

Fig. 1. Transmission images of (a) a "Silver Dollar" fish, and (b) holes drilled into an aluminum plate.

With the proper choice of the geometric shape and orientation of the input laser beam, we can obtain high-quality optical images of a cross section of the acoustic field from this diffracted light. Different cross sections are obtained by the simple translation of one lens along the axis of the optical system. This ability to obtain all cross sections implies that the acoustic field is reconstructed in its entire volume, a valuable asset in those applications where it may be desirable to scan through the different planes of the volume. This feature is directly analogous to holographic reconstructions.

A diagram of the imaging system is shown in Fig. 2 in a configuration to obtain a transmission-type image. In this arrangement the ultrasonic field insonifying the object is generated by a quartz transducer driven by the rf generator and amplifier. The sound field containing the object information then interacts with the laser light. The laser light is in the form of a wedge with the apex located to the right of the interaction, as in the figure. A vertically oriented cylindrical

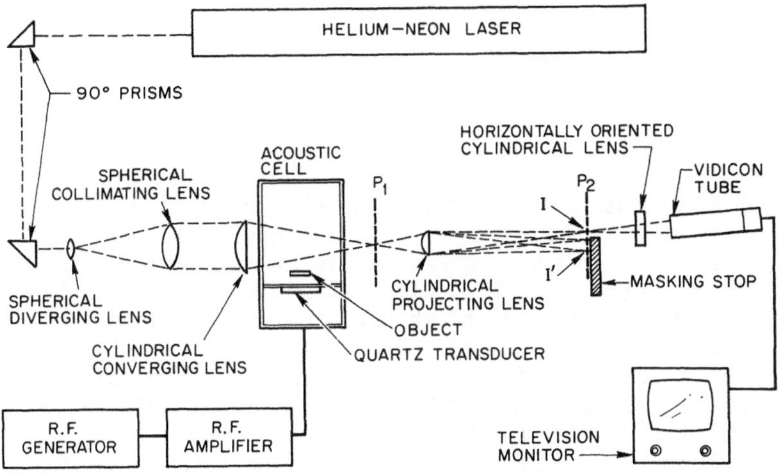

Fig. 2. Schematic diagram of the Bragg imaging system.

lens to the left of the acoustic cell acts on a collimated laser beam to form this wedge of light. (The plane of the figure is assumed to be the horizontal plane.) At approximately the plane P_1 two images of the sound field are formed, one to either side of the central order light. The image to the left of the central order light (looking along the axis of light propagation) is an upshifted (in frequency) virtual image of the sound field cross section. The image to the right of the central order component is a downshifted real image. Both images are demagnified in the horizontal direction by the ratio of the light wavelength to the sound wavelength. For rf sound frequencies and He-Ne laser light, this demagnification is of the order of 1/100. The vertical cylindrical projecting lens projects these images to the television camera face, restoring the horizontal dimension of the images to a useful size. At plane P_2 a stop removes the central order beam and one of the images. The remaining image is focused horizontally by the horizontally oriented cylindrical lens. An in-focus real image of the sound field is then picked up directly by a vidicon tube and displayed on a television monitor. Images such as those of Fig. 1 can then be displayed on the screen for real-time viewing or for being photographically recorded.

REVIEW OF BRAGG DIFFRACTION

The primary principle used, Bragg diffraction, is the diffraction of coherent light by a traveling acoustic wave when certain geometric conditions are met.[2,3] In this light-sound interaction the mechanism that diffracts part of the laser beam can be considered as an optical diffraction grating effect caused by a variation in the optical index of refraction due to the presence of the sound traveling through the medium. This effect was first predicted by Brillouin[4] in 1922.

One of the possible approaches to this phenomenon is based upon the interaction of a plane wave of sound with a plane wave of light. When a plane wave of sound intersects a plane wave of light it can be shown that, due to interference effects, the diffracted light wave will have significant amplitude only if the plane waves meet at the proper angle. Representing each wave by its propagation constant, the condition for diffraction may be written vectorially as

$$\vec{k} \pm \vec{K} = \vec{k}_{\pm} \tag{1}$$

where $\vec{k}$ and $\vec{k}_{\pm}$ are the propagation vectors of the incident and diffracted light, respectively, and $\vec{K}$ is the propagation constant of the sound wave. (It should be noted that here and throughout the analysis to follow we will be concerned only with weak interactions and first-order Bragg diffraction for the intended applications.)

An additional relation can be derived from parametric interaction theory to relate the frequencies of the diffracted waves to the frequencies of the incident waves.[5] This relation is

$$\omega_{\pm} = \omega \pm \Omega \tag{2}$$

where ω and $\omega_{\pm}$ are the frequencies of the incident and diffracted light, and Ω is the frequency of the sound. (The $\pm$ relates the frequency of the diffracted wave to the respective sign in the vector equation.) Because of this frequency behavior, the diffracted beams are called the upshifted beam and the downshifted beam. It should be noted from the above frequency relation that the diffracted light frequency is not very different from that of the incident light, since the typical sound frequency used for imaging ($\sim 10^6$ Hz) is very much less than the typical light frequency ($\sim 10^{14}$ Hz). This fact, together with the assumption of a weak interaction, implies that the magnitudes of $\vec{k}$ and $\vec{k}_{\pm}$ will be essentially equal.

The vector relation, Eq. (1), may then be drawn as closed isosceles triangles (called the Bragg triangles), as in Fig. 3.

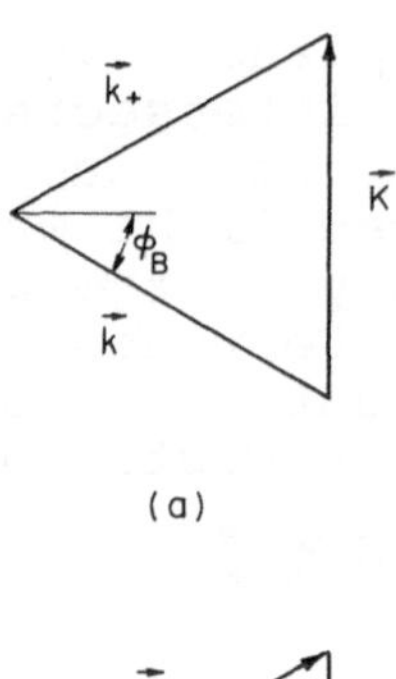

(a)

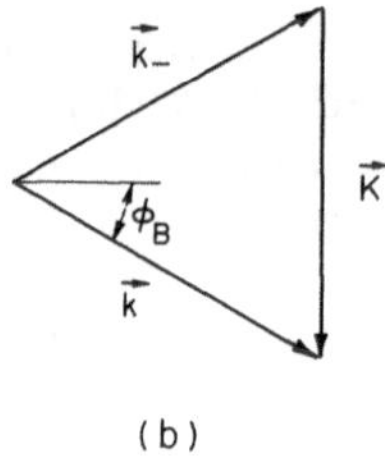

(b)

Fig. 3. Bragg triangles for (a) upshifted, and (b) downshifted Bragg diffraction.

From these trianges the required geometric condition for Bragg diffraction may be derived.

$$\sin \phi_B = \frac{|\vec{K}/2|}{|\vec{k}|} = \frac{\lambda}{2\Lambda} \tag{3}$$

where ϕ_B is the Bragg angle. By drawing the Bragg triangle in a different fashion, as in Fig. 4, the geometric requirements become clearer: the angle between the propagation vectors of the interacting light wave and sound wave must be $\pi/2 \pm \phi_B$ to permit Bragg diffraction.

When the geometrical conditions for diffraction are properly met, the amplitude of the diffracted light wave will be proportional to the product of the amplitudes of the incident sound and light plane waves.[5] The proportionality constant is a function of the interaction medium and the ratio of the acoustic wavelength to the light wavelength. The phase of the diffracted wave is dependent on whether the wave is upshifted or downshifted in frequency. If upshifted, the phase is equal to the sum of the phases of the light and sound plane waves; if downshifted, the phase of the diffracted wave is given by the difference in phase between the incident light wave and the sound wave.[5]

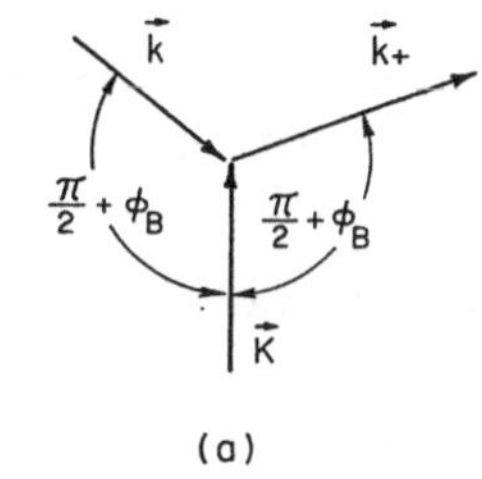

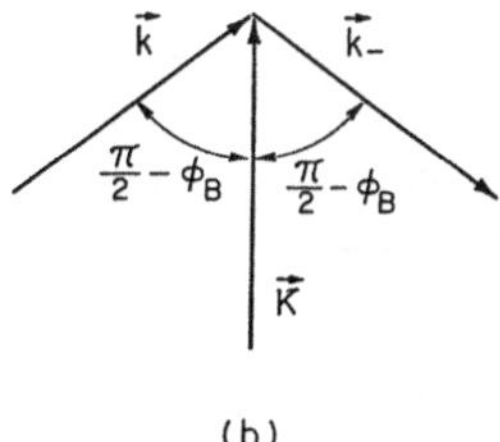

Fig. 4. Propagation vector diagrams for (a) upshifted, and (b) downshifted Bragg diffraction.

The amplitude and phase of the diffracted waves can be represented by complex notation. For example,

$$\underline{U}_+ = |U_+|e^{j\phi_+} \tag{4}$$

where $|U_+|$ is the amplitude and ϕ_+ is the phase of the upshifted plane wave. From the above statements we see that the upshifted and downshifted waves obey the following proportionalities

$$\underline{U}_+ \; \alpha \; \underline{U}_\ell \underline{U} \tag{5}$$

$$\underline{U}_- \; \alpha \; \underline{U}_\ell \underline{U}_s^* \tag{6}$$

where $\underline{U}$ and $\underline{U}_s$ are the corresponding complex forms for the incident plane waves of light and sound, respectively, and * represents complex conjugation.

Summarizing the results of the theory of first-order Bragg diffraction from this plane-wave approach:

1. Bragg diffraction can occur only when a plane wave of light and a plane wave of sound intersect so that their propagation vectors have an angular separation given by $\pi/2+\phi_B$ or $\pi/2-\phi_B$ where $\phi_B = \sin^{-1}\lambda/2\Lambda$.

2. The frequency of the diffracted light is shifted by an amount equal to the frequency of the sound where the sign of the shift depends on which angular condition is met.

3. The diffracted light's propagation vector will be directed at an angle of $\pi/2 \pm \phi_B$ from the propagation vector of the sound and will be in the plane determined by the propagation vectors of the input light and sound.

4. The amplitudes and phases of the diffracted waves are specified by the following proportionalities:

$$\underline{U}_+ \; \alpha \; \underline{U}_\ell \; \underline{U}_s$$

$$\underline{U}_- \; \alpha \; \underline{U}_\ell \; \underline{U}_s^*$$

where the quantities involved are the complex notation depicting the upshifted and downshifted diffracted plane waves.

SPATIAL FOURIER TRANSFORM REPRESENTATION OF ARBITRARY FIELDS

The utility of the plane-wave approach to Bragg diffraction becomes apparent when combined with the decomposition of an arbitrary sound or light field in terms of its planar-wave components.[6] This is a powerful tool in modern optics and is directly analogous to the Fourier series or transform approach of circuit and systems analysis.

In this method of analysis an arbitrary two-dimensional complex-valued scalar field distribution propagating in the +Z direction can be considered as a sum of plane waves of infinite extent with differing amplitudes and phases. Mathematically we can represent this sum as:

$$\underline{U}(x,y) = \int_{-\infty}^{\infty}\int U'(f_x,f_y)e^{+j2\pi(f_x x+f_y y)}df_x df_y \tag{7}$$

where

$$\underline{U}'(f_x,f_y) = \int_{-\infty}^{\infty}\int \underline{U}(x,y)e^{-j2\pi(f_x x+f_y y)}dxdy \tag{8}$$

Here each component $\underline{U}'(f_x,f_y)$ can be considered as a plane wave whose amplitude and phase are given by the amplitude and phase of this complex quantity. The propagation direction of this plane wave is simply related to the arguments of $\underline{U}'(f_x, f_y)$. The direction cosines of the propagation direction with respect to the x or y axis are equal to the wavelength times the value of f_x or f_y, respectively. Because of the assumed unidirectional propagation of the field, the direction cosines lie between +90° and -90°. Any values of f_x or f_y greater than λ^{-1} or Λ^{-1} where λ and Λ are the light and sound wavelengths (depending on whether a light field or sound field is being represented) are considered as evanescent waves. Thus the spatial distribution function

$\underline{U}'(f_x,f_y)$ can completely specify any sound or light field.

It should be noted that an equivalent representation of a spatial spectrum can be derived for any coordinate system that can express the propagation direction equally well. One such scheme is that specified by azimuth and inclination angles. Consider, for example, an arbitrary light field $\underline{U}(x,y)$ that propagates in the +z direction. Assuming that the geometric distribution is fairly simple, we can obtain the spatial Fourier transform, $\underline{U}'(f_x, f_y)$ from Eq. (7). Let us now define the inclination angle, β, as the angle that the light component makes with respect to the x-y plane and the azimuth angle, α, as the angle between the +x axis and the projection of the component on the x-y plane. Inclination angles are considered positive above the x-y plane; azimuth angles are positive in the counterclockwise direction.

To relate this notation to that of the spatial frequencies f_x and f_y, we consider a unit vector of arbitrary inclination and azimuth. We may then represent this vector in Cartesian coordinates as

$$\vec{1} = \hat{a}_z \sin\beta + \hat{a}_x \cos\beta\cos\alpha + \hat{a}_y \sin\alpha\cos\beta \tag{9}$$

The spatial frequencies are specified by relations such as

$$f_x = \frac{(\text{direction cosine})_x}{\lambda} = \frac{\cos\theta_x}{\lambda} \tag{10}$$

where θ_x is the angle between the vector and the x axis.

Thus
$$f_x = \frac{\vec{1}\cdot\hat{a}_x}{\lambda} \tag{11}$$

$$f_x = \frac{\cos\beta\cos\alpha}{\lambda} \tag{12}$$

Similarly,

$$f_y = \frac{\cos\beta \sin\alpha}{\lambda} \tag{13}$$

By substituting these expressions for the arguments of $\underline{U}'(f_x, f_y)$ we can obtain an expression for the spatial distribution function in terms of the angles of inclination and azimuth.

Conceptually the extension of the plane-wave/plane-wave viewpoint of Bragg diffraction to the general interaction of arbitrary light and sound fields is now evident. Simplistically, the approach is as follows: the interacting light and sound fields are each decomposed into plane-wave components. All components which meet the required angular conditions of Bragg diffraction interact to produce a diffracted component. After accounting for diffraction these diffracted components are assembled at any desired plane and the inverse transform performed to find the physical light distribution at that plane. The important contribution to Bragg imaging was Korpel's recognition[7] that the proper choice of the geometric shape of the incident light would cause the diffracted light to reproduce a scaled version of the sound field and hence to produce an image of that sound field.

TWO-DIMENSIONAL INTERACTION FOR IMAGING

To illustrate the simple imaging of a sound field consider a two-dimensional case. We assume that the light source is an infinite vertical line source of unit amplitude and that the sound source is also of infinite vertical extent, but not necessarily a line source (e.g., an infinitely long slit). The magnitudes of the one-dimensional angular spectra for this case are shown in Fig. 5. The spectrum of the line source of light has components of equal magnitude and phase for all angles. The infinite slit spectrum (with respect to the center-line) has a magnitude dependence of the $|\sin\phi/\phi|$ type and a phase variation which changes from 0 to π whenever $\sin\phi/\phi$ becomes negative.

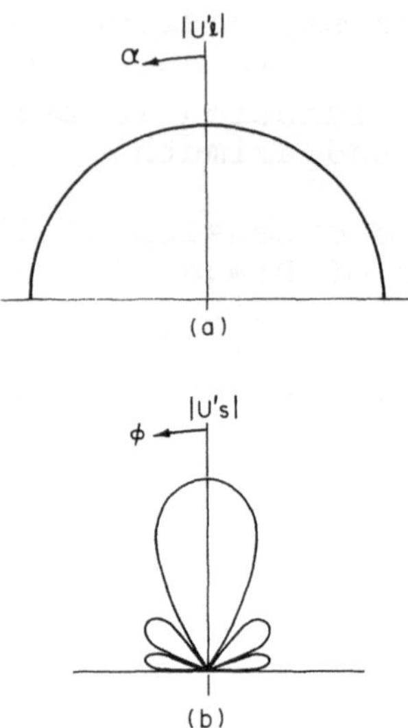

Fig. 5. Angular spectra of (a) an infinite line source of light radiating into only the +x half-space, and (b) an infinite slit source of sound.

The interaction generates a downshifted real image of the sound source in the manner shown in Fig. 6. Consider first the zero-order component of the sound field. Since it must interact at an angle $\pi/2-\phi_B$ we see geometrically that it must interact with the light component propagating at an angle $+\phi_B$ with respect to the x axis. We next consider a sound component at an angle ϕ. In order for it to intersect a light component at an angle $\pi/2-\phi_B$ it must select a light component which propagates at azimuth angle $\alpha = \phi+\phi_B$. The diffracted light component amplitude is directly proportional to the amplitude of the sound component. The phase is the difference of the

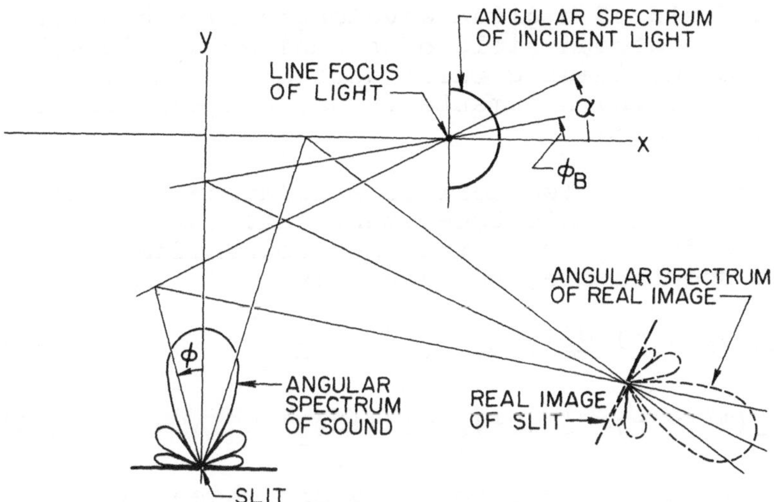

Fig. 6. Two-dimensional interaction to produce a real downshifted image of a slit.

incident light and sound phases. If the phase and amplitude of all incoming light components are equal, as in a line source, then the diffracted light duplicates the sound spectrum but with the propagation reversed.[8] Since all diffracted rays are in the plane of the figure and since the spectrum is reconstructed as the complex conjugate of the sound field, we produce a real image of the sound field at the image plane. Similar analysis of the upshifted case would have produced a virtual image of the field located symmetrically with respect to the undiffracted light beam.

Investigating the diffracted spectrum several observations can be made. First, the optical image is rotated by an angle $\pi/2+\phi_B$ from the acoustic field. Second, the sound spectrum is now reproduced in light. Because of the differences in wavelength this implies a demagnification of the width of the image by the scale of λ/Λ. This demagnification effect is common to all methods where a field of a given wavelength is reconstructed using a second field of a smaller wavelength. These are the imaging rules derived by Korpel from ray tracing arguments in his original paper[7] on Bragg imaging.

Korpel has also derived a relation between the spectra for this two-dimensional case.[9] The relations for the upshifted and downshifted diffracted spectra are, respectively,

$$\underline{U}'_{+}(\psi) = -jc\underline{U}'_{s}(\psi-\phi_B)\underline{U}'_{\ell}(\psi-2\phi_B) \tag{14}$$

$$\underline{U}'_{-}(\psi) = -jc\underline{U}'^{*}_{s}(\psi+\phi_B)\underline{U}'_{\ell}(\psi+2\phi_B) \tag{15}$$

where c is an interaction constant whose value depends on the medium and on the ratio of wavelengths λ/Λ and where * denotes the complex conjugate. The azimuth angles of both the incident and diffracted light spectra are measured with respect to the +x axis. The angle of the sound propagation is measured with respect to the +y axis. These formulae express analytically the heuristic geometrical description given above. From these formulae it is seen that in order to reproduce exactly the sound spectrum in the diffracted light spectrum the incident light spectrum should have constant amplitude and constant phase over all of its angular components. This in turn implies that the best choice of an incident light pattern to image the sound field is a line source. This conclusion was proven later in experimental practice to give the best images of several sources.

SPATIAL FILTERING IN THE TWO-DIMENSIONAL INTERACTION

Because the one-dimensional spectra are multiplied as in Eqs. (14) and (15), it should be possible to form a modified reconstruction of the sound spectrum by tailoring the incident light spectrum. This is similar to the principle used in spatial filtering of optical images.[10] Figure 7 shows one arrangement of optical elements to perform such spatial filtering. The point source S and lens L_1 serve to illuminate an object transparency in plane P_1 with a collimated light beam. Lens L_2 is a Fourier transform lens; the Fourier transform of the object in plane P_1 is formed at plane P_2. This spatial spectrum can then be manipulated by placing a "filtering" transparency in plane P_2 which can attenuate, eliminate, or change the phase of any plane-wave component of the object. This filtering operation is possible because the filter transparency and the Fourier transform light distribution multiply in the plane P_3. Lens L_3 then performs another Fourier transform and the resulting filtered image is displayed (in inverted fashion) in plane P_3. Some examples of the optical operations that can be performed in this fashion are phase contrast, noise removal, matched filtering, correlation, etc.

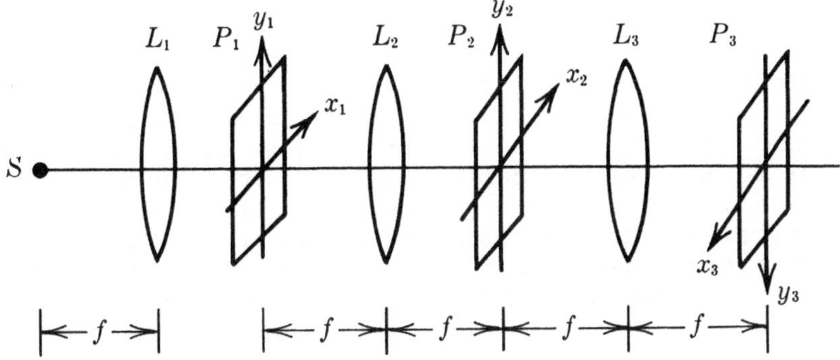

Fig. 7. Schematic diagram of an optical filtering setup.

One of the simplest filter functions is the removal of a component or components from the image spectrum. This is the principle used in contrast reversal (removal of the central order component), removal of periodic noise (removal of spatial component corresponding to periodicity of the noise), edge enhancement (removal of the lower spatial frequencies), etc. The elimination of a particular diffracted light component in the acoustic image can be accomplished by the elimination of the particular incident light component that interacts with the undesired sound component. Two possible methods for the removal of diffracted light components are the use of a modified geometry of the imaging system or the spatial filtering of the incident light.

The first of these methods makes use of the limited angular spectrum of the usual incident light pattern. Physical line sources used in the imaging system have to be introduced through a lens. The semiapex angle of the wedge forming the line image can be taken as a measure of the angular extent of the spectrum. This angle is related to the numerical aperture of the lens (assuming that the lens is fully illuminated) by

$$\alpha_m = \sin^{-1} (\text{N.A.}) \tag{16}$$

where N.A. is the numerical aperture. The angular spectrum of the incident light then can be assumed to have fan-shaped distribution with a sharp cutoff point at $\pm\,\alpha_m$. This distribution is an approximation to the more gradual cutoff of a physical laser beam. It is then possible to make use of this spectral cutoff feature to eliminate all spatial frequencies of the image above a certain frequency by changing the angle of incidence between the sound field and the incident light field.

The validity of this sharp angular spectrum cutoff feature of the incident light in two-dimensional imaging may be checked by measurement of the resolution obtained with different wedge semi-apex angles. In this experiment the central-order propagation components of the incident light and sound are aligned so that they interact to produce the downshifted central-order diffracted component as in Fig. 8a. Since the diffracted

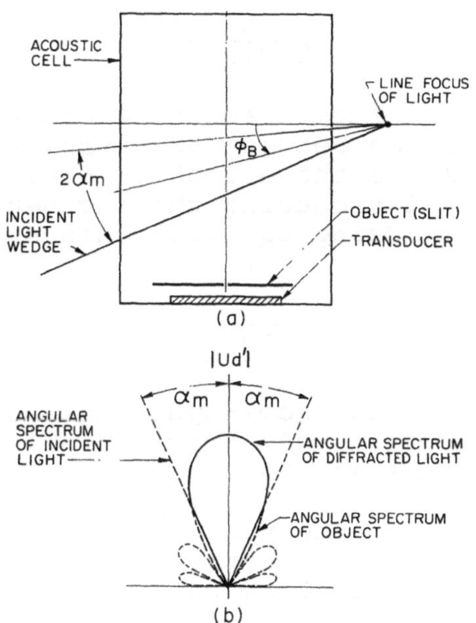

Fig. 8. (a) Physical orientation of the acoustic cell for a normal transmission image. (b) Diffracted light spectrum showing the effect of the limited semi-apex angle of the incident light wedge.

spectrum is proportional to the product of the input spectra, the angular width of the diffracted spectrum is equal to the width of the incident light spectrum as in Fig. 8b (assuming, as is usually the case, that the sound spectrum width is not the limiting factor). As the angular width of the light spectrum is changed, the behavior of the resolution of an acoustic image is an indication of the validity of the assumption of a sharp spectral cutoff of the incident light spectrum since a wider spectrum has better inherent resolution. Based essentially on the same type of analysis Korpel predicted that the image resolution would be given by[7]

$$\xi_h = \frac{\Lambda}{2(N.A.)} \tag{17}$$

where ξ_h is the minimum resolvable distance in the horizontal direction between two vertical-line sources, and N.A. is the effective numerical aperture of the lens introducing the incident light into the system. Using Eq. (16) this expression becomes

$$\xi_h = \frac{\Lambda}{2\sin\alpha_m} \tag{18}$$

This dependence on the semi-apex angle of the light wedge was confirmed experimentally[11] using long vertical wires as sound scatterers.

Two examples of the spatial filtering technique whereby components are eliminated from the image by using this angular cutoff of the incident light spectrum are "dark field" imaging and acoustic specular reflection imaging. Both of these techniques are characterized by the absence of a diffracted component corresponding to the central order sound component of the illuminating sound from the transducer. Only the sound scattered by or reflected from an object is imaged. Figure 9 shows typical images obtained from each method.[1] There is no image of the transducer sound field present.

To obtain "dark field" imaging it is necessary to rotate the cell through some angle η. The central order component of sound (which is largely the illuminating sound field) will have no component of light with which to interact if the magnitude of η is greater than α_m, the semi-apex angle of the light wedge. The experimental arrangement and the resulting diffraction pattern are sketched in Fig. 10. Note that not only is the central order sound component excluded from the diffracted pattern, but also more of the higher spatial frequencies are included in the image. This method provides images of the sound scattered into higher spatial frequencies from small scatterers. Figure 9a is the "dark field" image of a wire hook.

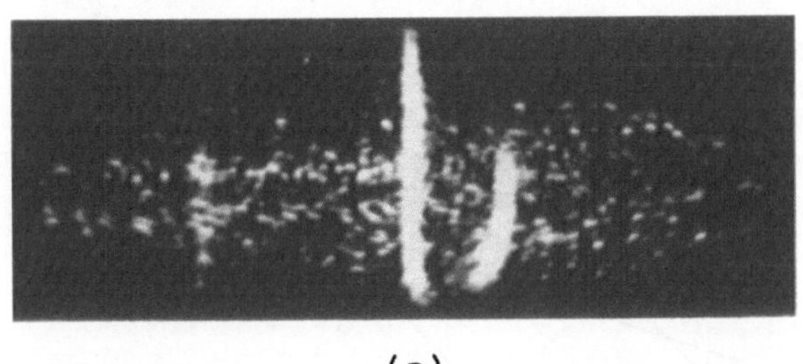

(a)

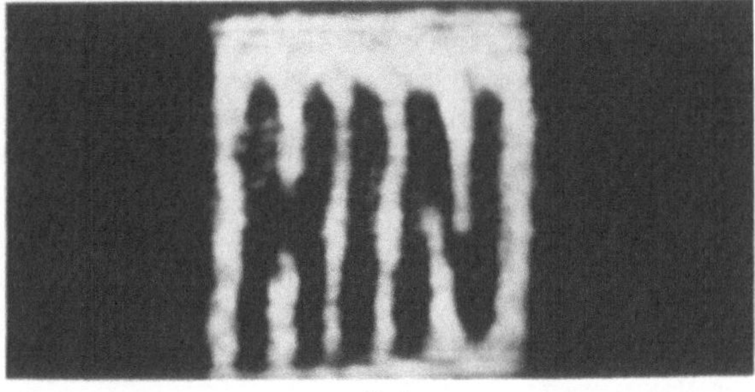

(b)

Fig. 9. Images obtained from (a) "dark field" imaging of a wire hook, and (b) specular sound reflection from a glass slide with masking tape letters.

When the cell is rotated to form a "dark field" image and there is no object to scatter sound energy there will be, of course, no image. It is possible, however, to specularly reflect enough sound energy from an object into the incident light field to form a good image of the reflected sound. As shown in Fig. 11, it is possible to orient the reflecting object so that the central order reflected wound component will interact with the central order light component to obtain a standard (i.e., not "dark field" image.)

Fig. 9b shows an image obtained from the sound reflected from a glass slide with masking tape letters which absorb the ultrasound.

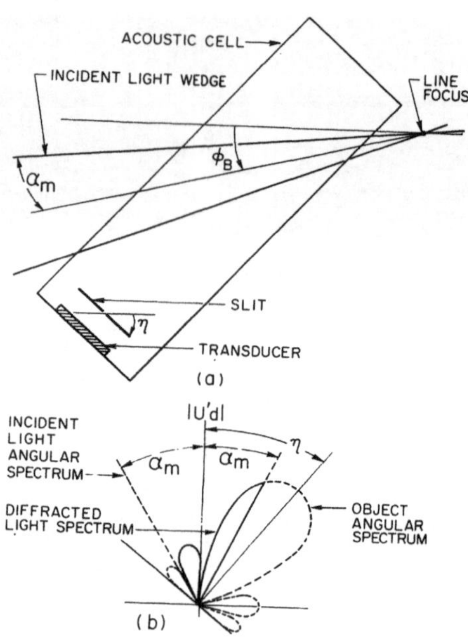

Fig. 10. (a) Physical orientation of the acoustic cell for "dark field" imaging. (b) Angular spectrum of the diffracted light shown as the overlapping portion of the incident light and sound spectra.

The second method of removing an undesired spatial component from the acoustic image is the removal of a particular component from the incident light spectrum by the means of optical spatial filtering. The spatial component that is removed is the component that would interact with the component of sound whose elimination is desired. In a physical imaging system the pseudo-infinite line source of light is formed by focusing a collimated laser beam with a cylindrical lens. The line focus will be located in the back

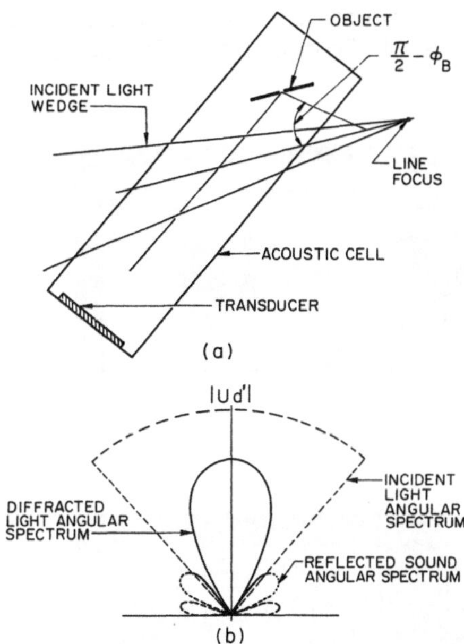

Fig. 11. (a) Physical orientation of the acoustic cell for imaging specularly reflected sound. (b) Diffracted light spectrum shown as the overlapping portion of the incident light and reflected sound spectra.

focal plane of the lens. The front focal plane of the same lens can serve as the filtering plane for the incident light. In this plane we are free to place one-dimensional filters (long strips of opaque stops) to achieve the desired results. Figure 12 shows the image of three vertical wires before and after spatial filtering of the incident light for contrast reversal. In this method the light component that interacts with the central order component of sound is removed by a stop placed in the filter plane of the incident light. This implies that the diffracted spectrum duplicates all of the sound spectrum except for the missing central order component and thus the contrast of the image is reversed. Similar techniques can be used for other filtering problems.

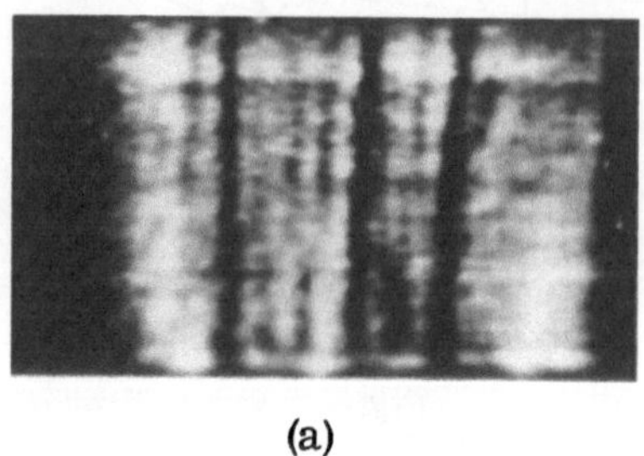

(a)

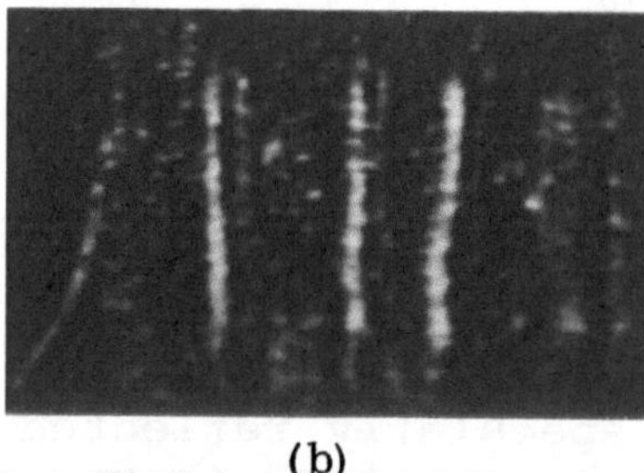

(b)

Fig. 12. Acoustic images of vertical wires (a) with normal transmission method, and (b) with spatial filtering of the incident light to achieve contrast reversal.

It should be noted that this method has greater flexibility than that previously described since attentuating and phase shifting filters can be used, as well as stops. The first technique of rotating the cell has only limited applications where it might be desirable to eliminate all spatial frequencies above or below a certain cutoff. Both methods are limited to the two-dimensional interaction, or equivalently, long vertical objects in the sound field.

THREE-DIMENSIONAL INTERACTION

The previous two-dimensional analysis predicted images of infinite vertical objects with good horizontal resolution. It is now of interest to consider the interaction for horizontal objects with vertical detail. To investigate this case we will assume that the sound source is infinitely long in the horizontal direction and that the incident light still forms an infinite vertical line source. These assumptions imply that all light components will be horizontal and all sound components will lie in the vertical plane.

As shown in Fig. 13, we represent the azimuth angle (with respect to the +x axis) of a general light component as the angle, α. The inclination of a sound component from the x-y plane is given by angle, θ. We now seek a relation between these angles that will hold when the components intersect at an angle $\pi/2-\phi_B$, insuring a Bragg interaction. A detailed view of the interacting

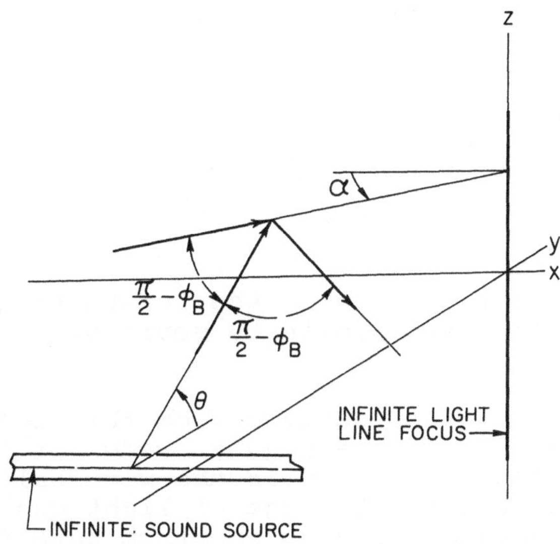

Fig. 13. Geometry for three-dimensional interaction involving infinite sources.

components relative to the intersection point is shown in Fig. 14. Applying Napier's rule to the right spherical triangle shown,[11] we obtain the required relation between the propagation directions of light and sound components

$$\sin\phi_B = \sin\alpha\cos\theta \tag{19}$$

From this type of relation one can predict the light component, specified by α, that would interact with any given sound component, specified by θ.

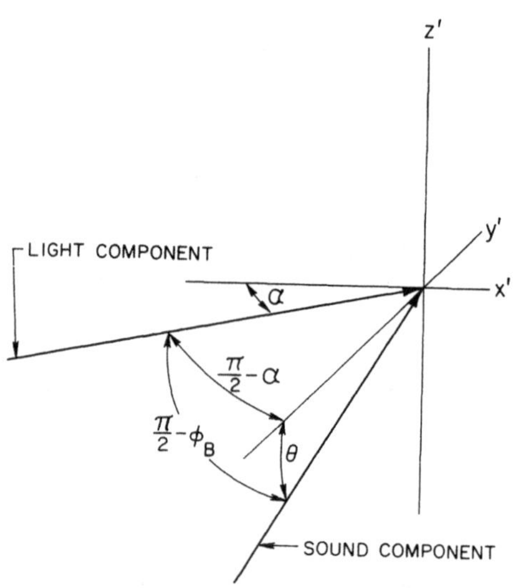

Fig. 14. Detailed view of the interacting components from infinite sources.

One useful fact derived from this relation is that the size of the angular wedge of light determines the resolution in the vertical dimension. Since the input wedge of light can be assumed to have an angular spectrum cutoff at some angle α_m (usually just the semi-apex angle of the wedge), there is a related maximum value of sound component inclination, θ_m. Since no sound components of greater inclination will be imaged, it is this maximum value of inclination that determines the resolution of vertical detail, according to the relation

$$\xi_v = \frac{\Lambda}{2\sin\theta_m} \tag{20}$$

where ξ_v is the minimum resolvable detail in the vertical direction. Experimental results[11] confirm this expression when the height of the line source can be considered infinite. In some experimental situations the physical height of the noninfinite light wedge can be the limiting factor of resolution since it has an effect similar to the pupil function of a lens, limiting the spatial frequencies of the image to those spatial components intersecting the interaction region. When compared with the vertically oriented sound source, it has been found that the size of the resolution element for horizontal objects is significantly less (2/3 of an acoustic wavelength to 10 acoustic wavelengths) than that for vertically oriented objects.

We now wish to consider the general case of interaction between arbitrary components of light and sound. We seek a relation that would identify all light components that could interact with a given specific sound component. Since the angular condition that components intersect at $\pi/2\pm\phi_B$ must be satisfied, this implies that all light components that lie in a cone whose axis is the given sound component and whose semi-apex angle is $\pi/2\pm\phi_B$ can interact with that sound component. Such a situation is illustrated in Fig. 15. In some special cases geometrical considerations will reduce the number of light components that can interact. In the simplest cases (such as those discussed previously) only one component of light can interact with one component of sound because of the planar nature of the light spectrum. This one-to-one correspondence is a desirable feature for imaging since there would be no redundant information in the diffracted light. Of course it is also required that the aberrations of the image not be too severe when introduced by such factors as nonfocussed diffracted components or phase differences due to differing path lengths for various components.

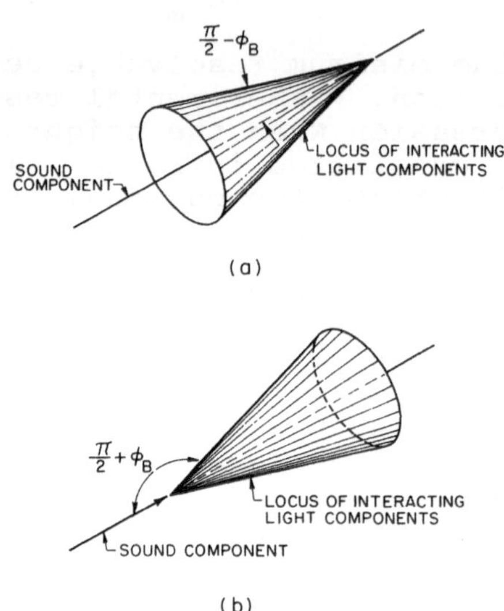

Fig. 15. Angular condition showing all possible interacting light components for a given acoustic propagation vector for (a) downshifted diffraction, and (b) upshifted diffraction.

For the purpose of specifying a component we again use an azimuth angle and an inclination angle coordinate system. Figures 16a and 16b show the coordinate system for the incident light and the sound, respectively. For the incident light distribution the inclination angle β measures the angle that the component makes with respect to the x-y plane; the azimuth angle, α, is the angle between the +x axis and the projection of the component on the x-y plane. Similarly the angle θ measures the inclination of the sound component with respect to the x-y plane; the aximuth angle ϕ is the angle between the +y axis and the projection of the sound component on the x-y plane.

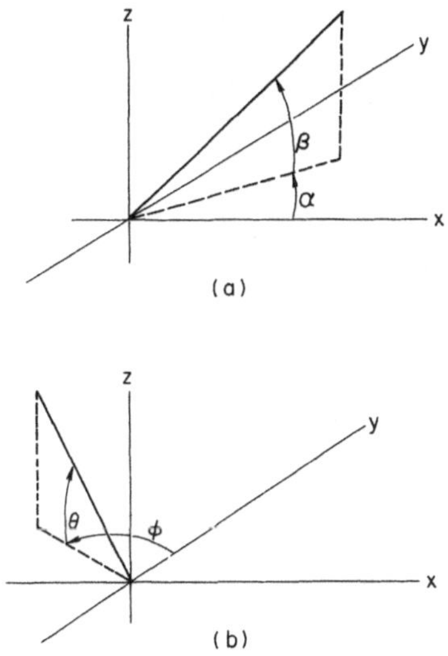

Fig. 16. Angular coordinates for (a) the incident light, and (b) the sound field.

Inclination angles are considered positive above the x-y plane; azimuth angles are positive in the counterclockwise direction.

To derive the relation between the components that can interact we assume that a sound component at an inclination θ and an azimuth ϕ is interacting with a light component at inclination β and azimuth α. As in Fig. 17 these components must intersect an an angle of $\pi/2-\phi_B$ in order to interact. (A similar analysis holds for the upshifted case where the angle of intersection is $\pi/2+\phi_B$.) Figure 18 shows the interaction point with the angular quantities noted. In order to obtain a relation between the quantities we must relate the sides of the sperical quadrilateral formed by the angular arcs (shown in Fig. 18 as BCDE).

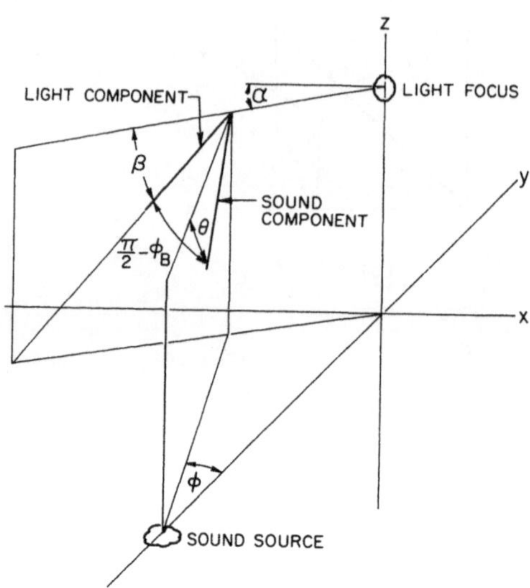

Fig. 17. Interaction between arbitrary sound and light components.

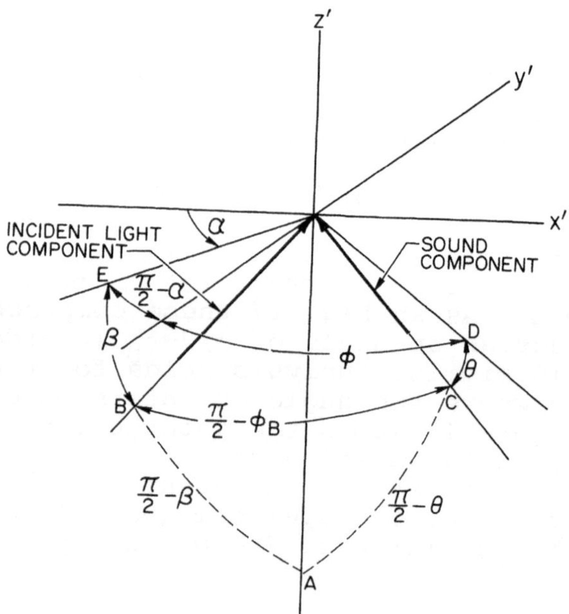

Fig. 18. Detail view of component intersection point.

Because of the geometry of the quadrilateral it is possible to find any one side in terms of the other three. One useful solution is to find β in terms of the other quantities. This is equivalent to determining the interacting ray's inclination while assuming knowledge of its azimuth and the interacting sound ray's azimuth and inclination.

To solve for β we first construct the spherical triangle ABC of Fig. 18 by continuing the meridian lines specifying β and θ to the pole of the hypothetical sphere on which the arcs are drawn. The arcs AB and AC will be complementary to β and θ. The angle A is equal to the angle measured along the equator, i.e., $\pi/2-(\alpha-\phi)$. Solving this oblique spherical triangle for β in terms of the other quantities we find

$$\beta = \pi/2-\cos^{-1}\left[\frac{1}{\sqrt{1+\frac{\sin^2(\alpha-\phi)}{\tan^2\theta}}}\right]$$

$$-\cos^{-1}\left[\frac{1}{\sqrt{1+\frac{\sin^2(\alpha-\phi)}{\tan^2\theta}}}\cdot\frac{\sin\phi_B}{\sin\theta}\right] \qquad (21)$$

This is the desired relation giving the inclination of the light component in terms of its azimuth and the angular coordinates of the given sound component. As noted previously, the amplitude and phase of the diffracted wave are determined respectively by the product of the amplitudes and the difference of the phases of the interacting components specified by this equation.

From this equation we note that for a specified sound component α serves as a parameter in determining the inclination angle β. Thus as α varies, the value of β will also change indicating that several light components will interact with the given sound component. This interaction with several light rays is undesirable for imaging because of the resulting ambiguity of the information. Thus we conclude that for imaging purposes the best light source is the infinite light source or an approximation since this restricts the spectrum to be one dimensional. With this type of source the components are restricted to being coplanar and the ambiguity of information is removed. Aberrations may still be present, however, that limit the usefulness of the image.

IMAGING WITH INFINITE LINE SOURCES OF LIGHT

With the assumption that an infinite light source produces an unambiguous image (although not necessarily without aberration) it is instructive to consider some special cases involving various orientations of the infinite light source. When the infinite light source is oriented vertically the expression of Eq. (21) simplifies to

$$\sin(\alpha-\phi)\cos\theta = \sin\phi_B \tag{22}$$

$$\alpha = \sin^{-1}\left[\frac{\sin\phi_B}{\cos\theta}\right] + \phi \tag{23}$$

This is a particularly important configuration since current Bragg imaging systems use vertical line sources or images (although not infinite) to obtain the best images of arbitrary sound fields. From Eq. (23) it is seen that a given component of the planar light spectrum specified by the azimuth angle α interacts with an infinity of sound components whose angular coordinates satisfy the above relation. Here it is implicitly assumed that the input laser beam is intense enough to interact with all possible sound components. Because of this one-to-many interaction for a general sound field the possibility is reduced for spatially filtering the image by modifying the input light beam as in the two-dimensional interaction unless the imaged object approximates an infinitely long vertical object. If not, removal or modification of a light component which interacts with the undesired sound component now implies that other spatial components of the image will be removed. While it is possible to predict which components will be affected (by Eq. (23)) this feature is generally undesirable for spatial filtering where a one-to-one correspondence is desired for precise control of the filtered image. Of course it is still possible to spatially filter the optical image of the acoustic field by conventional optical techniques before viewing the filtered image.

More information regarding verification of Eq. (21) can be obtained by making assumptions about the size and orientation of the sound source. In particular it is instructive to consider an infinite sound source oriented either vertically or horizontally. When the infinite sound source is vertical, then the spectrum varies only in azimuth and we can assume $\theta = 0$. Equation (22) then reduces to

$$\sin(\alpha-\phi) = \sin\phi_B \tag{24}$$

or

$$\alpha = \phi_B+\phi \tag{25}$$

Referring to the geometric two-dimensional analysis we see that this expression is the same as that previously obtained. This is expected since the case of two infinite parallel sources is exactly the same as the two-dimensional case.

When the infinite sound source is oriented horizontally, the sound spectrum varies only in inclination, and thus $\phi = 0$. Then Eq. (22) becomes

$$\sin\alpha \cos\theta = \sin\phi_B \tag{26}$$

This case is the same as that used to introduce the three-dimensional interaction, and the relation of Eq. (26) is the same as that derived geometrically (Eq. 19)).

Another case of interest is the interaction of a general sound field with the light from an infinite horizontal light source. For this imaging system the light spectrum varies only in inclination but not in azimuth (α=0). For the special case of the infinite horizontal light source, Eq. (21) reduces to

$$\beta = \frac{\pi}{2} - \cos^{-1}\left[\frac{1}{\sqrt{1 + \frac{\sin^2(-\phi)}{\tan^2\theta}}}\right]$$

$$- \cos^{-1}\left[\frac{1}{\sqrt{1 + \frac{\sin^2(-\phi)}{\tan^2\theta}}} \cdot \frac{\sin\phi_B}{\sin\theta}\right] \qquad (27)$$

Again the one light component β can interact with all sound components whose angular coordinates satisfy the above relation.

When further restrictions are made on the sound source limiting it to an infinite source, further simplifications of the relation occur. When the infinite sound source is vertically oriented (θ=0) the relation becomes

$$\beta = \cos^{-1}\left[\frac{\sin\phi_B}{\sin(-\phi)}\right] \qquad (28)$$

When the infinite sound source is oriented horizontally (ϕ=0), the inclination of the interacting light ray is given by

$$\beta = \sin^{-1} \frac{\sin\phi_B}{\sin\theta} \qquad (29)$$

From these relations we have verified that the choice of a line source of light is the best when considered on the basis of avoiding a redundancy of information in the diffracted light. This choice has been derived without consideration of possible aberrations in the image which depends in part on the orientation of the line source. Preliminary analyses[8] of some cases involving infinite sound sources show that two vertical infinite sources will give an essentially undistorted image while a vertical light source and a horizontal sound source give an aberrated image. Consideration of both facets of the imaging problem would be necessary to determine the light distribution and orientation that would give the best image for any general three-dimensional sound field.

SUMMARY

Consideration of Bragg diffraction imaging using a plane-wave analysis yields much information about this imaging technique. As might be expected the two-dimensional analysis gives, in a fairly simple fashion, a great deal of information about the formation of images and such diverse effects as the observed resolution, dark field imaging and novel spatial filtering techniques of the acoustic image. These results are based on the present Bragg diffraction imaging systems that use a quasi-infinite vertical line image of incident light. The geometrically more complicated three-dimensional analysis gives less quantitative information about the image. Angular relations between the interacting light and sound components have been derived. From this relation some simple qualitative predictions about image quality can be made.

ACKNOWLEDGEMENTS

This study is part of a thesis submitted in partial fulfillment of the requirements for the degree of Doctor of Philosophy at the University of California, Santa Barbara.

This work was supported by the Office of Naval Research through the Foundation Research Program at the Naval Postgraduate School and the National Institute of Health.

REFERENCES

1. J. Landry, J. Powers, and G. Wade, "Ultrasonic imaging of internal structure by Bragg diffraction," Applied Physics Letters 15(6):186-188 (1969).

2. R. Adler, "Interaction between light and sound," IEEE Spectrum 4(5):42-54 (1967).

3. C. F. Quate, C.D.W. Wilkinson, and D. K. Winslow, "Interaction of light and microwave sound," Proc. IEEE 53(10):1604-1628 (1965).

4. L. Brillouin, "Diffusion de la lumiere et des rayon X par un corps transparent homogène," Annals of Physics (Paris), 9th Ser., 17:88-122 (1922).

5. A. Korpel, The interaction of sound and light fields of arbitrarily prescribed cross-section," Zenith Radio Corporation Research Report No. 66-2, September 1966.

6. J. A. Ratcliffe, "Some aspects of diffraction theory and their application to the ionoshere," Reports on Progress in Physics, V. XIX, A. C. Strickland, Ed., (The Physical Society, London, 1956) p. 188-267.

7. A. Korpel, "Visualization of the cross section of a sound beam by Bragg diffraction of light," Appl. Physics Letters 9 (12):425-426 (1966).

8. J. P. Powers, R. Smith and G. Wade, "Phase Aberrations in Bragg Imaging," Acoustical Holography, V. 3, (Proc. Third International Symposium on Acoustical Holography), A. F. Metherell, Ed., (Plenum Press, N.Y., 1971) Chap. 5, p. 71-91.

9. A. Korpel, "Acoustic imaging by diffracted light. I - Two dimensional interaction," IEEE Trans. on Sonics and Ultrasonics, SU-15(3):153-157 (1968).

10. L. J. Cutrona, E. N. Leith, C. J. Palermo, and L. J. Porcello, "Optical data processing and filtering systems," IRE Trans. on Information Theory, 6(2):386-400.

11. R. Smith, G. Wade, J. Powers and J. Landry, "Studies of resolution in a Bragg imaging system," Acoustical Society of America, 49, (3):1062-1068, 1971.

GAIN AND PHASE VARIATIONS IN HOLOGRAPHIC ACOUSTIC IMAGING SYSTEMS

Jim Thorn

Naval Undersea Research and Development Center

San Diego, California 92132

ABSTRACT

The effect of random gain and phase errors in the hydrophones and associated amplifiers in a uniformly-filled square array of a holographic acoustic imaging system is theoretically analyzed. Temporal noise introduced by the amplifiers is also considered. The contrast of the image reconstructed from the holographic information is shown to depend on the statistics of the array, on the dynamic range of the amplifiers, and on the intensity distribution of the target. It is also shown that image contrast depends on field-of-view location when the width of the sensing elements in the array is an appreciable fraction of the separation of the elements. A method of compensating for known gain and phase variations is presented. The theory is used to evaluate the performance of some existing systems.

INTRODUCTION

This paper uses probability theory and Fourier transform theory to analyze the effect of gain and phase errors on the performance of a holographic Acoustic Imaging System. The system considered here is designed for long-range underwater viewing, so all targets are assumed to be in the far-field. The size, position, and intensity distribution of a target, as well as the statistics and noise characteristics of the receiving array, are found to affect the contrast of the reconstructed image of that target. The theoretical result is used to predict the image contrast attainable from hardware currently being developed.

THE ACOUSTIC IMAGING SYSTEM AND ITS MATHEMATICAL MODEL

The Acoustic Imaging System consists of a uniformly-filled two-dimensional array of MxM hydrophone receivers arranged in a plane square aperture of width L. Each hydrophone is connected to its own signal processor, and the combination is a "channel." An acoustic signal arriving at the hologram plane from a target is received by the hydrophones, mixed with an appropriate reference signal, and filtered to yield DC voltages for channel output signals, which make up the hologram of the target. The reconstructed image of the target may be obtained by performing a two-dimensional spatial Fourier transform on the array of channel output signals. The "ideal" channel outputs form a hologram which, when Fourier transformed, yields a perfect image of the target, subject only to wavelength, aperture, and sampling limitations. The "actual" channel outputs differ from the ideal by the introduction of gain and phase errors unique to each channel. It is the nature of a hologram that each of many small samples of the hologram contains information from the entire target and hence will contribute information to the entire image; therefore, it is logical to analyze the statistics of the reconstructed image in terms of the statistics of the gain and phase errors.

The following definitions will be used for this analysis.

S_{nm} - ideal channel output signal from the channel located at the n-th row, m-th column of the receiving array.

M^2 - total number of hydrophones in the MxM array.

L - linear dimension of the square array; rows and columns are therefore L/M apart, and the array density in hydrophones per unit area is M^2/L^2.

d - width of each hydrophone in the array; d must be less than L/M.

Γ_{nm} - factor (multiplicative) by which the signal from the channel at the n-th row, m-th column differs from the ideal signal, due to gain errors and/or phase shifts. Γ is assumed to be a random variable.

σ_Γ - rms value of Γ when averaged over all the channels.

T_{nm} - electronic noise term added by each channel.

H_{nm} - $\Gamma_{nm}S_{nm} + T_{nm}$ = actual channel output, forming the hologram.

I_a - the actual reconstructed image, found by Fourier transforming the hologram H.

I_i - the ideal reconstructed image, found by Fourier transforming the ideal channel outputs.

E( ) - statistical "expected value of" or "ensemble average of" a random variable.

$\exp(j\theta_{nm})$ - an abbreviation of the two-dimensional Fourier transform factor $\exp j2\pi(f_x x_n + f_y y_m)$, where f_x and f_y are the two components of spatial frequency in the hologram plane and correspond to spatial position in the reconstructed image plane.

F - Fourier transform.

GAIN-PHASE NONUNIFORMITIES AND RANDOM "SPATIAL NOISE"

The actual hologram as seen at the channel outputs is

$$H_{nm} = \Gamma_{nm}S_{nm} \,, \tag{1}$$

disregarding electronic noise. The Fourier transform of equation 1 is the reconstructed image:

$$I = F(H) = L^2/M^2 \sum_{nm} \Gamma_{nm}S_{nm}\exp(j\theta_{nm}) \,, \tag{2}$$

A spatially constant normalizing factor has been ignored. An average image may be defined as:

$$E(I) = L^2/M^2 \quad E(\Gamma) \quad \sum_{nm} S_{nm}\exp(j\theta_{nm}) \,. \tag{3}$$

The actual image differs from the average image by

$$B = I - E(I) \,. \tag{4}$$

B may be considered to be spatial noise introduced by the random gain-phase errors.

There is an analogy between this "spatial" noise, consisting of unwanted random signal fluctuations with position on the hydrophone array, and ordinary electronic "temporal" noise, consisting of unwanted random signal fluctuations with time. The power dissipated in a resistive load by a temporal noise signal at any given time is proportional to the square of the noise signal amplitude at that time; the total energy dissipated over a given interval is the integral, over time, of the instantaneous power. Similarly, the power or intensity of the spatial noise B at any point in the image plane is BB* (where * indicates complex conjugate). Since all parts of the hologram contribute equally to all parts of the image, the noise energy will be distributed equally over the image plane, so that the noise intensity P_N at any point in the image will be, on the average, E(BB*), which can be found from Eqs. 2 to 4:

$$P_N = E(BB^*) = \left[E(\Gamma^2) - E^2(\Gamma)\right] L^4/M^4 \sum_{mn} |S_{nm}|^2 \; . \quad (5)$$

The sum in Eq. 5 is evaluated by using the temporal noise analogy. The intensity of the ideal (noiseless) image is $I_i \, I_i^*$, and the total ideal image energy is the integral of $I_i \, I_i^*$ over the field of view in the image plane. However, Parseval's theorem[1] equates the energy in a function with the energy in the Fourier transform of the function. Therefore, the total ideal image energy must be equal to the total energy in the ideal channel output signals, since the image and hologram are Fourier transforms of each other. Hence,

$$U_i = \sum I_i \, I_i^* = L^2/M^2 \sum_{nm} |S_{nm}|^2 = L^2 \, (S_{rms})^2 \; , \quad (6)$$

where U_i is the total ideal image energy, and S_{rms} is the root mean square value of the ideal channel outputs. From Eqs. 5 and 6:

$$P_N = L^2/M^2 \left[E(\Gamma^2) - E^2(\Gamma)\right] U_i \; . \quad (7)$$

Equation 7 embodies three important qualitative relationships:

1. The intensity (energy per unit area) of the spatial noise is proportional to the total energy (intensity integrated over field of view) of the ideal image.

2. The intensity of the spatial noise is constant over the entire image plane.
3. The intensity of the spatial noise decreases as the density of hydrophones in the array (M^2/L^2) increases, if the statistics of the array is not changed.

The dependence of noise intensity on array density may be explained in terms of the sampled nature of the acoustic hologram. If a function is sampled at a rate Δ in the spatial domain, then the Fourier transform of the sampled function repeats itself at a rate $1/\Delta$ in the frequency domain.[2] Since the hologram is sampled at the rate L/M, which is the distance between rows and columns in the receiving array, the reconstructed image will repeat itself every M/L. The field of view is the two-dimensional area between repeated images, which is a square of dimension M/L x M/L. Thus the total noise energy is the noise intensity P_N times the field of view M^2/L^2. Equation 7 shows that this total noise energy must be independent of the array density. However, a greater array density would result in a greater field of view and a larger area for noise energy distribution. In this case, since the image size is not dependent on the array density, the noise is less intense relative to the image.

TEMPORAL NOISE

Channel gain and phase errors in the hologram cause background noise of uniform intensity in the image. Electronic noise (temporal noise) present in the preamplifiers and mixers of each channel, though not necessarily related to gain and phase shift of each channel, nonetheless affects the image in a similar manner by adding background noise.

The temporal noise term is T_{nm}, which is added to the signal at each channel. The amplitude contributed by this term to the image plane is

$$I_T = F(T) = L^2/M^2 \sum_{mn} T_{nm} \exp(j2\pi\theta_{nm}) \quad . \tag{8}$$

The total energy contributed by this term to the image plane is (from Parseval's theorem)

$$U_T = L^2/M^2 \sum_{mn} |T_{nm}|^2 = L^2 (T_{rms})^2 \quad , \tag{9}$$

where

$$T_{rms} = \sqrt{1/M^2 \sum_{mn} (T_{nm})^2} \quad , \tag{9}$$

and the corresponding intensity is

$$P_T = L^2/M^2 \quad U_T = L^4/M^2 \quad (T_{rms})^2 \quad . \tag{10}$$

Equations 7 and 10 may be added to yield a total noise intensity P caused by imperfections in the channels:

$$P = P_N + P_T = L^4/M^2 \left\{(S_{rms})^2 \quad [E(\Gamma^2) - E^2(\Gamma)] + (T_{rms})^2\right\}. \tag{11}$$

Note that the gain-phase noise P_N is multiplicative and depends on signal strength, while the temporal noise P_T is additive and is hence independent of signal strength.

Although T_{rms} has been derived as an average over space of noise terms added by the various channels, the result should be the same for a T_{rms} derived as an average over time of the actual electronic noise added by any one channel, since T_{nm} is merely a random sample of a time series. Therefore, Eq. 11 could be made to include ambient water noise and other medium anomalies if T were appropriately re-defined, and if such anomalies were of sufficient temporal and spatial incoherence.

ARRAY STATISTICS AND AMPLIFIER DYNAMIC RANGE

The parameter Γ in Eq. 11 can be defined in terms of array statistics. The receiving array contains a large number of mass-produced hydrophones, and electronic amplifiers and mixers. Assume that when all these components are assembled into channels and an identical signal is applied to every channel, the outputs of the channels, in decibels, will be distributed in a Gaussian manner about a mean of 0 with a standard deviation of σ_Γ. Since Γ_{nm} is the voltage gain factor of the n-m-th channel and since a decibel is 20 log (voltage gain), the probability distribution of the N factors will be the curve

$$Q(\Gamma) = \frac{1}{a\Gamma \, \sigma_\Gamma \sqrt{\pi/2}} \quad \exp\left(\frac{-2 \ln^2 \Gamma}{a^2 \sigma_\Gamma^2}\right) \quad , \tag{12}$$

where a = ln 10 / 10 = 0.23. The standard definition of probability and a standard table of definite integrals yield

$$E(\Gamma) = \int_0^\infty \Gamma Q(\Gamma)\, d\Gamma = \exp(\tfrac{1}{8} a^2 \sigma_\Gamma{}^2)$$
$$E(\Gamma^2) = \int_0^\infty \Gamma^2 Q(\Gamma)\, d\Gamma = \exp(\tfrac{1}{2} a^2 \sigma_\Gamma{}^2) \quad . \qquad (13)$$

Equation 13 relates inter-hydrophone voltage gain variations to the corresponding decibel gain variations.

The temporal noise term T in Eq. 11 can be expressed in terms of a more common amplifier parameter – dynamic range. Dynamic range is defined as the ratio of the maximum signal an amplifier can pass without clipping to the zero-signal noise level. If acoustic signals returning from a target are strong enough to overdrive any of the channel processors, harmonics of the spatial frequencies in the hologram are generated, causing spurious images to appear in the reconstruction of the hologram. The position and nature of the spurious images are highly dependent on the position and nature of the target. As a result it is not possible to analyze the effects of clipping in general. Instead, the sound transmitter may be adjusted so that the signal reflected from the target is not great enough to overdrive any of the channels; in this case the dynamic range D_A of any channel is the ratio of its maximum output signal S_{max} to its internal noise T, so that $D_A = 20 \log (S_{max}/T_{rms})$, or

$$T_{rms} = S_{max} \exp(-\tfrac{1}{2} a\, D_A) \quad . \qquad (14)$$

Equations 13 and 14 may be combined with Eq. 11 to yield

$$P = P_N + P_T = L^4/M^2 \quad (S_{rms})^2 \quad \Big[\exp(\tfrac{1}{2} a^2 \sigma_\Gamma{}^2) -$$
$$\exp(\tfrac{1}{4} a^2 \sigma_\Gamma{}^2) + \left(\frac{S_{max}}{S_{rms}}\right)^2 \exp(-aD_A)\Big] \quad , \qquad (15)$$

which relates total image noise intensity to signal strength (image energy), array density, dynamic range, and decibel gain variation. This noise intensity is the background noise which may be used to define image dynamic range.

CONTRAST AND IMAGE DYNAMIC RANGE

There are two aspects to acoustical holographic imaging which make the problem of defining an image contrast different from the similar problem in optical imaging. First, the image background noise in acoustic holography is proportional to the image energy, as shown in Eq. 15. If the contrast were defined, as it often is in optical images, as the ratio of the difference and the sum of the picture's maximum and minimum intensities, then contrast would not be a true measure of the system performance in all cases. For example, an image consisting of two "bright" objects would have, in the acoustic system, twice as much background noise and hence much less contrast than an image consisting of a single bright object; whereas an optical system would show both images as having the same contrast. Second, most targets of interest to a user of long-range acoustic holography will exhibit specular reflection of sound waves, so that images will appear as intense central spots surrounded by dim object outlines. A viewer is not so interested in the contrast between the intense spots and the background noise as in the contrast between the dim outlines and the background noise.

The parameter which accurately describes the performance of the acoustic system must take special account of specularity. For cases in which the target is so specular that those portions of the image outside the specular highlights do not have enough energy to contribute significantly to the background noise, a reasonable indication of contrast can be obtained from the parameter "image dynamic range":

$$D_I = \text{image dynamic range}$$

$$= 10 \log \left(\frac{\text{intensity of image highlight}}{\text{intensity of background noise}} \right) . \quad (16)$$

Image dynamic range is a measure of how dim a portion of a specular target may be, relative to the brightest portion of the target, without being buried in the background noise of the reconstructed image. For instance, a dynamic range of 10 dB indicates that the background noise is 10 dB lower than the image highlight. Thus, those portions of the object outline which are 10 dB lower than the highlight will exhibit a signal-to-noise ratio of unity.

The image dynamic range can be found using the expression for noise intensity in Eq. 15, and the expression for image energy in Eq. 6: image intensity is image energy per unit area, which is image energy divided by image size. The size of an image highlight can be no less than (but may be greater than) the size of a resolution element. The area of a resolution element for a sampled holographic system is equal to the field of view of that system divided by the number of samples;[3] since the field of view is M^2/L^2, the resolution element area must be $1/L^2$. Most of the image energy is concentrated into the image highlight, so the highlight intensity is

$$P_h = U_i \;/\; (n/L^2) = \frac{L^2}{n} U_i = \frac{L^4}{n} S_{rms}{}^2 \quad , \tag{17}$$

where n is the size of the highlight in resolution elements. A combination of Eqs. 15 to 17 gives

$$D_I = 10 \log \left(\frac{M^2/n}{\left[\exp(\frac{1}{2}\, a^2\, \sigma_\Gamma{}^2) - \exp(\frac{1}{4}\, a^2\, \sigma_\Gamma{}^2)\right] + \left[(\frac{S_{max}}{S_{rms}})^2 \exp(-aD_A)\right]} \right). \tag{18}$$

For images whose highlights occupy more of the field of view than a single resolution element, the area over which the highlight energy is spread is greater than for smaller highlights, and the dynamic range between the highlight intensity and the background noise intensity is less.

FINITE HYDROPHONE SIZE

Gain-phase variations determine the background noise level relative to the image highlight; hydrophone size determines the level of the dim portions of the image relative to the highlight. Therefore, an analysis of image contrast should include a consideration of hydrophone size.

In a holographic imaging system, spatial frequency in the hologram plane corresponds to position in the image plane — high spatial frequency corresponds to large distance from the center of the field of view. A high spatial frequency means the signal varies rapidly with position across the face of the receiving array. Because an individual receiving hydrophone does not have infinitesimal width, its output signal will be an integral of the signal appearing across its face, and rapid spatial variations may get averaged out, with the net effect that images of objects near the edges of the field of view will be attenuated relative to images of objects near the center of the field of view.[4]

To evaluate this attenuation quantitatively, consider the averaging action of the hydrophone to be a convolution of the acoustic signal with a weighting function given by the hydrophone's sensitivity over its face. If the hydrophone is uniformly sensitive over its entire width d, the weighting function is rect(x/d), where rect( ) = 1 when $-\frac{1}{2} \leq (\ ) \leq \frac{1}{2}$ and 0 otherwise. Since the Fourier transform of a convolution of two functions is the product of the transforms of the two functions, the amplitude of the image formed from the ideal hologram may be multiplied by

$$F(\mathrm{rect}(x/d)) = \sin(\pi x d) \ / \ \pi x \tag{19}$$

to get the image formed from the convoluted hologram. For an image highlight located at x_1 in the image and a dim outline located at x_2, the ratio R in decibels of the respective intensities of the dim outline in the ideal and in the attenuated cases is

$$R = 10 \log \left(\frac{x_2^2 \sin^2(\pi x_1 d)}{x_1^2 \sin^2(\pi x_2 d)} \right) \quad . \tag{20}$$

For example, if D_I is 10 dB and R is 3 dB, then the background noise level in the image is 10 dB below the image highlight level; and a portion of the target whose signal strength <u>in the array plane</u> was 10 dB below the highlight's signal strength will, because of the hydrophone convoluting effect, appear <u>in the image plane</u> to be 3 dB below the noise or 13 dB below the highlight. However, the hydrophone convoluting effect is not always detrimental to image contrast. If the highlight is closer to the edge of the field of view than the portion of the dim outline in question, then R will be negative and the signal-to-noise ratio of the dim outline will be enhanced in the image.

The effect on the image of hydrophone signal convolution increases as the hydrophone width increases. R assumes its maximum absolute value when the hydrophone width equals the hydrophone separation L/M, and when the image highlight and the dim outline are separated by half the field of view ($x_1 = 0$, $x_2 = M/2L$, or vice versa), in which case substitution into Eq. 20 yields

$$R_{max} = \pm 10 \log(\pi^2/4) = \pm 4 \text{ dB} \quad . \tag{21}$$

PRACTICAL PREDICTIONS

Equation 18 relates the dynamic range of the image of a specular target to the dynamic range of the channel amplifiers, the number of hydrophones in the array, the nonuniformity of the channel gains, and the size of the image highlight. Figure 1 shows image dynamic range versus gain nonuniformity for various amplifier dynamic ranges, under the following conditions: the array contains 100x100 hydrophones, signal strength is sufficient to take full advantage of the amplifier dynamic range ($S_{max}/S_{rms} = 1$), and the image highlight occupies a single resolution element (the highlight is as sharp and narrow as possible). Similar curves can be drawn for other array sizes and highlight sizes by simply shifting the Fig. 1 curves up or down an appropriate amount, since Eq. 18 shows that array size and highlight size do not affect the shape of the curves. According to Eq. 21, finite hydrophone width will also shift the curves up or down, in

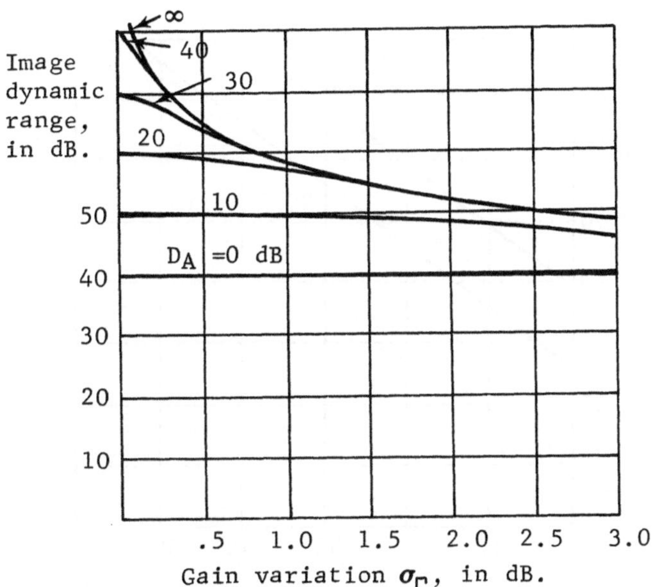

FIG. 1. Image dynamic range is shown as a function of array uniformity for various amplifier dynamic ranges D_A. These curves assume a 100x100 element array, a sharp image highlight, and signal strength sufficient to utilize the full amplifier dynamic range.

some cases by as much as 4 dB, depending on the hydrophone width-to-separation ratio and on the position of the specular highlights in the field of view.

A larger amplifier dynamic range will allow a larger image dynamic range, but there is a point of diminishing returns. For any given array nonuniformity, there is a maximum value which image dynamic range cannot be made to exceed by any increase in amplifier dynamic range. For an array of 100x100 elements with a 1 dB gain variation (see Fig. 1), no increase in amplifier dynamic range above 30 dB will significantly improve the image.

It may be possible to apply an individualized matrix of correction factors to eliminate array nonuniformities altogether, in which case amplifier dynamic range would be the contrast-limiting factor. Figure 2 shows image dynamic range versus amplifier dynamic range for two perfectly uniform arrays, again assuming sufficient signal strength and sharp highlights.

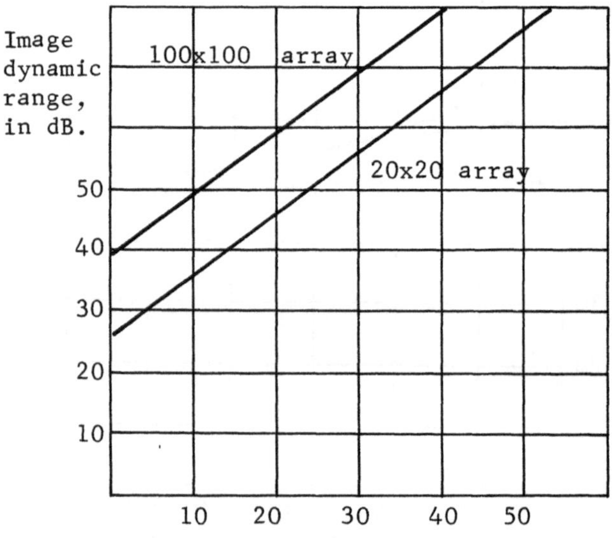

FIG. 2. Image dynamic range is shown as a function of amplifier dynamic range. These curves assume uniform channels, maximum unclipped signal strength, and sharp highlights.

SUMMARY

This paper reduces the consideration of array nonuniformities and noise characteristics to a simple graph of image dynamic range. Many assumptions and approximations are made. Specularity is assumed to be severe enough that nonspecular portions of the target contribute nothing to background noise. Randomly selected array elements are assumed to come from a gain distribution represented by an untruncated decibel Gaussian curve. The effective statistics of the array may be drastically different for cases such as near-field targets which shed most of their reflected energy onto relatively few array elements.

The theory presented here, though incomplete, may be helpful to those doing further analysis or designing hardware.

REFERENCES

1. Goodman, J. W. Introduction to Fourier Optics, McGraw-Hill, 1968. Page 277.

2. Goodman. Page 21.

3. Penn, W. A., and Chovan, J. L. "The Application of Holographic Concepts to Sonar," in Acoustical Holography, Volume II, Plenum Press, New York - London, 1970. Page 141.

4. Smith, J. M., and Moody, N. F. "Application of Fourier Transforms in Assessing the Performance of an Ultrasonic Holography System," in Acoustical Holography, Volume I, Plenum Press, New York, 1969. Page 102.

ACOUSTIC BRAGG IMAGING WITH AN OPTICAL POINT SOURCE

Justin L. Kreuzer

The Perkin-Elmer Corporation

Norwalk, Connecticut

ABSTRACT*

Most acoustic imaging by optical Bragg diffraction uses a coherent line source of light to illuminate the acoustic field and cylindrical lenses to form one conventional optical point image for each point in the acoustic field. Likewise it is generally realized that an optical point source of illumination produces an optical ring image for each acoustic field point [Ref. 1]. In general, the diameter of the ring is larger than the theoretical acoustic resolution, thus providing a poor image of the acoustic field. I will discuss three possible techniques for producing optical point images of acoustic field points with a point source of illumination. (1) The desired optical point images are formed directly when the acoustic wavelength equals one half the optical wavelength. (2) If the optical point source coincides with the acoustic point, the ring image reduces to a point image. (3) The optical ring image can be changed into a point image by a second imaging operation through a second acoustic field or a spatial filter.

*This work has been supported by the U.S. Army Materials and Mechanics Research Center, Watertown, Mass., under contract DAAG46-69-C-0010 and by the Perkin-Elmer Corporation IR&D program.

INTRODUCTION

Most acoustic imaging by optical Bragg diffraction uses a coherent line source of light to illuminate the acoustic field and cylindrical lenses to form one conventional optical point image for each point in the acoustic field. One of the major objections to the use of this line source technique is the high cost of appropriate optics. The resolution and the size of the field of view depend directly upon the quality of the cylindrical optics. It would be nice if we could get the same resolution with spherical optics. In this paper I will discuss some of the problems of using spherical optics. I will assume that the use of spherical optics includes the use of a point source of illumination.

An ordinary collimated beam formed from a point source cannot be used for high resolution imaging because it can interact only with a part of the acoustic field -- the acoustic plane wave components that satisfy the Bragg angle condition. The Bragg angle condition restricts the imaging interaction to only those acoustic plane wave components that lie on a right circular cone whose axis coincides with the direction of propagation of the incident spatial wave. The optical image of the acoustic field will be degraded because the optical image excludes the image of all the other plane wave components.

It is generally realized that an optical point source of illumination produces an optical ring image for each acoustic field point [Ref. 1]. In general the diameter of these rings is larger than the theoretical acoustic resolution and thus the rings provide a poor image of the acoustic field. Under some conditions an image at reduced resolution can be produced by using an optical point source of restricted angular subtense.

We may summarize the problem by noting that the line source technique can give a perfect optical image of an acoustic field because the technique produces one optical plane wave component for each acoustic plane wave component. The cylindrical lenses are needed to make the optical plane wave components have the proper relationships to form an accurate optical image of the acoustic field. A collimated point source or an optical point source of restricted angular subtense produces less than one optical plane wave for each acoustic plane wave component and thus cannot produce a perfect optical image.

On the other hand, a point source of light produces more than one optical plane wave component for each acoustic plane wave component. These extra plane wave components form an optical ring for each acoustic point source instead of a simple point. The problem is to find techniques in analogy to the use of cylindrical lenses with the line source to reduce the extra optical plane waves so that the ring becomes one point.

In order to find techniques to make points out of these rings I will first discuss the formation of these rings. Then I will discuss three possible basic techniques to produce optical point images of acoustic field points with a point source of illumination.

(1) The desired optical point images are formed directly when the acoustic wavelength equals one-half the optical wavelength.

(2) If the optical point source coincides with the acoustic point, the ring image reduces to a point image.

(3) The optical ring image can be changed into a point image by a second imaging operation through a second acoustic field or a spatial filter.

Lastly, I will show some experimental ring images and optical systems that use conventional high quality spherical optics.

THEORY

Acoustic Bragg imaging uses the linear Bragg angle diffraction of light by the acoustic field to form an optical "image" or map of the acoustic field [Ref. 1-5]. Light from a point source travels through a Bragg diffraction cell that is transparent to both the acoustic and optical field. Part of the light is diffracted by the optical index changes produced by the acoustic field. The acoustic field is weak enough so that only a small fraction of the incident light source is diffracted. The total amount of light diffracted is small enough so that the acoustic field is not changed significantly. This makes the interaction linear. Thus, the optical and acoustic fields can be expanded into sums (integrals) of plane waves, and pairs of differential plane-wave components can be treated alone. Each point source is expanded as a uniform distribution of plane waves propagating in all directions whose relative phases are the same at the center of the point. This is represented by the "rays" starting from the point sources. These rays are actually the propagation vectors of the plane-wave components of the point sources. Wherever rays intersect with equal optical path lengths the corresponding plane waves have similar phases. Where many plane wave components have similar phases, they interfere constructively to produce a bright point image.

Each plane-wave component diffracts a small portion of an optical plane wave only if the two waves are at the Bragg angle. The diffracted wave is also a plane wave propagating at the diffraction Bragg angle. The fraction of the optical plane wave diffracted is proportional to the acoustic plane wave component involved. The temporal frequency of the diffracted optical plane waves differs from the incident optical plane waves by the acoustic temporal frequency. This frequency shift will be ignored for now.

Figure 1 illustrates the interaction of an ideal point of light (ℓ) and an ideal point of sound (s) in the Bragg diffraction imaging cell. We need consider only the plane shown that contains the light and acoustic point sources because of the symmetry of the geometry. Thus we will consider 2-dimensional imaging first.

The following analysis of Figure 1 will show that the optical plane-wave components leaving the optical point source (ℓ) and diffracted by the acoustic point source(s), appear to have originated from the point (i). The larger circle (C_1) has a radius (d), where (d) is the distance from (ℓ) to (s), and is centered at the point (s). The small circle (C_2) has a diameter (g) and passes through the points (ℓ) and (s). The following geometric relation holds

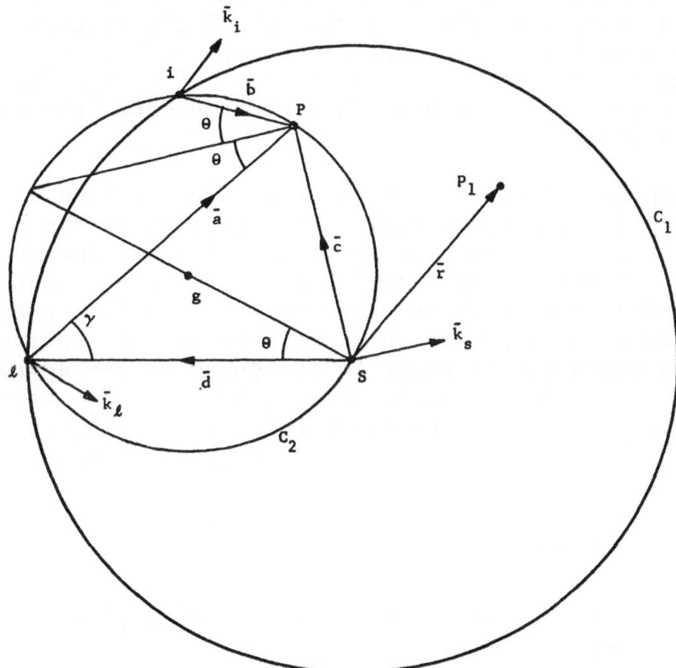

Fig. 1 Bragg Angle Imaging Cell Geometry

$$g = \frac{d}{\cos\theta} \tag{1}$$

where θ is the Bragg diffraction angle given by

$$|\sin\theta| = \frac{1}{2}\frac{k_s}{k} \tag{2}$$

Here, k_s and k are the acoustic and optical wave numbers, respectively. The magnitude is taken because it will be convenient to ignore the sign of the temporal frequency shift. The sign of the temporal frequency shift depends upon whether (ℓ) and (s) are converging or diverging sources or standing fields. This choice will not influence the conclusion.

The following notation will be useful. Bars over variables indicate vectors. The same variable without a bar is the magnitude (always positive) of the vector. The propagation vectors of the optical point-source, acoustic point-source, and optical image are denoted by $\bar{k}_\ell$, $\bar{k}_s$, and $\bar{k}_i$, respectively. A dot between vectors denotes the dot (scalar) product of the vectors.

The optical image's plane-wave components have a complex amplitude proportional to the product of the optical plane-wave component complex amplitude times the acoustic plane-wave component complex amplitude, subject to the condition that these three waves are at the Bragg angle. Thus, the relative phase of a diffracted optical plane-wave component is equal to the phase of the appropriate incident optical plane wave, plus the phase of the appropriate acoustic plane-wave component everywhere. With respect to the vectors shown in Figure 1, this phase relation at an arbitrary point (p_1) is

$$\bar{k}_1 \cdot (\bar{r}-\bar{c}+\bar{b}) = \bar{k}_\ell \cdot (\bar{r}-\bar{d}) + \bar{k}_s \cdot \bar{r} \qquad (3)$$

It is convenient to let the arbitrary point (p_1) lie on the circle (C_2). That is,

$$\bar{r} = \bar{c} \qquad (4)$$

and the point (p_1) coincides with the point (p_o). Equations (3) and (4) yield

$$\bar{k}_i \cdot \bar{b} = \bar{k}_\ell \cdot \bar{a} + \bar{k}_s \cdot \bar{c} \qquad (5)$$

Equation (5) can only be satisfied if the three interacting plane-wave components satisfy Bragg angle conditions. Because of the construction of circles (C_1) and (C_2), the three interacting plane-wave components of (ℓ), (s) and (i) will satisfy the Bragg angle condition if their respective propagation vectors ($\bar{k}_\ell$), ($\bar{k}_s$) and ($\bar{k}_i$) are parallel to the vectors ($\bar{a}$), ($\bar{b}$) and ($\bar{c}$), respectively; that is

$$\bar{k}_\ell = k_\ell \frac{\bar{a}}{a} \tag{6a}$$

$$\bar{k}_s = k_s \frac{\bar{c}}{c} \tag{6b}$$

$$\bar{k}_1 = k_i \frac{\bar{b}}{b} \tag{6c}$$

Also

$$k_i \approx k_\ell \equiv k \tag{7}$$

The incident optical wave number is approximately equal to the diffractive wave number because the velocity of light is approximately 10^5 times the acoustic velocity. Equations (5) and (6) yield

$$k\, b = k\, a + k_s c \tag{8}$$

From Figure 1 we can write the following geometric relations:

$$a = g\left|\cos(\theta-\gamma)\right|$$
$$b = g\left|\cos(\theta+\gamma)\right|$$
$$c = g\left|\sin(\gamma)\right| \tag{9}$$

Equation (8) is found to be true when Equations (2) and (9) are substituted into it with appropriate signs depending upon whether (p) lies between (i)(s), (s)(ℓ) or (ℓ)(i). This implies that (i) is the image of (s), which is what we wished to prove. This means that there is one acoustically diffracted optical plane-wave component for each acoustic plane-wave component. There is also a second similar image formed below the line $\bar{d}$ on the circle C_1 at the point analogous to (i).

The preceding shows that two perfect point images are formed in two dimensions. There are two cases of special interest:

(1) The light source (ℓ) is a coherent line-source perpendicular to the page emitting cylindrical waves, and

(2) The light source (ℓ) is a coherent point-source

emitting spherical waves.

First we will consider briefly the case where (ℓ) is a line-source. In the third dimension the image (i) is actually a slightly curved coherent line perpendicular to the plane of Figure 1. The acoustic point source (s) has been mapped into a line. A suitable set of high quality cylindrical lenses can be used to compress this line image into a point, to form a one-to-one ultrasonic imaging system [Ref. 6].

In the second case, (ℓ) is a coherent optical point-source emitting spherical waves. Figure 1 thus has rotational symmetry about the line (d). Therefore, we may rotate Figure 1 about line (d) as an axis to find that the image (i) becomes a circle centered about the line (d). Circle (C_1) has become a sphere, and circle C_2 has become a doughnut without a hole. The 3-dimensional optical image of an acoustic point-source is a circle centered on the straight line drawn through the illuminating optical point-source and the acoustic point-source, and lying on the intersection of the sphere (C_1) and the holeless doughnut (C_2). This imaging process is linear so that a set of acoustic points is mapped into a unique set of circles. Thus an acoustic image is uniquely mapped into a scrambled optical image.

Next I will discuss three techniques that will convert these optical rings into optical point images to provide an accurate image of the acoustic field.

First if the acoustic wavelength equals one half the optical wavelength a perfect optical poing image is directly formed for each acoustic point. This wavelength condition corresponds to a Bragg diffraction angle of 90°. The ring at the point (i) collapses into a point on the circle (C_1) opposite the optical source (ℓ). The optical image is exactly twice as large as the acoustic field as the ratio of the wavelengths would imply. Unfortunately this implies a very small acoustic wavelength and hence a very high acoustic frequency. For example if the optical wavelength is 0.5μm in the Bragg diffraction cell, the required acoustic frequency in water, where the acoustic velocity is 1.5 Km sec^{-1}, is 6 GHz. A frequency of 16 GHz is required for a typical solid with an acoustic velocity of 4 Km sec^{-1}. Under some conditions this frequency can be achieved in some solids but the frequency is too high for a liquid such as water. It is unfortunate that the required frequency is so high because this would provide the simplest direct Bragg acoustic imaging technique. A variation of the preceding would be to use a birefringent Bragg imaging cell in an appropriate geometry to produce a form of phase matching between the ordinary and extraordinary waves [Ref. 7]. In my opinion it does not seem possible to achieve the required phase matching over a large enough angle to achieve high acoustic resolution [Ref. 8].

A second technique is to let the acoustic point coincide with the optical point. Then the ring image collapses to a point. This is obvious by inspection of Figure 1. The optical image now coincides with the illumination point. The image point must be separated from the illumination point by the frequency, phase, or polarization shift associated with the image. The plane wave components in the image appear rotated with respect to the incident optical point source components. Thus if the source of optical illumination is a point source of limited angular extent, the image may be separated by spatial filtering [Ref. 9]. The light source for the second technique can be a temporally incoherent point source. The acoustic frequency is also arbitrary and may be broadband.

It is necessary to scan the optical point source to form the optical image one point at a time. The optical point is scanned in a TV raster scan over the surface of the acoustic field to be imaged. If the acoustic field to be imaged represents an object, it may be necessary to use acoustic optics to form an acoustic image coinciding with the optical point source scan plane. This technique should be capable of good images because you can scan rapidly and can easily discriminate against the scattered but nondiffracted light that bothers many Bragg imaging techniques.

The third technique is to use the Bragg imaging cell to form the rings and to compress the rings to points in a second operation. The rings can be converted to point images by imaging the rings through a second sound field in a second Bragg cell or by imaging through a spatial filter. The second (compression) acoustic point-source located at (ℓ) (or preferably at a one-to-one image of (ℓ)

with the source (ℓ) blocked out) can be used to compress the ring image (i) back into a point image. The second point-source of sound must satisfy the Bragg relation

$$\sin\theta_2 = \frac{1}{2}\frac{k_{s2}}{k} \tag{10a}$$

where

$$\theta_2 \equiv \left(\frac{\pi}{4} - \frac{\theta_1}{2}\right) \tag{10b}$$

and

$$\sin\theta_1 = \frac{1}{2}\frac{k_{s1}}{k} \tag{10c}$$

Subscripts 1 refer to the illuminating ultrasonic wavelength, and subscrpts 2 refer to the second compressing acoustic source.

For the correct intensity compression acoustic source, one-half of the ring energy is compressed to a point image [Ref. 10]. The image is correctly scaled by the illuminating ultrasonic and light wave numbers at a distance $\left(\frac{k_{s1}}{k}\right)$ d from (ℓ) along $\bar{d}$. The other half of the energy goes into a new ring that surrounds the point. This result may be derived by a reciprocity relation. Instead of starting with the sources (ℓ) and (s), start (s) and the ring image represented by (i) and compute the acoustic wavelength so that all points on the ring (i) form the same point image. If the intensity of this second point source is too low, only part of the energy will be diffracted. Thus part of the original ring will remain surrounding the compressed point.

Unfortunately the acoustic frequency of the second compression point source turns out to be too high to be useful in many cases. It would often be just as simple to use a higher frequency and form an image directly by the first technique. However, there are several variations that might avoid the problem of very high frequencies. The acoustic frequency is not important. Any "onion ring like" structure of the correct index variation will do. It might be possible to form a suitable spherical index variation by using a nonlinear optical process to record optical index variations produced by a standing optical field. A suitable optical field could be formed between two concentric spherical reflectors.

It would be very convenient if we could replace the second sound field by a two-dimensional diffracting plate instead of the three-dimensional index gradient produced by the acoustic field. We can consider this diffracting plate to be a spatial filter of a hologram in a general sense. It appears that the compression acoustic point-source can be replaced by a holographic transparency. The compression acoustic point-source has spherical symmetry, whereas the holographic transparency does not. That is, the optical plane-wave components of (i) see the same index gradient independent of their direction of propagation because of the spherical symmetry of the compression acoustic source. The holographic transparency does not have this symmetry and looks different from different directions because of an obliquity factor that states, in effect, that $\sin\theta \approx \theta$ only for small angles. However, if we magnify the plane perpendicular to $\vec{d}$, the angular extent of the plane-wave components of (i) is reduced. Enough magnification is required so that the optical plane-wave components of the ring images (i) (from the volume to be imaged near (s)) fall within a small angle of $\vec{d}$, such that $\sin\theta \approx \theta$ by conventional optical tolerances. The hologram would appear to consist of concentric rings; however, I have not yet worked out the details of the filter.

EXPERIMENTS

The following optical system can be used with any of the three imaging techniques previously discussed.

Figure 2 is a schematic diagram of a transmission Bragg imaging system. Lens (L_1) provides the optical point-source (ℓ). The Bragg diffraction takes place in water in a tank with two optical windows. Lens (L_2) corrects optical spherical aberration when the light leaves the tank. Lens (L_2) has a center of curvature approximately at the focus of lens (L_1). Lens (L_3) is a relay lens which images the (virtual) image from the tank into lens (L_4). Lens (L_4) is a short focal length lens that provides a magnified view for observation and photographs. A holographic filter could be placed in plane H, and then followed by further magnifying optics. The

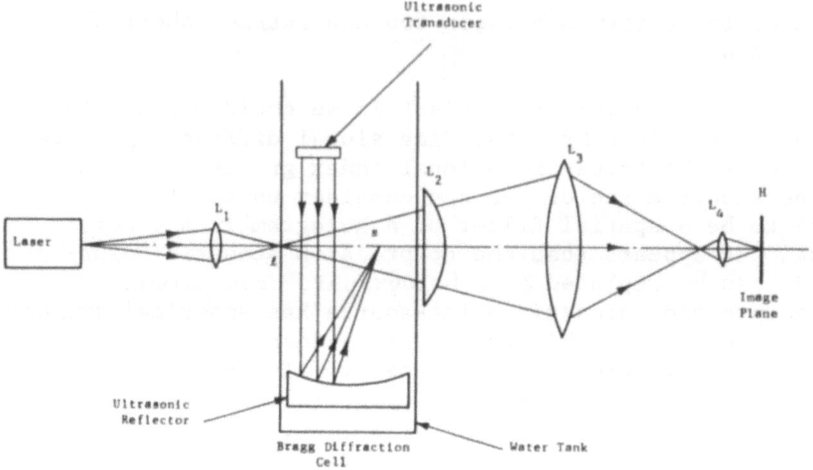

Fig. 2 Transmission Bragg Imaging Cell

optical system represented by Figure 2 can use conventional spherical optics, microscope objectives and transfer or relay optics in the way in which they were designed to be used.

Figure 3 is a photograph of the ring pattern produced by a point acoustic source on the optical axis. Only part of the diffraction ring image shows because the acoustic point-source is not a true point-source emitting spherical waves, but only a small part of a spherical wave. Only sound rays with angles corresponding to the waves emitted by the transducer are present.

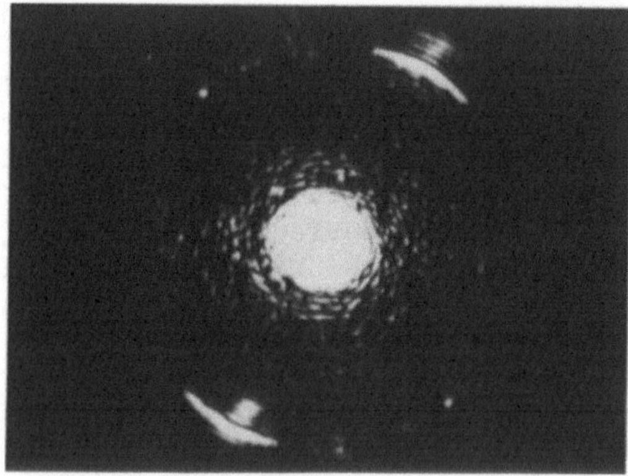

Fig. 3 Optical Ring Image of an Acoustic Point

The acoustic point source was formed by reflecting the wave from a plane transducer off an acoustical spherical mirror in the Bragg cell as shown in Figure 2.

Figure 4 illustrates a folded version of the optical system shown in Figure 2. A single spherical mirror replaces the large aperture transfer lens L_3. The return path through the Bragg cell would appear to give a second undesired image. A slight tilt of the return optical axis by tilting the mirror M should prevent overlap of the two images. The spherical mirror is located with its center-of-curvature at (ℓ). Lens L_2 could be eliminated if the mirror is placed in the cell or made on the outer surface of the cell. The beamsplitter separates the incident and image (return) light beams.

Figure 5 is a photograph of a system of the type shown in Figure 4. Lens L_4 is omitted. The ring image surrounding the beamsplitter was formed as shown in Figure 2.

I would like to thank A. Korpel for several valuable discussions.

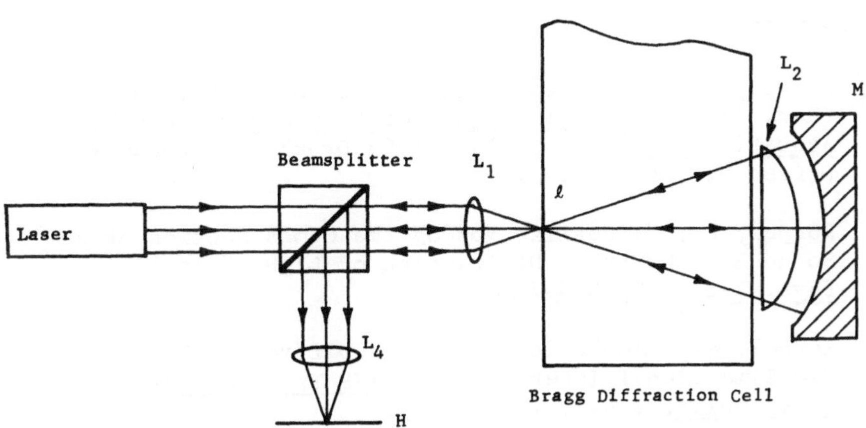

Fig. 4 Reflection Bragg Imaging Cell Schematic

Fig. 5 Reflection Bragg Imaging Cell with Ring Image

CONCLUSIONS

I have discussed three basic techniques to form Bragg images that use conventional point sources and spherical optics instead of cylindrical optics. It will be interesting to see if these techniques can be pursued to form images of extended acoustic fields.

REFERENCES

1 Korpel, A., Ph.D. thesis, University of Delft, Holland (Dec. 1969), Optical Imaging of Ultrasonic Field by Acoustic Bragg Diffraction, 1969 Drukkerij Bronder-Offset N.V. Rotterdam.

2 Korpel, A., "Visualization of the Cross Section of a Sound Beam by Bragg Diffraction of Light," Appl. Phys. Lett., 9, 425-7 (1966).

3 Korpel, A., "Acoustic Imaging by Diffracted Light 1. Two-Dimensional Interaction," IEEE Tran. Sonics Ultrasonics, SU-15, 153 (1966).

4 Tsai, Chen S. and Hance, A.V., "Optical Imaging of the Cross Section of a Microwave Acoustic Beam in Rutile by Bragg Diffraction of a Laser Beam," J. Acoust. Soc. Am., 42, 1345 (1967).

5 Landry, J., Powers, J., and Wade, G., "Ultrasonic Imaging of Internal Structure by Bragg Diffraction," Appl. Phys. Lett., 15, 186-8 (1969).

6 Korpel, A., "Astigmatic Imaging Properties of Bragg Diffraction," J. Acoust. Soc. Am., 49, 1059-61 (1971).

7 Dixon, R.W., "Acoustic Diffraction of Light in Anisotropic Media," IEEE J. Quant. Elect., QE-3, 85-93 (Feb. 1967).

8 Quate, C.F., Havlice, J., and Goodman, J., "Acoustic Microscope," Stanford University Microwave Lab Report #1701, 1-13 (Nov. 1968).

9 Korpel, A. and Kessler, L.W., "Acoustical Holography by Optically Sampling a Sound Field in Bulk," Chapter 9 in Acoustic Holography, Vol. 2, Ed. A.F. Netherell and L. Larmore, Plenum Press, New York (1970).

10 Fried, D.L. and Kleinhaus, W., "Efficient Diffraction of Light from Acoustic Waves in Water," Appl. Phys. Letters, 7, 19-20 (1 July 1965).

THE ULTIMATE SENSITIVITY OF SOKOLOV IMAGE CONVERTER TUBES

H. W. Jones

Physics Department
University of Calgary
Calgary 44, Alberta, Canada

ABSTRACT

The gridded Sokolov tube has been the subject of an analytical and experimental study to determine its ultimate sensitivity. The tube is treated as an electron coupled piezoelectric receiver and detailed consideration of the electron coupling leads to a complete expression for the signal to noise ratio at the output terminal, provided the operating frequency and bandwidth are specified. The factors which lead to the sensitivity are shown to depend on secondary emission characteristics of the piezosurface, the presence or lack of space charge noise suppression, the electron gun characteristics, the electrode spacing, the tube amplifier matching as well as the dielectric and piezoelectric properties of the receiving element. The part that charging and discharging processes play at the surface is discussed as is the effect of mechanical energy storage. Transit time effects do not normally arise from electron interaction in these tubes but depend on geometrical effects and are consequently excluded from the discussion.

It is shown that there is an optimum current and optimum properties for the piezoelectric material. It is also shown that the maximum sensitivity for a tube operating at 1 MHz would occur for a titanate ceramic receiver and is better than 8 x 10^{-11} watts/cm^2. The study required the determination of the complete secondary emission properties of powdered MgO and a new apparatus was designed for the purpose. A demountable image converter was used to measure performance under different conditions and comparison between experimental and theoretical results are reported.

INTRODUCTION

There have been a series of attempts to discuss the sensitivity of ultrasonic image converter tubes; the earliest are those attributable to Semmenikov[1] and Prokhorov[2] (1957 and 1958) who expressed somewhat discordant points of view. Since that time there has been a small but steady series of publications on the topic embracing the three types of electronic image converters (i.e., cathode potential stabilized, collector potential stabilized, and electron mirror). If any particular authors should be mentioned then perhaps they might be Turner[3] and Karplus et al[4] for their work on collector potential stabilized tubes, Smythe et al[5] for their work on anode potential stabilized tubes, and Szentesi[6] for his work on electron mirror tubes. Halstead[7], in his M.Sc. Thesis, provides a detailed review of the literature on the topic to 1967. Unfortunately many unanswered questions remain and it is an attempt to deal with these problems in connection with the collector potential stabilized tube that the work I shall describe was undertaken.

DISCUSSION

Difficulty arises from the many ways of describing tube performance in that various treatments have involved subsequent amplifier noise in different ways. In an attempt to avoid this difficulty the discussion in this paper considers the signal to noise ratio at the tube output terminal when it is connected to a single noiseless resistor. A bandwidth has to be specified and this is the only amplifier characteristic which is used.

The piezo receiving element is assumed to be relatively lossless and working at its resonant frequency; consequently the circuit shown in Fig. 1 can be used for a first consideration of the problem. The mechanical circuit has been transformed so that the sound pressure p produces a voltage V given by

$$V = \frac{p\ S_a}{2\alpha} \tag{1}$$

where S_a is the area of the piezo element in question given by

$$S_a = \frac{\pi l^2}{4} \tag{2}$$

l being the thickness of the transducer. The other components in the diagram are the acoustical radiation impedance Z_r of the coupling medium, the capacitive reaction of the element Z_c, the anode resistance of the image converter R_a, the external load Z_e (resistive in the following discussion) and a source of noise current In. The signal developed across the external load can be shown to be given

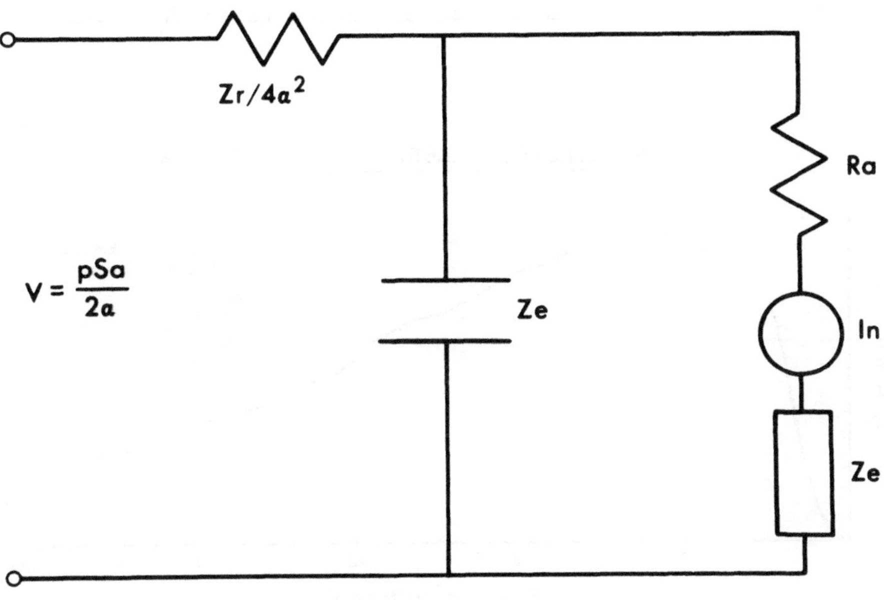

Figure 1

by

$$V_\delta = \frac{p\ Z_e\ S_a}{2\alpha}\left(1 - \frac{Z_r}{4\alpha^2\ Z_{tot}}\right)\bigg/\left(R_a + Z_e\right) \qquad (3)$$

where

$$Z_{tot} = \frac{\vec{Z}_r}{4\alpha^2} + \frac{\vec{Z}_c\ (\vec{Z}_e + \vec{R}_a)}{(\vec{Z}_c + \vec{Z}_e + \vec{R}_a)} \qquad (4)$$

It is not possible to proceed further at this point without considering the nature of the electronic coupling of the piezo element to the external load in more detail. The first difficulty which occurs is the apparent lack of data on the secondary emission processes which occur at the surface. A search of the literature did not provide the information required, consequently a side project was undertaken to obtain the necessary data. Because of the difficulty of obtaining the energy distribution of the emitted electrons from the apparatus used in previous measurements a new type of apparatus was designed[8]. Fig. 2 shows the electron energy distribution which was obtained for the emission from an amorphous magnesium oxide coating.

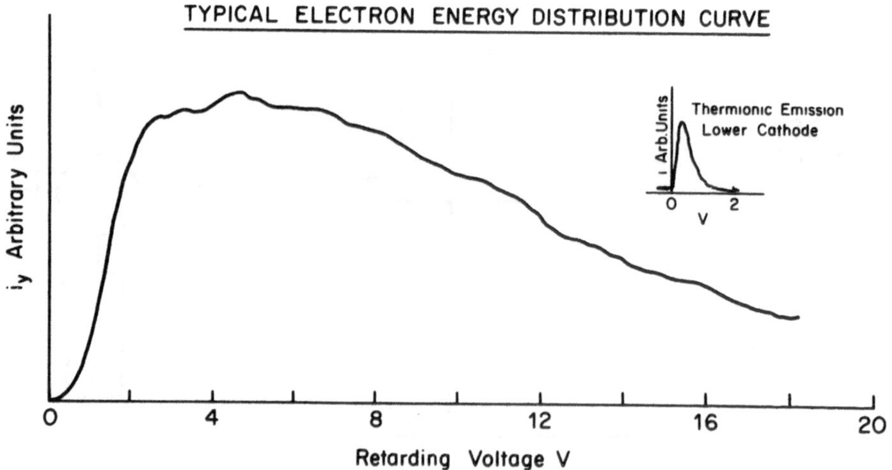

Figure 2

The same apparatus allowed the simultaneous determination of the secondary emission coefficient, Fig. 3.

The noise carried by the secondary emitted electrons is a problem which arises in electron multiplier tubes. This has been dealt with by several authors[9 10 11] who collectively reach the conclusion that the noise current is given by

$$i^2 = 2e\ I_b\ \delta(F_b\ \delta + H - \delta)\ \Delta f \tag{5}$$

where I_b is the image converter beam current

e is the electronic charge

Δf is the bandwidth of the amplifier

δ is the secondary emission coefficient.

The secondary emission coefficient can be related to a quantity β_n in the following manner, if we suppose that

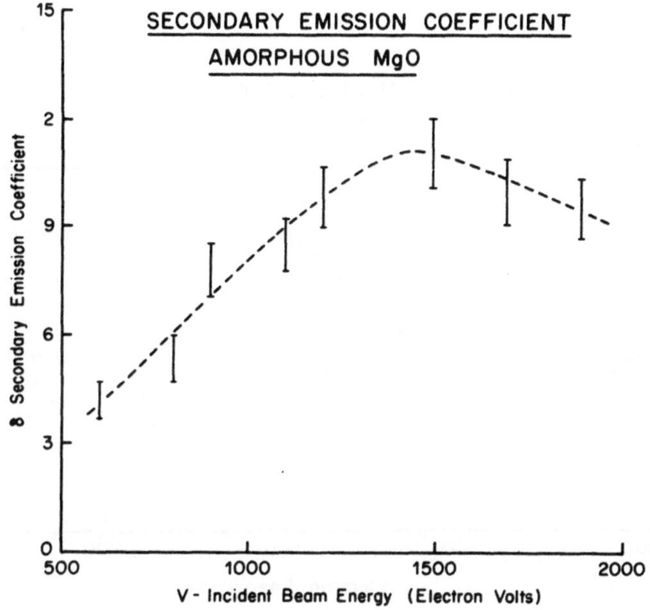

Figure 3

the probability of an electron releasing n electrons is β_n,

$$\sum_{n=0}^{\infty} n\,\beta_n = \delta \tag{6}$$

H is now defined as a moment of (6) such that

$$\sum_{n=0}^{\infty} n^2\,\beta_n = \delta H \tag{7}$$

Values of H are not readily available in the literature and its determination poses a difficult experimental problem as it requires detailed knowledge of the statistical processes of emission. The data of Bakker et al[12] (Fig. 4) was used in subsequent calculations[11].

F_b is the noise suppression factor accounting for the effects of the space charge quieting of the primary beam. This is a factor for which apparently no data exists (Smythe et al[5] assumed a value of unity) consequently a measurement

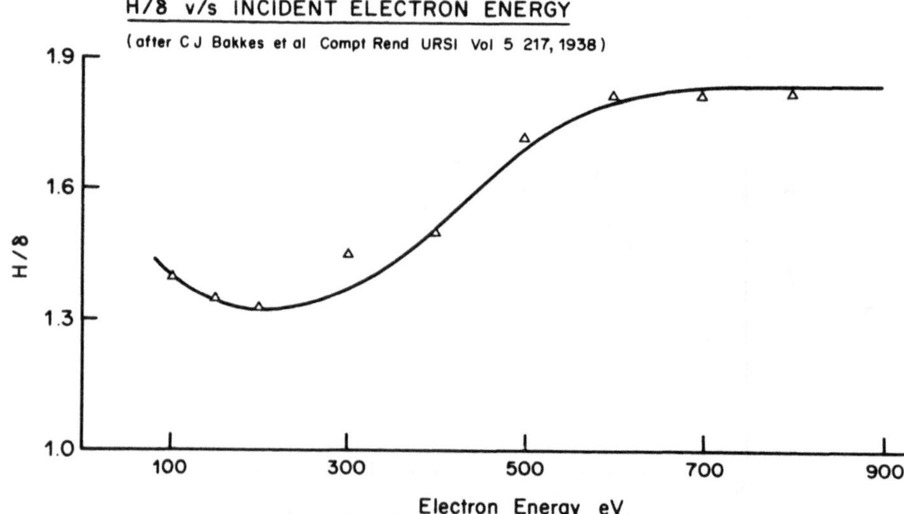

Figure 4

was made. The beam current noise was compared with a saturated 5722 noise diode. The result was surprising (Fig. 5). At a beam current of about 80 μA the noise voltage suddenly increased by nearly a factor of 10 above that of the noise diode. Further investigation showed an oscillation at 750 Kc/s approximately. The gun which was used was magnetically focussed and presumably some interaction with the space charge field produced this effect as a peculiar type of partition noise (the net cathode current remained constant over the oscillatory period). It was not possible to control the effect by external means except defocussing the beam.

If a space charge region exists near the piezo surface it follows that noise suppression can arise due to this effect and a further factor F_s needs to be introduced so that Eq. 5 needs to be rewritten as

$$i^2 = 2eI_b F_s (F_b \delta^2 + H - \delta)\, \Delta f \tag{8}$$

an expression that takes account of the fact that the average emitted current proceeding to the collector is 1/δ of

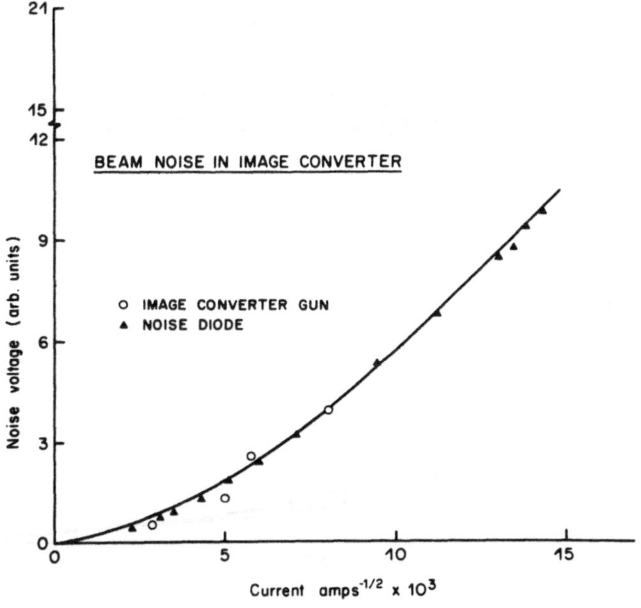

Figure 5

the total emitted current from the surface.

An examination of the possible existence of a space charge and its nature is required if F_s is to be obtained. The electron energy distribution (Fig. 2) is approximately Maxwellian and this allows an analytical solution of the problems in this region. With this assumption it is possible to treat the grid/piezo surface region as a diode if suitable boundary conditions are taken. Langmuir[13] obtained a solution to Poisson's equation for plane parallel electrodes under these circumstances and Fig. 6 shows the position of the space charge minimum for $\delta = 2$ and 15. When the space charge minimum occurs between the piezo surface and the grid it is possible to determine the value of F_s from well established theory[9 14]; Fig. 7 shows the values obtained.

The validity of the assumption of plane parallel geometry requires justification. It is clear that electrons emitted from the surface being interrogated must either reach the grid or collector or return to the surface. This is required by charge conservation. As $\delta > 1$ in general,

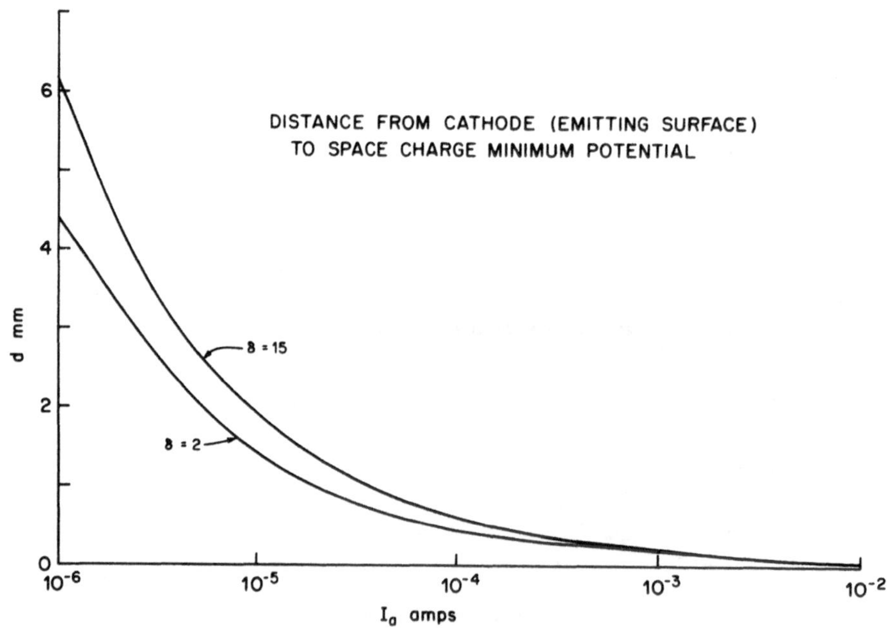

Figure 6

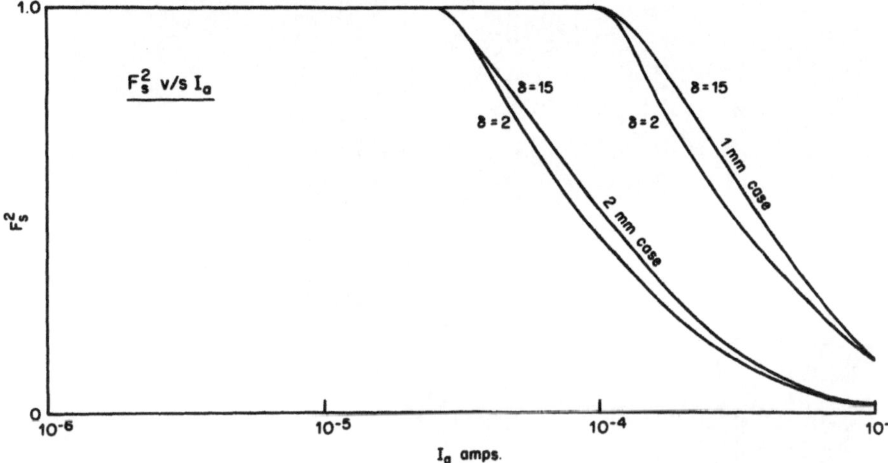

Figure 7

this implies that returning electrons must be focussed onto the correct part of the surface. This situation was simulated for various conditions by the use of a rubber model. Fig. 8 shows one of the profiles obtained in space charge limited conditions. Fig. 9 shows the nature of the potential profile on the piezo-surface. This charge distribution tends to limit the spread of emitted electrons and makes the parallel geometry reasonably justified, particularly in the case of large values of δ. When space charge is absent a similar effect occurs with the potential surrounding the emitting surface being at grid potential.

The diode resistance Ra between the piezo surface and the grid can be calculated from the following equations:

a) in the absence of space charge

$$Ra = \frac{V_T}{I_b} \tag{8a}$$

where V_T is the equivalent electron temperature (eV_T = kTe) and I_b is the scanning beam current, or

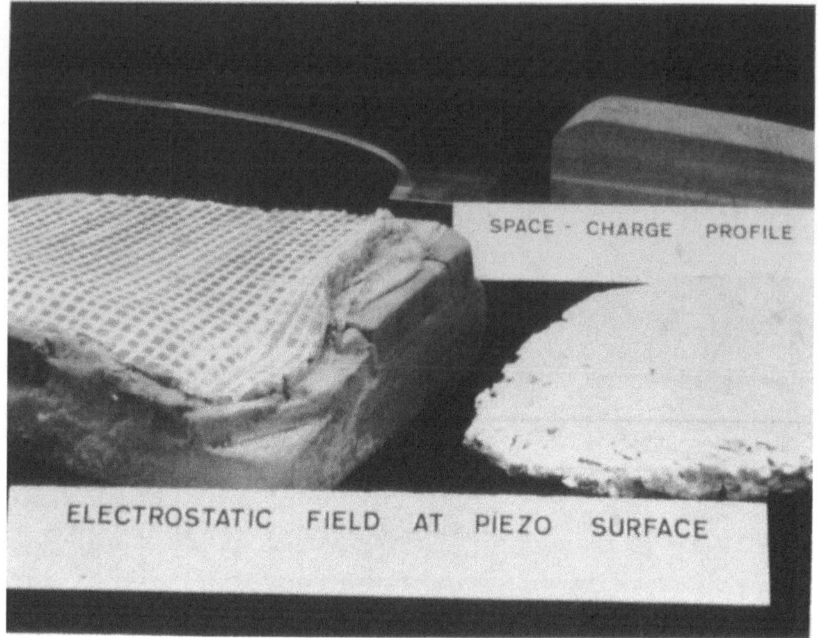

Figure 8

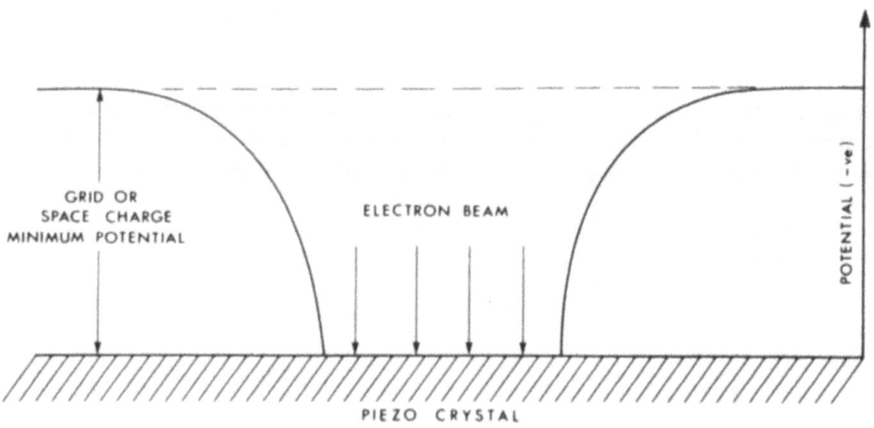

Figure 9 Potential distribution at piezo surface.

b) in the presence of full space charge

$$R_a = \frac{2(Vg')}{3\ F_b} \tag{8b}$$

where Vg' is the potential difference between grid and space charge minimum. When $Vg' \rightarrow 0$, R_a approaches the value given by 7. Figs. 10 and 11 show the values obtained for R_a vs. I_b and δ, for grid spacings of 1 and 2 mm.

With this information we are in a position to make a determination of the signal to noise ratio if it is recognized that the diode resistance R_a determines the dynamic impedance of the triode formed by the emitting surface grid and collector. That this is true can be seen from the fact that the collector is usually very far from the grid so that the penetration of the collector field into the grid/emitter space is negligible. This leads to μ (i.e. $\partial V_a/\partial V_g$) being ∞ and shows that the usual concept of plate resistance is inapplicable. Alternatively this can be demonstrated from [15]

$$I = 2.336 \times 10 \quad \frac{\left(V_g + \frac{V_a}{\mu}\right)^{3/2}}{\left(1 + \frac{1}{\mu} + \frac{4}{3\mu}\frac{d_{ga}}{d_{gk}}\right)^{3/2}} \tag{9}$$

the expression for a planar triode with grid/cathode spacing d_{gk} and grid/anode spacing d_{ga} [16] and

$$\mu = \frac{2\pi\ d_{ga}}{a\ \ln(2\ \sin \pi r/a)} \tag{10}$$

where r is the radius of the grid wire and a their spacing. Thus

$$I \doteqdot k\ V_g^{3/2} \tag{11a}$$

$$\text{and} \quad I \neq f\ (V_a) \tag{11b}$$

to a good order of approximation.

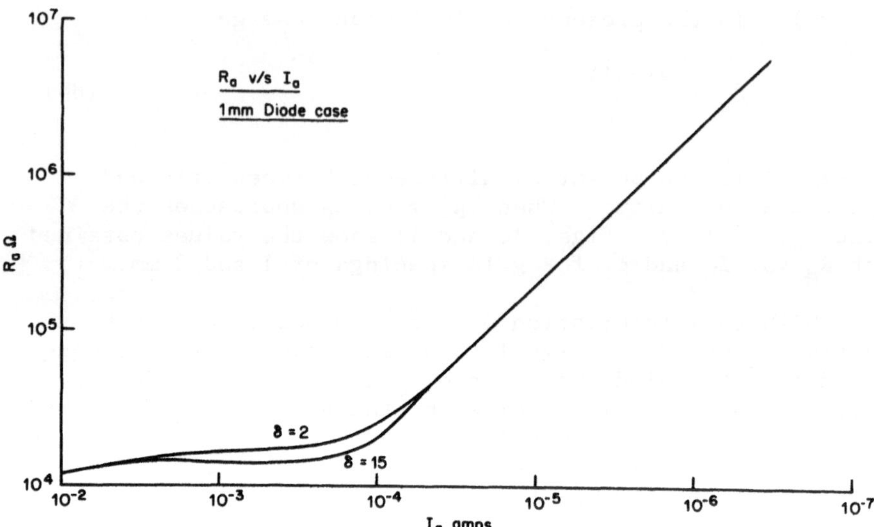

Figure 10

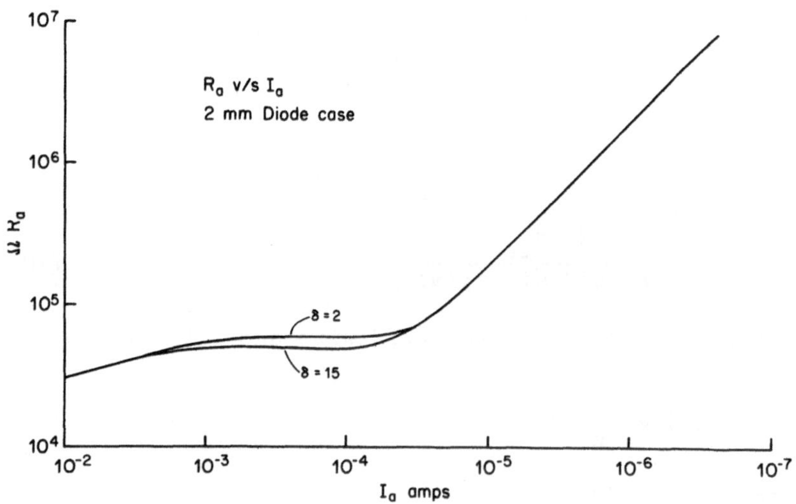

Figure 11

Equations 6 and 3 yield a signal to noise ratio of:

$$S = \frac{pS_a}{2\alpha}\left(1 - \frac{Z_r}{4\alpha^2 Z_{tot}}\right) \Big/ \left[(R_a+Z_e)(Z_e I_b F_s (F_b \delta^2 + H - \delta)\Delta f)^{\frac{1}{2}}\right] \quad (12)$$

putting S = 1 a value of p can be determined.

It should be noted that partition noise between grid and collector is sufficiently small to be neglected[16] as can be the noise in the radiation resistance[7].

Fig. 12 shows a variety of cases of a quartz receiver with a grid spacing of 2 mm. and a secondary emission coefficient of 2. For higher impedance operation there is a most sensitive region at about 15 μA which is only achieved again at currents of 1 mA. The effect of loading with impedances of 10^5, 10^4 and 10^3 Ω is shown.

Fig. 13 shows the effect of increasing δ and the associated considerable decrease in sensitivity. A comparison of grid spacing for d = 1 and 2 mm. is given.

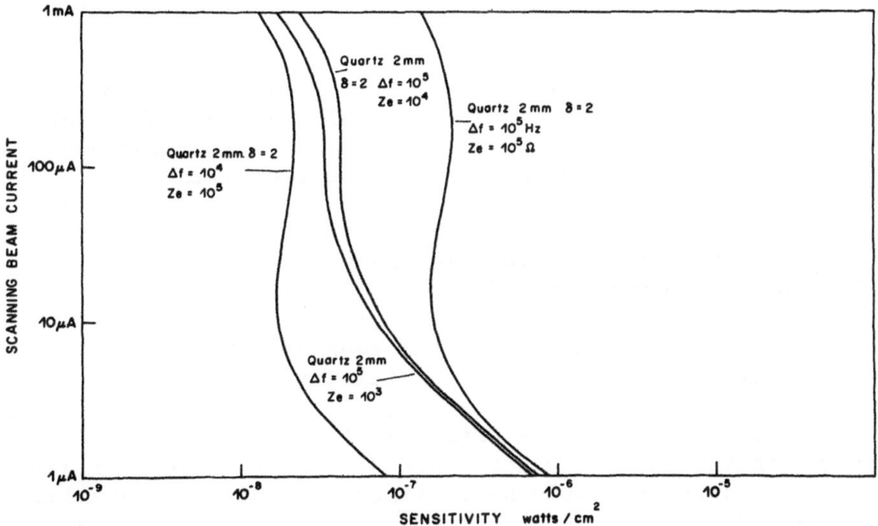

Figure 12

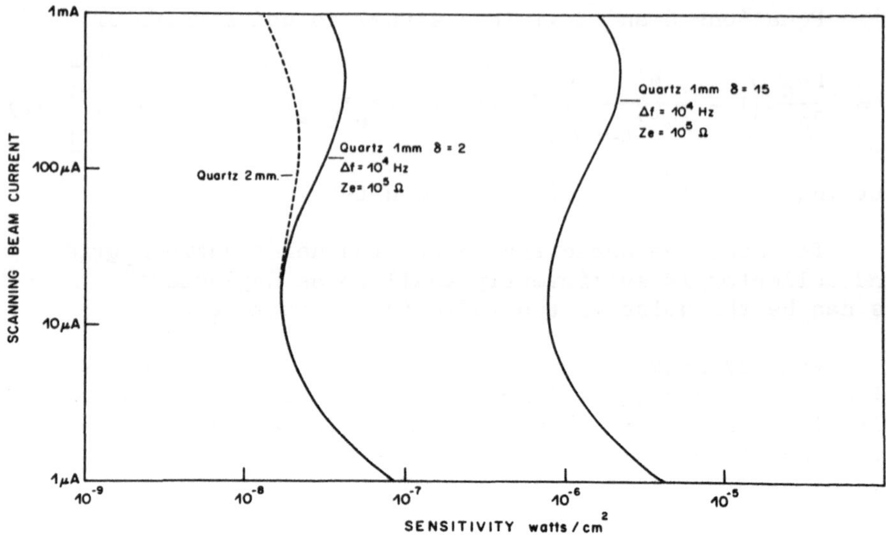

Figure 13

Fig. 14 shows the effect of space charge suppression in the gun, i.e. F_b = 1 and .1. It can be seen that such a quieting would produce a marked improvement in performance if it could be obtained.

Fig. 15 shows a $BaTiO_2$ case with Z_e as a variable (compared with one of the quartz cases). Values taken for the material are 1350 for the dielectric (ε) constant and the relevant piezo constant (e) of 15 coulombs/meters^2.

Finally (Fig. 16) a lead zirconate is compared with the barium titanate described in the previous cases. In this case ε = 1500 and e = 50.

All that has been said so far does not account for two effects which enhance the sensitivity of the tube.

Energy storage in the piezo-electric has been the subject of studies by Turner[17] and Halstead[7]. The original equivalent circuit should be modified to contain a parallel resonant branch shunting the input or output. It has been shown by the authors mentioned that the storage factor S which measures the mechanical energy stored is given by

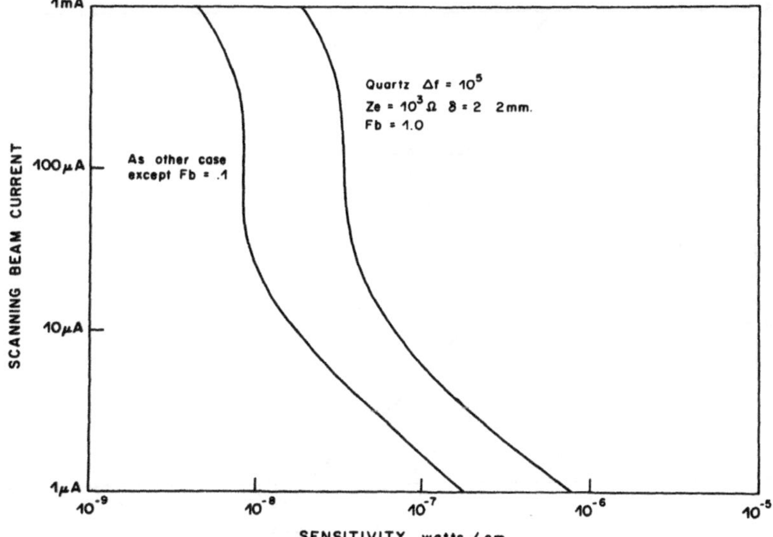

Figure 14

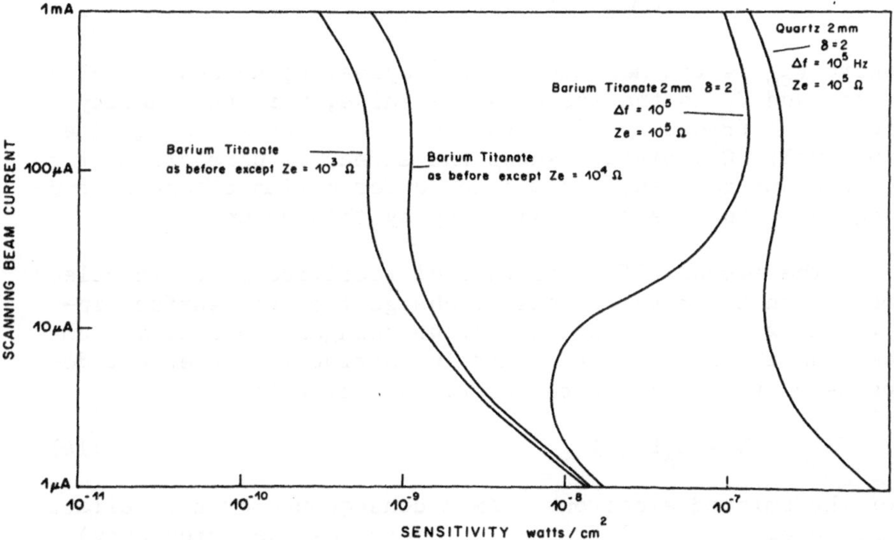

Figure 15

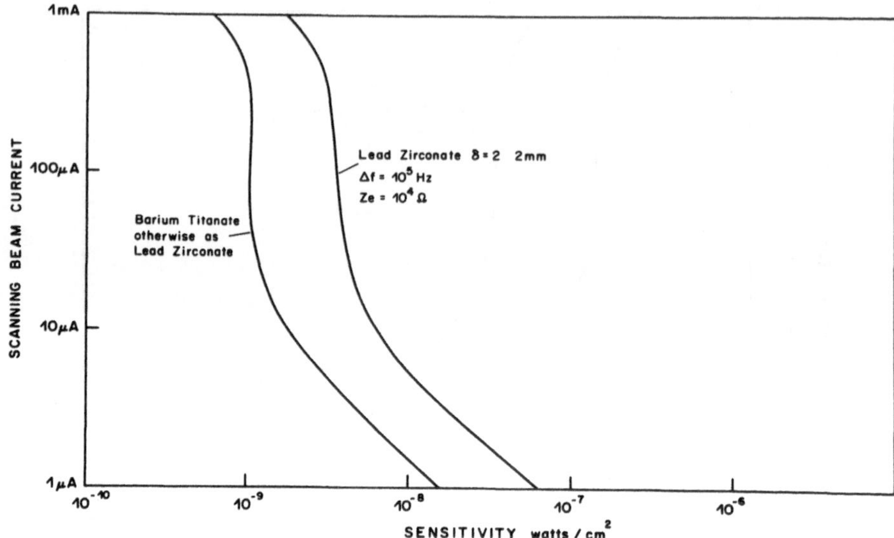

Figure 16

$$S = 1 + \left(\frac{(V_{ro})^2 - (V_r)^2}{(V_{ro})^2} \right) Q/2\pi f\tau \left(1 - e^{-2\pi f\tau/Q} \right) \tag{13}$$

where V_{ro} is the maximum voltage appearing across the element, and V_r is the steady state value, Q is the quality factor, f frequency of resonance, and τ the scanning time interval. Calculation shows the effect is unimportant for quartz but can range from 1 to 20 for barium titanate, consequently increasing sensitivity by this amount.

The second effect is that of rectification. In pulsed operation the rate of loss of charge from the surface increases when a modulating voltage changes the equilibrium between the grid and the emitting surface. The effect depends on the dN/dV of the energy distribution

$$N = N_o f\ (v) \tag{14}$$

of the emitted electrons. As a consequence of this effect the received signal frequency (piezo resonant frequency) gives an amplitude modulated wave form (Fig. 17). This modulation can be chosen to have a relatively small band-

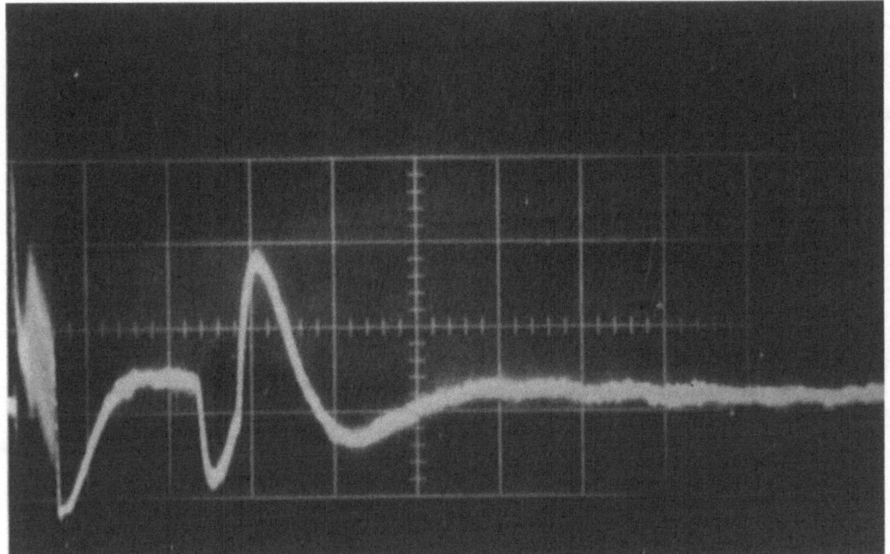

Figure 17 Amplitude modulation of pulsed signal.

width. Fig. 18 shows the output for such an envelope after going through the i.f. of a transistor radio with an effective bandwidth of about 5×10^3 Hz. This consequently is a high sensitivity arrangement.

The question of transit time effects is one which presents a difficulty in that it relates to the particular geometry of the tube considered. The scaling laws[16] for high frequency tubes show that an extremely high current density would be needed before the normal treatment[18,19] would be applicable to the sort of tubes generally discussed. For example, if we compared an image converter to a planar triode, then we would have the scaling law for current density as follows:

$$\text{If} \quad l_1/l_2 = m \tag{15}$$

where l_1 is the grid anode spacing in the image converter, l_2 that of a high frequency triode, and

$$f_1/f_2 = n \tag{16}$$

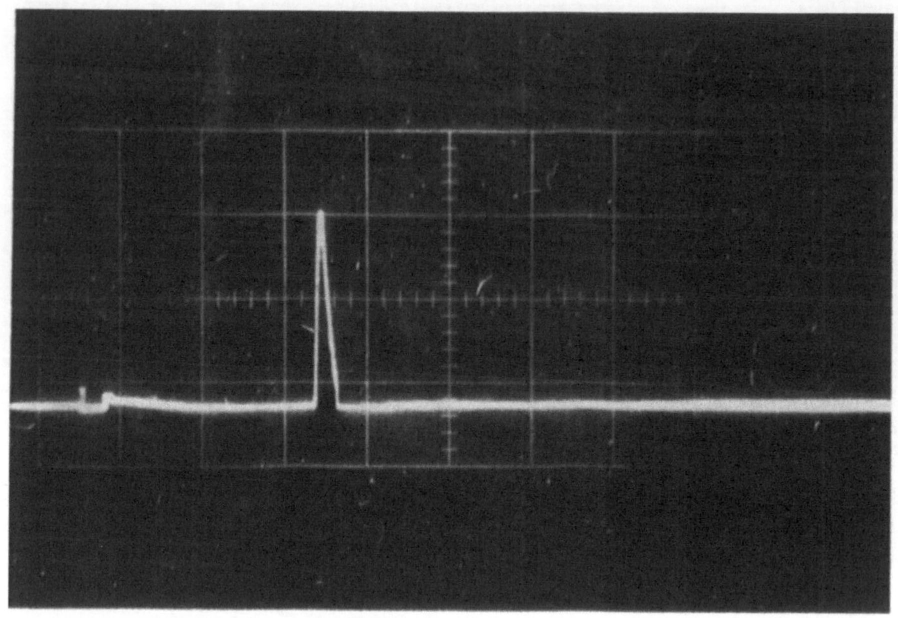

Figure 18 Gated amplified pulse from that shown in Fig. 17.

where f_1 is the operating frequency of the converter and f_2 the triode, then the ratio of current densities R is given by[16]

$$R = m^2/n \tag{17}$$

The question of the actual arrival time on an extended collector is another problem. Demodulation may occur depending on the geometry of the tube.

The theory which has been developed in this paper has been checked to some degree by experiment (Fig. 19) using a demountable tube with various grid spacings and piezo materials. Agreement is within the limits of the experimental error of the basic data and the measurement techniques used and appears to give general support to the theory advanced.

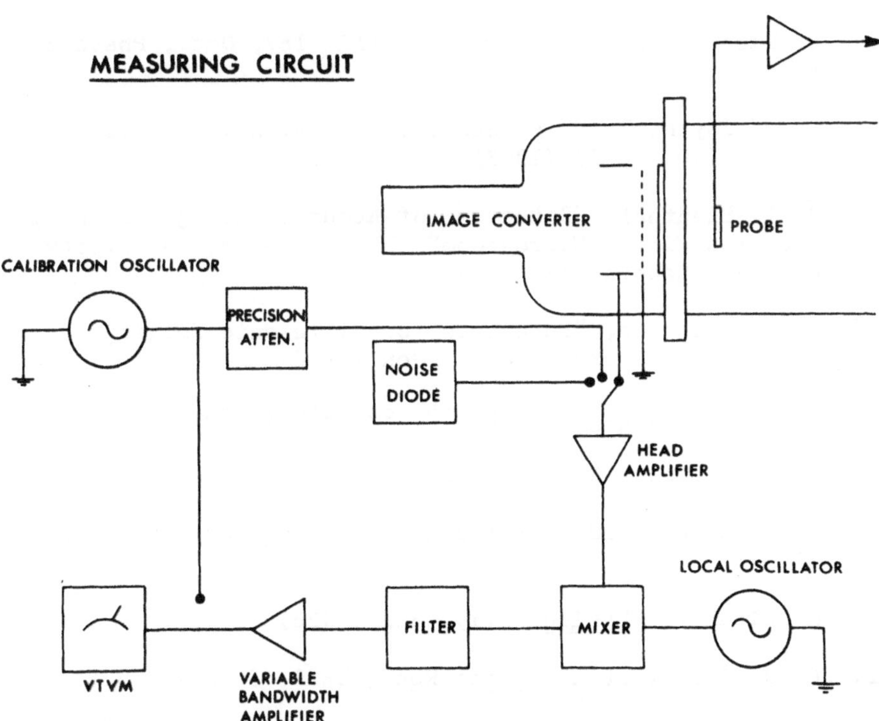
MEASURING CIRCUIT
IMAGE CONVERTER
PROBE
CALIBRATION OSCILLATOR
PRECISION ATTEN.
NOISE DIODE
HEAD AMPLIFIER
LOCAL OSCILLATOR
VTVM
VARIABLE BANDWIDTH AMPLIFIER
FILTER
MIXER

Figure 19

REFERENCES

1. Iu. B. Semennikov, Soviet Phys.-Acoustics, 4, 72 (1958)

2. V. G. Prokhorov, Soviet Phys.-Acoustics, 3, 272 (1957)

3. W. R. Turner, Ultrasonics, Oct.-Dec. 182 (1965)

4. H. Karplus and W. E. Lawrie, IITRI 187, Oct., Phase 3 (1963)

5. C. N. Smyth, F. Y. Poynton, and J. F. Sayers, Proc. I.E.E. 110, 1, 16 (1963)

6. O. I. Szentesi, "Detection of Acoustic Holograms using Electron Mirror Microscope", Ph.D. Thesis, University College, London, (July 1970)

7. J. R. Halstead, "Detectors for Ultrasonic Imaging Systems", University of London (1966)

8. H. W. Jones, J. Phys. E., 3, 997-999 (1970)

9. Van der Ziel, "Noise", Prentice Hall

10. L. J. Hayner and B. Kurrelmeyer, Phys. Rev., 52, 952 (1937)

11. M. Ziegler, Physica, 3(1), 307 (1936)

12. C. J. Babber et al, Compt Rend, URSI, 5, 217 (1938)

13. I. Langmuir, Phys. Rev., 21, 419 (1923)

14. O. O. North, RCA Review, 4, 441 (1940)

15. D. Tellegen, Physica, 5e, 301 (1925)

16. A. H. W. Beck, "Thermionic Valves", Cambridge University Press (1953)

17. W. R. Turner, NAVORD Report 4090, (2 July 1956)

A COMPARISON OF HOLOGRAPHIC VERSUS LENS TYPE ACOUSTIC IMAGE SYSTEMS BY COMPUTER SIMULATION

G. L. Sackman

Naval Postgraduate School

Monterey, California 93940

Ideal image systems of holographic or lens type should have similar performance, given similar aperture/wavelength ratios. However, practical acoustic image systems introduce amplitude and phase nonlinearities, noise, and geometric distortion at different points in the system. Computer simulation of a holographic and a lens type system of limited aperture/wavelength ratios has been made to emphasize the differences in subjective image quality introduced by such effects as diffraction, undersampling, saturation, and system noise.

INTRODUCTION

Acoustic image systems using either lens focusing or holographic techniques may have limitations in practical realization due to restricted apertures and finite sampling of the wave field, as well as amplifier saturation and self-noise. However, these limitations are manifest in different types of output image defects characteristic of lens type versus holographic systems. The fundamental reason for the difference is in the order of appearance of linear and non-linear processes.

From input object to output image, each system can be described as performing a minimum of two Fourier transformations in cascade as shown in Figure 1. In the course of proceeding thru the system from input to output, the

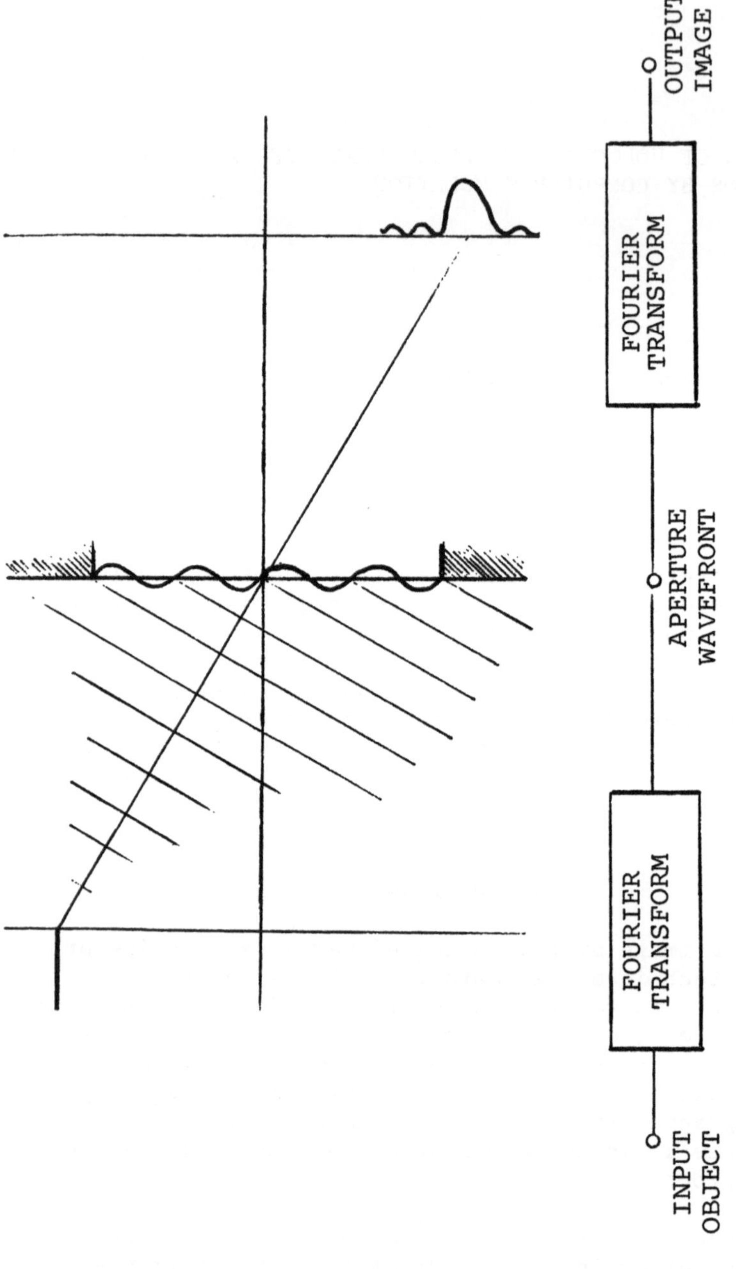

Fig. 1 An image system can be described as performing two Fourier transformations.

signal is degraded at each stage by linear and non-linear operations, which in general are not commutative and depend strongly on the order of occurrence. Since the non-linear operations associated with detection and amplification occur between the two Fourier transforms in the holographic system, the effect on the output is quite different from the lens system in which non-linearities are usually concentrated after the second transform as shown in Fig. 2

COMPUTER SIMULATION

A very simplified, one dimensional mathematical model of each type of system has been analyzed using a hybrid computer. The object is represented by an analog voltage, varying in time over a finite period T. Time here is analogous to angle within a finite field of view, the time origin representing the optic axis.

The input "object" function is then digitized by time sampling and amplitude quantizing. At this point the object signal can be described as a column vector with one element for each sample. Fourier transformation is done digitally using the Fast Fourier Transform (FFT) algorithm, which has the effect of performing a multiplication on the object vector by the Discrete Fourier Transform (DFT) matrix as shown in Figure 3. The output of this operation is another column vector with complex elements representing a finite number of Fourier coefficients, approximating the continuous Fourier transform of the object.

Other operations may also be described in the form of matrix multiplications, with the signal being represented always as a column vector. A typical amplifier gain characteristic including the effects of saturation and added noise is shown in Fig. 4. An amplifier bank can be represented by a square matrix with the gain factor for each amplifier appearing on the major diagonal, while crosstalk appears as off-diagonal elements, as shown in Fig. 5. Even non-linear operations can be done in this manner by using logic statements at each matrix element. For example, saturation effects can be simulated by the statement: " If the output product amplitude is greater than saturation, set the output product equal to the saturation constant."

Noise can be simulated by adding a random number to each element of the vector or matrix from a source with specified statistics.

After all operations of the system are accomplished, the output is displayed on a computer graphics terminal for inspection by the operator. Modifications to the operations can be made from a key set at the terminal, and the results are inspected in a short time so that the operator interacts with the computer directly.

The next two sections describe some experiments using the hybrid computer in which first a lens-type and then a holographic type image system were simulated. The experiments were intended to assist the operator in understanding the qualitative effect on the output image resulting from various systems restrictions and defects. For this reason, the effects are exaggerated for emphasis, and are not necessarily representative of any particular system, nor are they intended for quantitative measurement. Nevertheless, considerable insight into the nature of image systems can be gained, in order to identify potential sources of trouble and possible compromises.

LENS SYSTEM

The elementary lens system as shown in Fig. 6 can be described as a sequence of two Fourier transformations, the first between object and aperture (lens), the second between aperture and image (focal plane). Most system restrictions occur after the second transformation, although the effect of a finite aperture is to band-limit the spacial frequency spectrum after the first transform, as specified in part by the modulation transfer function (MTF). This operation is introduced in the experiment as a truncation of the aperture signal vector at some bound as suggested by the partially shaded block in the sketch. Aperture shading of any degree can be simulated if desired at this stage by multiplication with a diagonal matrix weighting factor. Lens defects and aberrations can be accounted for by a similar diagonal matrix with complex phase factors varying in a functional and/or stochastic manner. The signal is then displayed as the Real part of the aperture wavefront in order to inspect the spacial periodicity. The modulation transfer

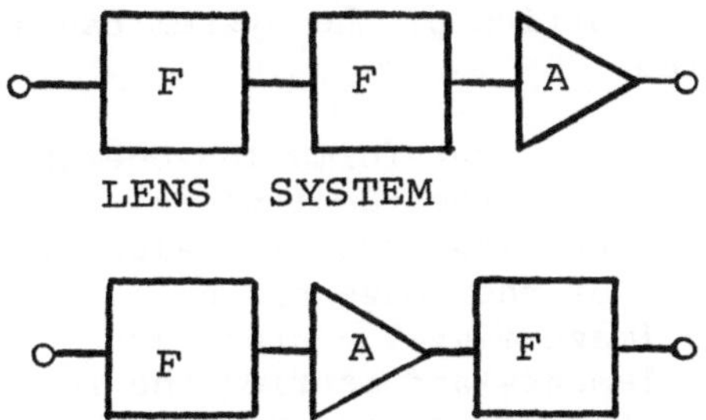

HOLOGRAPHIC SYSTEM

Fig. 2 Nonlinear operations occur between the transforms in a holographic system, whereas in a lens system they occur after the second transform.

$\overline{Y} = (F)\overline{X}$

(F) =

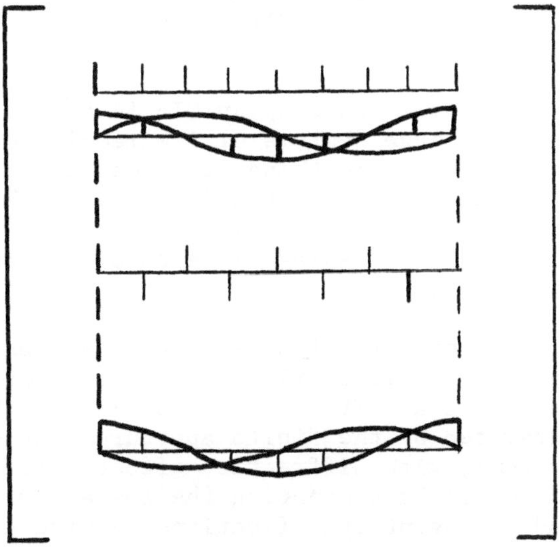

FOURIER TRANSFORM MATRIX

$$\overline{Y}(2\pi k/NT) = \sum_{n=0}^{N-1} \overline{X}(nT)\, e^{-j2\pi nk/N}$$

Fig. 3 The Discrete Fourier Transform is a matrix·operation upon a column vector composed of the input samples.

function for this portion of the system can also be derived if desired.

The second Fourier transformation operation yields the complex image signal at the focal plane. To simulate detection the absolute magnitude is calculated. The effect of finite sampling of the image can be simulated by thinning out the image signal vector, retaining more sparsely spaced elements and setting the rest to zero. The sampled vector is then operated upon by the matrix representing the amplifier array. Saturation is represented by the logic statements referred to earlier, while cross-talk can be introduced by inserting off-diagonal factors. Noise can be simulated by adding random numbers to the signal vector and/or amplifier matrix. The output signal is then displayed, either as discrete samples or smoothed by an appropriate filter. The output filter in this case corresponds to the MTF of the display optics in a real system.

An optimized display MTF is a critical requirement for sampled image systems of low resolution. Examples of poor display smoothing functions abound in the literature of digital image processing, giving the appearance of polka-dots or brick walls rather than low resolution images. The display MTF should not pass the quantization harmonics.

The effects of non-linearities such as amplifier saturation are familiar in the form of overexposed negatives and TV camera "blooming". These effects are agravated by the finite aperture size of acoustic image systems, with diffraction effects from specular objects in the field dominating the image. Drastic aperture shading and stringent specifications on side lobe levels are the only means of coping with these effects. Even then, regions of the image near specular points will be lost in overexposed diffraction fringes.

The effect of uniformly distributed noise ("snow") appears after smoothing primarily as a reduction in the effective contrast of the image, whereas noise peaks or dead channels can give the appearance of false objects.

Adjacent-channel cross-talk produces an affect similar to a poor MTF, reducing the resolution by broadening the

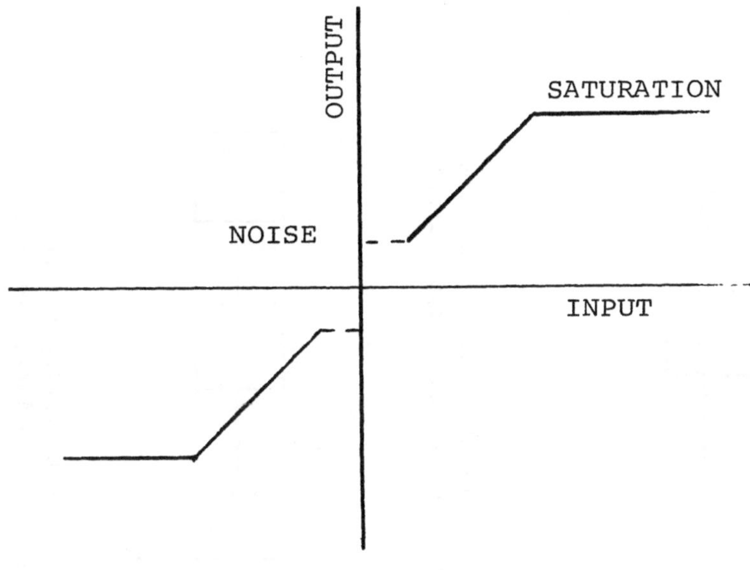

AMPLIFIER GAIN CHARACTERISTIC

SATURATION AND NOISE ADDED

Fig. 4 The amplifier gain region is bounded by saturation and noise.

$$\overline{Y} = (A)\overline{X} \qquad (A) = \begin{bmatrix} A & C & \cdot & \cdot & & & & \\ C & A & C & \cdot & \cdot & & & \\ \cdot & C & A & C & \cdot & \cdot & & \\ \cdot & \cdot & C & A & C & \cdot & \cdot & \\ & & & & & & & \\ & & & \cdot & \cdot & C & A & C & \cdot \\ & & & & \cdot & \cdot & C & A & C \\ & & & & & \cdot & \cdot & C & A \end{bmatrix}$$

AMPLIFIER

CROSSTALK MATRIX

Fig. 5 The amplifier matrix contains the gain factor on the major diagonal and crosstalk factors off-diagonal.

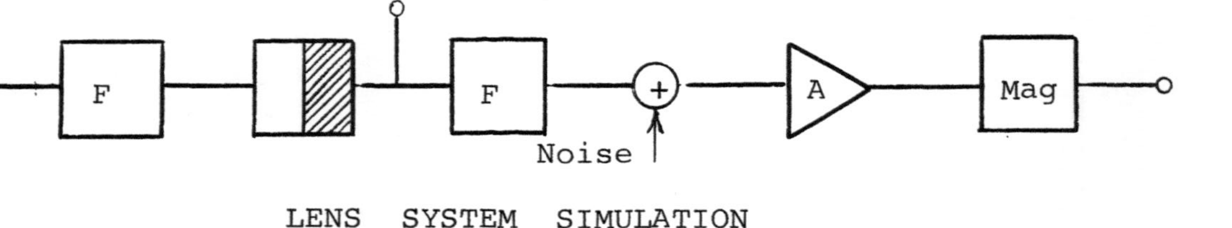

LENS SYSTEM SIMULATION

Fig. 6 Nonlinearities and noise enter the lens system after the second transform Detection is represented by displaying the magnitude of the output signal.

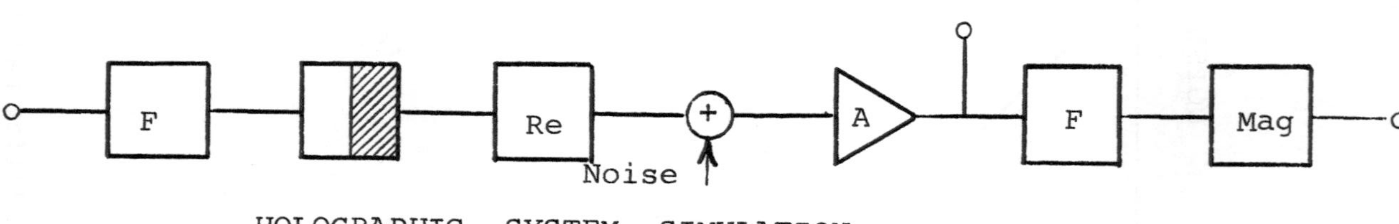

HOLOGRAPHIC SYSTEM SIMULATION

Fig. 7 In the holographic system, detection is represented by taking the Real part of the signal after the first transform. Noise and nonlinearities also enter the system at this point.

response to point objects. Rather large amounts of adjacent channel cross-talk can be accommodated in small-aperture systems.

However, non-adjacent channel cross talk (arising for example in cable bundles and along power supply lines) can cause the formation of spurious false objects at points scattered over the field.

Finite sampling of the image is equivalent to a band-limited MTF, since it sets an upper bound on the spacial frequency components at the image plane. Undersampling theoretically results in the possibility of folding or aliasing high frequency components into the lower spacial frequency passband due to a sort of stroboscopic effect. In the case of lens systems, sinusoidal or periodic objects are not normally encountered, except for test patterns. Therefore this effect is usually manifest only as anomalous resolution of bar charts.

HOLOGRAPHIC SYSTEM

An acoustical holographic system can be represented as a sequence of two Fourier transforms, but in this case sampling and detection of the signal occurs at the aperture plane between the two transforms as shown in Fig. 7. Most system restrictions appear in this sampling and detecting process, after which the image is reconstructed by a linear transformation.

This sequence is simulated by first multiplying the object signal vector by the DFT matrix. At this point, the Real part of the transform is retained, corresponding to a synchronously detected, on-axis hologram. The transform vector is then truncated to the aperture size, sampled, and multiplied by the amplifier matrix. The recorded wavefront signal is displayed for inspection, and then transformed. The absolute magnitude of the complex image signal is displayed for inspection as before, although the square of the magnitude could as easily be calculated and displayed if desired.

The effects of system restrictions now appear in rather interesting ways. Sampling the wavefront signal at the aperture plane has the effect of placing a limit on the

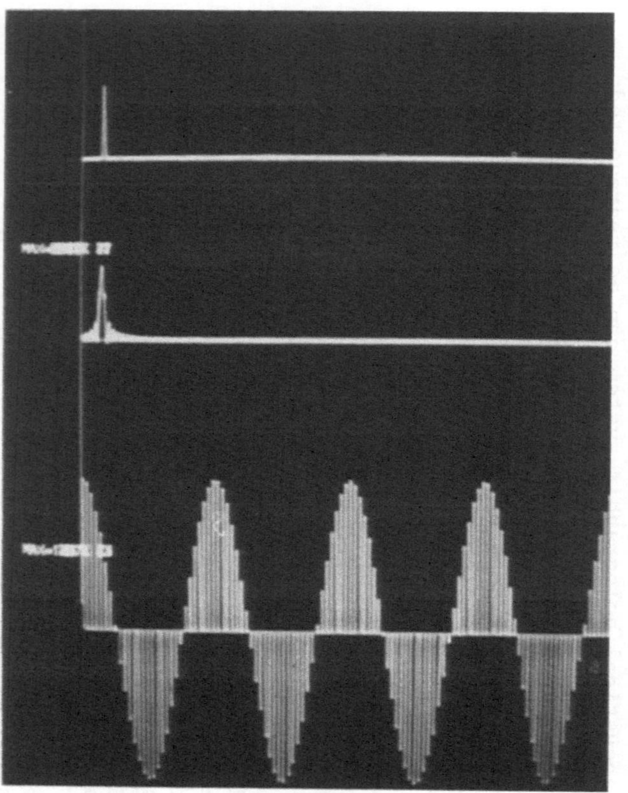

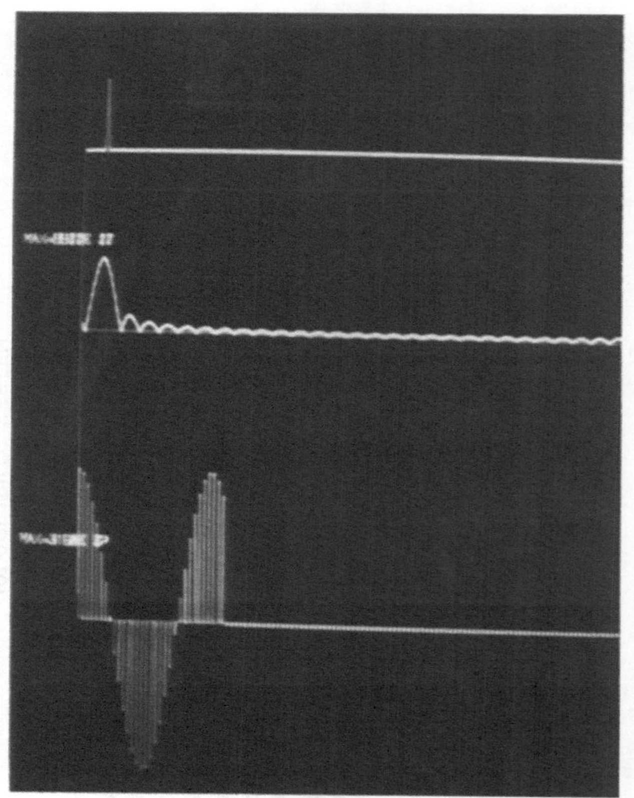

Fig. 8 The point object is shown on the first line, the image on the second, and the aperture wavefront on the third line. Bandlimiting is shown in the display on the right

unambigous field of view. Objects outside the field angle corresponding to the sampling limit are undersampled, and have aliases at lower spacial frequencies due to the stroboscopic effect. The result is ghost images inside the margin for objects outside the field angle limit. As specular objects would be especially troublesome, it would seem desirable to limit the field of view of a holographic system by controlling the radiation pattern of the insonifying transducer and the receiving elements, and/or use of an anechoic hood in front of the aperture.

Amplifier saturation is perhaps less likely in a holographic system due to the lack of the energy concentrating effect of a lens. Nevertheless, a specular object could cause saturation at periodic locations on the aperture plane corresponding to the spacial frequency of its field angle. These saturated fringes cause the other spacial frequencies to be blanked periodically, a form of amplitude modulation causing sidebands at the sum and difference frequencies about the blanking frequency and its harmonics. The spurious intermodulation products are equivalent to multiple ghost images all over the field of view, with intensities dependent upon the degree of clipping. A single brillant specular object could therefore cause considerable confusion of the image.

Adjacent channel cross-talk has the affect of reducing the response of the system at higher spacial frequencies in the aperture plane, which in this case results in a reduction in intensity of objects at wide field angles. Thus it produces an image with the appearance of vignetting.

Spurious signals due to non-adjacent channel cross-talk, noise, and dead channels are largely averaged out in the reconstruction process, with the primary effect of reducing contrast in the image.

CONCLUSION

The various restrictions and defects of lens and holographic systems result in various types of image degradation. Examples are shown in Figures 8 through 11, as described in the captions. It should be clear that neither system is free of potential problems, but their respective resistance to different types of disturbances

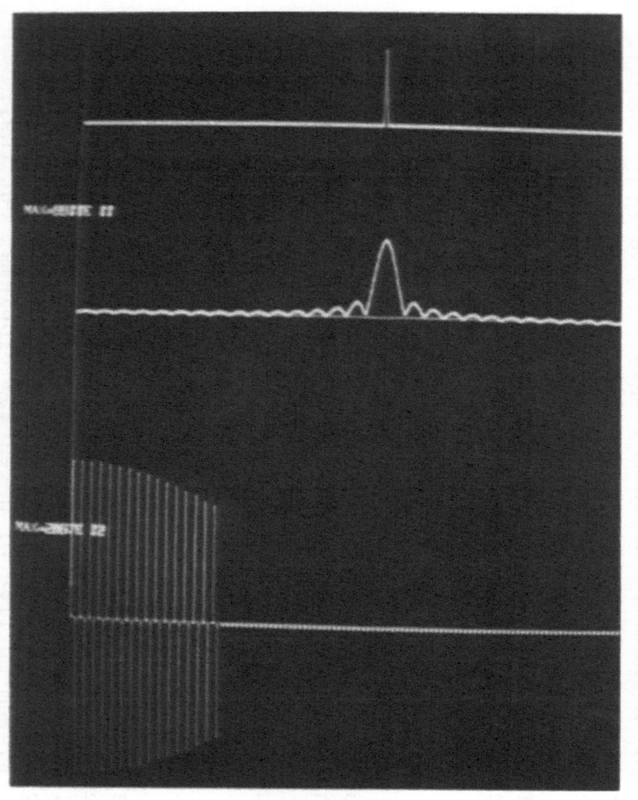

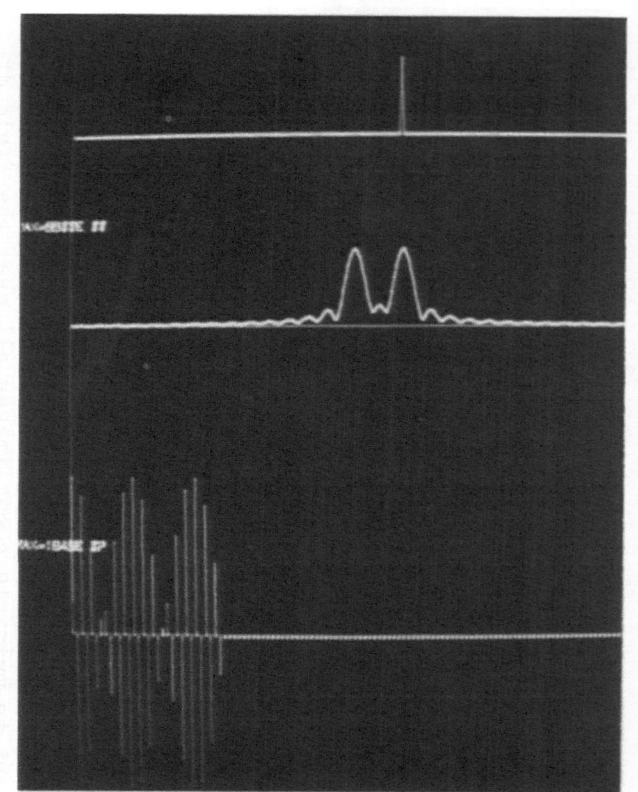

Fig. 9 Aliasing due to undersampling in a holographic system is shown on the right. The left display is barely within the sampling limit.

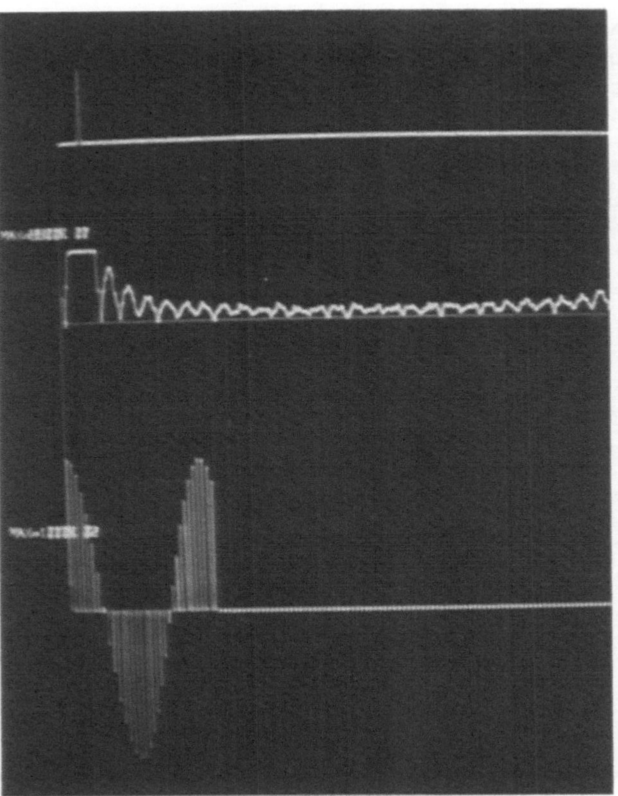

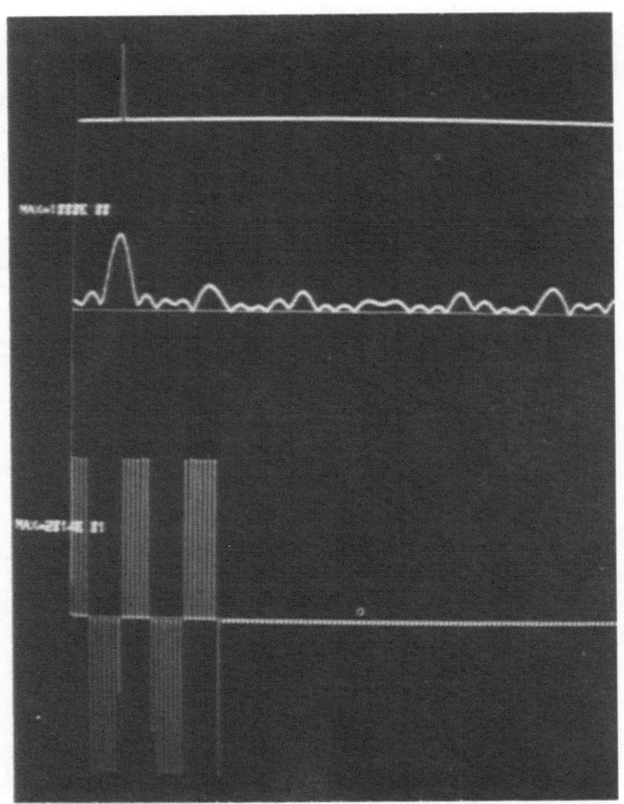

Fig. 10 Saturation and noise aggravate the diffraction effects in the lens system (1.) The saturated holographic system (R.) generates spurious ghost images.

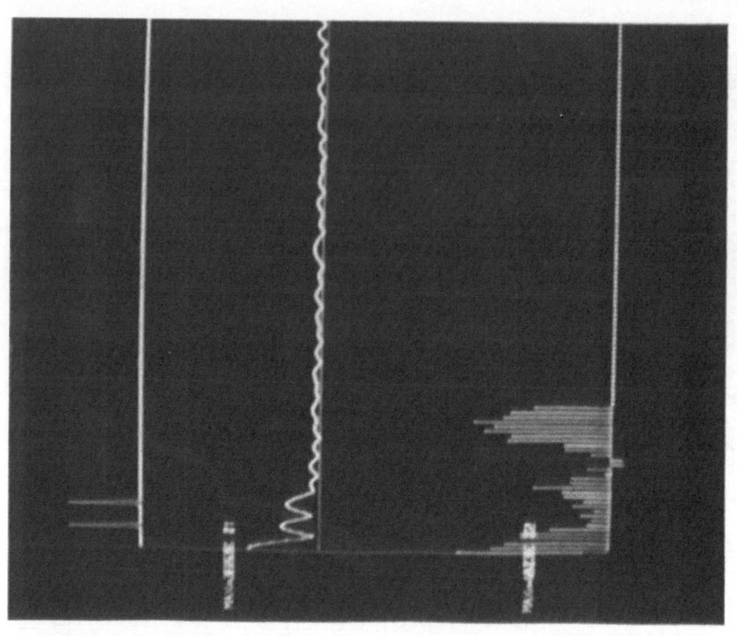

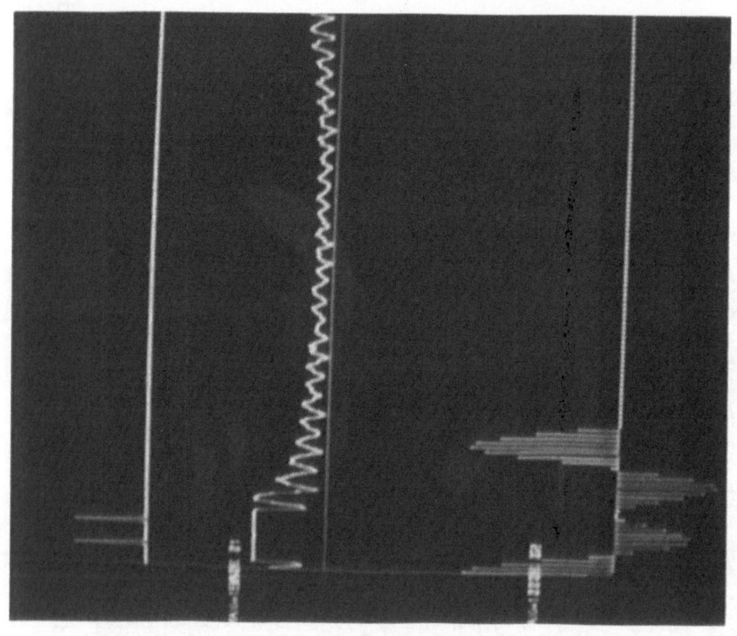

Fig. 11 Saturation and noise impair the resolution of the lens system (L.). Clipping the negative phase of the holographic recording (R.) causes a zero order response, but resolution is preserved with some loss of contrast.

DEGRADED IMAGE CHARACTERISTICS DUE TO SYSTEM DEFECTS

DEFECT	LENS SYSTEM	HOLOGRAPHIC SYSTEM
UNDERSAMPLING	Blurred Image, Anomalous Resolution of Bar Charts	Limited Field of View, Ghost Images due to Aliasing
SATURATION	Overexposed Highlights, Enhanced Diffraction Fringes	Ghost Images due to Intermodulation Products
ADJACENT CHANNEL CROSSTALK	Blurred Image	Vignetting of Image Field
NON ADJACENT CROSSTALK	Localize Bright Spots (Stars")	Reduced Contrast
DISTRIBUTED NOISE	Distributed Noise ("snow")	Reduced Contrast
LOCALIZED NOISE, DEAD CHANNELS	Bright Spots ("stars") Black Spots	Reduced Contrast
AMPLIFIER PHASE ERRORS	No effect	Blurred Image, Reduced Contrast
AMPLIFIER GAIN ERRORS	Distributed Noise ("snow")	Blurred Image, Reduced Contrast

can be compared to the performance cost in different system applications. A table is presented to summarize the results of this experiment, which although qualitative and far from exhaustive, should be of some benefit to the system designer as a kind of check-list.

ACKNOWLEDGEMENTS

The author would like to acknowledge his indebtdness to his colleague Prof. George A. Rahe who organized and developed the Electrical Engineering Computer Laboratory in the first place, and also translated the image system block diagrams into an efficient computational flow chart. Credit is due also to Mr. Robert Limes for writing and checking out the actual program.

CORRECTION OF PERCEIVED FIRST-ORDER DISTORTIONS IN HOLOGRAPHY*

Donald C. Winter

TRW Systems Group

Redondo Beach, California

INTRODUCTION

One of the fundamental problems in long wavelength holography is that of achieving undistorted three-dimensional imagery. To attain undistorted imagery, the longitudinal and lateral magnifications of the image must be equal. According to the classical analysis of holographic imagery, as formulated by Meier[1] and Champagne,[2] the lateral and longitudinal hologram magnifications will be equal if and only if the magnifications are also equal to the ratio of reconstruction wavelength to recording wavelength. If this rule were followed for long wavelength holography the resulting three-dimensional image, although undistorted, would, in most cases, be too small to be useful. If a larger image is obtained, the image will have unequal longitudinal and lateral magnifications, and therefore, first-order distortions.

As we are concerned here with a long wavelength hologram as a three-dimensional display to be viewed by an observer, an analysis of the distortion problem which included the observer in the system would seem appropriate. In particular,

*This work was performed while author was with the Radar and Optics Division, Willow Run Laboratories, Institute of Science and Technology, University of Michigan with partial support from NASA Langley Research Center under Contract NAS1-10810.

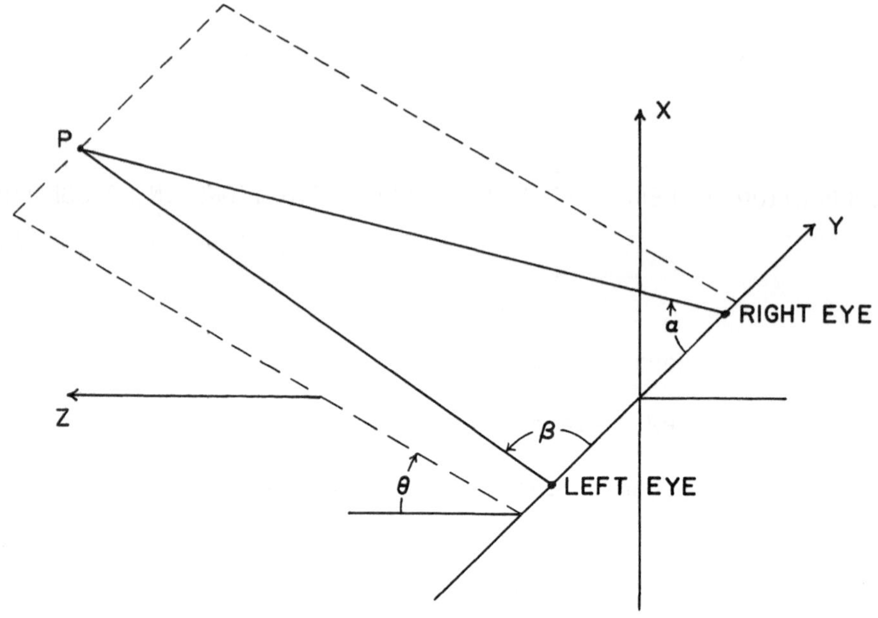

Figure 1 Bipolar Coordinate System

we are interested in examining the effects the holographic distortions have on the observer's perception of the image. The inclusion of the visual perception process into the description of holographic imagery necessarily complicates the analysis, but at the same time increases the degrees of freedom by which the image can be altered. We will see that these additional degrees of freedom enable us to overcome the problem of unequal longitudinal and lateral magnifications.

THEORETICAL ANALYSIS

To study the imaging process from the standpoint of an observer, it will be necessary to examine the mechanisms of visual space perception. Previous psychophysical studies have shown that the influence on visual perception of the level of accommodation, or focusing, of the eyes is negligible.[3] We are therefore justified in neglecting for the moment, the factor of accommodation. Under this assumption, the stimuli presented to the observer, for visual perception, can be specified by the angular distribution of the scene about the entrance pupils of the two eyes. This specification corresponds to the use of a bipolar coordinate system to provide an unambiguous localization of points in a three-dimensional space. The use of a bipolar coordinate system to model the binocular visual space perception of an observer can be attributed to Luneburg.[4,5,6]

The bipolar coordinate system is shown in Figure 1. By specifying the angles α, β, and θ we are able to locate uniquely the point P. To simplify our analysis, we can assume from here on that $\theta = 0$ without any loss in generality. Following Luneburg's approach, we will use the modified bipolar coordinates, ϕ and γ, shown in Figure 2, and defined by

$$\phi = 1/2(\alpha - \beta) \tag{1}$$

$$\gamma = \pi - \alpha - \beta \quad . \tag{2}$$

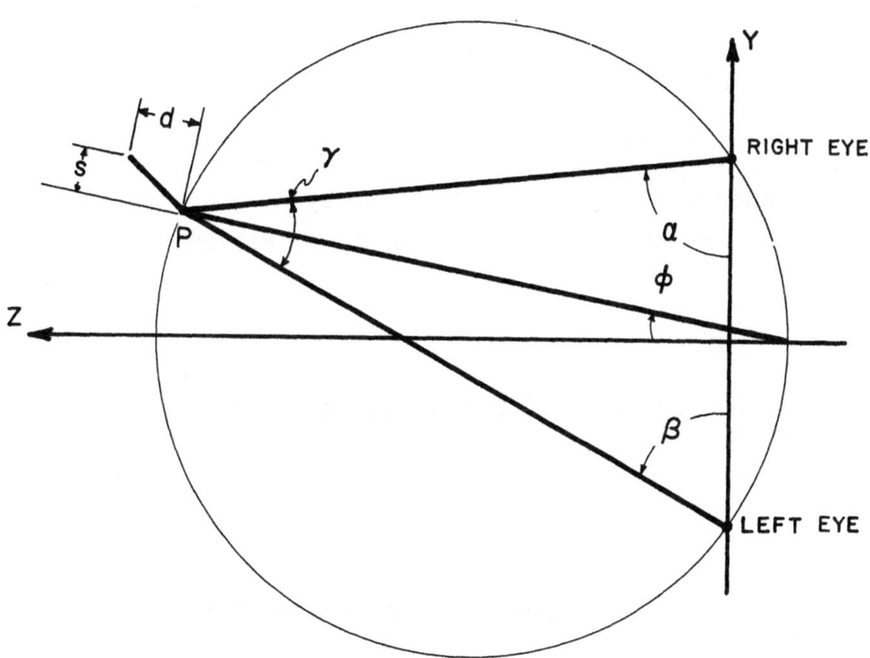

Figure 2 Modified Bipolar Coordinate System in Plane of Elevation of Point P

The bipolar coordinate system can be used to model an observer's binocular vision by placing the poles at the entrance pupils of the eyes. The coordinate γ can be identified as the angle of convergence, the angle at which the lines of sight intersect.

Let us now consider a small, arbitrary line element with longitudinal extent d and lateral extent s. That is, a small object with depth d and lateral size s. The object subtends an angle $\delta\alpha$ at the right eye, and $\delta\beta$ at the left eye. We define

$$\Gamma \equiv -\delta\gamma = \delta\alpha + \delta\beta \tag{3}$$

and

$$\varepsilon \equiv \delta\phi = 1/2(\delta\alpha - \delta\beta) \quad . \tag{4}$$

Γ is the angular disparity and represents the difference in the images seen by the left and right eyes; while ε is the visual angle and may be interpreted as the average angle subtended by the object at the observer's eyes. The seemingly misleading signs in these equations are due to our definition of the angles α and β.

We will denote the separation between the poles of the coordinate system by i and the distance from the object to the center of the coordinate system by R. Furthermore, we will assume that the small angle approximation holds so that R and γ will be related by

$$R = i/\gamma \quad . \tag{5}$$

Taking the depth, d, of the object as a differential of R, we obtain

$$d = \delta R = \frac{-i}{\gamma^2}\,\delta\gamma = \frac{\Gamma}{i} R^2 \quad . \tag{6}$$

Using the small angle approximation again, we write

$$s = \varepsilon R \quad . \tag{7}$$

The depth and size of the object can now be related to the modified bipolar coordinates by combining Equations 5, 6, and 7;

$$d/s = \Gamma/\gamma\varepsilon \quad . \tag{8}$$

The depth to size ratio is a convenient quantity to work with as we will see shortly. If an image is magnified equally in the longitudinal and lateral directions, the depth to size ratio will not be changed by the imaging process.

Using Equation 8 for the depth to size ratio, we can now examine how first-order distortions arise in holography. We know that, to first-order, the angular magnification of a hologram is a function only of the wavelengths used in the recording and reconstruction processes, and the hologram scaling factor. Therefore the angles $\delta\alpha$ and $\delta\beta$, and consequently Γ and ε, can be altered only by an equal multiplicative factor. We see, however, that if Γ and ε are multiplied by the same factor, the depth to size ratio is unchanged. Therefore, we can attribute any change in the depth to size ratio to a change in γ, the angle of convergence.

To enable us to utilize this observer oriented approach in a variety of situations, it is necessary to develop a more general formalism incorporating these techniques. To include the observer into a display system in a manageable way, we will need one additional concept, that of the space image. We define the space image to be the image that an observer would perceive, if he were able to interpret correctly the stimuli presented to him. By using the space image in place of the true perceived images we eliminate the impossible problem of analyzing the personal idiosyncrasies of the observers.

We can now analyze any holographic display system by calculating and comparing the depth to size ratios of the original object, the holographic image, and the space image. In defining the bipolar description of these images, we must specify the separation between the poles of the coordinate systems. In order to obtain a generalized analysis, three coordinate systems will have to be used, one for each image, with a specific relationship between the pole separations. To simplify the notation of our analysis, we will denote variables in the object space with the subscript "o" and variables in the hologram image space with the subscript "i." The variables which describe the stimuli to the observer and those which describe the space image will be primed. To model the binocular vision of the observer and thereby describe the space image, the separation between the poles must be equal to the separation between the observer's eyes--the interocular separation. When the observer views the holographic image by looking through the hologram, the lines of sight of each eye intersect the hologram at two points. These points will be taken as the poles of the coordinate system used to describe the holographic image. The separation between these poles is the effective separation of the observer's eyes in the image space. If the hologram is not scaled, we can use the same coordinate system to describe the original object and the holographic image. If the hologram is scaled, however, the separation between the poles, as well as all other dimensions on the hologram surface, will be scaled by the same factor. The pole separation for the coordinate system used to describe the original object will therefore be related to the pole separation used in describing the holographic image by the hologram scaling factor. That is,

$$i_o = \frac{i_i}{m} \tag{9}$$

where m is the hologram scale factor. By defining the coordinate systems in this manner, the ratio of the angular disparity, Γ, to the visual angle, ε, will be the same for all three systems.

The effective separation of the eyes can be altered by a set of four mirrors as shown in Figure 3. As both the angular disparity, Γ, and the angle of convergence, γ, are proportional to the separation of the poles, the depth to size ratio will not be changed. By varying the effective eye separation, however, we will be able to use holograms which are physically smaller than the normal interocular separation of about 6 cm. Furthermore, we can insure that the observer's angle of convergence will be within his range of comfort.

The depth to size ratio of the space image can be altered by correcting the angle of convergence of the observer. This correction can be accomplished by placing a pair of prisms between the hologram and the observer as shown in Figure 4.

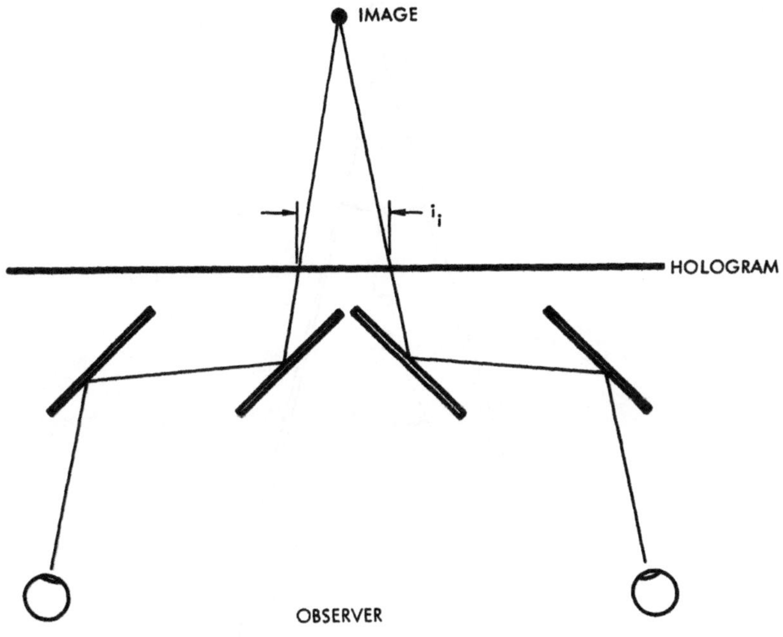

Figure 3 Mirror Assembly Used to Reduce the Effective Separation of the Eyes

The prisms add a constant angle Δ to the angle of convergence.

The prismatic deviation, Δ, needed to correct the first-order distortions, can be computed by requiring that the depth to size ratio of the space image be equal to the depth to size ratio of the object. We can write,

$$\frac{d'}{s'} = \frac{\Gamma'}{\varepsilon'\gamma'} \qquad (10)$$

for the space image depth to size ratio. As the ratio of Γ to ε is the same for all three coordinate systems, Equation 10 can be rewritten as

$$\frac{d'}{s'} = \frac{\Gamma_0}{\varepsilon_0\gamma'} \quad . \qquad (11)$$

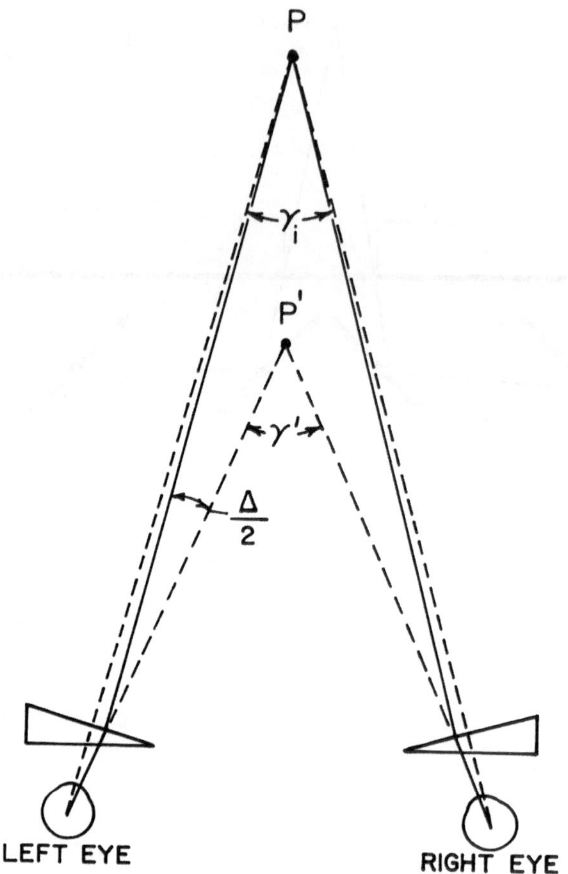

Figure 4 Use of Prisms to Increase the Observer's Angle of Convergence

In a similar fashion, the object depth to size ratio can be expressed as

$$\frac{d_o}{s_o} = \frac{\Gamma_o}{\varepsilon_o \gamma_o} \quad . \tag{12}$$

The requirement that $d'/s' = d_o/s_o$ can therefore be rewritten, using Equations 11 and 12, as

$$\gamma_o = \gamma' \quad . \tag{13}$$

If Equation 13 is satisfied, the space image will be free from first-order distortions. Placing a pair of prisms between the observer and the hologram, as shown in Figure 3, has the effect of adding a constant to the angle of convergence of the holographic image, approximately equal to the total prismatic deviation. That is

$$\gamma' \approx \gamma_i + \Delta \quad . \tag{14}$$

Substituting Equation 14 into Equation 13, we find that the condition for distortion free imagery is given by

$$\gamma_o = \gamma_i + \Delta \quad . \tag{15}$$

It is usually convenient to express Equation 15 as a function of the pole separations and the hologram recording geometry by using the appropriate versions of Equation 5 for the angles of convergence and Champagne's equations for the hologram image distance. When these operations are performed, we find that the required value of Δ is given by

$$\Delta = i_i \left[\frac{1}{mR_o} - \frac{1}{R_c} - \frac{\mu}{m^2} \left(\frac{1}{R_o} - \frac{1}{R_r} \right) \right] \tag{16}$$

where μ is the ratio of reconstruction wavelength to recording wavelength, and R_c and R_r are respectively the reconstruction beam and reference beam distances. From this equation we see that if the wavelength ratio, μ, is equal to the hologram scaling factor, m, Δ is independent of the object distance, R_o. In this case a single pair of prisms, of constant deviation, would be capable of correcting the first-order distortions for all points in the object space. Unfortunately, we are primarily concerned with holographic systems for which $\mu \neq m$. In that case, we see that the value of Δ necessary to compensate for the first-order hologram distortions will vary with R_o. We will have to choose one value of R_o and design the compensation system for that object distance. For object points with values of

R_o close to the selected value, the first-order distortions of the space image will be greatly reduced, but not eliminated.

As we cannot completely eliminate the first-order distortions of the space image we will need a measure of the distortions to predict a system's performance. We will use the ratio $(d'/s')/(d_o/s_o)$, which we denote by Λ, as a measure of the distortion due to unequal longitudinal and lateral magnifications. From Equations 11 and 12, we see that

$$\Lambda = \frac{\gamma_o}{\gamma'} \quad . \qquad (17)$$

As expected from our earlier analysis, the first-order distortions are determined only by the angle of convergence. If we now use a pair of prisms to eliminate the first-order distortions for objects at a distance $R_{o\Delta}$ from the hologram, we can show that the residual distortions for objects at any distance R_o, are given by

$$\Lambda = \frac{1}{\frac{\mu}{m} + \frac{R_o}{R_{o\Delta}} - \frac{\mu}{m}\frac{R_o}{R_{o\Delta}}} \quad . \qquad (18)$$

One very interesting fact is immediately evident from this equation. So long as the prisms are selected so that the distortions are eliminated for the object distance $R_{o\Delta}$, the variation in the distortions for different object distances is independent of the particular prismatic deviation and effective eye separations used. When the hologram wavelength ratio is much smaller than the hologram scale factor, we can approximate Equation 18 for the residual distortions as

$$\Lambda \approx \frac{R_{o\Delta}}{R_o} \quad . \qquad (19)$$

This equation provides a quick estimate of the performance of any given system. We see that the residual distortions will be reasonably small over a significant volume of the object space.

PSYCHOPHYSICAL EXPERIMENTS

From the preceding analysis, we have seen that it is theoretically possible to correct for the first-order distortions in long wavelength holography so long as we restrict our attention to the image perceived by an observer. The theoretical analysis indicated that the first-order distortions were due to changes in the angle of convergence. Our discussion centered about the space image, the image that the observer would perceive if, presented with the information necessary for a mathematical description of the image, he perceived the image as mathematically described. Our analysis has been limited in that we have not considered the extent to which an observer is able to use the information provided, in particular, by the angle of convergence, to obtain veridical perception. Several previous psychophysical studies [7,8,9] have examined the effect on the perception of an observer, of changes in the angle of convergence. These studies indicate that changes in the convergence angle of an observer are capable of causing changes in the observer's perception of image shape.

To examine experimentally an observer's ability to detect first-order distortions in holographic images, and to verify the possibility of correcting the distortions by altering the observer's angle of convergence, we prefer to limit the image degradation due to factors other than first-order distortion. In this manner the first-order distortions can be isolated from other factors which might affect the observer's perception. By using optically recorded holograms we can obtain high quality imagery while introducing the first-order distortions by variations in the reconstruction beam position. By varying the distance of the reconstruction beam point focus from the hologram as shown in Figure 5, the magnifications may be varied with the longitudinal magnification equal to the square of the lateral magnification. For convenience, lensless Fourier transform holograms were used. In addition to the prisms discussed previously, spherical lenses were used in some experiments to enable the observers to accommodate and converge for the same distance and thereby focus clearly on the holographic image. The test object consisted of two cards, separated in depth, with a scale ruled on the front card as shown in Figure 6. For each of several different magnification ratios, the observers were told to estimate the separation between the cards which they perceived, in terms of the units on the scale. This gives a

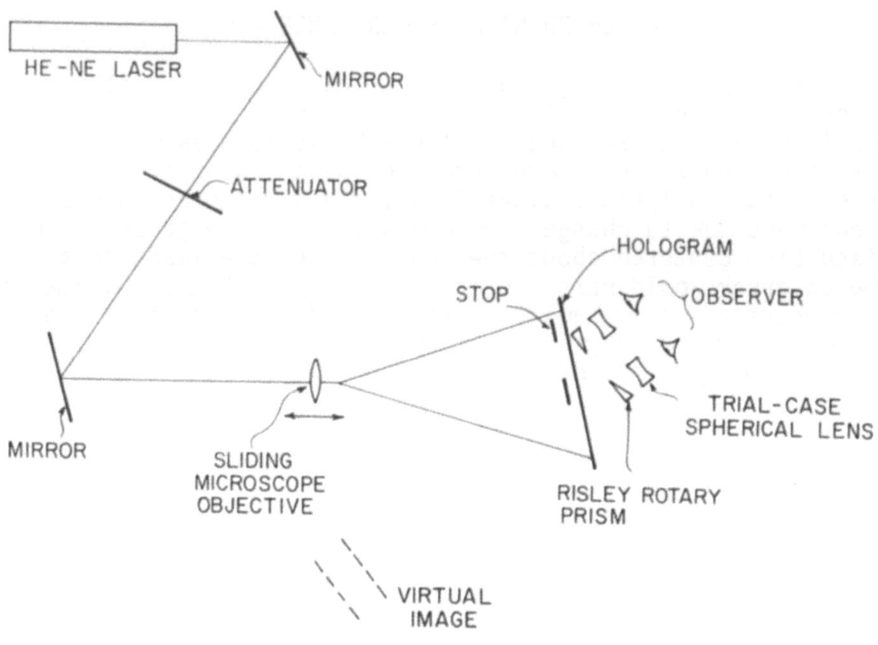

Figure 5 Configuration Used to Reconstruct Holographic Image

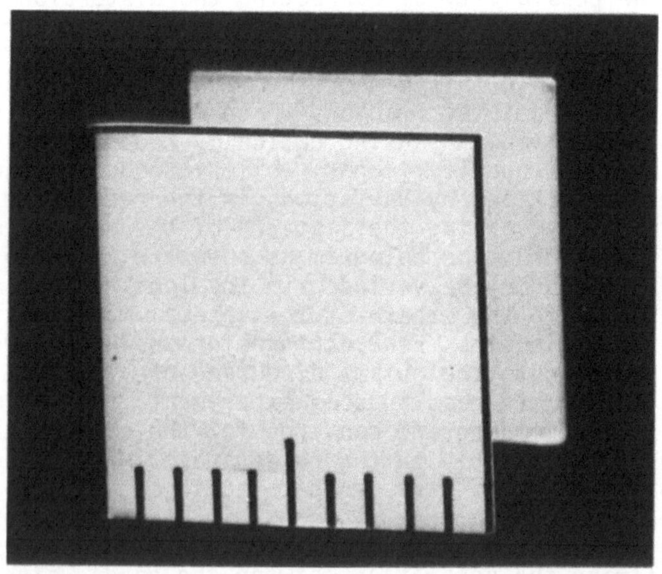

Figure 6 Object Used for Psychophysical Experiments

direct measure of the perceived depth to size ratio. Figures 7 and 8 show the responses of two of our observers. The perceived depth to size ratio has been normalized to the value perceived for the undistorted image. A normalized depth to size ratio of one, therefore corresponds to an undistorted perceived image. The normalized depth to size ratio is plotted as a function of the hologram image magnification ratio where each point represents the average of ten readings.

When the observers viewed the holographic image directly, they were able to perceive the distortions, although the sensitivity of the observers to the distortions differed. When a pair of prisms was used to correct the angle of convergence, the prismatic deviation necessarily varying with the magnification ratio, we see the elimination of the perceived distortions. Several studies showed that the additional use of opthalmic lenses did not affect the perceived distortions, but did improve the clarity of the perceived image. This experiment was repeated with several different observers and various test objects without any significant change in the results.

DISCUSSION OF POSSIBLE DIFFICULTIES

For the experiments described above, we may conclude that the perceived distortions were introduced by variations in the angle of convergence. We note, however, that when the holographic image was viewed directly the normalized depth to size ratio perceived by the observers was not identical to the magnification ratio. This discrepancy points out the complexities of the visual perceptual process, and, in particular, the presence of other factors, called empirical cues, which affect the observer's perception. The view of the test object shown in Figure 5, for example, provides information about the separation of the two cards even to a monocular observer; information which is not changed by variations in the angle of convergence. Fortunately, for a display system with an undistorted space image, we would expect that the empirical cues would reinforce the other visual cues, aid in the veridical perception of the image, and not induce distortions in the perceived image.

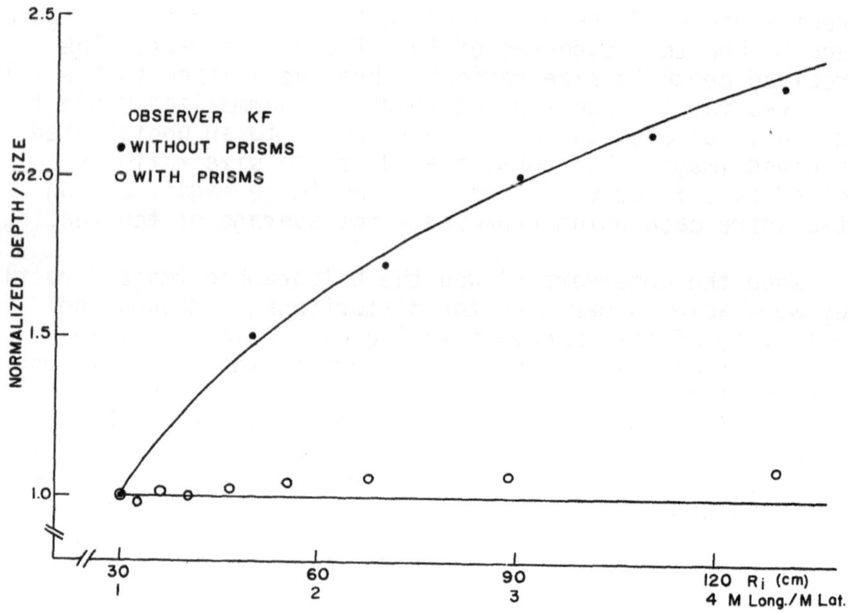

Figure 7 Responses of Observer KF for Psychophysical Experiment

Empirical cues are also capable of affecting the perception of distance. For example, when a binocular telescope is used to veiw people from a great distance, the known size of the people provides a strong cue for distance. Due to the relationship between the depth to size ratio of an object and its distance (or corresponding angle of convergence) described previously, there is reason to believe that the perceived depth to size ratio of an image will vary with the perceived distance, assuming that all other stimuli are constant. This effect has been noted during the use of binocular telescopes.[10] It is possible that the empirical factors for distance may be significant for some holographic systems, especially those for which the object and the image are in the distal region and the angles of convergence are consequently very small.

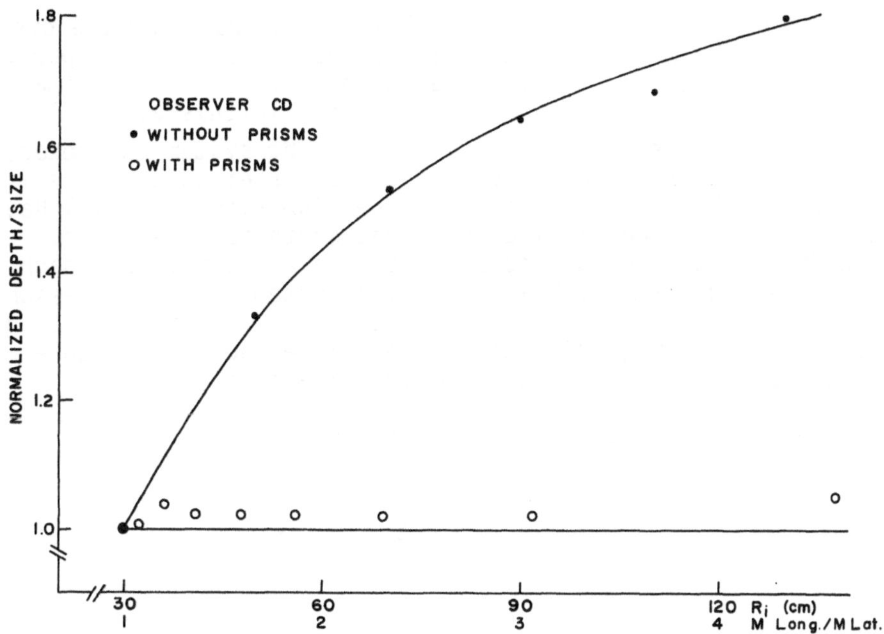

Figure 8 Responses of Observer CD for Psychophysical Experiment

It should be noted that the known size of an object will not affect the judgment of distance, however, if the observer believes that he is viewing a model, and not the original object. A photograph of a person, for example, is perceived as a model, and the known size of the person does not affect the observer's perception of the distance to the photograph. Although holographic images can be made with excellent quality, the monochromatic nature of the image, and the speckle caused by the coherent nature of the light make it evident that the holographic image is a model, distinctly different from the original object. This identification of holographic images, greatly reduces the effects of cues such as known size. As a test of this hypothesis, a hologram was made of a quarter (U. S. currency). When the distance to the image was varied, the perceived distance and size also varied. If the known size of the quarter played the determining role in the perceptual process, no variations in perceived distance or size would be expected.

Although it is unlikely that empirical cues for the distance of a holographic image will affect an observer's perception in most cases, it is conceivable that for certain images the empirical factors may be important. In this eventuality, it will be necessary to modify the viewing system parameters to obtain undistorted imagery. As the perceived distance is determined by empirical factors, the disparity, Γ', will have to be varied to adjust the perceived depth to size ratio. The following procedure is suggested as a possible approach to the design of suitable correction systems. We will assume that the perceived image can be approximated by a space image for which the image distance is determined not by the convergence angle, but by the perceived image distance.

From Equations 5 and 11 we can write

$$\frac{d'}{s'} = \frac{\Gamma_o}{\varepsilon_o} \frac{R'}{i'} \quad . \tag{20}$$

In Equation 20, R' is taken to be equal to the perceived image distance. Similarly from Equations 5, 9, and 12 we find that

$$\frac{d_o}{s_o} = \frac{\Gamma_o}{\varepsilon_o} \frac{R_o m}{i_i} \quad . \tag{21}$$

Consequently, the depth to size ratio of the space image will be equal to the depth to size ratio of the object if

$$i_i = \frac{R_o m\, i'}{R'} \quad . \tag{22}$$

By adjusting i_i, the disparity, Γ', is varied, thereby varying the depth of the space image and correcting the depth to size ratio.

The effects of empirical factors are highly dependent upon the subject of the image as well as the particular viewing systems. For this reason, the importance of the empirical factors will have to be examined for each situation.

CONCLUSIONS

The primary result of this study is the realization that a holographic, three-dimensional display system can be considered to be a binocular optical instrument. The various techniques used in the design of binocular instruments to vary the image presented to the observer can therefore be used with long wavelength holograms. In particular the effective separation of the eyes, and the angle of convergence can be varied such that the space image will have equal lateral and longitudinal magnifications. So long as we restrict our attention to the space image and perceived image, the first-order distortions in long-wavelength holography are removable and should not prevent the successful use of three-dimensional imagery. What is needed now, to evaluate this approach properly, is experimental data obtained by applying these techniques to practical long wavelength holographic systems.

REFERENCES

1. R. W. Meier, J. Opt. Soc. Am., 55, 987 (1965).

2. E. B. Champagne, J. Opt. Soc. Am., 57, 51 (1967).

3. C. H. Graham, Vision and Visual Perception, John Wiley and Sons, Inc., New York, 1965.

4. R. K. Luneburg, Mathematical Analysis of Binocular Vision, Princeton University Press, Princeton, 1947.

5. R. K. Luneburg in Courant Anniversary Volume, Interscience Publishers, Inc., New York, 1948.

6. R. K. Luneburg, J. Opt. Soc. Am., 40, 627 (1950).

7. E. G. Heineman, E. Tulving, and J. Nachmias, Am. J. Psychol., 72, 32 (1959).

8. H. Wallach and C. Zuckerman, Am. J. Psychol., 76, 404 (1963).

9. J. M. Foley, Percept. and Psychophysics, 2, 605 (1967).

10. J. Von Kries in Helmholtz's Treatise on Physiological Optics, Vol. III, edited by J. P. C. Southall, Optical Society of America, 1925.

ULTRASONIC HOLOGRAPHY IN A SOURCE- AND OBJECT-MOVABLE SYSTEM

T. Iwasaki and Y. Aoki

Department of Electronic Engineering
Hokkaido University
N 12, w 8, Sapporo, 060, Japan

ABSTRACT

In the acoustical holography by point-to-point mapping, various types of holographic systems are derived according to the relative movements of a wave-source, objects points and a receiver. We analyse such holographic systems in which a wave-source (or an object)moves rectilineally, while a receiver scans along the one-dimensional line, resulting in the equivalence of two-dimensional scanning by the receiver. From the analysis we find that an aberration of a constructed hologram is inevitable in this type of holography. We propose nonlinear reduction of acoustical holograms to eliminate the aberration. Experiment using 1 MHz ultrasonic-wave is conducted, where the source and object are allowed to move to one direction and a receiver scans repeatedly along the same line perpendicular to the source-movement. The experimental results verify the theoretical ones.

INTRODUCTION

Many investigations have been reported on the holographic imagery with sound-waves from the audio to ultrasonic frequency. The hologram-constructing technique is one of the important and interesting subjects in the holographic system in long-wavelength region. In ultrasonic holography, hologram-constructing techniques are classified as follows.[1]

1. Photographic or chemical method
2. Liquid surface deformation
3. Direct interaction of light and ultrasonic waves
4. Ultrasonic camera
5. Mechanical scanning

In those techniques, the mechanical scanning requires much time to construct a hologram. However, this holographic technique has advantages that the large hologram aperture is available compared with other techniques and that electronic components with high sensitivity and electronic circuits and systems can be utilized for hologram data processing. These advantages are so useful in long-wavelength holographies that this holographic technique should not be discarded by reason of spending much time.

Though two-dimensional arrays of receivers can be proposed for constructing a hologram rapidly[2,3], it seems to be difficult from the technical and economical points of view. In conventional acoustical holography, a receiver scans mechanically to map the two-dimensional acoustical fields[4,5]. The mechanic and electronic system for this scanning is rather complicated.Moreover, very close scanning is necessary for constructing fine hologram as the wavelength becomes shorter in ultrasonic region. If ultrasonic holograms are constructed by only one-demensional scanning along x-axis without moving the receiver along y-axis, this is a promising technique. In this paper, we propose such a holographic technique and conduct an experiment to confirm our proposal.

TWO-DIMENSIONAL SCANNING SYSTEM

In the conventional scanned-type holography, a receiver scans over the two-dimensional acoustical fields as shown in Fig.1, where the object and source are fixed. This is just the two-dimensional scanning system. Here, Gabor's coherent background holography is considered. The typical procedure to construct an acoustical hologram in the system of Fig.1 is as follows ; 1. Scanning the hologram plane with a receiver, 2. Displaying the holographic signals from the receiver with CRT, where the position of spot on the scope is synchronized to that of the receiver in the hologram plane, 3. Recording the hologram on the film by a camera and reducing the recorded hologram for optical reconstruction.

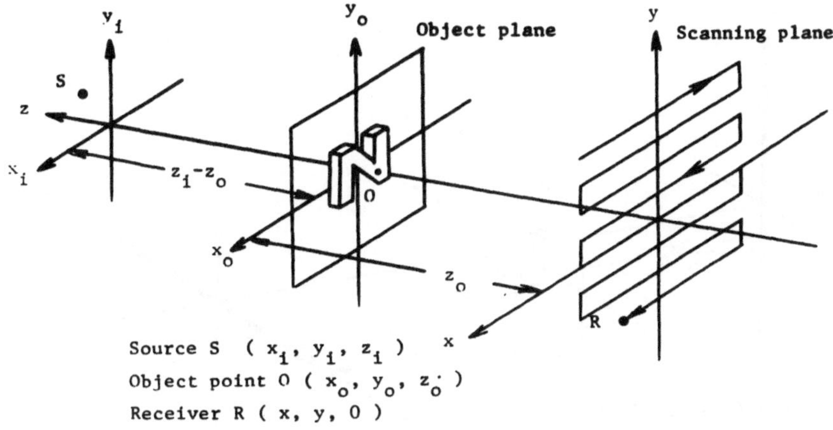

Fig. 1 Configuration of conventional two-dimensional scanning system.

These processes are described by following transformation,

$$\begin{aligned} x &= mx_1 \\ y &= my_1 \end{aligned} \qquad (1)$$

where m is the reduction ratio and (x, y), (x_1, y_1) are the coordinates of the scanning plane and final hologram plane, respectively. The transformation expressed by Eq. (1) is one of the various kinds of mappings from the scanning plane to the final hologram plane.

SOURCE-MOVABLE SYSTEM

Structure of the System

Two-dimensional scanning can be decomposed to the combination of two kinds of scannings with a source and a receiver along the crucial one-dimensional lines. This fact suggests us an another holographic technique as shown in Fig. 2. In Fig. 2 the vertical (y-axis) position of a receiver is fixed and the receiver scans repeatedly along x-axis. The source moves along y_i-axis perpendicular to the receiver scanning, resulting in the equivalent effect of receiver-scanning along y-axis. This system may be called

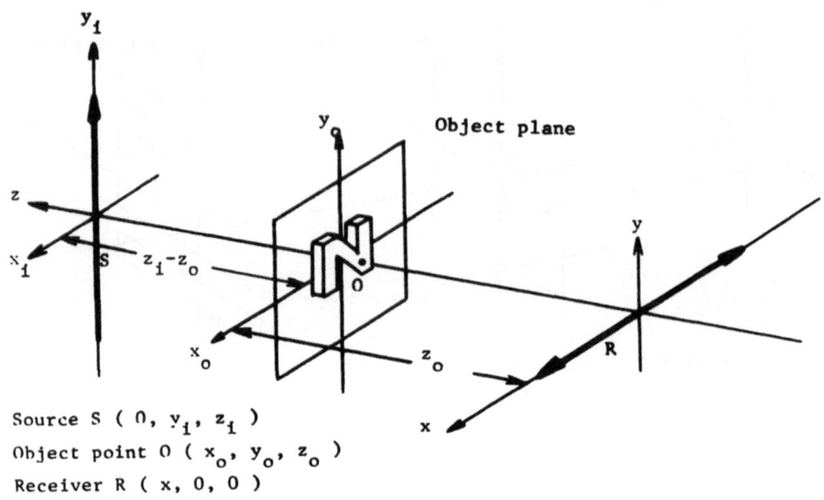

Fig. 2 Configuration of the source-movable system.

source-movable system. The horizontal position of the CRT-spot is synchronized with x-coordinate of the receiver, but the vertical position is synchronized with y_i-coordinate of the source. In this case, the reduction ratios with respect to x-and y-coordinate are not always equal. Therefor, the transformation of Eq. (1) is rewritten as follows,

$$x = mx_1$$
$$y_i = ny_1 \qquad (2)$$

where n is the reduction ratio with respect to y-coordinate.

An Analysis of the Hologram

For simplicity of analysis, the object-wave diffracted from a single point of the object is considered. A phase $\psi_f(x, y_i)$ is recorded under the condition that a source is at the position $(0, y_i, z_i)$ and a receiver at $(x, 0, 0)$. This phase arises according to the square-law detection of the supperposed fields of illuminating and object waves. The phase $\psi_f(x, y_i)$ may be written with the paraxial ray opproximation as follows,

$$\psi_f(x, y_i) = \pm \frac{k_1}{2} \left\{ \frac{x_o^2 + (y_i - y_o)^2}{z_i - z_o} + \frac{(x_o - x)^2 + y_o^2}{z_o} - \frac{x^2 + y_i^2}{z_i} \right\} \tag{3}$$

where $k_1 = 2\pi/\lambda_1$, λ_1 is the wavelength of the ultrasonic-wave employed in the construction of the hologram. Substituting Eq. (2) into Eq. (3), the phase $\psi_f(x_1, y_1)$ on the find hologram plane is obtained as follows,

$$\psi_f(x_1, y_1) = \pm \frac{k_1}{2} \left\{ A_x x_1^2 + A_y y_1^2 - 2(D_x x_1 + D_y y_1) + \frac{z_i\ (x_o^2 + y_o^2)}{z_o\ (z_i - z_o)} \right\} \tag{4}$$

where

$$A_x = m^2 \left\{ \frac{1}{z_o} - \frac{1}{z_i} \right\}, \qquad D_x = m \frac{x_o}{z_o}$$

$$A_y = n^2 \left\{ \frac{1}{z_i - z_o} - \frac{1}{z_i} \right\}, \qquad D_y = n \frac{y_o}{z_i - z_o} \tag{5}$$

In the case that A_x is not equal to A_y, Eq. (4) represents an ellipsoidal wavefront. This means that the final hologram is distorted.

The reconstructed image from this final hologram has an aberration. To eliminate this aberration, reduction ratios m and n must be chosen to satisfy the following relation.

$$\frac{m}{n} = \frac{z_o}{z_i - z_o} \tag{6}$$

The schematic explanation of Eq. (6) is shown in Fig. 3,

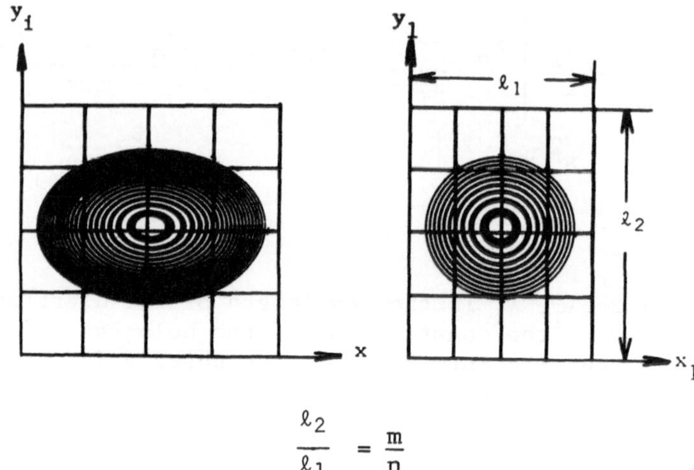

$$\frac{\ell_2}{\ell_1} = \frac{m}{n}$$

Fig. 3 Schematic explanation of the relation of Eq. (6).

where x-coordinate is reduced to the smaller size than that of y-coordinate (in the case m>n, that is, $z_o > z_i/2$). It should be noted that m/n of Eq. (6) varies with the position z_o of the object plane. This means that an aberration mentioned above is inevitable for the three-dimensional object. The linearly-propotional reduction, that is $m = n$, arises only when the planar object is located in the middle plane between $x_i - y_i$ and $x - y$ planes in Fig.2.

When Eq. (6) is satisfied, the position and magnifications of the reconstructed images are in agreement with those of the holographic system of Fig.1. Here, only analytical results are described. Applying the collimated coherent light to the final hologram reduced nonlinearly according to Eq. (6), the position Z_2 and magnifications $\partial x_2/\partial x_o$ and $\partial y_2/\partial y_o$ with respect to x- and y-coordinates of the reconstructed images are obtained as follows,

$$z_2 = \mp \frac{\mu}{m^2} \frac{z_i z_o}{z_i - z_o} \tag{7}$$

$$\frac{\partial x_2}{\partial x_o} = \frac{\partial y_2}{\partial y_o} = \frac{1}{m} \frac{z_i}{z_i - z_o} \tag{8}$$

where $\mu = \lambda_1/\lambda_2$, λ_2 is the wavelength of the light and (x_2, y_2) are the coordinates of an image point corresponding to the coordinates (x_o, y_o) of the object point. Negative and positive signs correspond to the true and conjugate images.

Experiment

Ultrasonic holograms in the object-movable system are constructed as in the experimental arrangement of Fig.4. Two PZT transducers, both with dismeter 5 mm, are used as a source and a receiver. The frequency is chosen at 1 MHz, that is the wavelength in water is about 1.5 mm. The carrier wave is modulated with 20KHz rectangular wave for convenience of amplifying the received signals. The receiver is allowed to scan about 20 cm in length. The nonlinear reduction is done, adjusting the gains of two D.C. amplifiers shown in Fig.4.

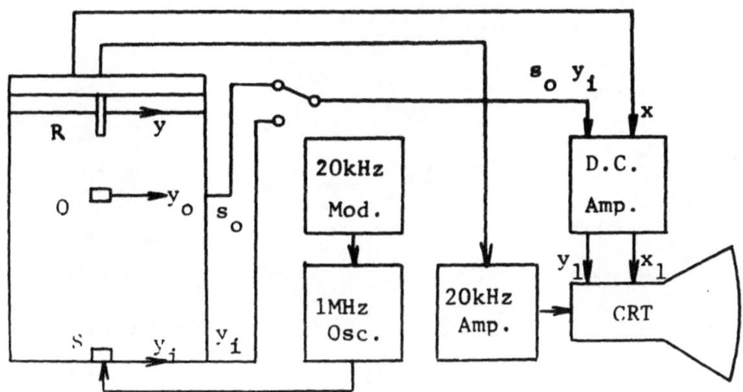

Fig. 4 Block diagram of the experimental arrangement.

The ultrasonic holograms constructed by moving the source are shown in Fig.5-(a), (b) and (c). The images reconstructed optically from the holograms of Fig.5-(a), (b) and (c) are shown in Fig.5-(d), (e) and (f). The object is a letter N made of air-contained spongy. The holograms of Fig.5 are constructed under the condition that $m/n = 1$, while the position z_o of the object is changing, resulting in $z_o/(z_i - z_o) = 0.6$, 1 and 1.56 for Fig.5- (a) (b) and (c) respectively. The hologram of Fig.5-(b) satisfies the Eq. (6), but other holograms do not. This means that the latter holograms are distorted and this distortion makes the reconstructed images unclear.

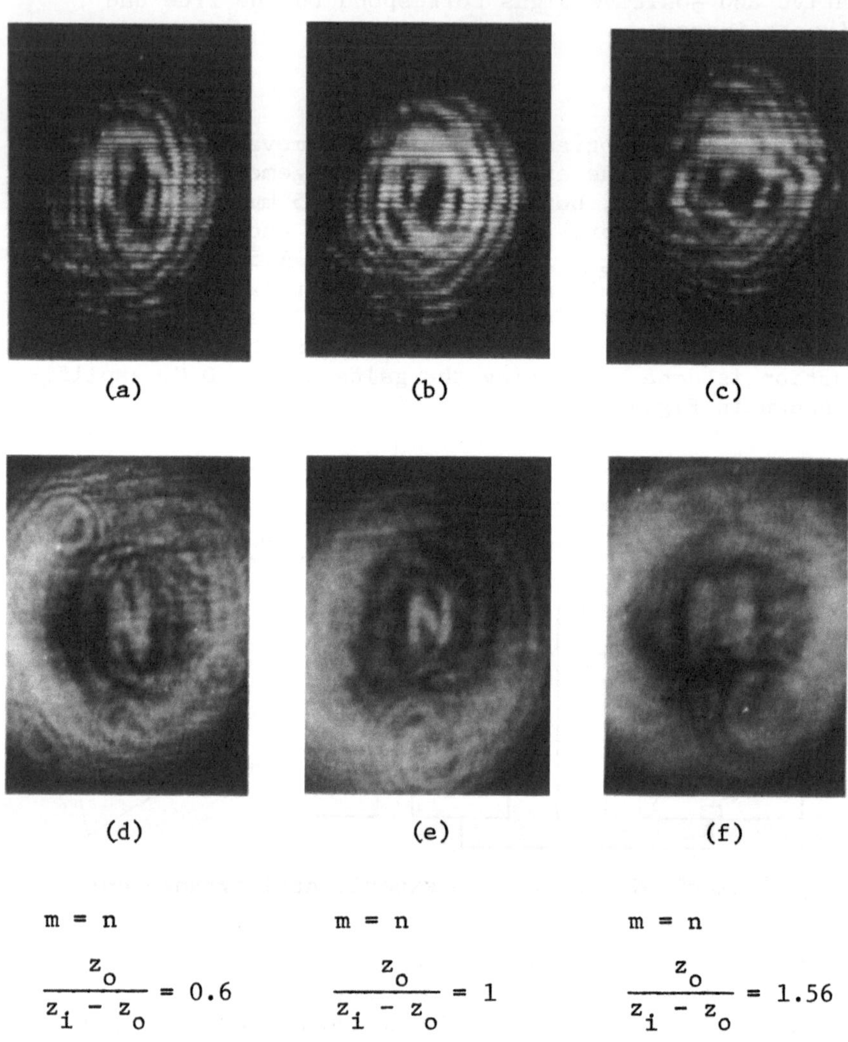

Fig. 5 Holograms (a), (b) and (c) are constructed in the system of Fig.2. The images (d), (e) and (f) are optically reconstructed from the holograms (a), (b) and (c) respectively.

This argument is clarified by the experimental results of Fig.5, where the best image is reconstructed for m = n (Fig.5-(b)), while the images of Fig.5-(d) and (f) reconstructed from the distorted holograms (Fig.5-(a) and (c)) are unclear. For the case of m = n, it is verified that the experimental values of image position and magnifications are in good agreement with that of theoretical ones of Eqs. (7) and (8).

OBJECT-MOVABLE SYSTEM

The movement of the source is relative with respect to the object. This fact suggests us an another holographic technique as shown in the schematic explanation of Fig.6. In Fig.6, an object is moved along y_o-axis, after every

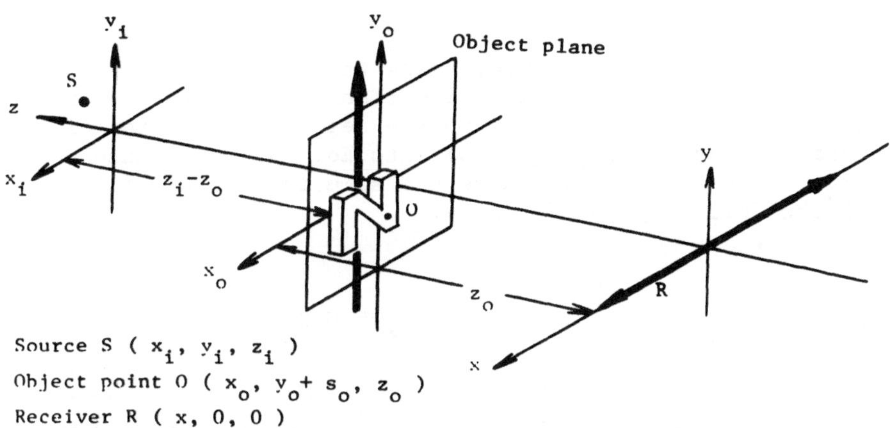

Fig. 6 Configuration of object-movable system.

time the receiver-scanning along x-axis finishes. This holographic system may be called object-movable system.

In this system the recorded phase $\psi_f(x, s_o)$ corresponding to Eq. (3) can be expressed as follows,

$$\psi_f(x, s_o) = \pm \frac{k_1}{2} \left\{ \frac{x_o^2 + (y_o + s_o)^2}{z_i - z_o} + \frac{(x_o - x)^2 + (y_o + s_o)^2}{z_o} - \frac{x^2}{z_i} \right\} \qquad (9)$$

In the deduction of Eq. (9), the source is assumed to be placed at the origin of the source plane, that is $x_i = y_i = 0$, for simplicity of discussion. In the system of Fig.6, the y-position of the CRT-spot is synchronized with the object-movement. Therefore the relation between the coordinates (x_1, y_1) of the find hologram and the coordinate x of the receiver and the displacement s_o of the object is expressed as following transformation.

$$x = mx_1$$
$$s_o = - ny_1 \qquad (10)$$

The negative sign of Eq. (10) means that the movement of the object point to the positive direction of y_o-axis is equivalent to that of the source to the negative direction of y_i-axis. Substituting Eq. (10) into Eq. (9), the phase $\psi_f(x_1, y_1)$ can be rewritten as the same form of Eq. (4) with the following coefficients.

$$A_x = m^2 \left\{ \frac{1}{z_o} - \frac{1}{z_i} \right\}, \qquad D_x = m\frac{x_o}{z_o}$$
$$A_y = n^2 \left\{ \frac{1}{z_i - z_o} + \frac{1}{z_o} \right\}, \qquad D_y = n\left\{ \frac{1}{z_i - z_o} + \frac{1}{z_o} \right\} y_o \qquad (11)$$

When A_x is not equal to A_y in Eq. (11), a distorted hologram is constructed. Putting $A_x = A_y$, a condition to produce the undistorted hologram is obtained as follows,

$$\frac{m}{n} = \frac{z_i}{z_i - z_o} \qquad (12)$$

Under the condition of Eq. (12), the positions and magnifications of reconstructed images in the object-movable system is obtained by Eqs. (7) and (8). Equation (12) indicates that nonlinear reduction is always necessary in the object-movable system as long as the illuminating wave is a spherical wave. The condition of m = n holds only when z_i approaches infinity, that is, illuminating wave becomes a plane wave.

The experimental results using the equipments of Fig.4 is shown in Fig.7. The ultrasonic hologram of Fig.7-(a) is constructed under the condition that $|z_o| = 23$ cm, $|z_i| = 64$ cm and m/n is adjusted to 1.56 to satisfy Eq. (12). The optically reconstructed image of Fig.7-(b) shows the original object of a letter N.

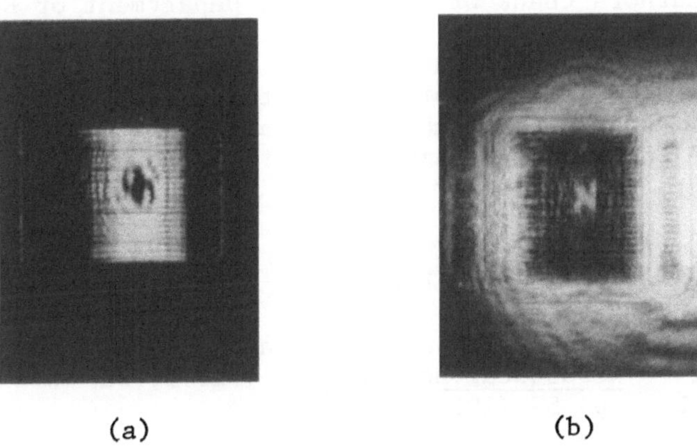

(a) (b)

Fig. 7 Hologram (a) is constructed in the system of Fig.6. The image (b) is optically reconstructed from the hologram (a).

CONCLUSION

A holographic technique in a source- and object-movable system is proposed. In this system holograms can be constructed with one-dimensional scanning of a receiver, resulting in the simplification of the scanning mechanism and electronic system compared with the two-dimensional scanning system. In the proposed system, however, distorted holograms are constructed. Condition to eliminate this distortion by nonlinear reduction is obtained theoretically. The experiment with 1MHz ultrasonic-wave is conducted according to the proposed method. The experimental results coincide with the theoretical ones and this verifies our proposal.

In holographic technique proposed here, the one-dimensional scanning can be replaced by a one-dimensional array of receivers switching them electronically[6]. In this case, the disadvantage of the conventional two-dimensional scanned-type holography, in which it takes much time to construct a hologram can be overcomed.

ACKNOWLEDGMENTS

The authors thank Dr. M. Suzuki, Department of Electronics, Hokkaido University for supporting the present work. They thank Dr. M. Onoe, Institute of Industrial Science, University of Tokyo, for presenting this work at the Symposium. They also thank Dr. G. Wade, Department of Electrical Engineering, University of California, for arranging to present their paper.

REFERENCE

1. B. B. Brenden, " A comparison of acoustical holography method, " Acoustical Holography, Vol. 1, Plenum Press, New York, : p. 57 (1969)

2. E. Marom and R. K. Mueller, " Design and preliminary test of an underwater viewing system using sound holography, " Acoustical Holography, Vol. 3, Plenum Press, New York, : p. 191 (1971)

3. G. L. Sackman and LT. R. J. Larkin, " An electronically scanned transducer array using microcircuit devices, " Acoustical Holography, Vol. 3, Plenum Press, New York,: p. 211 (1971)

4. Y. Aoki, " Acoustical holograms and optical reconstruction, " Acoustical Holography, Vol. 1, Plenum Press, New York, : p. 223 (1969)

5. A. F. Metherell, " The relative importance of phase and amplitude in acoustical holography, " Acoustical Holography, Vol. 1, Plenum Press, New York, : p. 203 (1969)

6. W. H. Wells, " Acoustical imaging with linear transducer arrays, " Acoustical Holography, Vol. 2, Plenum Press, New York, : p. 87 (1970)

ACOUSTIC INTERFERENCE IN SOLIDS AND HOLOGRAPHIC IMAGING

J. A. Cunningham and C. F. Quate

Stanford University

Stanford, California 94305

ABSTRACT

This paper reports a new technique for the visualization of ultrasonic interference in solids. The technique utilizes small (≈ 1 μm) polystyrene spheres suspended in a very thin liquid film formed on the surface of the solid. The nonlinear response of this film to acoustic waves is such that the spheres reveal the interference pattern present at the surface. There are two important advantages to this technique. The first is the simplicity of the technique, no elaborate equipment is required. The second is the easing of restrictions on ultimate resolution by our ability to visualize ultrasonic interference in solids.

We demonstrate the capability of obtaining a linear hologram resolution of 16 μm at a sensitivity on the order of 10^{-4} watts/cm^2 to 10^{-3} watts/cm^2. We also present the optical reconstruction of a mesh from an acoustic hologram. The resolution shown in the reconstructed image is approximately 380 μm with a magnification approaching 50 ×.

INTRODUCTION

Since its rudimentary beginnings, acoustic imaging has grown into a complex technology encompassing many distinct and often novel techniques. One such technique, which has spurred much recent investigation, is acoustical holography.[1] Holography involves the recording and subsequent reconstruction of a wavefront which contains both phase and amplitude information about an object. The information contained in this wavefront is recorded by interfering it with a known, phase related, reference wavefront. The resulting record, or hologram, may then be used to optically reconstruct an image.

To date, most successful acoustic holograms have been generated in a liquid, usually water, and recorded in the form of a surface relief pattern whose displacement was proportional to the acoustic intensity. This method has three limitations.[2] First, it is limited in highest usable frequency, and hence resolution, by the large acoustic attenuation in liquids at high frequencies. Second, any unwanted surface deformations produce distortions in the reconstructed wavefront which seriously degrade image quality. Third, for a given acoustic power density, the displacement decreases inversely as frequency squared.

We have developed a technique which should relieve some of these limitations and, after further development, provide a high resolution acoustic imaging system. This new technique permits the direct visualization of acoustic interference in solids by means of a nonlinear response of small particles suspended in a thin liquid film on its surface. Since solids are low acoustic loss materials, the limit on upper usable frequency, hence resolution, is greatly relaxed. Also, since the method of recording the interference pattern does not involve surface properties, degradations in the final image are expected to be less.

THE VISUALIZATION TECHNIQUE

The interference pattern generated in the solid is revealed in a thin liquid film which is formed on its surface. The film consists of a solution of water and photo-flo (for wetting purposes) in which are suspended 1.1 micron (μm) polystyrene spheres. The acoustic waves interacting in the solid are passed into this film via acoustic transformers. There, by means of a nonlinear process in the liquid, stationary radiation pressures are exerted on the spheres causing them to move either to regions of highest or lowest intensity dependent on their density relative to the liquid.

The nature and magnitude of these forces is most easily demonstrated for the simple model shown in Fig. 1.

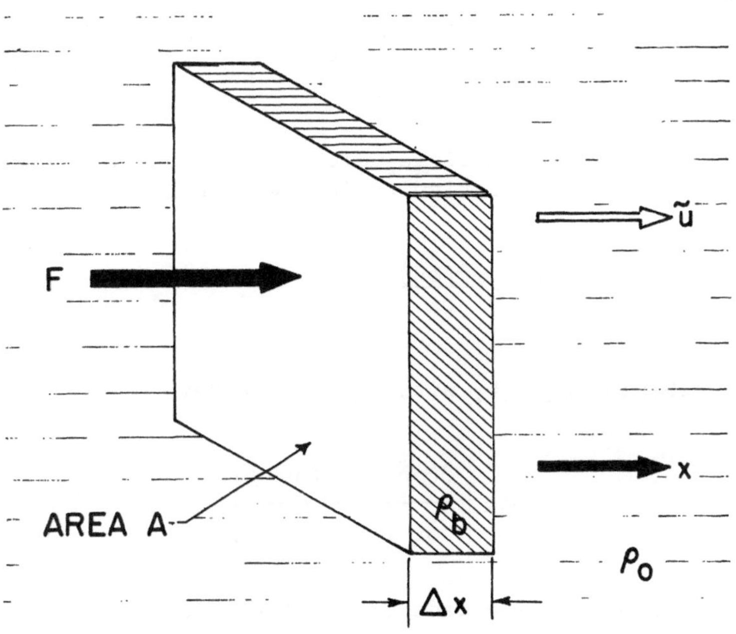

FIG. 1--Force on a differential volume element in a surrounding liquid due to a traveling plane acoustic wave.

We assume a thin slab of material of density ρ_b suspended in a liquid of density ρ_0 and oriented so that it is normal to the plane waves (wavevector $\tilde{k}$) traversing the liquid. If the thickness of the slab is Δx , the force per unit area is given by

$$f = (\rho \, \Delta x) \frac{d\tilde{u}}{dt} , \tag{1}$$

where upon using

$$\frac{d\tilde{u}}{dt} = \frac{\partial \tilde{u}}{\partial t} + \tilde{u} \frac{\partial \tilde{u}}{\partial x} \tag{2}$$

we obtain

$$f = \rho \, \Delta x \left(\frac{\partial \tilde{u}}{\partial t} + \tilde{u} \frac{\partial \tilde{u}}{\partial x} \right) . \tag{3}$$

Here $\tilde{u}$ denotes the particle velocity. In addition, we have the equation of continuity in the form

$$\frac{\partial \rho}{\partial t} + \frac{\partial (\rho \tilde{u})}{\partial x} = 0 , \tag{4}$$

and by combining Eq. (4) with Eq. (3) we have

$$f = \Delta x \left[\frac{\partial (\rho \tilde{u})}{\partial t} + \frac{\partial (\rho \tilde{u}^2)}{\partial x} \right] . \tag{5}$$

In Eq. (5) use has been made of the fact that

$$\frac{\partial [(\rho \tilde{u}) \tilde{u}]}{\partial x} = \rho \tilde{u} \frac{\partial \tilde{u}}{\partial x} + \tilde{u} \frac{\partial (\rho \tilde{u})}{\partial x} . \tag{6}$$

In linear acoustic problems we would use only the first term on the right of Eq. (5), equate this to the gradient of the pressure and proceed to the wave equation. Here we want to exploit the force arising from the second, or non-linear term on the right of Eq. (5). We assume that there are small amplitude acoustic waves at frequency ω moving through the fluid and, as a result of the interference term involving $\tilde{u}^2$ in Eq. (5), there will be a component of force at the sum frequency, $\omega_s = 2\omega$, and at the difference frequency, $\omega_d = 0$. We focus our attention on this term since it can be used to describe the effects presented in the experimental section. The first term on the right of Eq. (5) cannot give rise to a time independent force and we can therefore rewrite the equation without this term as

$$F = \frac{\partial(\rho\, \tilde{u}^2)}{\partial x} \Delta x\, A \quad , \tag{7}$$

where F is now used to denote the force resulting from the two interfering sound waves. The density ρ can be expanded as $\rho = \rho_b + \rho_1$, where ρ_b is the average density of the media and $\rho_1 = \rho - \rho_b$ is the excess density resulting from the sound wave. In a similar manner, $\tilde{u}$ can be expanded in the form $\tilde{u} = \tilde{u}_0 + \tilde{u}_1$, where $\tilde{u}_0 = 0$ since here we have no average particle velocity. In this case we may keep only the second-order term $\rho_b \tilde{u}_1^2$ and simplify Eq. (7) to read

$$F = \rho_b\, V \frac{\partial \tilde{u}_1^2}{\partial x} \quad . \tag{8}$$

Here $V = \Delta x A$ is the volume of the element, and it is understood that we are considering only the time independent portion of the expression on the right. In this form we recognize this as the translational force exerted on bubbles contained in the liquid.[3] We may apply the result to our case simply by considering $\rho_b > \rho_0$.

We are not yet finished with our calculation for the element of mass density ρ_b, for we must remember that there is a force on the volume V of magnitude $\rho_0 V\, \partial \tilde{u}_1^2/\partial x$ when the element is absent. It is the differential force which will move the element relative to the liquid and this differential force is simply expressed as

$$F_d = \left(\frac{\rho_b}{\rho_0} - 1\right) \rho_0 V \frac{\partial \tilde{u}_1^2}{\partial x} \quad . \qquad (9)$$

In this form we can now state that this force is appropriate for a body of arbitrary shape and we will use it in our discussion of the force on spheres. More exact calculations of the radiation pressure on spheres have been carried out by several authors,[4] but our approximation yields good results for the case we are considering.

For the case of a standing wave, we may consider $\tilde{u}_1$ as taking the form

$$\tilde{u}_1 = \tilde{u}_0[\cos(\omega t - kx) + \cos(\omega t + kx)] = \tilde{u}_+ + \tilde{u}_- \quad , \qquad (10)$$

and

$$\tilde{u}_1^2 = \tilde{u}_0^2[\cos^2(\omega t - kx) + 2\cos(\omega t - kx)\cos(\omega t + kx) + \cos^2(\omega t + kx)] \quad . \qquad (11)$$

The time independent part of this expression is

$$\tilde{u}_1^2 = \tilde{u}_0^2 (1 + \cos 2kx) \qquad (12)$$

and for the force as given by Eq. (9) we have

$$F_d = -\left(\frac{\rho_b}{\rho_0} - 1\right) \rho_0 V \, 2k \, \tilde{u}_0^2 \sin(2kx) \quad . \qquad (13)$$

This result is illustrated in Fig. 2. In Fig. 2(a) are shown the two oppositely directed waves at times t_1, t_2, and t_3. In each succeeding interval $\tilde{u}_+$ has progressively traveled to the right, while $\tilde{u}_-$ has traveled to the left. In Fig. 2(b) the interaction term $\tilde{u}_+\tilde{u}_-$ is shown at these same instants of time. This interaction term may be seen to have a bias superimposed on it, but it remains spatially stationary. Finally, in Fig. 2(c) we display the resultant force on the element. Superimposing the forces obtained at times t_1, t_2 and t_3 shows them to be identical. The resultant radiation pressure on the spheres is therefore stationary in both time and space.

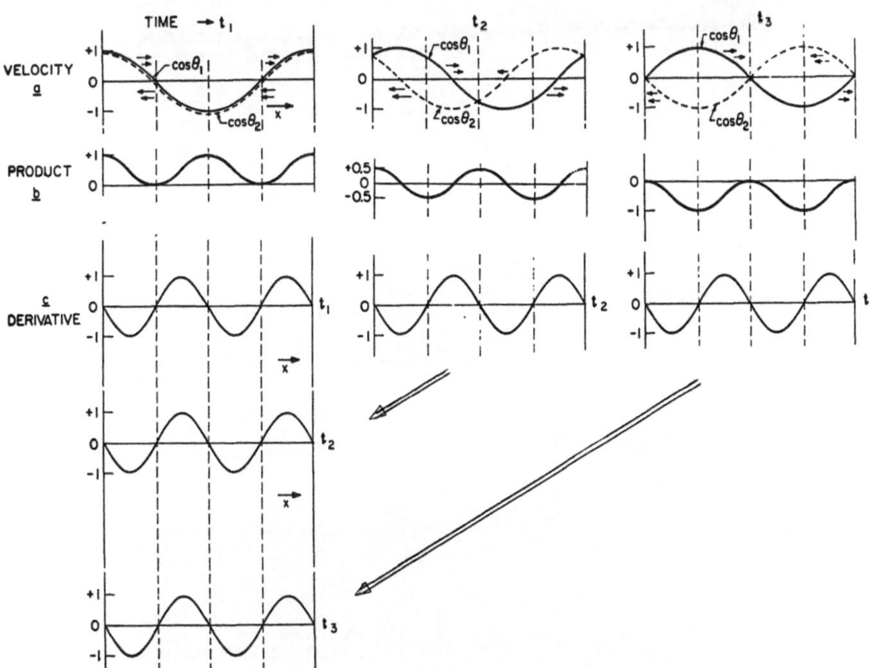

FIG. 2--Time independent stationary force field created by oppositely directed plane waves. (a) The two velocity waves and their relative phases at times t_1, t_2, and t_3; (b) The nonlinear response (product) at those corresponding times; (c) The resultant time independent stationary force on the volume element.

An elegant experiment illustrating this effect has been performed by Hanson, et al.[5] Their results are shown in Fig. 3. By means of a piezoelectric transducer seen at the bottom of the figure and a conical reflector at the top, an acoustic standing wave is created in the air column between. The stationary radiation pressure field thus produced is capable of supporting the liquid drops against gravity. The necessary force is 0.5 dynes.

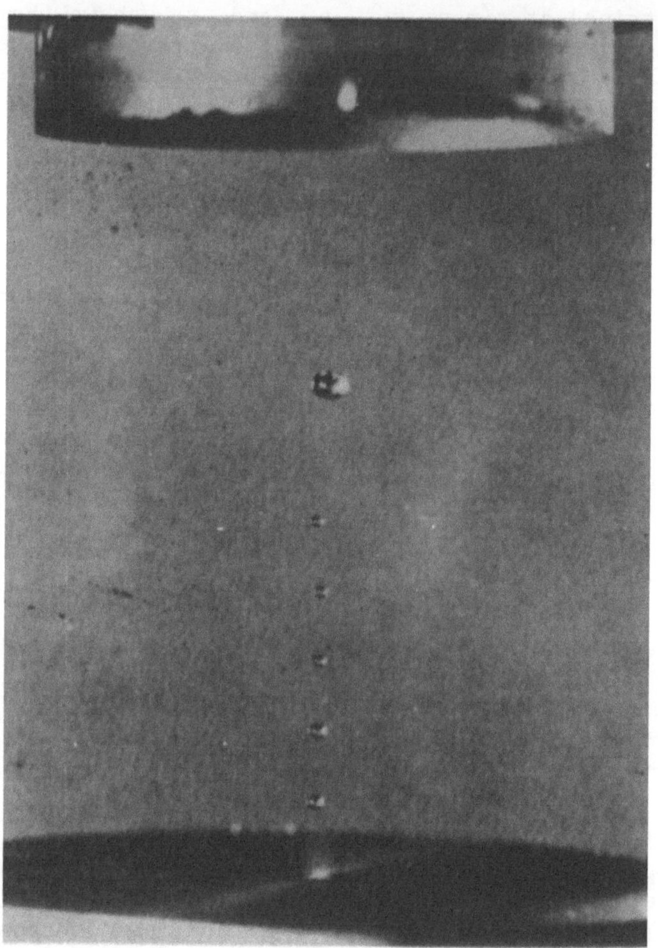

FIG. 3--Bench test of acoustical drop holder. Approximate diameter of top drop, 3 mm. (Courtesy of Hanson, et al.)

The situation of interest to us is shown in Fig. 4 where we have two plane acoustic waves intersecting at a solid-liquid interface (x-y plane). The waves enter the liquid at an angle θ from the normal. This yields a standing wave pattern in the x direction where now

$$\tilde{u}_1 = \tilde{u}_0[\cos(\omega t - k'x) + \cos(\omega t + k'x)] \quad , \qquad (14)$$

with $k' = k \sin\theta$. This creates a stationary force field along x of the form

$$F = -\left(\frac{\rho_b}{\rho_0} - 1\right) \rho_0 V(2k \sin\theta)\tilde{u}_0^2 \sin[2kx \sin\theta] \quad , \qquad (15)$$

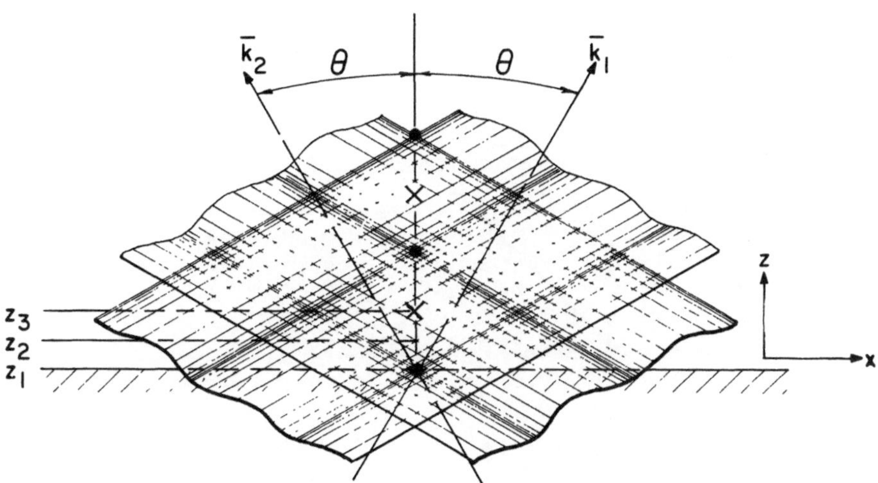

FIG. 4--Interference pattern generated by two crossed plane waves with wavevectors $\bar{k}_1$ and $\bar{k}_2$. The planes z_1 , z_2 , and z_3 correspond to the times t_1 , t_2 , and t_3 of Fig. 2.

which is independent of y and z . The periodicity of this force field is given by

$$S = \pi/k \sin \theta \quad . \tag{16}$$

If small spheres ($kr \sin \theta \ll 1$) are now placed in the liquid they will be forced to either velocity nodes when $\rho_b < \rho_0$ or to antinodes when $\rho_b > \rho_0$. The planes z_1 , z_2 , and z_3 of Fig. 4 correspond to the conditions of interference that would exist at the interface at the times t_1 , t_2 , and t_3 shown in Fig. 2.

The effect of this force field on small spheres is shown more clearly in Fig. 5. The planes of constant force which are normal to the surface and periodic with x will force the spheres to form linear fringes parallel to the y axis. The fringe spacing, S , as shown on the figure

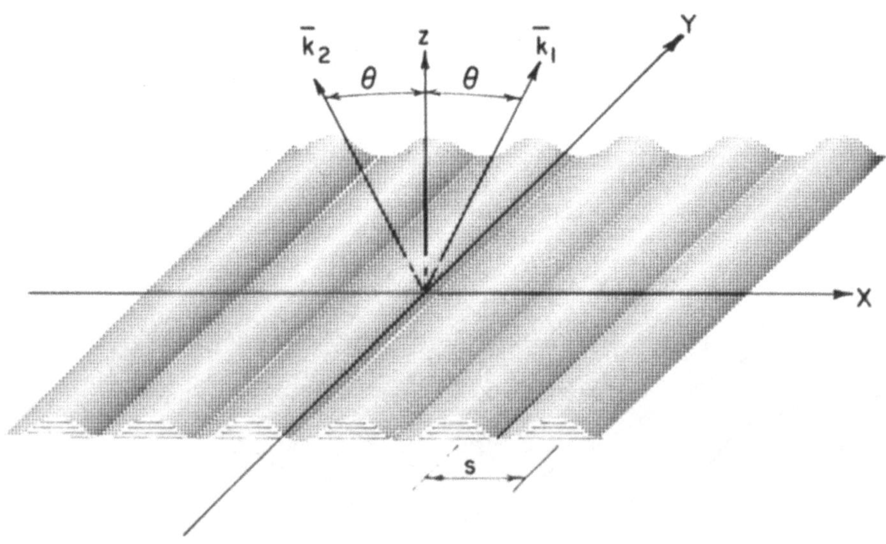

FIG. 5--Fringe pattern created in the x-y plane by the crossed plane waves.

is identical with the expected result of $\pi/k \sin \theta$. This pattern as recorded by the density of the spheres is the "hologram" of two plane waves.

The downward force exerted on the 1.1 μm polystyrene spheres used in our experiments as a result of gravity and buoyancy is 4.75×10^{-11} dynes. With an acoustic input into the liquid of 10^{-3} watts/cm^2 the sideways force exerted on the spheres by the radiation pressure is on the order of 5×10^{-12} dynes. The above analysis may be extended to include more complex wave patterns as exist when waves are scattered from an object.

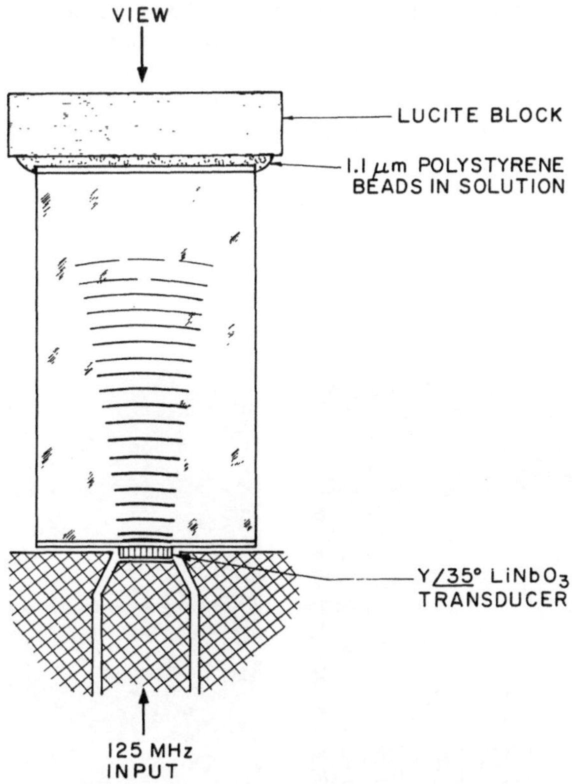

FIG. 6--Schematic diagram of apparatus for displaying the self-interference of a plane acoustic wave resonating in a quartz rod.

SELF-INTERFERENCE

One of the earliest experimental trials of our technique was in the visualization of the self-interference of a plane wave resonating in a quartz bar. The arrangement is shown in Fig. 6. A plane acoustic wave at 125 MHz (λ = 48 μm) is generated in the quartz rod by means of a Y $\angle 35^o$ $LiNbO_3$ transducer. Since there are no acoustic impedance transformers on the opposite end, the wave resonates in the rod and interferes with itself. This interference is revealed in the liquid film as described earlier. The lucite block has three functions: (1) it forms the thin liquid film by compression of a drop of liquid, (2) since its acoustic impedance closely matches that of water, it serves to eliminate unwanted reflections, and (3) it is a medium through which we can view the scene.

Two of the self-interference patterns thus generated are shown in Fig. 7. In Fig. 7(a) we see the pattern

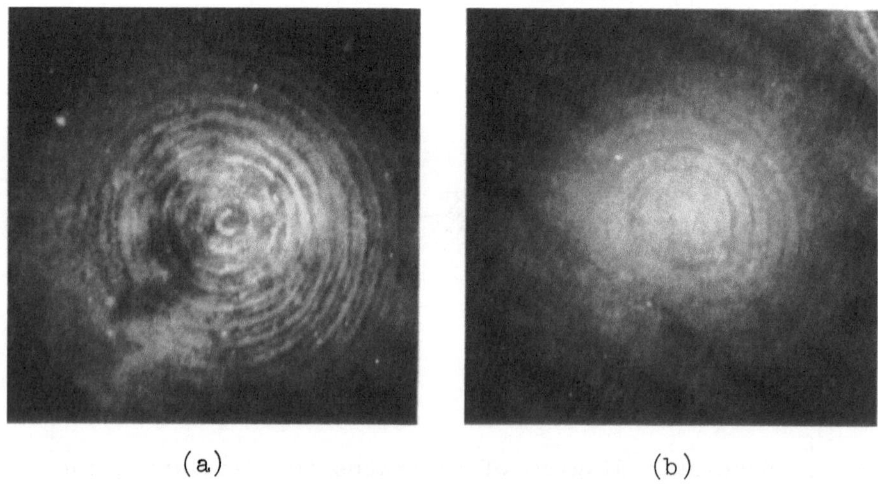

(a) (b)

FIG. 7--Direct visualization of the self-interference pattern. (a) Near anti-resonance; (b) near resonance.

generated when the rod is near an anti-resonant condition, the interfering waves being out of phase at the top surface. A large number of interference rings are thus formed. In Fig. 7(b) after a slight frequency shift (~ 200 kHz) the rod is near its resonant condition. The fringe spacing has correspondingly increased and the central region is now uniformly illuminated, characteristic of in-phase interference.

RESOLUTION AND SENSITIVITY

As a simple test of the resolution and sensitivity capabilities of our visualization technique, it was decided to try to visualize the fringe pattern created by the intersection of two plane waves as shown in Fig. 5. The experimental arrangement used to generate such a pattern is shown in Fig. 8. Two plane acoustic beams at 260 MHz

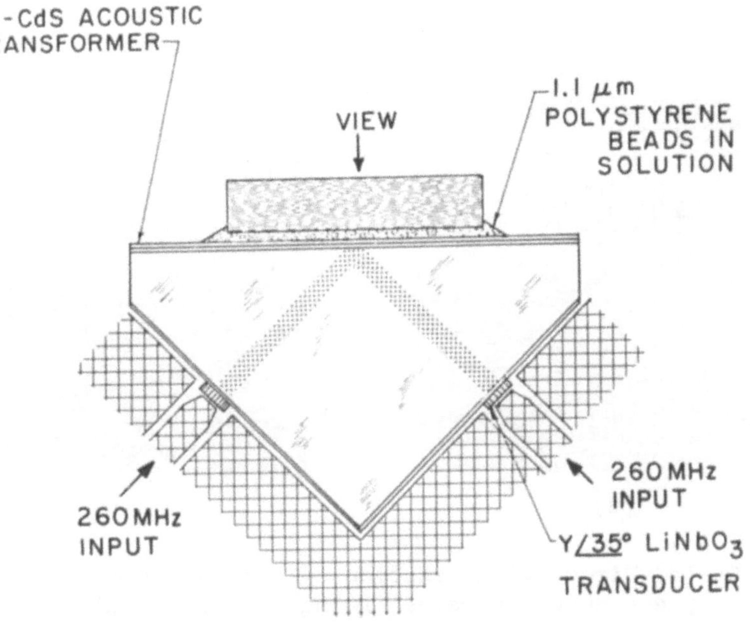

FIG. 8--Experimental arrangement for displaying the interference of two plane acoustic waves in a solid.

($\lambda = 23$ μm) are generated in the fused quartz prism and intersect at the surface at an angle of $\theta = 45°$. This determines a linear fringing in the interference pattern with a separation of 16.2 μm, which corresponds to the prediction of Eq. (16).

The interference pattern generated in the above procedure and recorded by the spheres in solution is shown in Fig. 9(a). The white lines (fringes) are the spheres collecting along the intensity maxima and the dark background is the acoustic transformer. The fringe spacing is 16.2 μm as expected. There are approximately 50 fringes across the figure. The rectangular dark spot in the center of the figure is actually a hole in the Au film of the acoustic transformer as shown in Fig. 9(b). Acoustic power is not

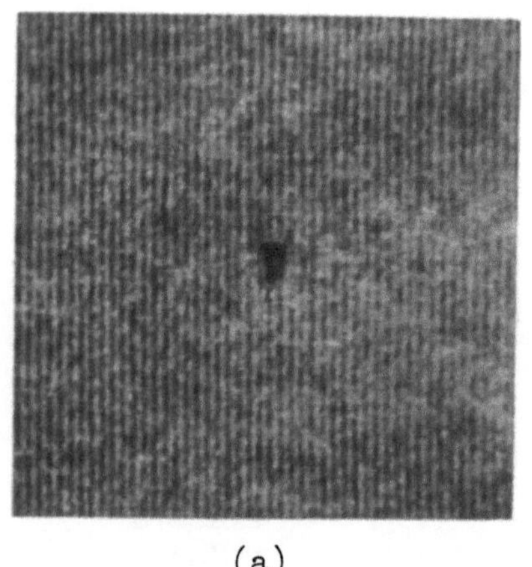

(a)

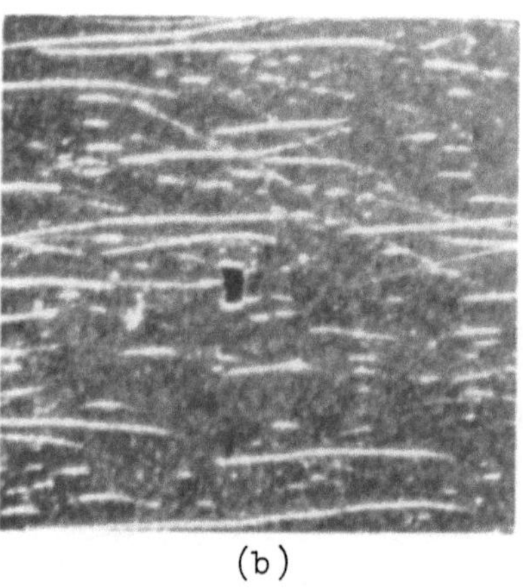

(b)

FIG. 9--(a) Direct visualization of the interference of two plane acoustic waves (fringe spacing 16.2 μm); (b) Optical photograph of hole in gold film of acoustic transformers (hole 40 μm × 60 μm). Scratches in CdS film are also evident.

passed through this region and hence no fringes are formed there. Although the linear resolution already demonstrated is good, we feel that the ultimate resolution capability of the technique has not yet been reached. It will be mainly determined by the angle of incidence of the two acoustic beams on the surface as well as their wavelength. The finest fringe spacing in the hologram will determine the ultimate resolution in the reconstructed image. However, as the fringe spacing becomes finer, resolution will be limited by the properties of the liquid suspension.

The sensitivity of the system, or the acoustic power per unit area required to reveal the interference pattern in the solid has also been experimentally determined. Present values range from 10^{-4} watts/cm^2 to 10^{-3} watts/cm^2 and place the sensitivity of this technique on a par with many of the imaging schemes now in use. Limits on sensitivity will be determined by the acoustic properties of the liquid suspension.

ACOUSTIC IMAGING

Application of the above technique to acoustic imaging is now, in principle, a simple step. The details of a system designed for this purpose are shown in Fig. 10. By means of a Y $\underline{/35^{\circ}}$ $LiNbO_3$ transducer, a plane acoustic beam is generated in the Z-axis quartz rod. It is passed into a water cell containing the object via acoustic transformers. By a similar set of transformers the wave which has been diffracted by the object is passed into a YAG crystal. A plane reference wave is also generated in the YAG crystal by a matched transducer. These two waves intersect at the surface to produce an interference pattern. As before, this interference pattern is displayed as a distribution of the spheres in the liquid film to form an acoustic hologram. A positive optical transparency of the fringe pattern may now be taken using a microscope camera, thereby generating an optical hologram of the object. It is possible, by passing a laser beam through this transparency, to reconstruct an optical image of the object.

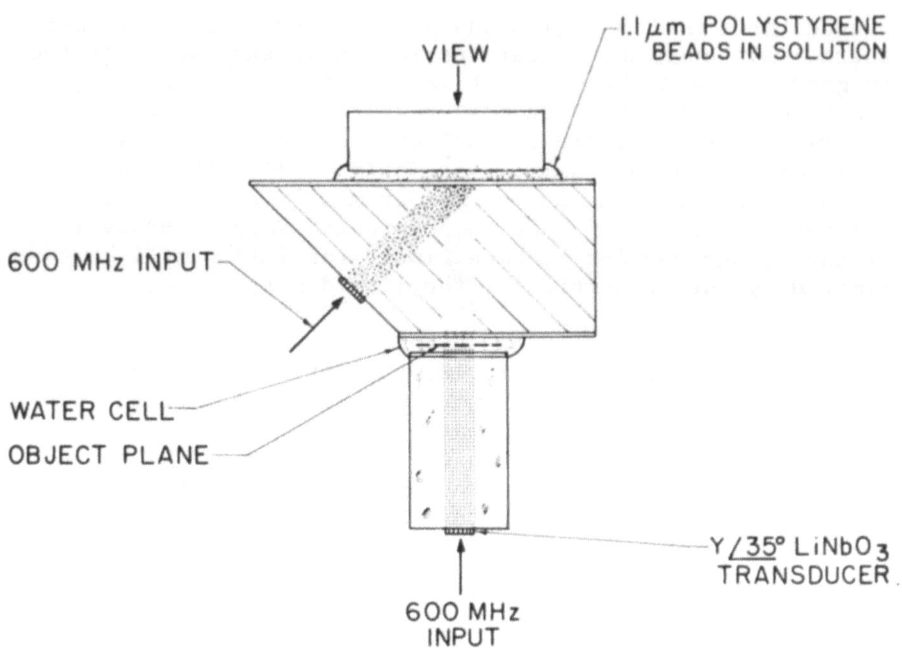

FIG. 10--Experimental apparatus for performing acoustic holography in solids.

One of the test objects used in this imaging system was a circular mesh, a portion of which is shown in the optical photograph of Fig. 11(a). The holes are 300 μm in diameter on 420μm centers. Note the ragged edges. A photograph of the acoustical hologram is shown in Fig. 11(c). It is essentially a shadowgraph of the mesh with superimposed linear interference fringes. Considerably more than four holes may be seen, although the fringe contrast in the outer holes is not good. Finally, in Fig. 11(b) is shown the optical reconstruction. Four holes can be distinguished. The total magnification is approximately 50 ×. The fringes apparent on the reconstructed image are due to noise in the optical system. The upper right hand hole in the image may be seen to display some of the raggedness of the edge shown in the optical photograph of the object.

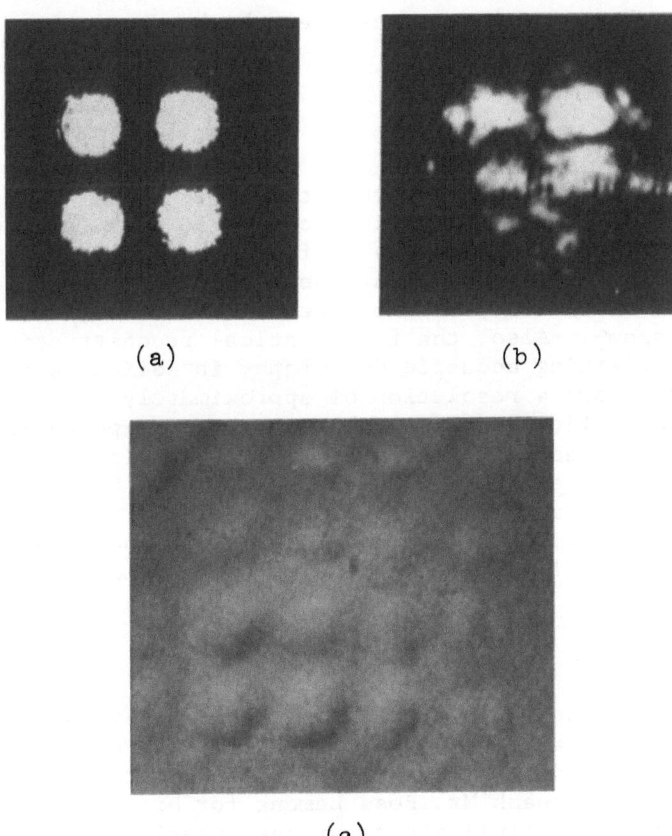

FIG. 11--Optical reconstruction from acoustic hologram obtained using apparatus shown in Fig. 10. (a) Portion of mesh used for imaging (300 μm holes on 420 μm centers); (b) Optically reconstructed image (actual 50 × magnification); (c) Photograph of acoustic hologram.

The resolution obtained here is not yet at the theoretical limit. The primary reason for this was our inability to optically reproduce the acoustic hologram with sufficient contrast. This will be corrected in future work by using high contrast film and low angle illumination. With this step our experimentally obtainable resolution should more closely match our theoretical expectations. Once we have accomplished this we can further increase resolution by going to progressively higher operating frequencies.

CONCLUSIONS AND SUMMARY

In this paper we have presented a technique which permits the visualization of acoustic interference in a solid via a thin liquid film on its surface. We have demonstrated the capability of obtaining a linear hologram resolution of 16 μm at a sensitivity on the order of 10^{-4} watts/cm^2 to 10^{-3} watts/cm^2. Also, the first optical reconstruction obtained utilizing acoustic holography in solids is presented and shows a resolution of approximately 300 μm with a system magnification approaching 50 ×. The potential for future development is clearly shown. By utilizing better optical techniques and going to progressively higher acoustic frequencies we can bring our experimental and theoretical resolution capabilities into better accord, and push those capabilities well beyond the results already obtained.

ACKNOWLEDGEMENTS

This work was supported by the John A. Hartford Foundation, Inc. The authors wish to thank Messrs. A. R. Hanson, E. G. Domich, and H. S. Adams for permission to use Fig. 3. We also wish to thank Mr. Ross Lemons for his technical assistance in the final stages of the experimental work presented here.

REFERENCES

1. Mueller, Rolf K., "Acoustic Holography," Proc. IEEE, vol. 59, p.1319 (1971).

2. El-Sum, H.M.A., "Progress in Acoustical Holography," in Acoustical Holography, vol. 2, A. F. Metherell and Lewis Larmore, Eds., New York: Plenum, p.7 (1970).

3. Eller, A., "Force on a Bubble in a Standing Acoustic Wave," J. Acoust. Soc. Amer., vol. 43, p.170 (1968).

4. King. K. V., "On the Acoustic Radiation Pressure on Spheres," Proc. Roy. Soc. (London), A147, p.212 (1934); Yosioka, K. and Kawasima Y., "Acoustic Radiation Pressure on a Compressible Sphere," Acustica, vol. 5, p. 167 (1955); Nyborg, W. L., "Radiation Pressure on a Small Rigid Sphere," J. Acoust. Soc. Amer., vol. 42, p.947 (1967).

5. Hanson, A. R., Domich, E. G., and Adams, H. S., "Acoustical Liquid Drop Holder," Rev. Sci. Inst., vol. 35, p.1031 (1964).

DETECTION OF BURIED GEOLOGICAL ORE-BODIES BY RECONSTRUCTED WAVEFRONTS

A.K. Kalra and P.W. Rodgers

Division of Engineering Geoscience
Department of Materials Science and Engineering
University of California
Berkeley, California 94720

ABSTRACT

A new seismic method based on concepts of reconstructed wavefronts is described. It employs a mono-frequency source of seismic energy and records the interference pattern between the scattered waves from the geological feature of interest buried at a certain depth and the direct waves travelling along the surface. The interfering direct waves that do not contribute to the characteristic signatures of the geological ore-body are removed by spatial filtering before reconstruction.

The interpretation procedures or the reconstruction then consists in determining the complex amplitude distribution pattern at planes parallel to but situated at different depths from the known pattern at the surface. This is achieved on the computer by using the reconstructing integrals which are nothing but the inverse Fresnel diffraction integrals. Using the magnitude of the projected complex amplitude, an unbiased estimator of the true depth to the scatterer is formed. The distribution pattern at the true depth is also diagnostic of the physical relief of the scatterer. The usefulness of this method is illustrated by computing the theoretical responses for a point scatterer and other examples and compared with other existing methods. The use of this method in petroleum exploration is also discussed.

INTRODUCTION

There are many applications of acoustical holography that have been suggested ranging from biology and oceanography to geological investigations. One considered here is of particular interest to an exploration geophysicist. Farr (1970) has discussed such an application in oil exploration but his experiments were confined to fluid media (water tank) and the reconstruction of the immersed object obtained optically. Optical reconstructions from acoustical hologram are known to produce highly distorted images. Moreover he has considered a direct application of holographical technique similar to Mueller and Sheridon (1966). The principles of acoustical holography are employed here to describe a new seismic method for detection of buried geological ore-bodies. First the 'seismic hologram' is formed at the recording surface and the complex amplitude pattern so obtained is then reconstructed on the computer at different depth estimates using the proper diffraction integrals. This method is, in a way, analagous to other potential-field methods and thus its interpretation procedures are compared with those of gravity and magnetic.

NEW SEISMIC METHOD

The new seismic method called Wavefront Reconstruction employs a mono-frequency source of seismic energy that illuminates the geological feature of interest buried at a depth, z^*, as shown in Figure 1. The interference pattern between the waves scattered from the target and those travelling along the surface is recorded at the surface. This constitutes the seismic hologram. The interfering direct waves that do not contribute to the characteristic signatures of the geological ore-body are removed by spatial filtering before reconstruction.

The reconstruction then consists in determining the complex amplitude distribution pattern at planes parallel to but situated at different depths from the known pattern at the surface. This is achieved on the computer by using the reconstructing integrals which are nothing but the inverse Fresnel diffraction integrals. Using the magnitude of the projected complex amplitude, an unbiased estimator

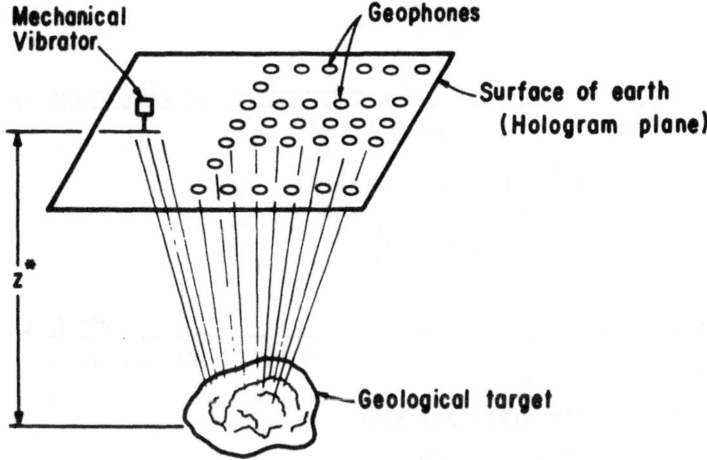

Figure 1. General Exploration Situation

of the true depth to the scatterer is formed. The distribution pattern at the true depth is also diagnostic of the physical relief of the scatterer. The main features of wavefront reconstruction method are illustrated for a two dimensional case of a point-scatterer as the target. Response for a scatterer of arbitrary shape can then be obtained by convolving it with that due to the point scatterer. This technique is extended to include the examples of horizontal and inclined scatterers of limited extend.

POINT SCATTERER

The method of reconstructed wavefronts for a point scatterer in a two dimensional seismic model is illustrated in Figure 2. The coherent source of seismic energy, S, is considered to be a mono-frequency horizontally polarized shear wave generator. This would eliminate wave conversion problems on scattering. S also defines the origin for the rectangular coordinate system in the recording plane X_0. The scatterer is contained in another plane X_1 which is parallel to plane X_0 and displaced by a distance z^* along the positive z - axis as shown.

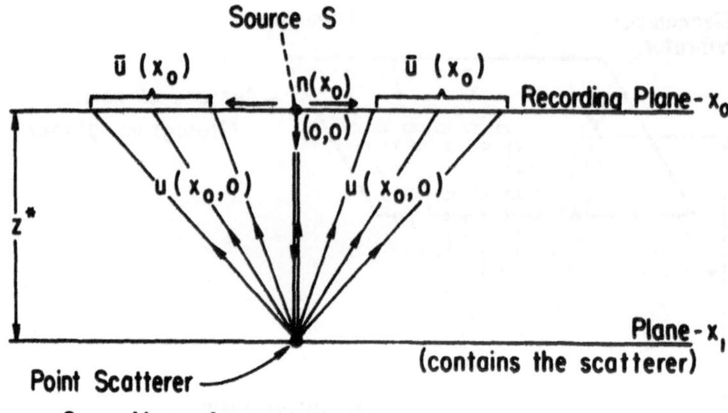

Figure 2. The Point Scatterer

Construction of the Hologram

The complex amplitude of the waves scattered from the object are denoted by $u(x_0, 0)$ in the recording plane X_0 which also defines the hologram plane. The direct or surface wave which does not contribute to the interpretation of the scatterer is represented by $n(x_0)$. The seismic hologram or the interference pattern, $\bar{u}(x_0)$, between the scattered signal $u(x_0,0)$ and the direct wave thus recorded at the surface of the model is mathematically written as

$$\bar{u}(x_0) = u(x_0, 0) + n(x_0) \tag{1}$$

$\bar{u}(x_0)$, $u(x_0, 0)$ and $n(x_0)$ are all complex signals so that any one of these has an amplitude $a(x_0)$ and phase $\phi(x_0)$ and is generally represented as

$$\bar{u}(x_0) = a(x_0)e^{j\phi(x_0)} \tag{2}$$

where the frequency term $e^{j2\pi f_s t}$ has been suppressed. The scattered signal $u(x_0, 0)$ is mathematically determined by using Fresnel diffraction integrals (Goodman, 1968). Accordingly, if we know the complex distribution function say $u(x_1)$ in plane X_1 containing the scatterer then the scattered field $u(x_0, z)$ in another plane X_0 at a distance z but parallel to the first plane X_1 is given, in one dimension, by the integral

$$u(x_0, z) = \frac{e^{jkz}}{\sqrt{j\lambda z}} \int_{-\infty}^{\infty} u(x_1)\, e^{\frac{jk}{2z}(x_0 - x_1)^2}\, dx_1 \qquad (3)$$

where $k = \frac{2\pi}{\lambda}$ and λ = wave length of the scattering wave.

For the case of the point scatterer and the coordinate system used as described above, we have

$$u(x_1) = \delta(x_1) \text{ and } z = -z^* \qquad (4)$$

Substituting these values in (3) and neglecting anelastic attenuation and multiple scattering of the elastic waves the complex amplitude $u(x_0,0)$ in observation plane X_0, after simplification, is given by

$$u(x_0,0) = \frac{e^{-jkz^*}}{\sqrt{-j\lambda z^*}}\, e^{-\frac{jk}{2z^*} x_0^2} \qquad (5)$$

The use of Fresnel integral (3) in arriving at the scattered field $u(x_0,0)$ requires that the depth of burial, z^*, for the scattering object should be far greater than its lateral extent and the observation region on the surface. This restriction which is referred to as the Fresnel approximation can be expressed mathematically for the two dimensional case as

$$z^{*3} >> \frac{\pi}{4\lambda} (x_0 - x_1)^4_{max}$$

The direct wave $n(x_0)$ for a two dimensional seismic model however is expressed as

$$n(x_0) = b \cos \left(\frac{2\pi}{\lambda} x_0 + \phi\right) \tag{6}$$

where x_0 is the distance of the recording geophone from the source S along the x-axis and λ is the wavelength of the seismic wave in the medium.

The direct wave $n(x_0)$ can be visualized as a surface wave which is only confined nearer to the recording surface and thus does not contain any information about the characteristic nature of the scatterer at depth. Consequently it is desirable to eliminate the direct wave $n(x_0)$ and recover the scattered signal $u(x_0,0)$ from the observed interference pattern $\bar{u}(x_0)$ before reconstruction. This is achieved by wavelength or spatial filtering techniques which is discussed below.

Wavelength Filtering

The wavelength filtering procedures are summarized in Figure 3. The scattered waves travelling through the medium would have a wavelength λ in the medium which is the same as that for the seismic waves originating from the source and illuminating the scattering object. However when these waves reach the surface of the seismic model they strike the free-boundary at certain angle of incidence θ as shown and appear to generate a different wave in the recording plane X_0 with a new wavelength dependent on the incidence angle θ. This apparent wavelength λ_a in the hologram plane X_0 is given by the expression

$$\lambda_a = \frac{\lambda}{\sin \theta} \tag{7}$$

Accordingly the recorded field $\bar{u}(x_0)$ at the surface of the seismic model can be regarded as the superposition of two sinusoidal waves with wavelengths λ and λ_a. The apparent wavelength λ_a associated with the signal $u(x_0,0)$ in the plane X_0 is larger than that of the surface wave $n(x_0)$ which has a wavelength λ, i.e.

$$\lambda_a >> \lambda \tag{8}$$

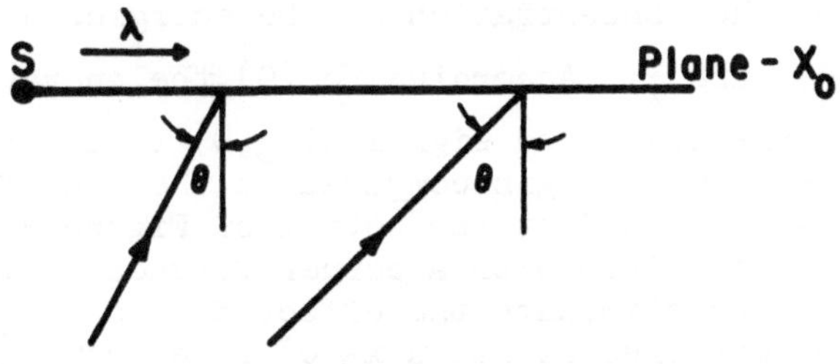

$$\text{Apparent Wave Length } \lambda_a = \frac{\lambda}{\text{SIN } \theta}$$

$$\lambda_a >> \lambda \text{ (for small } \theta\text{)}$$

$$\text{or } \frac{1}{\lambda} >> \frac{1}{\lambda_a} \text{ (spatial frequencies)}$$

Figure 3. Wavelength Filtering

for any angle θ except at $\theta = \frac{\pi}{2}$ when $\lambda_a = \lambda$. Moreover for the particular case where the incident scattered waves subtend a small angle with the normal to the free boundary, the inequality becomes greater.

In wavefront reconstruction method where vertical depth z* to the scatterer is larger than the horizontal extent of the observation plane X_0, θ is essentially small and the relation (8) holds. Expressing the relation (8) in terms of spatial frequencies we have

$$\frac{1}{\lambda} >> \frac{1}{\lambda_a} \quad \text{(for small } \theta\text{)} \tag{9}$$

Consequently the power spectrum of the observed signal in

(1) would show the concentration of the energies at two frequencies $\frac{1}{\lambda_a}$ and $\frac{1}{\lambda}$. According to (9) the energy peak mainly due to the scattered signal $u(x_0,0)$ should be located nearer to the origin compared to the peak for the direct wave as indicated at the bottom of Figure 3. A suitable low-pass filter with a corner frequency ν_c then can be employed to eliminate the effect due to $n(x_0)$ and obtain a true scattered signal $u(x_0,0)$ free from any interference from the direct wave. The corner frequency ν_c for such a filter for the observed data of finite length, say 2L, is difficult to determine analytically because the transforms of $u(x_0,0)$ rect. $(x_0/2L)$ and $n(x_0)$ rect. $(x_0/2L)$ are transcendental in nature. However ν_c in such a case can be determined approximately and is given by

$$\nu_c = \frac{\lambda_a L - \lambda L - \lambda_a \lambda - \nu_h \lambda_a \lambda L}{2\lambda_a \lambda L} \tag{10}$$

where ν_h is the highest frequency content for the noise spectrum.

<u>Sampling Interval</u>. In the recording of interference data $\bar{u}(x_0)$ we can essentially consider the process to be band limited to the highest frequency $\frac{1}{\lambda}$ and consequently we employ the sampling theorem for band limited functions to determine the sampling interval X_s. The sampling theorem states that the sampling frequency ν_s should be equal to or greater than twice the highest frequency ν_b present in the process, i.e.

$$\nu_s \geq 2\nu_b \tag{11}$$

In our case (11) becomes

$$\nu_s \geq 2\,\frac{1}{\lambda} \tag{12}$$

or the sampling interval

$$X_s \leq \frac{\lambda}{2} \tag{13}$$

where the sampling interval is defined, in relation to sampling frequency as

$$X_s = \frac{1}{\nu_s} \tag{14}$$

In order to demonstrate the usefulness of the wave-front reconstruction method in mining exploration we consider an ideal situation where the surface wave $n(x_0)$ is completely eliminated and thus the recorded seismic hologram consists purely of the scattered signal $u(x_0,0)$ given by (5). Thus the interpretation procedures consist in the reconstruction of this hologram.

Reconstruction

The scattered field recorded on the surface of the seismic model in the form of a hologram is sufficient to determine the depth and other characteristics of the buried scatterer. This is obtained as the reconstruction process by reconstructing the characteristic wavefronts at the plane containing the scattering object by using the reconstruction integral which are nothing but the inverse of Fresnel diffraction integrals as given in (3). However, for interpretation purposes, the true depth z^* to the plane containing the scatterer is not known and so we would substitute $\hat{z}$ for z^* which is an estimate of the true depth z^*. The outline of the reconstruction geometry is sketched in Figure 4. The reconstruction integral then becomes

$$u(x_1,\hat{z}) = \frac{e^{jk\hat{z}}}{\sqrt{j\lambda\hat{z}}} \int_{-\infty}^{\infty} u(x_0,0)\; e^{\frac{jk}{2\hat{z}}(x_1 - x_0)^2}\, dx_0 \tag{15}$$

Substituting the value of $u(x_0,0)$ from (5) and simplifying we obtain

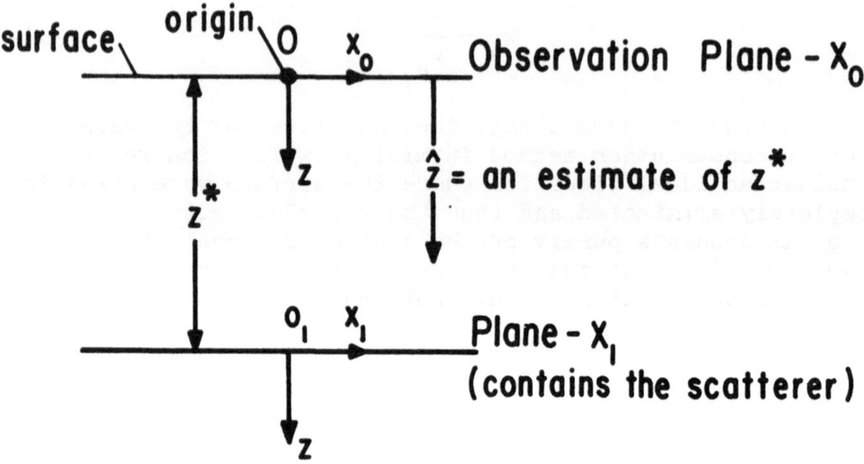

RECONSTRUCTION INTEGRAL

$$u(x_1, \hat{z}) = \frac{e^{jk\hat{z}}}{\sqrt{j\lambda\hat{z}}} \int_{-\infty}^{\infty} u(x_0, o)\, e^{\frac{jk}{2\hat{z}}(x_1 - x_0)^2}\, dx_0$$

WHERE $k = \frac{2\pi}{\lambda}$

λ = Reconstructing Wave-length

Figure 4. Reconstruction Geometry

$$u(x_1,\hat{z}) = \frac{e^{-jk(z^* - \hat{z})} e^{\frac{jk}{2\hat{z}} x_1^2}}{\lambda\sqrt{z^*\hat{z}}} \int_{-\infty}^{\infty} e^{j\alpha x_0^2} e^{-j\omega x_0} dx_0 \tag{16}$$

where
$$\alpha = \frac{k(z^* - \hat{z})}{2z^*\hat{z}}$$

and
$$\omega = \frac{kx_1}{\hat{z}} \tag{17}$$

The integral in expression (16) can be evaluated as Fourier transform of $e^{j\alpha x_0^2}$ and thus using the Fourier transform tables (Papoulis, 1968; p. 67) we finally obtain

$$u(x_1,\hat{z}) = \frac{e^{-jk(z^* - \hat{z})} e^{\frac{jk}{2\hat{z}} x_1^2}}{\sqrt{\lambda(z^* - \hat{z})}} e^{j\frac{\pi}{4}} e^{\frac{jkz^*}{2\hat{z}(z^* - \hat{z})} x_1^2} \tag{18}$$

Since $u(x_1,\hat{z})$ is a complex quantity, (18) is not very helpful for interpretation purposes. However, looking at the absolute magnitude of this reconstructed field at an arbitrary plane a distance $\hat{z}$ from the recording surface we determine that

$$|u(x_1,\hat{z})| = \frac{1}{\sqrt{\lambda(z^* - \hat{z})}} \tag{19}$$

We notice in expression (19) that the absolute magnitude of the reconstructed anomaly $|u(x_1,\hat{z})|$ goes on increasing with the increase in the value of the depth estimate $\hat{z}$ and at the correct value of $\hat{z}$, i.e., $\hat{z} = z^*$, $|u(x_1,\hat{z})|$ becomes infinite. For the values of $\hat{z}$ greater than the true depth to the scatterer z^* we observe that

$$|u(x_1,\hat{z})| = \frac{1}{\sqrt{-\lambda(\hat{z} - z^*)}} = \frac{1}{|j\sqrt{\lambda(\hat{z} - z^*)}|} = \frac{1}{\sqrt{\lambda(\hat{z} - z^*)}} \tag{20}$$

And so when the value of the estimator $\hat{z}$ exceeds the correct value, i.e. $\hat{z} > z^*$, the magnitude for the reconstructed anomaly $|u(x_1,\hat{z})|$ decays continuously and so the estimator $\hat{z}$ is unbiased in that sense.

Thus in the wavefront reconstruction process we determine that the magnitude of the downward projected complex amplitude goes on increasing with the increase in the value of the depth estimator $\hat{z}$, and at one particular value when the estimator equals the true value of the depth of burial for the scatterer this magnitude becomes infinite. For a value of the estimator larger than the true value the magnitude decreases continuously. We note that the correct value of this estimator is the one where the magnitude is infinite. Thus by comparing the reconstructed anomaly for different depth estimates, the wavefront reconstruction method provides us with an unbiased estimate of the true depth to the scattering body.

Furthermore, if we substitute the true value of the estimate $\hat{z} = z^*$ as determined from the above criteria into (16) we also obtain the shape of the original scatterer. Accordingly we have

$$u(x_1,z^*) = \frac{e^{j\frac{\pi x_1^2}{z^*\lambda}}}{\lambda z^*} \int_{-\infty}^{\infty} 1.\ e^{-j2\pi\frac{x_1}{\lambda z^*} x_0} dx_0 \tag{21}$$

where the values of α, ω and k have been substituted from (17) and (3). Again using the Fourier transform tables, (21) can simply be written as

$$u(x_1, z^*) = 2\pi \frac{e^{j\frac{\pi x_1^2}{z^*\lambda}}}{\lambda z^*} \delta\left(\frac{2\pi x_1}{\lambda z^*}\right) \tag{22}$$

And using the Fourier transform theorem for the change of scale we obtain the shape of the scatterer as

$$u(x_1, z^*) = \delta(x_1) \tag{23}$$

Expression (23) is an exact reproduction for the point scatterer we used as the original object in the plane X_1 at the correctly predicted depth z*. It has therefore been illustrated at least for the case of the point scatterer that the wavefront reconstruction method, as applied to the problem of delineating a scattering object, is very effective not only in determination of the true depth to the buried body but also in describing its exact shape.

The use of reconstruction integral (15) however, implies that the hologram (5) be recorded for an infinite observation aperture. However, in a more reasonable situation we are bound to restrict the observation aperture to some finite length say from $x_0 = -L$ to +L. The reconstruction (16) now becomes

$$u(x_1, \hat{z}) = \frac{e^{-jk(z^* - \hat{z})} e^{\frac{jk}{2\hat{z}} x_1^2}}{\lambda\sqrt{z^*\hat{z}}} \int_{-L}^{+L} e^{\frac{jk(z^* - \hat{z})}{2z^*\hat{z}} x_0^2} e^{-\frac{jkx_1}{\hat{z}} x_0} dx_0 \tag{24}$$

Note that we can no longer use Fourier transforms to evaluate the above expression as was done earlier for integral (16). In fact it is a very difficult integral to evaluate analytically. However, by proper substitutions, we can reduce it to Fresnel integrals (Goodman, 1968) which are tabulated and thus solve it numerically. Accordingly we obtain the final expression for the reconstruction integral as

$$u(x_1,\hat{z}) = \frac{e^{-jk(z^* - \hat{z})} e^{\frac{jk}{2\hat{z}} x_1^2} e^{-\frac{j\omega^2}{4\alpha}}}{\lambda\sqrt{z^*\hat{z}}}$$

$$\sqrt{\frac{\pi}{2\alpha}} \left\{ C\left[\sqrt{\frac{2}{\pi}} \left(\sqrt{\alpha}L - \frac{\omega}{2\sqrt{\alpha}}\right)\right] + C\left[\sqrt{\frac{2}{\pi}}\left(\sqrt{\alpha}L + \frac{\omega}{2\sqrt{\alpha}}\right)\right] \right.$$

$$\left. + j\left[S\left[\sqrt{\frac{2}{\pi}}\left(\sqrt{\alpha}L - \frac{\omega}{2\sqrt{\alpha}}\right)\right] + S\left[\sqrt{\frac{2}{\pi}}\left(\sqrt{\alpha}L + \frac{\omega}{2\sqrt{\alpha}}\right)\right]\right]\right\} \tag{25}$$

where α and ω are given by expression (17), and $C(t)$, $S(t)$ are the set of Fresnel integrals.

Substituting the value of α in the multiplying factor outside the brackets and taking the magnitude of both sides of (25) we get

$$|u(x_1,\hat{z})| = \frac{1}{\sqrt{2\lambda(z^* - \hat{z})}} \left| \left\{ C\left[\sqrt{\frac{2}{\pi}}\left(\sqrt{\alpha}L - \frac{\omega}{2\sqrt{\alpha}}\right)\right] \right.\right.$$

$$+ C\left[\sqrt{\frac{2}{\pi}}\left(\sqrt{\alpha}L + \frac{\omega}{2\sqrt{\alpha}}\right)\right]$$

$$\left.\left. + j\left[S\left[\sqrt{\frac{2}{\pi}}\left(\sqrt{\alpha}L - \frac{\omega}{2\sqrt{\alpha}}\right)\right] + S\left[\sqrt{\frac{2}{\pi}}\left(\sqrt{\alpha}L + \frac{\omega}{2\sqrt{\alpha}}\right)\right]\right]\right\} \right| \tag{26}$$

It is difficult to predict the behavior of $|u(x_1,\hat{z})|$ w.r.t. $\hat{z}$ because of the Fresnel integrals appearing in expression (26). A convenient aid in interpreting such an expression is the graphical construction known as Cornu's spiral (Goodman, 1968) which is a simultaneous plot of $C(t)$ and $S(t)$, vs. the parameter t.

The values of Fresnel integrals oscillate rather widly for the smaller arguments but stabilize for higher argument values reaching a constant value of one-half for large enough 't'. Thus the anomaly characteristics due to (26) are very much dependent on the magnitude of the quantity within the square brackets which in turn is determined by α, ω and L or using (17) more truly by z^*, x_1, $\hat{z}$, λ and L. Accordingly for large L and small value of ω or x_1, $|u(x_1,\hat{z})|$ increases with an increase in $\hat{z}$ as before. However when $\hat{z} = z^*$, the above expression becomes indeterminate and the original integral (24) reduces to

$$u(x_1,z^*) = \frac{e^{\frac{jk}{2z^*}x_1^2}}{\lambda z^*} \int_{-L}^{+L} 1.\ e^{-j(\frac{kx_1}{z^*})x_0} dx_0 \qquad (27)$$

or
$$u(x_1,z^*) = \frac{e^{\frac{jk}{2z^*}x_1^2}}{\lambda z^*} \quad 2L\ \frac{2\pi z^*}{k}\ \mathrm{sinc}(2L\cdot x_1) \qquad (28)$$

where
$$\mathrm{sinc}\ x = \frac{\sin\ \pi x}{\pi x} \qquad (29)$$

and so

$$|u(x_1,z^*)| = 2L\ \mathrm{sinc}\ (2L\cdot x_1) \qquad (30)$$

Thus the peak value of the reconstructed anomaly at $\hat{z} = z^*$ has its maximum value of 2L at the origin, i.e. $x_1 = 0$. In comparison with the case having an infinite aperture considered earlier, we find that the amplitude $|u(x_1,\hat{z})|$ does not become infinite for the correct depth estimate $\hat{z} = z^*$ but does attain its maximum value of 2L at $\hat{z} = z^*$ for a particular value of x_1 at the origin. In addition, the shape of the reconstructed anomaly is a sinc function as expected for the finite aperture case. The first zero crossing for this sinc function along the x_1-axis is at $\pm$ 1/2L as is obvious from (30). Clearly then the larger the value of the aperture L the higher the peak for the sinc-shaped reconstructed anomaly and smaller its width. Consequently when L becomes infinite the reconstructed anomaly is a delta function as was predicted for such a case in (23). Again for $\hat{z} > z^*$, it is apparent from (26) that $|u(x_1,\hat{z})|$ decreases monotonically.

Thus for finite but large values of the observation aperture L, the interpretation for determining the true depth to the scattering body remains virtually the same as that for an infinite aperture function.

Computer Reconstructions for the Point Scatterer. The above results were verified numerically for a two dimensional seismic model. The computer model consists of an aluminum plate and the horizontally polarized shear wave generator has a frequency of 300 kHz and thus produces elastic waves having a wavelength, λ, of 1.06 cms. The point scatterer is assumed to be situated at a depth z^* of 127.0 cms directly below the source. The scattered signal $u(x_{0n},0)$ is generated on the computer using the discrete form of (5) as expressed below

$$u(x_{0n},0) = \frac{e^{-jkz^*}}{\sqrt{-j\lambda z^*}} e^{-\frac{jk}{2z^*} x_{0n}^2} \quad (31)$$

$$\text{for } n = \pm 1, \pm 2 \ . \ . \ . \ N$$

where x_{0n} is the nth x-coordinate in the observation plane X_0 and 2N is the total number of data points. In relation to (5) we have

$$x_0 = N\Delta x_o \quad (32)$$

Δx_0 is the increment in the value of x_0. The corresponding surface wave, $n(x_{0n})$, from (6) is added to (31) to obtain the hologram data $\bar{u}(x_{0n})$. Accordingly the hologram was obtained from x_0 = -25.55 to 25.55 cms at an interval of 0.1 cm giving a total length, 2L, for the observation aperture as 51.1 cms.

The effect of the surface wave is minimized by wavelength filtering techniques as discussed earlier, and the signal so recovered is denoted by $u_f\,(x_{0n},0)$ to distinguish it from the pure signal $u(x_{0n},0)$.

The reconstruction for the estimated depth $\hat{z}$ = 40, 80, 120, 127, 134 and 250 cms were obtained on computer using the corresponding discrete form of the reconstruction integral (15) for finite observation aperture, i.e.

$$u(x_{1m},\hat{z}) = \frac{e^{jk\hat{z}}}{\sqrt{j\lambda\hat{z}}} \sum_{-N}^{N} u_f(x_{0n},0)e^{\frac{jk}{2\hat{z}}(x_{1m} - x_{0n})^2} \Delta x_{0n}$$

$$\text{for } m = \pm 1, \pm 2 \; . \; . \; . \; M \qquad (33)$$

where x_{1m} is the mth x-coordinate in the reconstruction plane X_1 and 2M is the total number of reconstruction points defining the total linear dimensions of the plane X_1. Again

$$x_1 = M \cdot \Delta x_1$$

where Δx_1 is the increment in the value of x_1. In the example considered we have

$$\Delta x_1 = \Delta x_0 = 0.1 \text{ cm}$$

and N = M = 256 and therefore 2L = 51.1 cm.

Figure 5 shows the plot of the magnitude of this downward projected complex amplitude against the lateral distance x_1. The plot for $\hat{z}$ = 40 shows a number of small peaks but lacks the presence of any prominent anomaly because the estimated depth is far below true depth. At $\hat{z}$ = 80 it is interesting to note that most of the small peaks on the sides have disappeared and there is a concentration of an overall broad peak near the center, though it is not well developed yet. At the next depth estimate of 120.0 cms the central peak is very prominent and does indicate the presence of the point scatterer. At the true depth, i.e. at $\hat{z}$ = 127.0 the anomalous peak is very well developed and has attained the maximum peak value of 0.43 as against 0.421 at $\hat{z}$ = 120 and obviously much lesser values for $\hat{z}$ = 40 and 80. The reconstruction at $\hat{z}$ = 134 which is a larger estimate for the true depth, the prominent peak though present shows a sharp decrease in its peak value of 0.412 and at

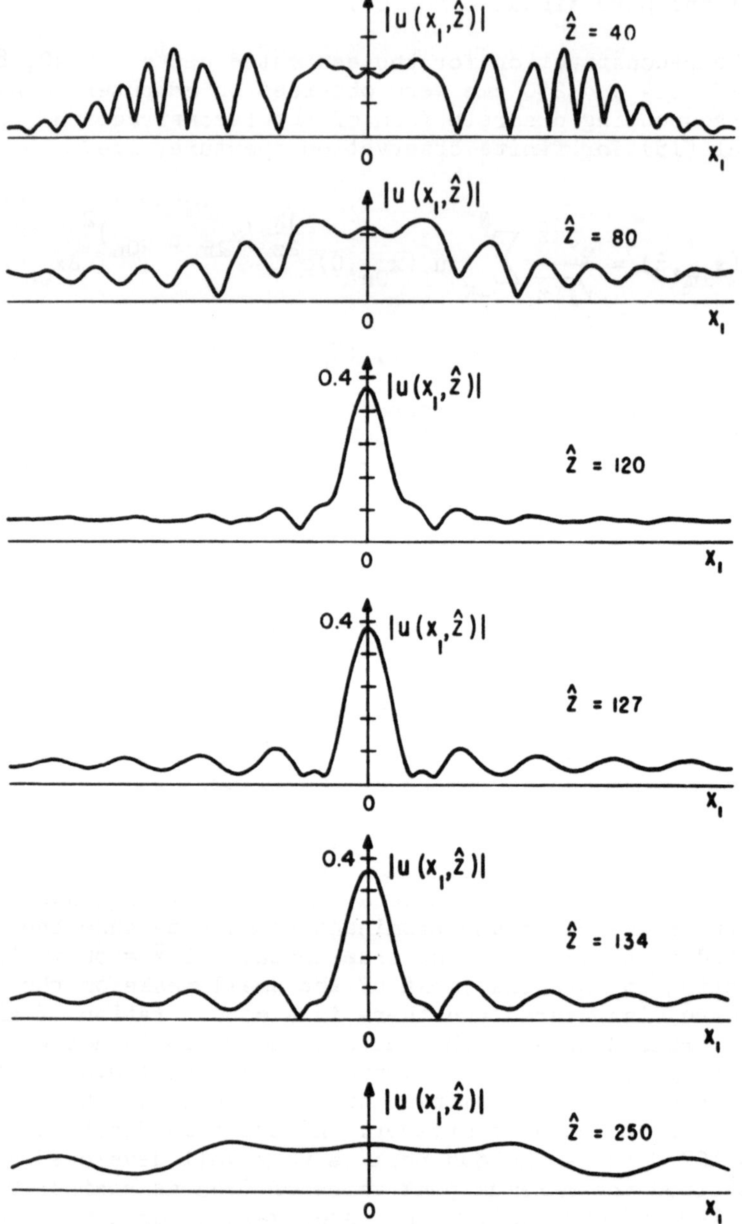

Figure 5. Reconstructed Anomalies For Point Scatterer

$\hat{z} = 250$, it has completely subsided.

Thus, the magnitude of the prominent peak does show a steady increase in its amplitude on approaching the true depth of burial for the scatterer and then it drops rapidly for depth estimates greater than the true depth as was predicted earlier. Moreover the shape of the anomaly is best developed at the correct estimate and is in fact a sinc function as expected from (30). Also most of the small peaks that are present at lower depth estimates have the tendency to shift laterally with depth estimates nearing true depth.

OTHER EXAMPLES

The details of the wavefront reconstruction method and its usefulness in determining the true depth to the scattering body have been demonstrated above with the example of a point scatterer. The use of the point scatterer for such an illustration did help in dealing with considerably simpler expressions and therefore it proved to be easier to handle the different aspects of the method under consideration. Such an example might appear to be an over simplification of an otherwise complicated problem but is very helpful in extending this technique to any arbitrary shaped scatterers. The response due to any scattering object at the surface can be obtained by convolving the aperture function $u(x_1)$ of the object with that due to the point scatterer as given in (5). Considering the case of a line scatterer, for example, having a total length 'a' and centered at the origin in the scattering plane X_1, we have

$$u(x_1) = \text{rect}\left(\frac{x_1}{a}\right) \tag{34}$$

where

$$\text{rect}\left(\frac{x_1}{a}\right) = \begin{cases} 1 & |x_1| \leq \frac{a}{2} \\ 0 & \text{otherwise} \end{cases} \tag{35}$$

And we obtain the response, $u_\ell(x_0,0)$, due to the line scatterer at the surface by convolving (35) with (5), i.e.

$$u_\ell(x_0,0) = \text{rect}\left(\frac{x_1}{a}\right) \star \frac{e^{-jkz^*}}{\sqrt{-j\lambda z^*}}\, e^{-\frac{jk}{2z^*} x_0^2}$$

$$= \frac{e^{-jkz^*}}{\sqrt{-j\lambda z^*}}\left[\text{rect}\left(\frac{x_1}{a}\right) \star e^{-\frac{jk}{2z^*} x_0^2}\right] \qquad (36)$$

Expanding the square bracket in the form of the convolution integral we finally obtain that

$$u_\ell(x_0,0) = \frac{e^{-jkz^*}}{\sqrt{-j\lambda z^*}} \int_{-\infty}^{\infty} \text{rect}\left(\frac{x_1}{a}\right) e^{-\frac{jk}{2z^*}(x_0-x_1)^2}\, dx_1 \qquad (37)$$

Using (34) we can also write (37) in a more general form as

$$u_\ell(x_0,0) = \frac{e^{-jkz^*}}{\sqrt{-j\lambda z^*}} \int_{-\infty}^{\infty} u(x_1) e^{-\frac{jk}{2z^*}(x_0-x_1)^2}\, dx_1 \qquad (38)$$

Comparing (38) with the construction integral (3) after making the necessary modifications for the coordinate system used, i.e. $z = -z^*$, we find that the two integrals are identical and therefore (38) does represent the correct expression for the formation of hologram in plane X_0. Indeed an expression similar to (38) can also be written to define the hologram $u_a(x_0,0)$ due to any arbitrary scatterer as given below, i.e.

$$u_a(x_0,0) = \frac{e^{-jkz^*}}{\sqrt{-j\lambda z^*}} \int_{-\infty}^{\infty} u_a(x_1) e^{-\frac{jk}{2z^*}(x_0-x_1)^2} dx_1 \tag{39}$$

where $u_a(x_1)$ is the aperture function for the arbitrary scatterer in plane X_1. Expression (39) can be rewritten as

$$u_a(x_0,0) = \frac{e^{-jkz^*}}{\sqrt{-j\lambda z^*}} \left[u_a(x_1) \star e^{-\frac{jk}{2z^*} x_0^2} \right] \tag{40}$$

and using (5) we obtain that

$$u_a(x_0,0) = u_a(x_1) \star u(x_0,0) \tag{41}$$

where $u(x_0,0)$ is the hologram for a point object as given by (5).

Computer Reconstructions for the Line and Inclined Scatterers

Horizontal Line Scatterer. Once we have generated the hologram due to the line scatterer or for that matter any arbitrary scatterer, the reconstruction is obtained by computing the complex distribution function in the plane X_1 at an estimated depth of $\hat{z}$ below the surface according to the layout shown in Figure 4. Thus the reconstruction integral is exactly similar to the one used for the reconstruction of the point hologram with finite aperture dimensions and therefore the criteria for determining the true depth to the scatterer follow the same procedures as outlined in the case of the point scatterer. However, the shape of the reconstructed anomaly, for example its peak-height and width, etc. is directly dependent on the value of the observed complex function on the surface and therefore would differ for different scatterers. In this respect the reconstructed anomaly in the true plane containing the scatterer represents the characteristic features of the original scatterer. This aspect of the wavefront reconstruction method is demonstrated by considering examples of a horizontal line scatterer and that of an inclined interface.

The horizontal scatterer was assumed to be located in plane X_1 centered at the origin measuring from $x_1 = -1.55$ to $+1.55$ cms with a total extent of 3.10 cm and the hologram $u_\ell(x_0,0)$ at the surface of the seismic model was computed by using expression (38) above. The scattering aperture $u(x_1)$ for the horizontal scatterer, due to the coherent seismic source located at origin on the surface of model, is given by

$$u(x_1) = \frac{e^{jkr}}{\sqrt{r}} ; \qquad k = \frac{2\pi}{\lambda} \tag{42}$$

where λ is the seismic wavelength in the model and r is the distance from the seismic source (origin) to any point contained in the line scatterer and is given by

$$r = (z^{*2} + x_1^2)^{\frac{1}{2}} \tag{43}$$

z^* is the perpendicular distance between the two parallel planes X_0 and X_1 as before. For computation purposes the whole scatterer is thought to consist of 32 points each 0.1 cm apart and the value of $u(x_1)$ determined from (42) for the coherent source operated at a frequency of 300 kHz as in the previous examples. Accordingly, the hologram at the surface is computed by using the corresponding discrete form of (38) as given below

$$u_\ell(x_0,0) = \frac{e^{-jkz^*}}{\sqrt{-j\lambda z^*}} \sum_{-1.55}^{+1.55} u(x_{1m}) e^{-\frac{jk}{2z^*}(x_{0n}-x_{1m})^2} \Delta x_{1m}$$

$$\text{for } m = \pm 1, \pm 2 \ldots \pm 16 \tag{44}$$

where x_{0n} and x_{1m} were defined for (31) and (33) respectively. The hologram is obtained for $n = \pm 1, \pm 2 \ldots 256$ for a total aperture $2L = 51.10$ cm as before. The

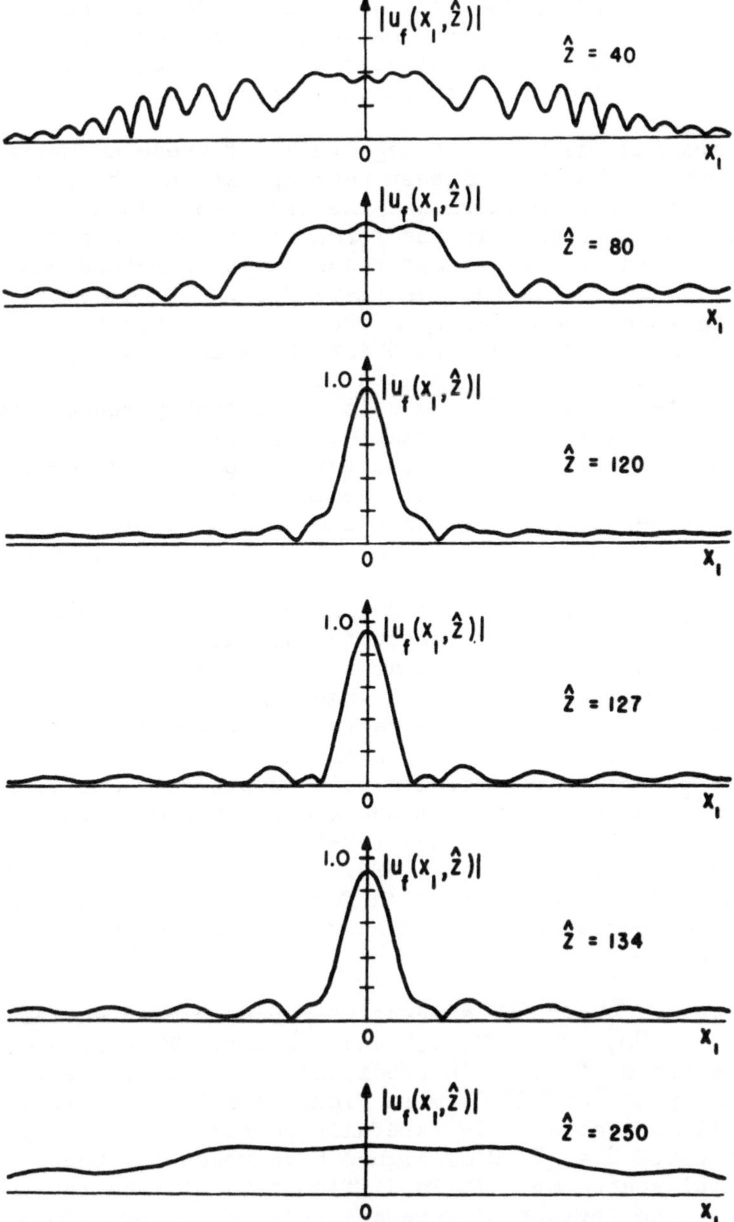

Figure 6. Reconstructed Anomalies For Line Scatterer

corresponding reconstruction integral is the same as that given in (33) for the point scatterer. The reconstructed anomalies obtained for $\hat{z}$ values of 40, 80, 120, 127, 134 and 250 cms on computer are shown in Figure 6.

Comparing the relative magnitudes of these anomalies, it is obvious that the highest peak appears at the depth estimate of 127.0 cm which was the true depth z^* assumed for the line scatterer in the formation of the hologram. In this regard, the wavefront reconstruction method has again provided the correct estimate for the true depth. In comparing these anomalies with the corresponding reconstructed anomalies of Figure 5 for the same values of depth estimate $\hat{z}$, it should be pointed out that the vertical scale in the plots of Figure 6 is considerably reduced and so the amplitudes of the anomalies for the case of line scatterer are higher than those for the point scatterer. For example the characteristic anomaly at $\hat{z} = 127.0$ for the line scatterer has a peak amplitude of approximately 0.961 units as against 0.43 in Figure 5 for the point scatterer.

Inclined Scatterer. To demonstrate that the characteristic features of the reconstructed anomaly at the correct estimated depth, in our case at $\hat{z} = 127.0$ cm, are truly representative of the physical shape and size of the scatterer, the wavefront reconstruction technique was also extended to include the example of an inclined scatterer. For this purpose the horizontal line scatterer considered above is assumed to have rotated clockwise around the origin through an angle $\phi = 5°$. Scattered field $u(x_1)$ is similarly computed by using (42). The construction of the hologram and its reconstruction for the inclined interface is again computed as detailed above for the horizontal interface.

Figure 7 shows the successive reconstructed anomalies for $\hat{z} = 40, 80, 120, 127, 134$ and 250 cms. The value of the true depth z^* is again predicted correctly to 127.0 cm. The anomaly at $\hat{z} = 127.0$ cm in Figure 7 which represents the inclined scatterer is carefully compared with the one also at depth $\hat{z} = 127.0$ of Figure 6 representing the horizontal scatterer. It is difficult to distinguish and visualize the obvious differences between the two anomalies as presented in Figures 6 and 7 because of the small scales used in drawing the sketches. Nevertheless some of the important features are as follows. One of the main

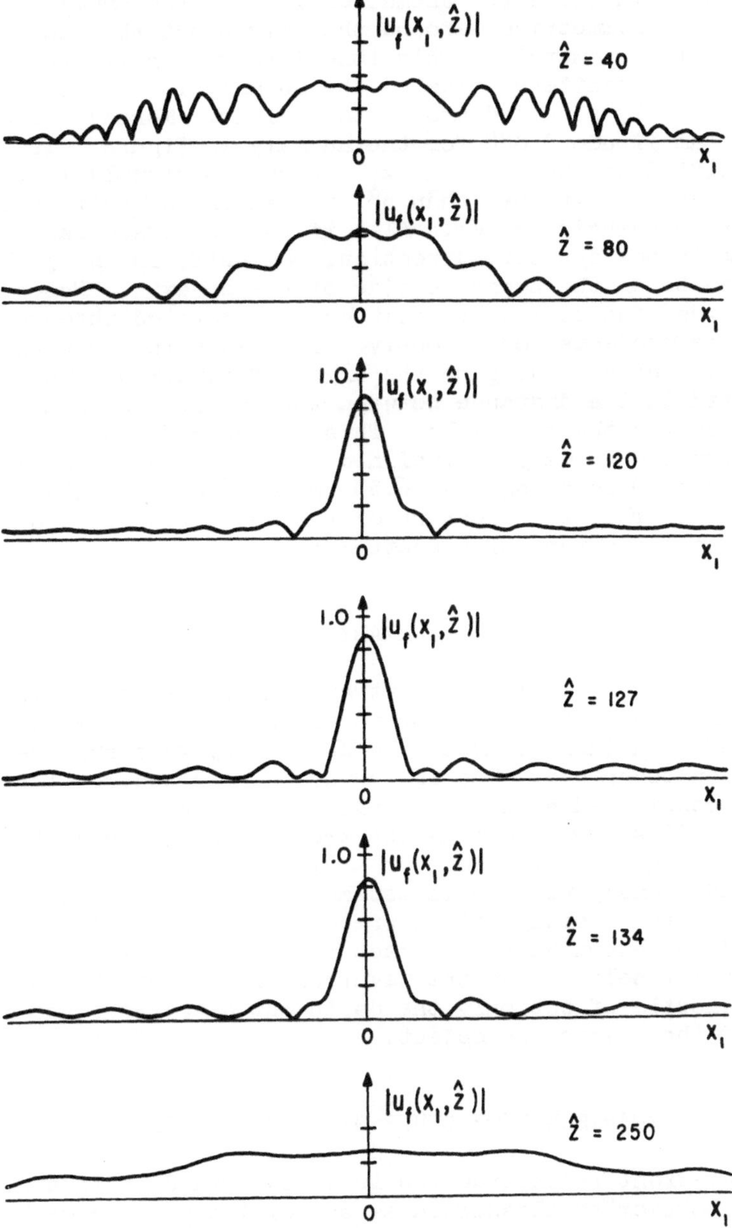

Figure 7. Reconstructed Anomaly For Inclined Scatterer

differences in these two anomalies is that the anomaly of Figure 6 is symmetrical around the origin but that in Figure 7 is asymmetric. This immediately suggests that the shape of the scatterer causing the anomaly in Figure 7 is asymmetrically situated around the origin. Moreover the highest amplitude 0.888 for the anomaly in Figure 7 is in fact shifted to the right at $x_1 = 0.25$. Obviously if the dip angle θ was greater than 5°, the amount of shift would be correspondingly larger. Also if the interface was dipping in the opposite direction, the shift in the peak ought to be on the negative side of the x_1-axis. Essentially when the horizontal scatterer is rotated through an angle the projected or effective dimensions in the plane X_1 is reduced depending on the angle of rotation. This is reflected in the distance between the two points of minimum amplitude for the main lobe. This distance for the reconstructed anomaly of inclined interface in Figure 7 measures 6.10 cm as against 6.50 cm for the horizontal interface. Thus this feature of the reconstructed anomaly indicates the extent of a scatterer.

SUMMARY

The above examples demonstrate the procedures involved in determining the true depth to the scatterer from the observed data (hologram) at the surface by wavefront reconstruction technique. The interpretation procedures for the method considered here can be thought to consist of two stages. First we determine the true depth to the scatterer by observing the variations in amplitude of the reconstructed anomaly with depth estimates. Later the particular reconstructed anomaly at the predetermined true depth is studied to determine the characteristic features of the scatterer itself. Thus the wavefront reconstruction method is diagnostic of not only the depth but also the shape and size of the scattering object.

COMPARISON WITH OTHER GEOPHYSICAL METHODS

Wavefront reconstruction is a new method in itself. It can however be classified as one of the seismic methods in a broad sense though it differs considerably from the conventional reflection and refraction techniques. In comparison with the existing seismic methods of geophysical exploration, the method considered here employs a monofrequency source of seismic energy and records the scattered field at the surface in its entirety as against a multifrequency source used in reflection and refraction surveys. In addition, their interpretation is based on the phenomenon of specular reflection or refraction of seismic energy and thus involves the recognition of some definite events on the seismogram. Consequently, the seismic reflection method has been very successful in mapping the relatively smooth interfaces like gently dipping beds and anticlinal structure, etc. However, in the mapping of salt domes that are frequently associated with large oil reserves the currently available seismic methods perform very poorly. This is because most of seismic energy from salt domes and other structures like stratigraphic traps and fault zones, etc. is not specularly reflected back to the surface but experiences a scattering phenomenon. For such a situation, of course, there are no well defined events in the seismogram and so the reflection method fails. The wavefront reconstruction as presented here is designed to provide an accurate interpretation for the scattered seismic signal and it is hoped this method fulfills a long required need to cover this aspect of seismic interpretation. Thus the method introduced here occupies a new dimension in geophysical prospecting and it in no way replaces the seismic reflection or refraction techniques.

The interpretation technique employed in determining the true depth to the geological body in this method is similar to the downward continuation procedures used in the interpretation of potential field data. The reconstruction integral (15) for example can be regarded as the corresponding downward continuation integral for the wave propagation case.

The downward continuation technique as a valid interpretation method for potential field data was first suggested by Peters (1949). The criteria for determination of the true depth to the body as postulated by him for the potential data and later extended for interpreting d.c. telluric, self-potential and electromagnetic field data by Roy (1963 and 1966) is summarized here in comparison to the one obtained for the new seismic method under discussion.

The criteria for depth determination by continuing the

observed potential field downwards is qualitative in nature as compared to the unbiased estimate obtained by wavefront reconstruction method discussed earlier. Accordingly, there are three criteria that would indicate that the depth to the anomalous body has been reached. First is the basic change in the shape of the continued anomaly. It seems that the downward continued anomaly tends to retain its shape up to the true depth in a manner that resembles the observed anomaly at the surface. Of course, the anomaly becomes sharper and sharper on approaching the correct depth; however this trend continues even beyond this depth with the appearance of extra peaks and troughs as side lobes. The first appearance of such side lobes is taken to be an indication of the true depth. As the second criteria, the existence of the so-called violent oscillations in the downward continued anomaly where some of the magnitudes attain very large values is understood to mean that the depth of the body has been crossed. Thirdly, the point of maximum curvature or the fastest change of slope on the semi-log plot of the peak-amplitude for the primary anomaly vs. the depth of continuation is said to represent the true depth of body.

In contrast to the qualitative determination of the depth for these geophysical methods, the depth determination by wavefront reconstruction technique is exact, unbiased and quantitative in nature. Once the true depth of the scatterer is known, the shape and size of the reconstructed anomaly at the above determined true depth can be interpreted to define the characteristic features of the anomalous body.

CONCLUSIONS

The wavefront reconstruction method of geophysical prospecting as described here thus involves the following three steps:

1. Recording of the interference pattern $\bar{u}(x_0)$ between the complex scattered field $u(x_0,0)$ and the direct wave $n(x_0)$.

2. Wavelength filtering of the observed data $\bar{u}(x_0)$ to eliminate the effect of the direct wave $n(x_0)$ and obtain the complex distribution pattern purely due

to the scattered field of the anomalous body $u(x_0,0)$.

3. Reconstruction of the anomalies at different depth estimates to determine the true depth of the scattering body and later to interpret the corresponding characteristic anomaly in terms of the physical relief of the scatterer.

The usefulness of this technique has been demonstrated for the detection of mineral ore bodies of any arbitrary shape; however, the examples of horizontal and inclined scatterers considered illustrate its application to a two layer seismic case. Thus, the wavefront reconstruction as applied to the problem of geophysical exploration can be used for mineral as well as oil prospecting. Moreover, the examples presented were purposely selected to be of two dimensional nature for computational simplicity and to correspond with the 2-D laboratory model experiments. It can be easily extended for a three dimensional situation in which case the reconstructed anomaly would be a two dimensional contour map of the anomalous body.

REFERENCES

Farr, J. B., Acoustical holography experiments using digital processing: Acoustical Holography, v. 2, p. 225, Plenum Press, New York (1970).

Goodman, J. W., Introduction to Fourier optics: McGraw-Hill Book Co., Inc., New York (1968).

Mueller, R. K. and N. K. Sheridon, Sound holograms and optical reconstruction, Appl. Phys. Letters, v. 9, p. 328 (1966).

Papoulis, A., Systems and transforms with applications in optics: McGraw-Hill Book Co., Inc., New York (1968).

Peters, L. J., The direct approach to magnetic interpretation and its practical application: Geophysics, v. 14, p. 290 (1949).

Roy, A., New interpretation techniques for telluric and some direct current fields: Geophysics, v. 28, p. 250 (1963).

Roy, A., Downward continuation and its application to electromagnetic data interpretation: Geophysics, v. 31, p. 167 (1966).

LIST OF PARTICIPANTS

Robert C. Addison, Jr.
American Optical
P.O. Box 2267
Framingham Ctr.,Mass.

Mahfuz Ahmed
Zenith Radio Corporation
6001 W. Dickens Avenue
Chicago, Illinois 60639

Pierre Alais
University Paris VI
Laboratoire de Mecanique
2pl de la gare de Ceinture
78 - Saint Cyr l'ecole
Paris, FRANCE

George A. Alers
North American Rockwell
Science Center
1049 Camino Dos Rios
Thousand Oaks, Cal. 91360

B. A. Auld
W.W. Hansen Laboratory
Stanford University
Stanford, Calif. 94305

LeRoy Barncastle
Santa Barbara Res. Ctr.
75 Coromar Drive
Goleta, Calif. 93017

William Bertrando
2685 W. Puesta Del Sol
Santa Barbara, Cal.93105

Helge Bodholt
(MIT-Cambridge)
Simrad A.S.
Horten, NORWAY

Newell O. Booth
U.S. Naval Undersea Ctr.
Code 6513
San Diego, Cal. 92132

Byron B. Brenden
Holosonics, Inc.
2950 Go. Washington Way
Richland, Wash. 99352

Peter S. Bringham
University of California
Lawrence Berkeley Lab.
Berkeley, Calif. 94720

Thomas Calkins
115 Deerboro Apt.102
Goleta, Calif. 93017

John C. Campbell
Santa Barbara Res. Ctr.
75 Coromar Drive
Goleta, Calif. 93017

Robert F. Carlson
Channel Industries
P.O. Box 3680
Santa Barbara, Cal. 93106

Henry H. Chaskelis
U.S. Naval Research Lab.
Code 8435
Washington, D. C. 20390

C. L. Coates
University of Illinois
Coordinated Science Lab.
Champaign, Ill. 61820

Dale Collins
Holosonics, Inc.
2950 Go. Washington Way
Richland, Wash. 99352

James A. Cunningham
W.W. Hansen Laboratories
Stanford University
Stanford, Calif. 94305

William J. Dallas
University of California
Applied Physics and Info.
Science Dept. San Diego
LaJolla, Calif. 92037

Dominick J. DeBitetto
Philips Laboratories
Briarcliff Manor, NY

John A. Edward
General Electric Co.
Court St. Plant Bldg. 4
Syracuse, NY

Kenneth R. Erikson
Actron Industries, Inc.
700 Royal Oaks Drive
Monrovia, Calif. 91016

James Ewing
Bendix Corp.
15825 Roxford
Sylmar, Calif. 91345

N. H. Farhat
University - Pennsylvania
200 South 33rd Street
Philadelphia, Pa. 19104

John B. Farr
Amoco Production Co.
(Research Center)
P.O. Box 591
Tulsa, Oklahoma 74102

Wayne R. Fenner
The Aerospace Corp.
Bldg. 120, Rm. 1217
P.O. Box 95085
Los Angeles, Cal. 90045

B. W. Finney
General Electric
P. O. Drawer QQ
Santa Barbara, Cal. 93102

Wolfgang K. Fischer
U.S. Naval Underwater Ctr.
New London Laboratory
New London, Conn. 06320

Gerald L. Fitzpatrick
U.S. Bureau of Mines
Denver Mining Res. Ctr.
Bldg 20, Denver Fed. Ctr.
Denver, Colorado 80225

John J. Flynn, M.D.
5322 Franklin Avenue
Los Angeles, Calif. 90027

Martin D. Fox
Duke University
4216 Garret Rd. Apt F18
Durham, N.C. 27707

G. Roger Gathers MS1-504
Lawrence Livermore Lab.
P.O. Box 808
Livermore, Calif. 94550

Fred G. Geil
U.S. Naval Undersea Ctr.
Bldg 128, Rosecrans St.
San Diego, Calif. 92109

I. Lee Gelles
Kennecott Copper Corp.
Ledgemont Laboratory
128 Spring Street
Lexington, Mass. 02173

P. -A. Grandchamp
Hoffmann-La Roche
Ch-4002 Basle
SWITZERLAND

Lou Ann Granger
Bendix Corporation
Electrodynamics Division
15825 Roxford
Sylmar, Calif. 91345

Philip S. Green
Stanford Research Inst.
333 Ravenswood Avenue
Menlo Park, Cal. 94025

Lt. Carlton A. Griggs
US Naval Postraduate Sch.
380A Bergin Drive
Monterey, Calif. 93940

W. R. Guard
Newark College of Engrg.
323 High Street
Newark, N.J.

Francis J. Henry
Res. & Tech. Directorate
(ORD-035C2B)
Naval Ordnance Systems
Washington, D.C. 20360

Richard W. Hicks
680 Arundle Road
Goleta, Calif. 93017

B. P. Hildebrand
Battelle-Northwest Lab.
Box 999
Richland, Wash. 99352

David R. Holbrooke, M.D.
Children's Hospital-S.F.
3700 California Street
San Francisco, Ca. 94119

John B. Hough
U.S. Naval Ship Engrg.Ctr.
Code 6126, Ctr. Bldg.
Prince George's Center
Hyattsville, Md. 20782

Jen-Shu Hsieh
University of California
Dept. of Radiological Sci.
Center for Health Sciences
Los Angeles, Calif. 90024

Charles S. Ih
CBS Laboratories
227 High Ridge Road
Stamford, Conn. 06905

Robert J. Johnston
Three-D Systems Inc.
408 Grove Lane
Santa Barbara, Cal. 93105

H. W. Jones Rm 714
University of Calgary
Physics Department
Calgary, Alberta, CANADA

David W. Jorgensen
McDonnell Douglas
Astronautics Company
5301 Bolsa Avenue
Huntington Beach, Calif.

Finn Jorgensen
6173 La Goleta Road
Goleta, Calif. 93017

William J. Joyce
Philips Medical Systems
710 Bridgeport Avenue
Shelton, Conn. 06484

A. K. Kalra
University of California
Hearst Mining Bldg.
Berkeley, Calif. 94720

H. B. Karplus
Argonne National Lab.
Argonne, Ill. 60439

William E. Katzenmeyer
D/472-G3
Goodyear Aerospace Corp.
1210 Massillon Road
Akron, Ohio 44315

Dr. Leslie Kay
University of Canterbury
Christchurch, N. ZEALAND
or 108 Lake Shore Road
Brighton, Mass. 02135

Dr. Patrick N. Keating
The Bendix Corporation
Research Laboratories
Bendix Center
Southfield, Mich. 48076

L. W. Kessler
Zenith Radio Corp.
6001 W. Dickens Avenue
Chicago, Illinois 60639

H. (Tom) Keyani
Dept. of Elec. Engrg.
University of California
Santa Barbara, Cal.93106

Yoshimitsu Kikuchi
Tohoku University
Research Institute of
Electrical Communication
Sendai, Miyagi-Pref,
JAPAN

Adrianus Korpel
Zenith Radio Corp.
6001 W. Dickens Avenue
Chicago, Ill. 60639

Justin L. Kreuzer
Perkin Elmer Corp.
Main Avenue
Norwalk, Conn. 06852

M. J. Kuhn
Atlantic Richfield
P.O. Box 2819
Dallas, Texas 75221

John Landry
University of California
Dept. of Elec. Engrg.
Santa Barbara, Cal.93106

Pearl Lattaker
U.S. Naval Undersea Ctr.
Code 6513
San Diego, Calif. 92132

Achille Leblanc
Universite du Quebec
Bldg. Des Forges
Trois-Rivieres
Quebec, CANADA

Chin-Hwa Lee
University of California
Dept. of Elec. Engrg.
Santa Barbara, Cal.93106

A. W. Lohmann
University of California
San Diego-A.P.I.S.
La Jolla, Calif. 92037

David H. MacAnlis
U.S. Navy
CNO-Op967
Pentagon
Washington, D.C. 20360

Edward M. McCurry
Children's Hospital-S.F.
3700 California Street
San Francisco, Ca. 94119

James W. McHarg
Naval Ship Systems Comm.
Attn: Code PMS 395-A46
Washington, D.C. 20360

James A. McKinnis
U.S. Air Force
Rocket Propulsion Lab.
Edwards AFB, Cal. 93523

Kung-Shiang Ma
University of California
San Diego, A.P.I.S.
La Jolla, Calif. 92037

Dr. Max G. Maginness
Stanford Electronics Lab.
Stanford, Calif. 94305

F. H. Mahoney, Jr.
3339 Richland Drive
Santa Barbara, Ca. 93105

O. K. Mawardi
Case Western Reserve Univ
University Circle
Cleveland, Ohio 44106

Albert A. Melkonian
726 Juanita Avenue
Santa Barbara, Ca. 93105

Alexander F. Metherell
Actron Industries, Inc.
700 Royal Oaks Drive
Monrovia, Calif. 91016

Lyle Minkler
Yavapai College
17 Walking Diamond
Prescott, Arizona 86301

W. Duane Montgomery
University of Saskatchewan
Mathematics Department
Renina, Saskatchewan
CANADA

Robert G. Morris
Office of Naval Research
Code 421
Arlington, Va. 22217

Gerald Mott
Westinghouse Electric
Beulah Road
Pittsburgh, Pa. 15235

Dr. R. K. Mueller
Bendix Research Lab.
Bendix Center
Southfield, Mich. 48076

Dr. Anant K. Nigam
Bell Telephone Labs.
MH 15-208
600 Mountain Avenue
Murray Hill, N.J. 07974

Dr. Charles P. Olinger
University of Cincinnati
3064 Victoria Avenue
Cincinnati, Ohio 45208

Morio Onoe
Institute of Industrial
Science
University of Tokyo
7-22-1, Roppongi, Minato-k
Tokyo, JAPAN

Peter R. Palermo
Zenith Radio Corp.
6001 W. Dickens Avenue
Chicago, Illinois 60639

Soo-Chang Pei
University of California
Dept. of Elec. Engrg.
Santa Barbara, Ca. 93106

Jerome L. Pfeiffer
The Bendix Corporation
Research Laboratories
Bendix Center
Southfield, Mich. 48076

Alonzo W. Phillips
Aerojet ElectroSystems
P.O. Box 296
Azusa, Calif. 91702

William L. Pitt
Children's Hospital-S.F.
Rm 612-OPR
3700 California Street
San Francisco, Ca. 94119

John P. Powers
U.S. Naval Post. School
Elec. Engrg. Dept.
Monterey, Cal. 93940

David W. Prine
General American Res.Div.
7449 N. Natchez Avenue
Niles, Illinois 60648

Calvin F. Quate
Stanford University
W.W. Hansen Laboratories
Stanford, Calif. 94305

C. G. Roberts
Stanford University
96A Escondido Village
Stanford, Calif. 94305

Henry A. F. Rocha
General Electric
Bldg. 37-Rm 679
Schenectady, NY 12305

G. L. Sackman
U.S. Naval Post. School
Code 52 Sa
Monterey, Calif. 93940

Ben Saltzer, Head
Code 6513
U.S. Naval Undersea Ctr.
San Diego, Calif. 92132

Louis Schaefer
Stanford Research Inst.
333 Ravenswood Avenue
Menlo Park, Cal. 94025

John A. Scopatz
International Transducer
640 McCloskey Place
Goleta, Calif. 93017

Joseph T. Siedlecki
Chief of Naval Operations
OP 967 - Pentagon
Washington, D.C. 20350

Dr. Daniel Silverman
Technology Management
5969 S. Birmingham
Tulsa, Okla. 74105

Bruce D. Sollish
Columbia University
School of Engineering
New York, N.Y. 10027

Jeffrey M. Speiser
U.S. Naval Undersea Ctr.
San Diego, Cal. 92132

Gordon E. Stewart
The Aerospace Corporation
Bldg. 120, Room 1221
P.O. Box 95085
Los Angeles, Cal. 92667

Jerry L. Sutton
U.S. Naval Undersea Ctr.
San Diego, Cal. 92132

Mikio Takagi
University of California
Dept. of Elec. Engrg.
Santa Barbara, Ca.93106

Nobuyuki Takagi
Tokyo Shibaura Elec. Co.
(Toshiba R&D Center)
1 Komukai Toshiba-cho
Kawasaki, 210, JAPAN

Dr. Donald O. Thompson
North American Rockwell
Science Center
1049 Camino Dos Rios
Thousand Oaks, Ca. 91360

R. B. Thompson
North American Rockwell
Science Center
P.O. Box 1085
Thousand Oaks, Ca. 91360

Jim Thorn
U.S. Naval Undersea Ctr.
San Diego, Calif. 92132

Neal Tobochnik
University of California
Dept. of Radio. Sciences
Ctr. for Health Sciences
Los Angeles, Calif. 90024

William Trousdale
Wesleyan University
Physics Department
Middletown, Conn. 06457

Nie-But Tse
University of California
Dept. of Elec. Engrg.
Santa Barbara, Ca. 93106

Francis X. Urrico
475 Camino La Guna Vista
Goleta, Calif. 93017

David Vilkomerson
RCA Laboratories
Princeton, N.J. 08540

W. E. Wallace
Computer Command and
Control Co. Suite 1305
1717 Pennsylvania Avenue
Washington, D.C. 20006

Glen Wade
University of California
Dept. of Elec. Engrg.
Santa Barbara, Ca.93106

Keith Wang
University of California
Dept. of Elec. Engrg.
Santa Barbara, Ca. 93106

E. Eugene Watson
Ordnance Research Lab.
Pennsylvania State Univ.
Box 30
State College, Pa. 16801

James L. Weaver
Lockheed Research Lab.
5221 Bldg. 204
3251 Hanover Street
Palo Alto, Calif. 94304

Harper John Whitehouse
U.S. Naval Undersea Ctr.
Code 6003
San Diego, Calif. 92132

Robert L. Whitman
Zenith Radio Corp.
6001 W. Dickens Avenue
Chicago, Illinois 60639

Dr. D. E. Willcox
Doric Corporation
120 Park Avenue
Oklahoma City, Okla.

Dr. Donald C. Winter
TRW Systems
One Space Park
Redondo Beach, Ca. 90278

S. V. Yadavalli
Stanford Research Inst.
333 Ravenswood Avenue
Menlo Park, Cal. 94025

Gene Zilinskas
Bendix Electrodynamics
11600 Sherman Way
N. Hollywood, Calif.

1

2

3

SOME OF THE PARTICIPANTS AND SPEAKERS AT THE FOURTH INTERNATIONAL SYMPOSIUM ON ACOUSTICAL HOLOGRAPHY.

View 1 Left to right - G. L. Fitzpatrick, Denver Mining Research Center; J. P. Powers, USN, Naval Postgraduate School; A. K. Nigam, Bell Telephone Laboratories; R. K. Mueller, Bendix Research Laboratories; P. N. Keating, Bendix Research Laboratories; G. L. Sackman, USN, Naval Postgraduate School; F. G. Geil, US Department of the Navy; G. Wade, University of California at Santa Barbara; H. Keyani, University of California at Santa Barbara; A. F. Metherell, McDonnell Douglas Actron Industries, Inc.; H. W. Jones, University of Calgary; A. Korpel, Zenith Radio Corporation; J. L. Kreuzer, Perkin-Elmer Corporation, J. A. Cunningham, Stanford University; O. K. Mawardi, Case-Western Reserve University; A. K. Kalra, University of California at Berkeley; N. H. Farhat, University of Pennsylvania; L. W. Kessler, Zenith Radio Corporation; W. R. Guard, University of Pennsylvania; and P. Alais, Université de Paris.

View 2 Left to right - A. W. Lohmann, University of California at San Diego; Y. Kikuchi, Tohoku University; W. J. Dallas, University of California at San Diego; J. L. Pfeifer, Bendix Research Laboratories; J. L. Sutton, USN, Naval Undersea R&D Center; H. D. Collins, Holosonics, Inc.; B. P. Hildebrand, Battelle Memorial Institute; M. D. Fox, Duke University; B. B. Brenden, Holosonics, Incorporated; B. A. Saltzer, USN Naval Undersea R&D Center; D. C. Winter, University of Michigan (T.R.W. Systems); M. Onoe, University of Tokyo; J. Thorn, USN Naval Undersea R&D Center; M. Takagi, University of Tokyo (University of California at Santa Barbara); B. D. Sollish, Columbia University; P. R. Palermo, Zenith Radio Corporation; M. Ahmed, Zenith Radio Corporation; J. L. Weaver, Lockheed Research Laboratory; D. Vilkomerson, RCA Laboratories; B. A. Auld, Stanford University; K. Wang, University of California at Santa Barbara; and J. Landry, University of California at Santa Barbara.

View 3 Left to right - P. S. Green, Stanford Research Institute; O. N. Wilde, receptionist, G. Mott, Westinghouse Electric Corporation; and M. G. Maginness, Stanford Electronics Laboratories.

Not Pictured: C. F. Quate, Stanford University; C. G. Roberts, Stanford University; and L. F. Schaefer, Stanford Research Institute.

INDEX